Catalysis of
Organic Reactions

CHEMICAL INDUSTRIES

A Series of Reference Books and Textbooks

Consulting Editor

HEINZ HEINEMANN
Berkeley, California

ADDITIONAL VOLUMES IN PREPARATION

Catalysis of Organic Reactions

John R. Sowa, Jr.

Department of Chemistry and Biochemistry
Seton Hall University

CRC Press
Taylor & Francis Group
Boca Raton   London   New York

CRC Press is an imprint of the
Taylor & Francis Group, an **informa** business
A TAYLOR & FRANCIS BOOK

CRC Press
Taylor & Francis Group
6000 Broken Sound Parkway NW, Suite 300
Boca Raton, FL 33487-2742

First issued in paperback 2019

© 2005 by Taylor & Francis Group, LLC
CRC Press is an imprint of Taylor & Francis Group, an Informa business

No claim to original U.S. Government works

ISBN-13: 978-0-8247-2729-1 (hbk)
ISBN-13: 978-0-367-39311-3 (pbk)

Library of Congress Cataloging-in-Publication Data

Catalog record is available from the Library of Congress

Visit the Taylor & Francis Web site at
http://www.taylorandfrancis.com

and the CRC Press Web site at
http://www.crcpress.com

Contents

Board of Editors

Chronology of Organic Reaction Society Conferences

Conf	Year	Chair	Location	Proceedings Publisher
1	1967	Joseph O'Connor	New York City	NYAS
2	1969	Joseph O'Connor	New York City	NYAS
3	1970	Mel Rebenstrot	New York City	NYAS
4	1973	Paul Rylander	New York City	NYAS
5	1974	Paul Rylander & Harold Greenfield	Boston	Academic Press
6	1976	Gerry Smith	Boston	Academic Press
7	1978	Bill Jones	Chicago	Academic Press
8	1980	Bill Moses	New Orleans South	Academic Press
9	1982	John Kosak	Charleston	Dekker
10	1984	Bob Augustine	Williamsburg	Dekker
11	1986	Paul Rylander & Harold Greenfield	Savannah	Dekker
12	1988	Dale Blackburn	San Antonio	Dekker
13	1990	Bill Pascoe	Boca Raton	Dekker
14	1992	Tom Johnson & John Kosak	Albuquerque	Dekker
15	1994	Mike Scaros & Mike Prunier	Phoenix	Dekker
16	1996	Russ Malz	Atlanta	Dekker
17	1998	Frank Herkes	New Orleans South	Dekker
18	2000	Mike Ford	Charleston	Dekker
19	2002	Dennis Morrell	San Antonio	Dekker
20	2004	John Sowa	Hilton Head	CRC Press

Preface

This volume of Catalysis of Organic Reactions is a collection of 63 papers that were presented at the 20th Conference on Catalysis of Organic Reactions, March 21–25, 2004, in Hilton Head Island, South Carolina. The conference was organized by the Organic Reactions Catalysis Society (ORCS, http://www.orcs.org) and also marks the 20th publication of a proceedings from ORCS. This volume documents recent and important developments in the field of catalysis as related to organic synthesis with a special emphasis on processes with actual and/or potential applications in industry.

Each paper has been expertly processed by an Editorial Board consisting of academic and industrial experts in catalysis. Furthermore each paper was reviewed by peers within and outside of the ORCS community. The volume is organized into five major symposia as follows:

I-Catalytic Hydrogenation
II- Novel Concepts and Approaches to Catalysis of Organic Reactions
III-Acid-Base Catalysis
IV-Catalytic Oxidation
V-Catalysis in Organic Synthesis

Where possible, subsections have been organized within each symposium. In the lead-off symposium, **Catalytic Hydrogenation**, the first subsection consists of five papers on *Hydrogenation on Raney-type Catalysts*. This subsection includes the 2004 Murray Raney Award address by Professor Jean Lessard, University of Sherbrooke. Additional subsections include *Anchored and Supported Hydrogenation Catalysts* (9 papers) and five papers on *Hydrogenation of Interesting Substrates and Renewable Sources*. The symposium on **Novel Concepts and Approaches to Catalysis of Organic Reactions** includes five papers in *Combinatorial and Parallel Methods in Catalyst Design, Optimization and Utilization*, which was a Featured Technology Area for the Conference. The *Symposium on Catalysis in Organic Synthesis* includes the 2004 Paul N. Rylander award address by Professor Richard C. Larock and also includes subsections on *Deprotection* reactions and *Asymmetric Catalysis*. This organization is intended to help the reader find a number of interesting and related papers grouped together.

The success of the 20th Conference was greatly enhanced by financial contributions from the following organizations: *W. R. Grace-Davison, Parr Instrument, North American Catalysis Society, ACS-PRF, Engelhard, Degussa, AMC-PMC, Merck & Co., Eli Lilly & Co., CRI Catalysts, The Catalyst Group, Umicore, Süd Chemie, Nova Molecular Technologies, Inc., and Rohm & Haas.* I am also grateful to the ORCS Board of Directors (Frank Herkes, Robert DeVries, Alan Allgeier, Anne Gaffney, Yongkui Sun, Steve Jacobson, Mike Ford) and the Executive Committee (Vic Mylroie, Steve Schmidt, Dennis Morrell) for time and energy to running the 20th Conference. A special acknowledgement is made to Prof. Donna G. Blackmond, as the 2003 Paul N. Rylander winner. Finally, I am forever indebted to my wife Sonia for her patience and love during the process of organizing the 20th Conference and assembling these Proceedings.

To Sonia, Julian and Ariana

I. Symposium on Catalytic Hydrogenation

a. Hydrogenation on Raney-Type Catalysts

1.

2004 Murray Raney Award: Electrocatalytic Hydrogenation of Organic Compounds at Raney Metal Electrodes: Scope and Limitations

Jean Lessard

Laboratoire de chimie et électrochimie organiques, Département de chimie, Université de Sherbrooke, Sherbrooke, Québec, J1K 2R1 Canada

Jean.Lessard@USherbrooke.ca

Abstract

The interest of electrocatalytic hydrogenation (ECH) of organic compounds stems from the fact that: i) higher selectivities than in catalytic hydrogenation (CH) can be achieved by using milder conditions and by using catalysts that have a low activity in CH such as Co and Cu catalysts; and ii) the use of volatile and persistent organic solvents can be avoided. Typical examples are presented: the electrohydrogenation of polycyclic fused aromatic hydrocarbons and of glucose, at Raney Ni electrodes, at atmospheric pressure and at 80-110°C; the electrohydrogenation of lignin models and of lignins at Raney Ni electrodes; and the highly chemoselective electrohydrogenation of nitro groups (nitroaryl groups as well as primary, secondary, and tertiary nitroalkyl groups) to the corresponding amine in neutral and basic medium at Raney Co and/or Raney Cu electrodes in the presence of easily reducible functions such as nitriles (conjugated or not), alkynes, C–I bonds, etc. The limitations of the ECH method are also pointed out and some mechanistic aspects are discussed.

Introduction

The electrocatalytic hydrogenation (ECH) of an unsaturated organic substrate (Y=Z) in aqueous or mixed aqueous-organic media (eqs [2] to [4] where M represents an adsorption site of the catalyst, and M(H) and M(Y=Z), chemisorbed hydrogen and the adsorbed substrate respectively) involves the same hydrogenation steps as those of classical catalytic hydrogenation (CH) (steps [2] to [4]: stoichiometry for adsorbed species only, not for surface M) (1):

$$2H_2O(H_3O^+) + 2e^- + M \longrightarrow 2M(H) + 2OH^-(H_2O) \qquad [1]$$

$$Y{=\!=}Z + M \rightleftharpoons M(Y{=\!=}Z) \qquad [2]$$

$$M(Y{=\!=}Z) + 2M(H) \rightleftharpoons M(YH{-\!-}ZH) \qquad [3]$$

$$M(YH{-\!-}ZH) \rightleftharpoons YH{-\!-}ZH + M \qquad [4]$$

$$M(H) + 2e^- + H_2O \longrightarrow H_2 + OH^- + M \qquad [5]$$

$$H_2 + M \rightleftharpoons 2M(H) \qquad [6]$$

The same catalysts can be used for both types of hydrogenation. ECH differs from CH in the way to generate M(H), respectively by reduction of water (or hydronium ion depending on the pH) in ECH (eq. [1]) and by thermal dissociation of dihydrogen in CH (eq. [6]). The main *advantages* of ECH over CH are: 1) the kinetic barrier due to the splitting of H_2 is completely bypassed in ECH (therefore, catalysts that have a low activity in CH can be used successfully in ECH even at room temperature and atmospheric pressure); 2) the mass transport of insoluble hydrogen gas through the solution and its manipulation are also completely bypassed in ECH (1). The main *disadvantage* of ECH over CH comes from the competition between hydrogenation of the organic substrate (steps [2]-[4]) and electrochemical desorption of M(H) (eq. [5]). As a result of this competition, ECH of substrates very difficult to hydrogenate either does not occur or is very inefficient (low current efficiencies) (1). When compared to electro-hydrogenation by successive electron transfer and protonation steps (electronation-protonation, EP), which is the mechanism of electrohydrogenation at high hydrogen overvoltage cathodes (Hg, glassy carbon, etc.), ECH has the advantage 1) of requiring a less negative potential (less energy), water being more easily reduced at transition metal electrodes than most of the organic substrates (with the exception of the nitro group family) and 2) of avoiding the generation of radical species (radical anions and/or neutral radicals).[1]

In this paper, results on the ECH of representative substrates are presented i) in order to illustrate the advantages and the limitations of the ECH at Raney metal electrodes as a method of hydrogenation of organic compounds and ii) in order to discuss some synthetic and mechanistic aspects. The Raney metal electrodes consisted of Raney metal particles embedded in a nickel matrix (1) or dispersed in a lanthanum polyphosphate matrix (2), or of pressed Raney metal alloy of which the outmost layer only has been leached (3).

ECH of Polycyclic Aromatics

Under the best conditions determined after a thorough investigation of the various factors which may have an influence on the competition between hydrogenation (steps [2]-[4]) and reduction of M(H) to H_2 (eq. [5]), the hydrogenation of phenanthrene (**1**) at a RNi cathode (in a two-compartment H-cell with a Nafion-324 membrane) gives a 1:1 mixture of octahydrophenanthrenes (**2** and **3**) in 92% yield and 77% current efficiency after complete conversion (Q = 10 F/mol: Q is the amount of electricity (charge = integral of current x time) used for the electrolysis and F is the Faraday constant (94487 coulombs/mol)) (Scheme 1) (4). These optimum conditions are: i) T = 50°C; J = 6.25 mA/cm^2; ii) periodic current control (T_1 = 1 s (potential applied), T_2 = 8 s (open circuit)); (iii) H_3BO_3 0.1 M and NaCl 0.1 M in ethylene glycol-water 95:4 (v/v); (iv) pH = 3.5-6.8; (v) pressure of an inert gas (P(N_2) = 3.5 atm) (4). These conditions are much milder than those required for the CH of **1** to **2** and **3** at a RNi catalyst: T = 110-120°C, P(H$_2$) = 175-260 atm (5). This represents an advantage of ECH over CH.

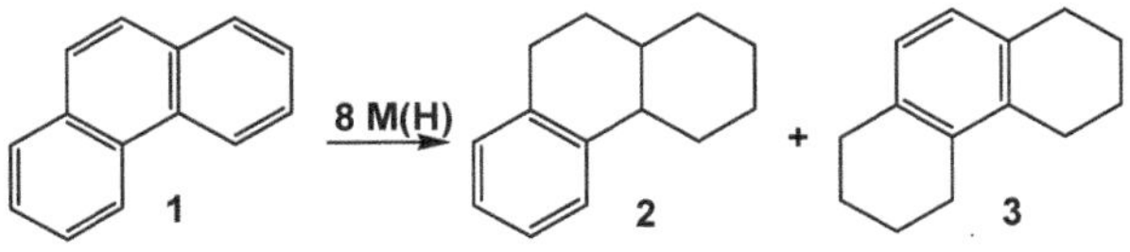

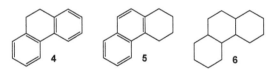

Scheme 1

On the other hand, the hydrogenation of phenanthrene (**1**) to perhydrophenanthrenes (**6**) (Figure 1) by ECH at a RNi cathode is very inefficient because, at RNi, the rate of hydrogenation (eq. [3]) of the π bonds of a single aromatic ring of octahydrophenanthrenes (**2** and **3**) (stronger than the π bonds of **1** itself and of the intermediate dihydro- (**4**) and tetrahydro-phenanthrene [**5**]) (Figure 1) is much slower than that of the electrochemical desorption of hydrogen (eq. [5]) (V$_3$/V$_5$ << 1). This represents a limitation of ECH over CH. Perhydrophenanthrenes (**8**) have been obtained in good yields (68-85%) by CH at RNi at T = 175-200°C and P(H$_2$) = 175-250 atm (5).

Figure 1 Products/intermediates of the hydrogenation of phenanthrene (**1**).

ECH of Glucose

The ECH of glucose at a RNi cathode in aqueous medium illustrates the advantage of ECH over CH. Under the more stringent conditions used in CH (T = 100°C), hydrogenation of glucose to sorbitol is quantitative but isomerisation of glucose to mannose also proceeds to the extent of 25-30%. The yield of hydrogenation is quantitative. The ECH of a 1.6 M aqueous solution of glucose containing Na$_2$SO$_4$ 0.1 M as supporting electrolyte, carried out in a two-compartment flow-cell (Nafion-324 membrane, geometrical area of the RNi electrode of 36 cm^2) at T = 60°C and J = 33 mA/cm^2, gave a 100% yield of sorbitol contaminated by only 3-6% of mannitol (hydrogenation of mannose) (Scheme 2) (6). The current efficiency was 50-60%, the electrochemical desorption of M(H) occurring to the extent of 50-40%.[6] The ECH method also gives better yields than electrohydrogenation by the electronation-protonation (EP) method because the formation of pinacol-type dimer is a competing process in EP electrohydrogenation (7) due to the formation of radical intermediates as already pointed out.

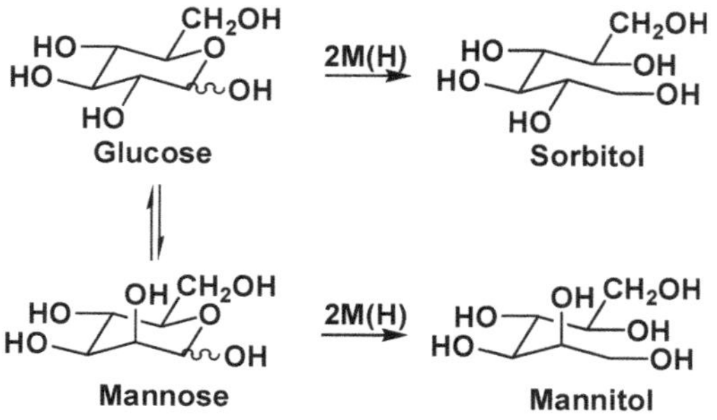

Scheme 2

The ECH of glucose serves also to illustrate one of the disadvantages of any electrochemical method when working with organic compounds soluble in water: the use of a supporting electrolyte which is also soluble in water and must be separated from the organic product. In the present case, electrodialysis had to be used to separate sodium sulphate from sorbitol, which added to the cost of the process. As a result, the cost of producing sorbitol from glucose by ECH was estimated to be about twice that of the classical CH process, and there is no market for a sorbitol containing little mannitol (or none).

ECH of Lignin Models and Lignins

The ECH of β-O-4 lignin models have been carried out at constant current (J = 0.2 mA/cm^2) at a Raney nickel cathode in an aqueous 1 M NaOH solution and at 75°C (two-compartment flow cell with a Nafion-324 membrane) (2). It was found that the hydrogenolysis of phenolic β-arylethyl-aryl ethers with one β-O-4 linkage takes place to give a mixture of monomeric phenolic compounds. For instance, the ECH of model **7** (Scheme 3) gives acetovanillone (**9**) (12%), α-methylvanillyl alcohol (**10**) (18%), guiacol (**11**) (37%), and products derived from their further hydrogenation (phenol (**12**) (7%) and five other products (18% total)). Compound **7** is first dehydrogenated to ketone **8** (Scheme 3). The hydrogenolysis of the C-O bond of **8** is much faster than that of alcohol **7** because, in ketone **8**, the C-O bond is conjugated to both an aromatic ring and an arylketone (2). The ratio of compounds derived from the ArC$_2$ moiety (ketone **9** and alcohol **10**) to those derived from the ArC$_1$ moiety (guaiacol (**11**) and phenol (**12**)) is close to 1 as expected. The β-O-4 bond is more difficult to cleave upon replacing one hydrogen of the methylene group of **9** by a hydroxymethyl group (see **13** in Figure 2). However, model **14** (with two β-O-4 linkages) (Figure 2) obtained by the introduction of an α-methylvanillin group at the phenolic position of **7** was hydrogenolyzed at the same rate as **7** (2).

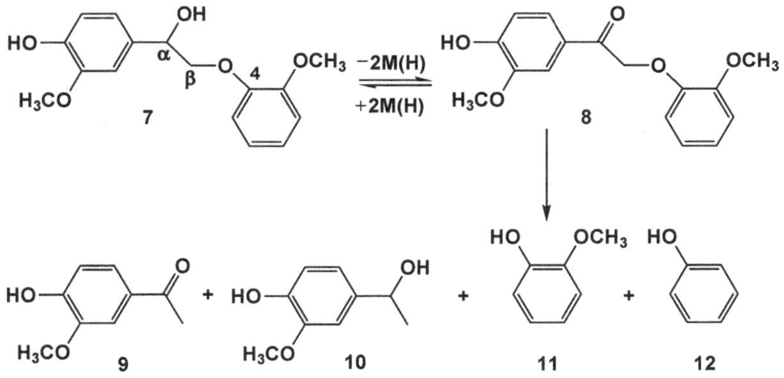

Scheme 3

illustrated by the two following examples (8). In the first example, the ECH of a modified lignin (Kraft Indulin) at 130°C and 3 atm gave 16 monomeric phenolic products (including **10**, **11**, **12**, **13**, and **14**) in a total yield of 0.7% only (near zero current efficiency). In the second example, the ECH, in alkaline (1 M NaOH) ethylene glycol-water at 140°C and $P(N_2) = 3$ atm, of a native lignin extracted *in situ* from maple tree sawdust, gave 14 different monomeric phenolic products with a total yield of 8% (near zero current efficiency). By comparison, CH of the same native lignin, under the same conditions of temperature and solvent but under a pressure of H_2 of 100 atm, was reported to give 20% of monomeric phenolic products (40% of total phenolic products) (9).

The inefficiency of the ECH of lignins can be readily explained by the fact that, the lignins being polymeric, very few hydrogenolysable β-O-4 linkages can access the chemisorbed hydrogens, M(H), at the surface of the electrode. Therefore, the hydrogenation step [3] is very slow and the main reaction occuring is electrochemical desorption of M(H) (eq. [5]). In CH, such a competition between reactions [3] and [5] does not exist and the hydrogenolysis of C-O bonds occurs to a much greater extent. The lignin case constitutes another example of the drawback of the ECH method when the hydrogenation step [3] is much slower than the electrochemical desorption of M(H) (eq. [5]).

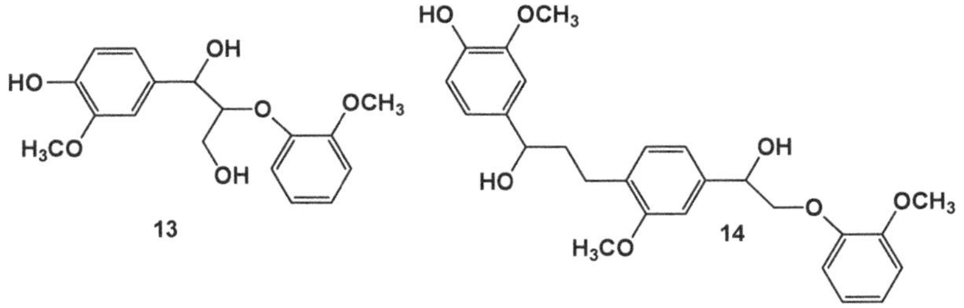

Figure 2 Lignin models.

Synthetic Aspects of Electrohydrogenation of Nitro Compounds at Raney Metal (RM) Electrodes in Aqueous Alcohols

The electrohydrogenation of nitro compounds at Raney metal (RM) electrodes in aqueous methanol or ethanol, in slightly acidic (pH = 3-5), neutral (pH = 6-8) or basic (pH 10 to >13) solution, is very efficient and gives the corresponding amines in high yields (80-100%) and current efficiencies (80-95%) for arylnitro as well as alkylnitro groups (1). By comparison, the electrohydrogenation at a high hydrogen overvoltage cathode (Hg, glassy carbon) in neutral (pH > 5) and basic medium stops at the hydroxylamine because unprotonated hydroxylamines are not reducible by electron transfer (EP mechanism) (10). For example, in methanol-water (93:7 w/w) and at RNi, RCo, or RCu electrodes, the controlled potential electrohydrogenation (working potential (E_w) ~ -1.0 V *vs* SCE in basic medium, ~ -0.7 to -0.9 V in neutral medium, H-cell with a Nafion-324 membrane) of nitrobenzene, nitrohexane, nitrocyclohexane, and *t*-nitrobutane gives exclusively the corresponding amine (aniline, hexylamine, cyclohexylamine, and *t*-butylamine respectively) both in basic (0.14 M KOH) and neutral (acetate or borate buffer) medium (11,12). It is noteworthy that electrohydrogenation at RCo or RCu electrodes is one of the most efficient methods to reduce *t*-alkylnitro groups to the corresponding amino goups.

The electrohydrogenation at RCo and RCu cathodes can be highly selective as shown by the following examples carried out under controlled potential (E_w *vs* a saturated calomel electrode (SCE) as reference electrode). E_w is referred to as the "zero-current potential" and is determined in the following way (1,11). The cathode is placed in the cell containing the electrolysis medium without the organic substrate and the potential is lowered in order to obtain a current density $J = 16$ mA / cm^2 (hydrogen evolution) for a total charge $Q = 100$ coulombs. Then the potential is slowly increased to get the zero-current point and the substrate is added at once, this addition causing a sudden increase in current which afterwards decreases as the substrate is hydrogenated (1,11).

Arylnitro groups

1) The electrohydrogenation of *p*-nitro-acetophenone (**15**) at a RCu cathode in an alkaline (0.28 M KOH) MeOH-H$_2$O (1.5% of H$_2$O) solution (E_w = -0.72 V) gives exclusively *p*-amino-acetophenone (**16**) (79% yield of isolated product, 85% current efficiency) (Scheme 4) (11).

15   **16**

Scheme 4

2) The conversion of *p*-nitro-nitrocumene (**17**) to *p*-amino-nitrocumene (**18**) at a RCo or RCu cathode (E_w = -0.9 V) is quantitative (by vapor phase

chromatography (VPC)) (70% current efficiency) in neutral methanol-water 95:5 (v/v) at pH = 6 (acetate buffer) (Scheme 5) (13).

Scheme 5

3) Under the same conditions as in example 2) above, *p*-cyanonitrobenzene (**19**) is quantitatively converted (by VPC) to *p*-cyanoaniline (**20**) at a RCo or RCu cathode (E_w = -0.9 V) (98% current efficiency) (Scheme 6) (14).

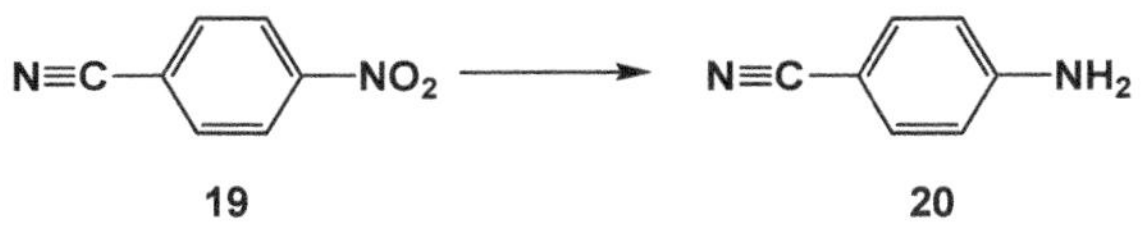

Scheme 6

4) Under the same conditions again but at a less negative potential (E_w = -0.7 V), nitro compound **21** is converted to quinolone **22** at a RCu cathode (53% yield of isolated product, 46% current efficiency), and nitro compound **23** to N-hydroxyquinolone **24** at a RCo cathode (85% yield, 64% current efficiency) (Scheme 7) (15). Interestingly, in the electrohydrogenation of **21**, the hydrogenolysis of the intermediate hydroxylamine to the amine is faster than its cyclisation to the corresponding N-hydroxyquinolone (which was synthesized by an independent method and was found unreactive under the electrohydrogenation conditions), whereas, in the electrohydrogenation of nitro compound **23**, the cyclisation of the intermediate hydroxylamine to N-hydroxyquinolone **24** is faster than its hydrogenolysis to the corresponding amine.

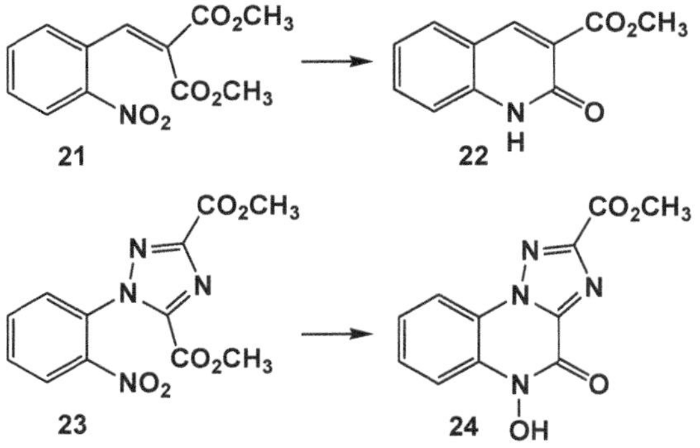

Scheme 7

5) The electrohydrogenation of nitro-dioxolane **25** in ethanol-water 60:40 (v/v) at a RCo (E_w = -1.06 V) or a RCu (E_w = -1.06 V) cathode gives the amino derivative **26** quantitatively (by VPC) (95% current efficiency) (Scheme 8) (15).

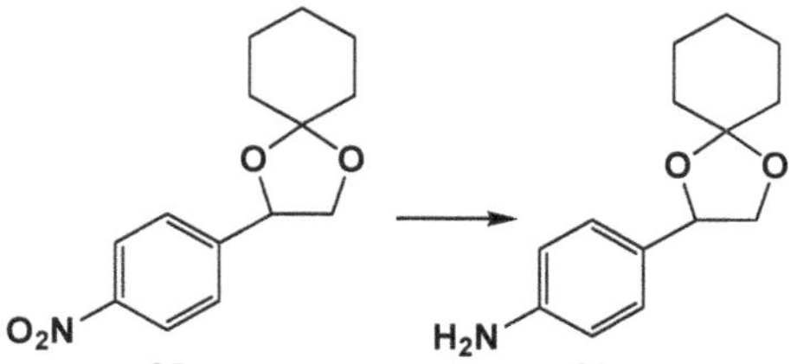

Scheme 8

With the more active RNi cathode, the electrohydrogenation is much less selective and hydrogenolysis of the benzylic C-O bond of **26** occurs to an appreciable extent to give *p*-aminophenethyl alcohol (**28**) and cyclohexanol (**30**) in 70% yield (30% yield of aminodioxolane **28**) after 8.9 F/mol (15). The hydrogenolysis of the benzylic C-O bond gives the intermediate hemiketal **27** which cleaves to *p*-aminophenethyl alcohol (**30**) and cyclohexanone (**29**) (Scheme 9). Cyclohexanone is electrohydrogenated further to cyclohexanol (**30**).

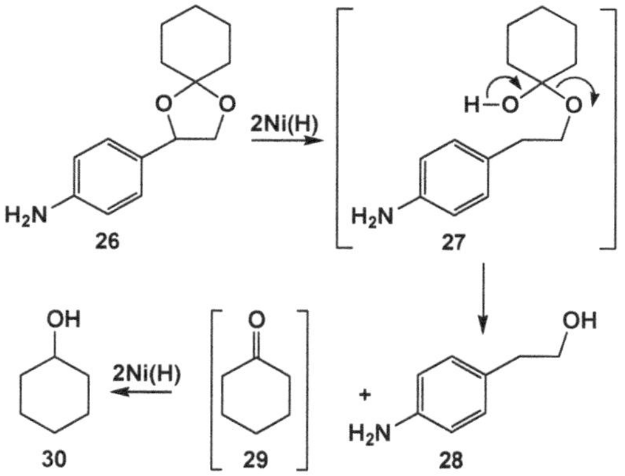

Scheme 9

6) A last example of a very selective electrohydrogenation of an arylnitro group is that of *o*-iodonitrobenzene (**31**) to *o*-iodoaniline (**32**) at pH 3 (pyridine·HCl buffer) and at a RCu cathode (E_w = -0.7 V) in methanol-water 95:5 (v/v) Scheme 10. The yield of *o*-iodoaniline (**32**) is 97% (by VPC) (14). Electrohydrogenation by classical electroreduction (EP mechanism) cannot be selective because of the rapid cleavage of the radical anion formed upon the first electron transfer. To our knowledge, such a high selectivity for the hydrogenation of **31** to **32** has never been reported before.

Scheme 10

Alkylnitro groups

1) The electrohydrogenation of *p*-cyano-nitrocumene (**33**) in methanol-water 95:5 (v/v), at a RCo or RCu electrode (E_w = -0.9 V) and at pH 5 (NaCl 0.1 M in the catholyte, acetate buffer in the anolyte), is highly selective giving exclusively *p*-cyano-aminocumene (**34**) (91% yield of isolated product, 33% current efficiency) (Scheme 11) (14).

Scheme 11

2) The nitroacetylenic acetal **35** is electrohydrogenated selectively to the corresponding amine **36** in 72% yield (by VPC) at a RCu electrode under the same conditions as those specified in 1) above (13). This hydrogenation could not be achieved by other methods (Scheme 12) (16).

Scheme 12

3) The electrohydrogenation of 3-nitro-2,3-dimethylbut-1-ene (**37**) to 3-amino-2,3-dimethylbut-1-ene (**38**), at a RCu electrode (E_w = -0.9 V) and in methanol-water 95:5 (v/v) at pH 6 (acetate buffer), is quantitative (by VPC) (Scheme 13) (17). Other methods of hydrogenation were unsuccessful (18).

Scheme 13

Mechanistic Aspects of Electrohydrogenation of Nitro Compounds at Raney Metal (RM) Electrodes in Aqueous Alcohols

Experimental facts and some mechanistic consequences

There are three categories of important experimental facts about the electrohydrogenation of nitro groups at large surface-area electrodes (such as Raney metal electrodes) in an aqueous alcoholic medium and these facts have mechanistic consequences.

1) The so-called "zero-current" potential (E_{zcp}) used in potentiostatic electrohydrogenations (E_w = -0.7 to -0.9 V in slightly acidic and neutral medium, -1.0 to -1.2 V in basic medium), which corresponds to the potential of reduction of water to chemisorbed hydrogen (eq. [1]), is either the same as or more negative than that of the reduction potential of nitro groups to the corresponding radical anion (eq. [7]). E_{zcp} is the same in the case of alkylnitro groups (for example, E_p for the cyclic voltammetric reduction of nitrocumene (**39**) (see Scheme 15 for its structure) at Hg (19) is -0.75 V at pH 3, -0.9 V at pH 7, and -1.1 V vs. SCE in basic medium (pH >10)) and it is more negative in the case of arylnitro groups (for example, E_p for the electroreduction of *o*-iodobenzene (**31**) at Hg (20) is -0.58 V at pH 3, -0.74 V at pH 7 and -0.90 V at pH > 13). As a consequence, there are, a priori, two possible mechanisms for the reduction of a nitro group to the hydroxylamine: an electronation-protonation (EP) mechanism (eqs. [7] to [1]) or an electrocatalytic hydrogenation (ECH) mechanism (eq. [1)] then [1] to [17]). Interestingly, since the hydrogenolysis of hydroxylamines to amines (eq. [18]) occurs readily at Raney metal electrodes under the various conditions studied, there is the possibility of direct hydrogenolysis of the adsorbed dihydroxylamine to the adsorbed hydroxylamine in an ECH mechanism as shown in equation [14]. This could be the main pathway if the hydrogenolysis of the adsorbed dihydroxylamine (eq. [14]) is faster than its dehydration to the nitroso derivative (eq. [15]) and(or) its diffusion away from the electrode surface.

2) Unprotonated hydroxylamines are not reducible by electron transfer (10) as already mentioned and the electrohydrogenation of nitro compounds at Raney metal electrodes in neutral and basic aqueous alcoholic media gives the corresponding amines as shown by the numerous examples illustrated above. Therefore, in these media, hydroxylamines must be reduced to the amine by an ECH mechanism (eq. [1] followed by eqs. [18] and [19]).

3) Nitro groups attached to a primary and secondary alkyl group in a highly basic (pH > 13) medium exist as the nitronate (enolate) anions. These anions must be very difficult to reduce by electron transfer and are surely much more difficult to reduce than water. Since the electrohydrogenation of such nitro compounds to the corresponding amines is very efficient at Raney metal cathodes in 0.1 to 0.15 M KOH (or NaOH) aqueous alcohol (pH > 13) (12), as

already pointed out, there is only one possible mechanistic pathway for their hydrogenation to the corresponding hydroxylamine, the ECH pathway illustrated in Scheme 14. Hydrogenolysis to the amine proceeds necessarily by ECH as pointed out above (eqs. [18] and [19]).

$$RNO_2 + e^- \rightleftharpoons RNO_2^{\cdot -} \qquad [7]$$

$$RNO_2^{\cdot -} \xrightarrow{H_2O} RNO_2H^{\cdot} + OH^- \qquad [8]$$

$$RNO_2H^{\cdot} + e^- \rightleftharpoons RNO_2H^- \qquad [9]$$

$$RNO_2H^- \rightleftharpoons RNO + OH^- \qquad [10]$$

$$RNO + 2e^- \xrightarrow{H_2O} RNHOH + OH^- \qquad [11]$$

$$RNO_2 + M \rightleftharpoons M(RNO_2) \qquad [12]$$

$$M(RNO_2) \xrightarrow{2M(H)} M(RN(OH)_2) \qquad [13]$$

$$M(RN(OH)_2) \xrightarrow{2M(H)} M(RNHOH) + H_2O \qquad [14]$$

$$M(RN(OH)_2) \rightleftharpoons M(RNO) + H_2O \qquad [15]$$

$$M(RNO) \xrightarrow{2M(H)} M(RNHOH) \qquad [16]$$

$$M(RNHOH) \rightleftharpoons RNHOH + M \qquad [17]$$

$$M(RNHOH) \xrightarrow{2M(H)} M(RNH_2) + H_2O \qquad [18]$$

$$M(RNH_2) \rightleftharpoons RNH_2 + M \qquad [19]$$

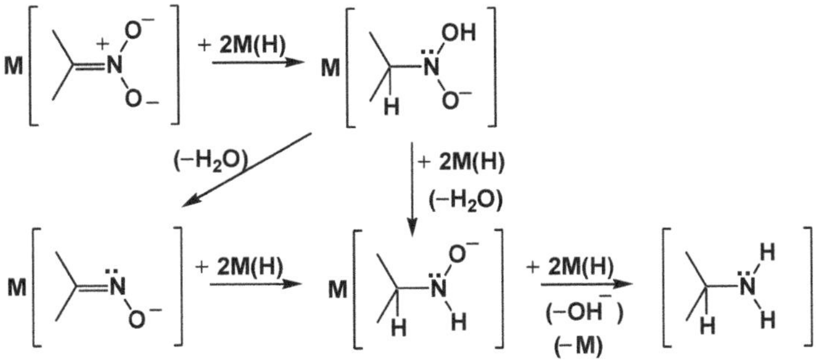

Scheme 14

Nitrocumene (39) to probe the mechanism of electrohydrogenation of alkylnitro groups

In order to determine whether the electrohydrogenation of alkylnitro groups at Raney metal (large surface-area electrodes) occurs by an ECH mechanism or by an EP mechanism, or if both mechanisms could be involved simultaneously (could be in competition), nitrocumene (39) was used as a probe. The cleavage of the nitrocumyl radical anion (40) to the cumyl radical (41) and the nitrite anion is fast enough (k_c = 3 x 10^6 s^{-1} in acetonitrile containing trietlhylammonium perchlorate (21)) to compete with its protonation in basic (0.15 M KOH, pH > 13) ethanol-water 40:60 (w/w). Indeed, as shown in Scheme 15, when nitrocumene (39) was reduced at a Hg electrode (E_w = -1.13 V) under those conditions, bicumyl (43) was obtained in 85% yield along with 2% cumene (42) (22). The other products (about 10%) were derived from hydroxylaminocumene (not shown) so that the ratio of the rate of cleavage to the rate of protonation of radical anion 40 is about 9 at pH > 13 in aqueous ethanol. At pH values below 10, protonation competes effectively with the cleavage of radical anion 40 (19).

When nitrocumene (39) was electrohydrogenated in the same basic medium as above but at Raney metal electrodes, aminocumene (44) was obtained as expected from an ECH mechanism pathway (eq. [1] followed by eqs. [12] to [19]) but bicumyl (43), resulting from electronation of nitrocumene (39) (Scheme 2), was formed also (Scheme 15) (22).

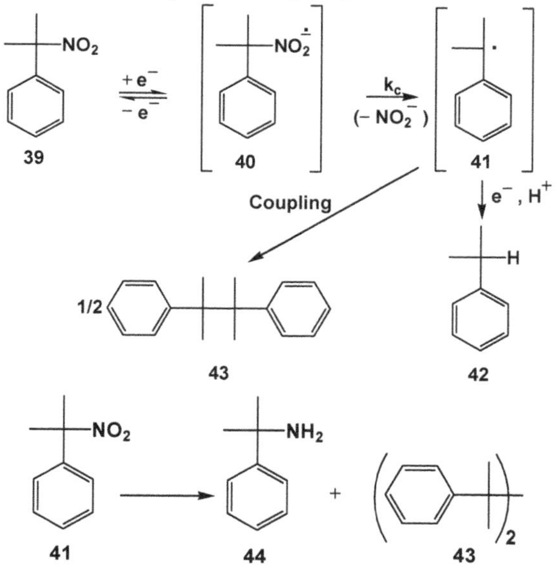

Scheme 15

The results of Table 1 show clearly that electronation of nitrocumene (**39**) (Scheme 15) does compete with reaction with chemisorbed hydrogen, M(H), at some stage in the electrohydrogenation process. The simplest interpretation is a direct competition between electronation of the nitro compound (eq. [7]) and reaction of the adsorbed nitro compound with chemisorbed hydrogen, M(H) (eq. [13]). However, it is quite possible that the electronation of the adsorbed nitro compound (eq. [20]) could be faster than its reaction with M(H) (eq. [13]) and the competition would then be between the cleavage of the adsorbed radical anion (eq. [21]) and its reaction with M(H) (eq. [22]).

$$M(RNO_2) + e^- \rightleftharpoons M(RN\dot{O_2^-}) \qquad [20]$$

$$M(RN\dot{O_2^-}) \longrightarrow M(R^\bullet) + NO_2^- \qquad [21]$$

$$M(RN\dot{O_2^-}) \xrightarrow{2M(H)} M(RNO_2H^-) \qquad [22]$$

It is noteworthy that the relative proportion of amine **44** and bicumyl (**43**) which reflects the ratio of the rate of electronation to the rate of reaction with M(H) (the competition between electronation and reaction with M(H)), varies with the Raney metal (compare entries 1 and 3 of Table 1, and entries 2 and 4) and with the electrode potential (compare entries 1 and 2). The more negative is the potential, the faster is the rate of electronation and the higher should be the proportion of bicumyl (**43**) as observed (entries 1 and 2). The less active the Raney metal as hydrogenation catalyst, the slower is the rate of reaction with M(H) (the lower is the amount of M(H) at the surface of the electrode) and the lower is the amount of aminocumene (**44**). RCu is the least active catalyst and the proportion of aminocumene (**44**) is indeed the lowest at the RCu cathode (entry 4).

Table 1 Electrohydrogenation of nitrocumene (**39**) at Raney metal (RM) electrodes in EtOH-H$_2$O 40:60 (w/w) (22).

Entry	RM	E_w (V)[a]	Amine **44** (%)[b]	Bicumyl **43**(%)[b]
1	RNi	-1.10[c]	97	2
2	RNi	-1.20	75	20
3	RCo	-1.10[c]	75	20
4	RCu	-1.20[c]	19	76

[a] Working potential *vs.* SCE.
[b] Yield determined by VPC.
[c] Zero-current potential (E_{zcp}, see text).

Conclusions

Electrohydrogenation at Raney metal electrodes is a mild method of hydrogenation, the advantages and disadvantages of which have been pointed out in the Introduction and have been illustrated in the paper with selected examples. In particular, electrohydrogenation of nitro compounds, at Raney copper electrodes and in neutral and basic aqueous alcohols, is a very efficient and very selective method of hydrogenation to the corresponding amines. There is a competition between two mechanistic pathways in the electrohydrogenation of nitrocumene (**39**) to the corresponding hydroxylamine at Raney metal electrodes in basic medium, electrocatalytic hydrogenation (ECH) and electronation-protonation (EP). Since arylnitro groups are more easily electronated than nitrocumene (**39**) by 200 mV, their electrohydrogenation to arylhydroxylamines at Raney metal electrodes must proceed mainly through an EP mechanism. Conversion of the hydroxylamine to the amine involves an ECH mechanism.

Acknowledgements

I gratefully acknowledge the contribution of M.Sc. and Ph.D. students, postdoctoral fellows, professional researchers, and colleagues. Their names can be seen from the list of references. I am most thankful also to the NSERC of Canada, the Fonds FCAR of Quebec, the Ministry of Energy and Natural Resources of Quebec, and the Université de Sherbrooke for financial support.

References

1. J.M. Chapuzet, A. Lasia, and J. Lessard, *In* Electrocatalysis, J. Lipkowski and P.N. Ross, Eds., Wiley-VCH, 1998, p. 155 and references therein.
2. A. Cyr, F. Chiltz, P. Jeanson, A. Martel, L. Brossard, J. Lessard, and H. Ménard, *Can. J. Chem.*, **78**, 307 (2000).
3. B. Mahdavi, P. Chambrion, J. Binette, E. Martel, and J. Lessard, *Can. J. Chem.*, **73**, 846 (1996).
4. R. Menini, A. Martel, H. Ménard, J. Lessard, and O. Vittori, *Electrochim. Acta*, **43**, 1697 (1998).
5. M. Hudlicky, Reduction in Organic Chemistry, Wiley & Sons, New York, 1984, p. 52.
6. G. Belot, S. Desjardins, and J. Lessard, unpublished results.
7. M.L. Wolfrom, W.W. Binkley, C.C. Spencer, and B.W. Lew, *J. Am. Chem. Soc.*, **73**, 3357 (1951).
8. P. Jeanson, *M. Sc. Dissertation*, Université de Sherbrooke, 2001.
9. B. Loubinoux, French Pat. 8015272 (1980).
10. H. Lund, *in* Organic Electrochemistry, H. Lund and O. Hammerich, Eds, Marcel Dekker, Inc., New York-Basel, 2001, p. 379.
11. G. Belot, S. Desjardins, and J. Lessard, *Tetrahedron Lett.*, **25**, 5347 (1984).
12. P. Delair, A. Cyr, and J. Lessard, *in* Electroorganic Synthesis, Festschrift for Manuel Baizer, R.D. Little and N. L. Weinberg, Eds., Marcel Dekker, Inc., New York-Basel-Hong Kong, 1991, p. 129.
13. J. M. Chapuzet, B. Côté, M. Lavoie, E. Martel., C. Raffin, and J. Lessard, *in* Novel Trends in Electroorganic Synthesis, S. Torii, Ed., Kodansha, Tokyo, 1995, p. 321.

14. J.M. Chapuzet, R. Labrecque, M. Lavoie, E. Martel, and J. Lessard, *J. Chim. Phys.*, **93,** 601 (1996).
15. J.M. Chapuzet, C. Godbout, and J. Lessard, unpublished results.
16. J. Goré, personal communication. We thank Professor Goré (Université de Lyon, France) for a generous gift of compound **35**.
17. J. M. Chapuzet and J. Lessard, unpublished results.
18. P.S. Engel, personal communication. We thank Professor Engel (Rice University) for a generous gift of compound **37**.
19. E.S. Chan-Shing, *Ph.D. Thesis*, Université de Sherbrooke, 1999.
20. B.J. Côté, D. Després, R. Labrecque, J. Lamothe, J.M. Chapuzet, and J. Lessard, *J. Electroanal. Chem.*, **355**, 219 (1993).
21. Z-R. Zheng, D.H. Evans, E.S. Chan-Shing, and J. Lessard, *J. Am. Chem. Soc.*, **40**, 9429 (1999).
22. E.S. Chan-Shing, D. Boucher, and J. Lessard, *Can. J. Chem.*, **77**, 687 (1999).

2.   Low Pressure Slurry Hydrogenation Process for the Production of Dimethylaminopropylamine Using Sponge Nickel Catalyst

Gregory J. Ward and Bryan C. Blanchard

Solutia, Inc., P.O. Box 97, Gonzalez, FL 32560-0097

Abstract

A process for the production of 3-dimethylaminopropylamine (DMAPA) in high (>99%) purity from N,N-dimethylaminopropionitrile (DMAPN) utilizing a low pressure slurry hydrogenation process is described. The basic process comprises contacting the nitrile with hydrogen at low pressure in the presence of a sponge nickel catalyst under conditions sufficient to effect the conversion of the nitrile to the primary amine product. The improvement in the process resides in a combination of carrying out the hydrogenation process at very low pressure in the presence of an optimum amount of caustic and sponge nickel catalyst in order to give an improved selectivity of greater than 99.9 % of DMAPN to DMAPA. This process can be readily adapted to the established Solutia low pressure hydrogenation process, which employs a continuous gas lift slurry reactor. In this process, a decanter is used to perform the bulk catalyst separation, and entrained catalyst and fines in the decanter effluent are removed by hydroclones. Complete removal of catalyst from the crude amine can be obtained with this system with minimal maintenance and high reliability. This system has a variety of advantages over other competing technologies including high selectivity, low catalyst usage, low capital cost, high reliability, established sources of catalyst, and no need to completely recharge catalyst to reactor. Solutia has been practicing this technology for over 30 years for the production of hexamethylenediamine, and considers it the best technology available to convert nitriles to primary amines.

Introduction

The most common and least expensive catalyst for producing primary amines from nitriles is sponge nickel. The generalized reaction, carried out in the presence of sponge nickel catalyst, is the following:

$$R\text{-}C\equiv N + 2H_2 \ \rightarrow \ R\text{---}CH_2\text{-}NH_2$$

Solutia has been producing hexamethylenediamine via low pressure slurry hydrogenation of adiponitrile since 1973. This process can also been adapted for the production of other amines such as DMAPA. The catalyst employed for this process is a promoted sponge nickel catalyst, and the reaction is carried out in a continuous gas lift reactor as shown in Figure 1.

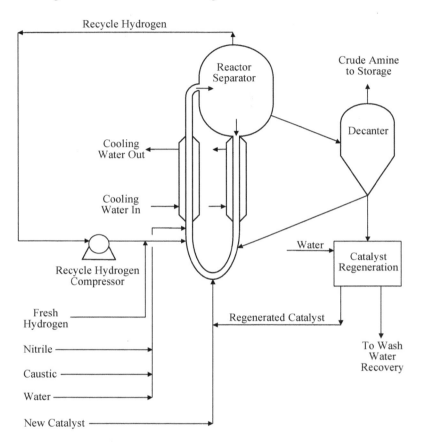

Figure 1 Solutia low pressure amine synthesis process.

Recycle hydrogen is injected into the bottom of the upleg. This produces a very high circulation rate of the reaction mass without high maintenance equipment such as pumps or agitators. Unreacted hydrogen is separated from the reaction mass in the separator at the top of the reactor. A compressor is used to recycle the unreacted hydrogen back to the base of the upleg. By maintaining steady recycle hydrogen flow, good circulation rates of the reaction mass are assured, even at low production rates. Nitrile is injected into the reactor leg just above the hydrogen inlet. It reacts quickly with the hydrogen on the surface of the promoted sponge nickel catalyst. The reaction goes essentially to completion producing high purity crude amine product with typically less than one percent organic byproducts. The heat of reaction is removed by tempered water in cooling jackets around the legs.

The reaction mass consists of two liquid phases and one solid phase; no solvent is required. The major liquid phase is the crude amine product itself. The solid phase is promoted sponge nickel catalyst. Surrounding the catalyst is a second liquid phase consisting of concentrated caustic and water. Water and caustic are added continuously to make up for losses leaving in the crude product. The ratios of water, caustic, and catalyst in the reaction mass are controlled to produce high yields of product amine and very low catalyst usages. High catalyst concentrations are employed in the reaction mass to keep the concentration of unreacted nitriles very low; the upper limit on the catalyst concentration is the point where the circulation rate is inhibited.

The crude amine product is taken from the reactor through a decanter as shown in Figure 1. In the decanter, the clarified crude amine product is continuously separated from the catalyst, and the catalyst is returned to the reactor through the decanter under flow line as shown in Figure 1. Entrained catalyst and fines in the decanter effluent are removed by hydroclones. Complete removal of catalyst from the crude amine is important to prevent fouling and product degradation during downstream processing.

Periodically, a portion of the catalyst slurry is purged from the reactor for regeneration. In the catalyst regeneration process, the catalyst is washed with water to remove impurities that accumulate in the caustic phase. Most of the regenerated catalyst is returned to the reactor along with new catalyst. With this configuration, fresh catalyst can be added as required to maintain acceptable catalyst activity without the need to replace the entire reactor charge in a batchwise manner.

Recent publications and patents have suggested selectivity enhancements over the process practiced by Solutia (1, 2) are possible by adjusting the chemical additives used in the process (3, 4) or that fixed bed catalysts are better suited for continuous processes than slurry catalyst (5, 6). These claims have prompted workers at Solutia to re-evaluate the technology employed for production of hexamethylenediamine and to evaluate Solutia low pressure slurry hydrogenation technology for chemistries other than hexamethylenediamine. Specifically, hydrogenation of DMAPN to produce DMAPA (9) was selected to study due to a relatively large number of recent publications pertaining to the production of DMAPA (3, 4, 7, 8).

Experimental Section

Semi-batch hydrogenation involves feeding the nitrile to an autoclave, containing a slurry of catalyst in the reaction product, for a specified time period after which time the nitrile feed is stopped. After the nitrile feed is stopped, the reaction will continue for a short period of time while the residual unreacted nitrile is consumed. Close monitoring of the hydrogen uptake during the time period after the nitrile feed is stopped provides insight into the rate at which catalyst deactivation is occurring, since the hydrogen uptake rate between cycles will decrease noticeably if the catalyst is experiencing significant deactivation. The period from the time the nitrile feed is started until the time the nitrile feed is stopped is referred to as a cycle. After the cycle is completed, a quantity of reaction product equal to the quantity of nitrile fed to the reactor is withdrawn from the autoclave. The feed is then resumed under the same conditions as before. Semi-batch operation is an excellent tool for evaluating catalysts and the effects of process changes.

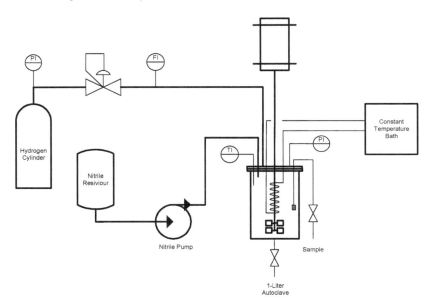

Figure 2 Experimental apparatus.

The experimental apparatus is shown in Figure 2. The reaction vessel is an Autoclave Engineers one-liter autoclave reactor equipped with double turbine blades, dispersimax-type agitator, a coil extending to the bottom to circulate the transfer fluid from a temperature controlled bath for temperature control, and a fritted, stainless steel metal sample port below the liquid level. Hydrogen is fed from a cylinder equipped with a pressure gauge and a regulator. Hydrogen is continuously added to the reactor as hydrogen is consumed by the reaction to maintain a constant reactor pressure. The hydrogen flows through a mass flow meter. The DMAPN (obtained from Acros®) is pumped to the autoclave with an Isco Model 500D syringe pump.

Caustic preparation begins with obtaining distilled water that has been boiled to remove dissolved carbon dioxide. Caustic solutions are prepared in 100 gram batches containing about 25% caustic by weight. The caustic (KOH,

NaOH, etc.) is added to the degassed water (~ 60 ml) with stirring. After complete dissolution of the caustic, additional water is added to bring the weight of the solution to a total weight of 100 grams. The solution is filtered, and stored in a closed container until use in order to minimize adsorption of CO_2 from the air.

The catalyst used for this reaction is sponge nickel catalyst (Degussa MC502) that contains iron and chromium to promote the hydrogenation reaction (the catalyst contains approximately 86% nickel, 10% aluminum, 2% chromium, and 2% iron). 37.5 grams of catalyst is washed 3 times with water and 3 times with DMAPA (Acros; contaminated with 72 ppm TMPDA by GC analysis), each wash consisting of mixed catalyst and material in a 100 ml graduated cylinder, settling the catalyst, and decanting the top 50 ml of clear liquid.

The catalyst slurry in DMAPA is then charged to the autoclave. Additionally, 265 ml of 100% DMAPA and 6 - 8 mL of 25% (wt.) caustic solution in water is charged. The agitator is turned on, and the autoclave heated to 60 °C. The autoclave is then purged three times with nitrogen, and then three times with hydrogen, before being pressurized to 100 psig with hydrogen. The autoclave is then heated to 90 °C, and the pressure is monitored for 5 minutes with no hydrogen addition to confirm there are no leaks.

After the autoclave is charged with catalyst slurry, DMAPA and aqueous caustic solution, the feed of DMAPN is then started to the autoclave at a rate of 5 ml/minute using the syringe pump. Pressure and temperature are maintained at 100 psig and 90 °C, respectively, during the entirety of the run. After 27 minutes, the feed is stopped, and a 150 g sample is withdrawn from the autoclave for analysis. The feed is then resumed under the same conditions as before. This procedure is then repeated for a total of 6 - 8 cycles.

Analytical Section

The reaction mixture is sampled after each reaction cycle and analyzed for purity, reaction progress, and the presence and amount of by-products (if any) formed. Analysis is by gas chromatography (HP 5890 Series II; Phenomenex Zebron ZB-1 capillary column, Phenomenex Cat. No. 7HK-G001-36) with flame ionization detection in order to quantify the by-product impurities. The quantity of the byproducts is determined using an external standard calibration method.

Results and Discussion

In the initial test, 6 ml of 25% (wt.) caustic solution in water was charged to the autoclave with 37.5 grams of catalyst. The caustic used in this run was a blend containing 50 wt. % sodium hydroxide and 50 wt. % potassium hydroxide. The feed of DMAPN containing 0.04 wt. % water was then started to the autoclave at a rate of 5 ml/minute using the syringe pump. Pressure and temperature were maintained at 100 psig and 90 °C, respectively, during the run. After 27 minutes, the feed was stopped, and a 150 g sample was withdrawn from the autoclave for analysis. The feed was then resumed under the same conditions as before. This procedure was then repeated for a total of 7 cycles. The reaction mixture was sampled after each reaction cycle, and analysis of the impurities in the reaction product, with the balance being DMAPA, is given in Table 1.

Table 1 Product analysis.

Cycle	NPA (ppm)	DAP (ppm)	DMAPN (ppm)	TMPDA (ppm)	2°Amine (ppm)	Water Wt. %
1	89	ND	ND	43	200	7.26
2	123	ND	ND	29	267	4.72
3	143	ND	ND	17	284	3.47
4	174	ND	ND	14	308	2.65
5	190	ND	ND	7	265	2.11
6	209	ND	ND	5	236	1.71
7	223	ND	ND	ND	214	1.37

ND = Not Detected
NPA = n-propylamine
DAP = 1,3-diaminopropane
DMAPN = dimethylaminopropionitrile
TMPDA = N,N,N',N'-tetramethyl-1,3-propanediamine
2° Amine = 3,3'-iminobis(N,N-dimethylpropylamine)

The data in Table 1 shows that the amount of 2° Amine remains generally at or below 300 ppm over the course of the seven reaction cycles indicating that this particular blend of caustic is very effective at suppressing secondary amine formation. NPA is produced via decyanoethylation of the DMAPN followed by hydrogenation of the liberated acrylonitrile. Also, TMPDA is not formed in this process. It is introduced into the system with the DMAPA used to slurry the catalyst charged to the autoclave, and it is purged out by the seventh cycle. Conversion is 100% on all of the cycles as evident by no DMAPN detected in the reaction product. Although not reported in Table 1, no evidence of catalyst deactivation was detected between cycles as measured by the hydrogen uptake after the nitrile feed was stopped. No water was added to the system other than the water that was added with the aqueous caustic solution charged to the autoclave with the catalyst. As the data in Table 1 shows, the water concentration decreased with each cycle as the contents of the autoclave are turned over. If additional cycles had been run, it would have been necessary to add water to the nitrile feed to maintain a constant water concentration in the system since some amount of water is required to maintain good activity and selectivity. Finally, as the data in Table 1 shows, the crude DMAPA product is produced with a molar yield of approximately 99.95%, no TMPDA production, and less than 300 ppm of the secondary amine present in the final product.

Next, a series of runs was conducted to determine the effect of various alkali metal hydroxide additions along with the sponge nickel catalyst. The 50 wt. % sodium hydroxide and 50 wt. % potassium hydroxide caustic solution used in the initial test was replaced with an aqueous solution of the alkali metal hydroxide at the level indicated in Table 2. After the reaction number of cycles indicated in Table 2, a sample was removed for analysis. The conditions and results are shown in Table 2. The results reported in Table 2 show the level of 2° Amine in the product from the final cycle. The level of NPA in all of the runs was comparable to the level observed in the initial test. No significant levels of other impurities were detected.

Tables 1 and 2 clearly show that the use of such alkali metal hydroxides as KOH, CsOH, and mixtures of KOH/NaOH allowed the reaction to proceed to a high DMAPN conversion with a very high selectivity for the primary amine. These results suggest that the highest selectivity in the hydrogenation of DMAPN to DMAPA is obtained with KOH, and mixtures of KOH/NaOH.

Table 2 Effect of alkali metal hydroxide on activity and selectivity.

Run	Alkali Metal Hydroxide	Metal Hydroxide Quantity	Temperature (°C)	Pressure (PSIG)	Number of Cycles	2° Amine (ppm)
1	CsOH	8 ml of 25 % (wt) aqueous CsOH	90	500	6	15
2	LiOH	80 ml of 10 % (wt) aqueous LiOH	90	500	6	2567
3	NaOH	8 ml of 25 % (wt) aqueous NaOH	90	110	6	1440
4	KOH	6 ml of 25 % (wt) aqueous KOH	90	100	6	9
5	RbOH	7 ml of 25 % (wt) aqueous RbOH	90	100	6	1463
6	KOH/NaOH	6 ml of 25 % (wt) aqueous 50/50 NaOH/KOH	90	100	6	9

Another important feature of this reaction is the low pressure at which the reaction proceeds. Unlike hexamethylenediamine or other amines produced with this process, the hydrogenation of DMAPN to DMAPA proceeds at very low pressures. High catalyst activity and high selectivity are obtained at 100 psig for NaOH, KOH, RbOH and blended NaOH/KOH. Testing with CsOH and LiOH was only conducted at 500 psig, and these tests were not repeated at 100 PSIG for CsOH and LiOH due to time constraints.

Acknowledgements

The authors would like to thank Solutia, Inc. for permission to publish work that is the basis for a US patent (9). We also wish to thank our numerous colleagues at Solutia for developing the methods and establishing the technology that are the basis for this study. We are indebted to those who have worked for many years to establish this technology as the premier technology for converting nitriles to primary amines.

References

1. Charles R. Campbell and Arthur D. Hufford, *Montifibre-Monsanto Hexamethylenediamine Process*, 1984.
2. G. Bartalini and M. Giuggioli, to Montedison Fibre S. p. A., US 3,821,305, June 28, 1974.
3. Thomas A. Johnson and Douglas P. Freyberger, Catalysis of Organic Reaction (82), Marcel Dekker, Inc., New York, 201-227 (2001).
4. Thomas Albert Johnson, to Air Products and Chemicals, Inc., US 5,869,653 February 9, 1999.
5. John D. Super, Catalysis of Organic Reaction (82), Marcel Dekker, Inc., New York, 35-49 (2001).
6. Barbara Bender, Monika Berweiler, Konrad Moebus, Daniel Ostgard, and Gernot Stein, to Degussa-Huels A.-G., US 6,284,703, September 4, 2001.
7. Andreas Ansmann, Christoph Benisch, to BASF AG, US Patent Application 20030120115, June 26, 2003.
8. Jiri Krupka, Josef Pasek, Marketa Navratilova, *Coll. Czech. Chem. Commun.*, Vol. 65 (11), 1805-1819 (2000).
9. Gregory J. Ward and Bryan C. Blanchard, to Solutia Inc., US Patent Application Serial No. 10/327,765, filing date December 23, 2002.

3. The Conversion of Ethanolamine to Glycine Salt Over Skeletal Copper Catalysts

Dongsheng Liu, Noel W. Cant, Liyan Ma, and Mark S. Wainwright

School of Chemical Engineering and Industrial Chemistry, The University of New South Wales, UNSW Sydney 2052, Australia

m.wainwright@unsw.edu.au

Abstract

The oxidative dehydrogenation of ethanolamine to sodium glycinate in 6.2 M NaOH was investigated using unpromoted and chromia promoted skeletal copper catalysts at 433 K and 0.9 MPa. The reaction was first order in ethanolamine concentration and was independent of caustic concentration, stirrer speed and particle size. Unpromoted skeletal copper lost surface area and activity with repeated cycles but a small amount of chromia (ca. 0.4 wt%) resulted in enhanced activity and stability.

Introduction

The oxidative dehydrogenation of ethanolamines to their corresponding aminocarboxylic acids is of considerable commercial interest for the production of agricultural and other chemicals. The reactions can be carried out in an alkaline environment without a catalyst but the reaction is slow with low yields (1). The first attempts to develop a catalyst were carried out with water absent. Chitwood (2) disclosed that cadmium oxide exhibited the highest activity but yields were moderate and the use of cadmium oxide is highly undesirable due to its toxicity. Copper was found to exhibit an activity only slightly less than that of cadmium oxide but the duration of maximum catalytic activity was shorter. Goto et al. (1) invented a new method in which copper, or a mixture of copper and zirconium compounds, was used in the presence of aqueous NaOH or KOH to manufacture aminocarboxylic acid salts more quickly with very high yields under milder reaction conditions. More recent patents carry claims for improved copper catalysts and for their use to produce other target compounds in a similar way (3-6).

Recently, a novel process for the preparation of chromia promoted skeletal copper catalysts was reported by Ma and Wainwright (8), in which Al was selectively leached from $CuAl_2$ alloy particles using 6.1 M NaOH solutions containing different concentrations of sodium chromate. The catalysts had very high surface areas and were very stable in highly concentrated NaOH solutions at temperatures up to 400 K (8, 9). They thus have potential for use in the liquid phase dehydrogenation of aminoalcohols to aminocarboxylic acid salts.

The oxidative dehydrogenation of aminoalcohols has received little attention in the non-patent literature. Yang et al. (7) recently made a kinetic study of the dehydrogenation of ethanolamine to glycine salts

$$NH_2CH_2CH_2OH + OH^- \rightarrow NH_2CH_2COO^- + 2H_2 \tag{1}$$

using a Cu-ZnO catalyst and reported a first order dependence on ethanolamine concentration. The activation energy for the reaction was found to be 148kJ/mol and a two-step reaction mechanism was proposed. The purpose of the current investigation is to investigate unpromoted and chromia promoted skeletal copper catalysts for the liquid phase dehydrogenation of ethanolamine to sodium glycinate. The reaction orders, the effects of stirrer rate and catalyst particle size on the reaction rate, and the effect of chromia on catalyst performance have all been determined.

Results and Discussion

Chitwood (2) found that copper compounds exhibited only a short period of maximum catalytic activity for the dehydrogenation of ethanolamine to glycine salt. In this study, the catalytic activity of a skeletal copper catalyst was tested in repeated use. The catalyst used was prepared by selectively leaching CuAl$_2$ particles in a 6.1 M NaOH solution at 293 K for 24 hours. Figure 1 shows the profiles of hydrogen evolved versus reaction time.

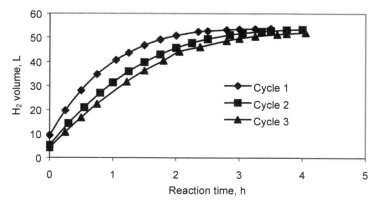

Figure 1 Hydrogen evolution during repeated cycles of ethanolamine dehydrogenation over an unpromoted skeletal copper under standard conditions.

Clearly the rate of reaction declines in repeated cycles. However the decrease between cycle 2 and cycle 3 was considerably smaller than that between cycle 1 and cycle 2, indicating that loss of activity occurred primarily in the first cycle. This is due to loss of surface area as discussed later.

The effect of reactant concentrations on reaction rate was studied using unpromoted skeletal copper catalysts initially leached at 278 K and then pretreated in 6.2 M NaOH at 473 K for 6 hours. The experiments were performed with varying the initial concentrations of ethanolamine and NaOH while keeping the total liquid volume constant by varying the amount of water accordingly. Figure 2 indicates that when the concentration of ethanolamine was halved, the initial rate (L_{H2} h^{-1}) was also approximately halved, so the reaction is first order with respect to ethanolamine concentration. However, when the initial concentration of NaOH was halved, the rate was almost unchanged, so the reaction is zero order with respect to NaOH concentration. Therefore, the rate equation for this reaction is given by $r = k$ C_A, where C_A is the concentration of ethanolamine in the reactant mixture. Hence a first order plot of $-\ln(1-X)$, where X is the conversion of ethanolamine, versus reaction time t is expected to be linear in experiments carried our with the concentration of NaOH equal to or greater than that of ethanolamine. It was not possible to carry out experiments with ethanolamone in excess, to further test the dependence on NaOH, since such conditions led to the formation of oligomeric byproducts.

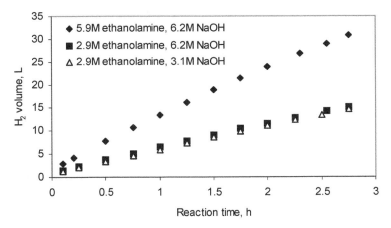

Figure 2 The effect of different reactant concentrations on hydrogen evolution during ethanolamine dehydrogenation over unpromoted skeletal copper at standard conditions.

Two techniques were available to monitor the extent of the dehydrogenation reaction. Multiple sampling of the liquid phase and measuring the glycine salt by HPLC has the disadvantages that it changes the volume of the reaction mixture during the experiment and is only available post run. Measurement of H_2 evolution is continuous but can provide an accurate measure of conversion only if the stoichiometry follows reaction (1) throughout. In order to validate the use of the H_2 evolution method, experiments were conducted over unpromoted skeletal copper under standard reaction conditions with liquid samples being withdrawn and analyzed whilst continuous measurements of H_2 evolution were being made. Equation 2 was used to estimate the conversion of ethanolamine from the volume of H_2 observed at time t, V_{H2}^t, and the volume of hydrogen expected for complete conversion of the starting weight of ethanolamine according to reaction (1), V_{H2}^{max}, i.e.

$$X = V_{H2}^t / V_{H2}^{max} \tag{2}$$

As may be seen from Figure 3, first order plots based on the conversions obtained by both methods were linear with slopes in close agreement. The apparent rate constant is 0.57 h^{-1} based on H_2 evolution and 0.54 h^{-1} based on the formation of glycine salt thereby indicating that it is valid to use the simpler H_2 evolution measurements when measuring the effects of process variables on the reaction.

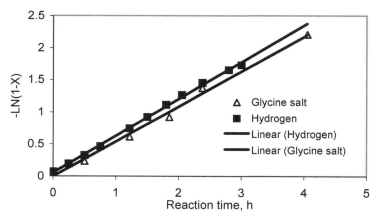

Figure 3 Comparison of first order plots for the formation of hydrogen and of glycine salt during ethanolamine dehydrogenation over unpromoted skeletal copper under standard conditions.

In order to investigate the relationship between the surface area of skeletal copper and activity, the same sample of catalyst was tested in four successive runs. Rate constants was compared with that of another sample prepared in the same way but pretreated in 6.2 M NaOH at 473 K before use. Figure 4 shows that the first order rate constants, calculated so as to take into account the mass of catalyst relative to the volume of solution, decreased in the first three cycles but then stabilised. The surface areas, measured on small samples taken after reaction, mirrored this pattern. The rate constant, and the surface area, for the pretreated catalyst was similar to those obtained in cycles 3 and 4. It is apparent that activity and surface area are closely related for the unpromoted skeletal copper catalyst and that the pretreatment in NaOH at 473 K is approximately equivalent to three repeated reactions in terms of stabilising activity and surface area.

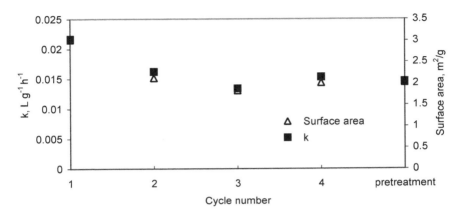

Figure 4 Effect of repeated reaction cycles and catalyst pretreatment in 6.2 M NaOH at 473 K on rate constant and catalyst surface area for ethanolamine dehydrogenation over unpromoted skeletal copper under standard conditions.

Dehydrogenation of ethanolamine produces large quantities of hydrogen, which may assist in mass transfer processes. In order to investigate the effect of stirrer rate on reaction rate, two skeletal copper catalysts were prepared with particle sizes of 106-211μm and 355-420μm, respectively, and each was pretreated at 473K for 6 hours in 6.2 M NaOH to stabilise activity. After pretreatment, the catalysts were used in the reaction at varying stirrer rates. For the catalyst with particle size of 106-211μm, the stirrer rate was changed in the order 80, 40, 160, 0 and 80 rpm every 48 minutes, and for the catalyst with particle size of 355-420μm, the stirrer rate was changed in the order 80, 40, 0, 160 and 80 rpm every 48 minutes. Figure 5 shows that the first order plots are continuous for both sizes of catalyst particles despite major step changes in stirrer speed. This indicates that the evolution of H_2 during the reaction is probably sufficient to provide the agitation necessary to overcome any mass transfer limitations existing for this range of stirrer speeds. The first order rate constants estimated from the plots were 0.011 L g^{-1} h^{-1} for the 106-211 μm particles and 0.012 L g^{-1} h^{-1} for the 355-420 μm particles indicating that there is no effect of particle size in this range. This is to be expected due to the large pores that exist in skeletal copper (8).

To investigate the effect of chromia on catalyst behaviour for the dehydrogenation of ethanolamine to glycine salt, four catalysts were prepared using the method described by Ma and Wainwright (8). The activities of each catalyst, the properties of which are reported in Table 1, were evaluated in three or four successive cycles. The first order rate constants obtained are plotted as a function of cycle number in Figure 6 from which it can be seen that the activity of the catalyst decreased with increasing chromia content at high chromia levels. For skeletal copper catalysts with the two higher chromia contents (designated

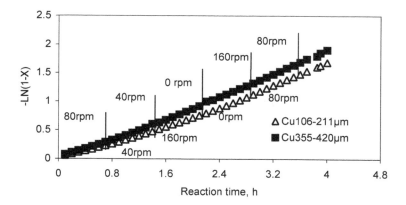

Figure 5 Effect of stirrer speed during ethanolamine dehydrogenation over unpromoted skeletal copper under standard conditions for particles with different sizes.

CuCr0.01 and CuCr0.1) there was no obvious deactivation, but due to their lower initial activities they had no advantage compared with the unpromoted skeletal copper catalyst. For the low chromia content skeletal copper catalyst (CuCr0.002), and the unpromoted skeletal copper catalyst, the deactivation in the first cycle was significant. However, CuCr0.002 had both a higher initial activity and a higher, stable residual activity than the unpromoted skeletal copper catalyst.

   A previous study (8) showed that low concentrations (between 0.001 and 0.008M) of Na_2CrO_4 were sufficient to stabilise the total surface area and the Cu surface area, but at Na_2CrO_4 concentrations above 0.008M, the Cu surface area decreased, presumably due to blocking by excess chromia deposited on it. The current investigation confirmed this conclusion. The catalysts with high chromia contents had lower activities than unpromoted skeletal copper whereas the catalyst with a low chromia content had higher surface area and activity than unpromoted skeletal copper after operating in the severe reaction conditions.

Table 1 Compositions and surface areas of unpromoted and chromia-promoted skeletal copper catalysts

Catalysts	Cr_2O_3[a] content (%)	Surface areas (m^2g^{-1}) [b]				
		Fresh	Cycle 1	Cycle 2	Cycle 3	Cycle 4
Cu	0	13	3.0	2.1	1.8	2.0
CuCr0.002	0.4	29	17	18	14	8.7
CuCr0.01	3.7	61	25	32	24	21
CuCr0.1	7.0	72	45	63	60	-

[a] The Cr_2O_3 contents were obtained for fresh catalysts.
[b] The surface areas for cycles 1-4 are for the catalyst recovered after each cycle.

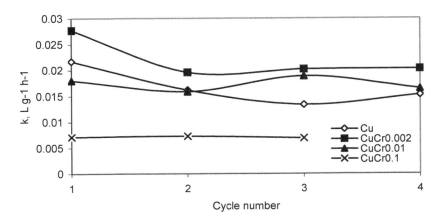

Figure 6 First order rate constants for repeated cycles of ethanolamine dehydrogenation over chromia-promoted skeletal copper catalysts under standard conditions.

The results in Table 1 and Figure 4 show that the surface area of unpromoted skeletal copper decreases to around 2 m^2 g^{-1} after exposure to 6.2 M NaOH during successive dehydrogenation reactions at 433 K and by pretreatment at 473 K in 6.2 M NaOH solution. This coarsening of the copper structure has been extensively discussed in the literature (11, 12) and results from dissolution of copper into the highly concentrated caustic solution and subsequent redeposition by precipitation. Recently it has been shown that leaching in 6.1 M NaOH solution containing Na_2CrO_4 slows this coarsening process and increases the overall and specific copper surface areas (13). In that study copper crystallite sizes were measured after unpromoted and chromia promoted skeletal copper had been exposed to 6.1 M NaOH solutions at 423 K for times up to 100 hours. The results showed that whereas the copper crystallite size in unpromoted skeletal copper grew from 13.8 nm to 39 nm after 100 hours, the crystallite size of chromia promoted skeletal copper produced by leaching $CuAl_2$ in 6.1 M NaOH containing 0.01 M Na_2CrO_4 only grew from from 9.6 to 11.9 nm. The results in Table 1 suggest that leaching in a 6.1 M NaOH solution containing 0.002 M Na_2CrO_4 produces a significant reduction to the consuming of skeletal copper whilst still maintaining a relatively high level of active copper surface which appears to be lost at higher chromia loadings. Optimization of the Na_2CrO_4 concentration in the leach liquor should lead to further improvements to skeletal copper catalysts for use in this and similar processes.

Conclusions

The oxidative dehydrogenation of ethanolamine over skeletal copper catalysts at temperatures, pressures and catalyst concentrations that are used in industrial processes has been shown to be independent of the agitation rate and catalyst particle size over a range of conditions. A small content of chromia (ca. 0.7 wt %) provided some improvement to catalyst activity and whereas larger amounts provided stability at the expense of activity.

Experimental Section

The preparation of unpromoted and chromia promoted skeletal copper catalysts was similar to that reported in previous studies (8-10). $CuAl_2$ alloy (52.17 wt. % Al and 47.83 wt. % Cu) was crushed and screened to various size fractions before leaching. The leach liquor was a large excess of 6.1 M sodium hydroxide containing various concentrations of sodium chromate in order to deposit different amounts of chromia on the surface of skeletal copper. The study mainly involved unpromoted catalyst particles (106-211 μm) which were prepared by leaching for 24 hours at 278 K or 293 K. Chromia promoted skeletal copper catalysts were produced using concentrations of 0.002 M, 0.01 M and 0.1 M Na_2CrO_4 in the 6.1 M NaOH leach liquor. All leached catalysts were washed thoroughly in distilled water until the pH was neutral and then refrigerated in distilled water at around 275 K prior to use. In some experiments catalysts were pretreated in NaOH at the same concentration used in reaction (6.2 M), but at 40 K higher temperature, prior to use. The mixture of catalyst and NaOH solution was heated in the closed autoclave used for reaction to 473 K over a period of 1 hour with a stirrer rate of 80 rpm and the temperature was then maintained constant for 6 hours. The temperature was then decreased to below 343 K and the catalyst washed again with distilled water after recovery.

The Cr_2O_3 content of each catalyst was determined by atomic absorption spectroscopy (Varian/Spectr AA-20 plus) on acid-digested samples. Total surface areas were determined by a single point BET method (nitrogen adsorption-desorption at 77.5 K) using a mixture of 29.7% N_2 in helium. Samples were wet-loaded into the flow tube and dried at 423 K in a hydrogen flow for 15 minutes and then for another 30 minutes at 513 K before cooling in helium.

Ethanolamine dehydrogenation, reaction (1), was carried out in a 300ml stainless steel autoclave (Parr Inst. Co., series 4560) equipped with stirrer, liquid sampling valve, gas inlet and pressure release valves, thermocouple and pressure gauge. The gas inlet was connected to a cylinder containing nitrogen which was used to flush air from the autoclave and to pressurise the reaction system with nitrogen. The exit line was connected to an electronic pressure controller (Rosemount Inst., model 5866) set at 0.9 MPa and a wet gas flow meter (G.H. Zeal Ltd., model DM3A) was used to follow the release of gas by reaction. The autoclave was equipped with a heating mantle housed in an aluminum shell that provided uniform heat distribution to the wall and bottom of the autoclave. The mantle was connected to a temperature controller (Parr Inst. Co., series 4843) and a digital readout of temperature, stirrer speed and system pressure was provided. The temperature of the reaction mixture was controlled to better than ± 2 K as indicated by a thermocouple in a well in the autoclave

In a typical experiment, 50 g of NaOH flakes were first dissolved in 120 g of water. The required amount of catalyst was then wet loaded into the caustic solution and 72 g of ethanolamine added. Before heating the autoclave was closed and the air inside purged out with N_2. The autoclave was then heated with the set temperature (433 K) reached after about 80 minutes. This time is taken as zero in the plots that follow. The standard operating conditions used for catalyst evaluation unless otherwise stated were as follows: temperature 433 K; pressure 0.9 MPa; ethanolamine concentration 2.9 M; NaOH concentration 6.2 M; stirrer speed 80 rpm; catalyst 8 g with particle size 106-211 μm.

The course of reaction was determined by continuously monitoring hydrogen evolution and by post-run detemination of the glycine content of small samples of liquid taken at intervals during the run. Glycine concentrations were determined using a Shimadzu LC-10AT*VP* HPLC fitted with a Shimadzu SPD-10A*VP* UV-VIS detector operating at 203 nm. The column was an alphaBond C18 125A 10 μ column (300×3.9 mm i.d.); the mobile phase was a 30 vol% methanol in water mixture set at pH 6.7 using phosphate buffer. The mobile phase flow rate was 1 ml/min under 19.5-20 MPa pressure. The liquid samples from the reactor were diluted 50 times with mobile phase before analysis.

Acknowledgements

One of the authors (NWC) is grateful to the Petroleum Research Fund of the American Chemical Society for assistance with travel expenses.

References

1. T. Goto, H. Yokoyama and H. Nishibayashi, U.S. Patent No. 4,782,183 (1988).
2. H. C. Chitwood, U.S. Patent No. 2,384,817 (1945).
3. Y. Urano, Y. Kadono and T. Goto, U.S. Patent No. 5,220,055 (1993).
4. J. Ochoa-Gomez, J. Martin-Ramon, J. Sanchez-Sanchez and A. De Diego-Zori, U.S. Patent No. 5,225,592 (1993).
5. T. Franczyk, U.S. Patent No. 5,292,936 (1994).
6. T. Franczyk, Y. Kadono, N. Miyagawa, S. Takasaki and H. Wakayama, U.S. Patent No. 5,739,390 (1998).
7. Y. Yang, Z. Duan, W. Liu, G. Li, and Y. Xiong, *Huaxue Fanying Gongcheng Yu Gongyi* **17**, 210 (2001).

8.   L. Ma and M. S. Wainwright, *Appl. Catal. A: General* **187**, 89 (1999).
9.   L. Ma, D.L. Trimm and M.S. Wainwright, *Topics in Catal.* **8**, 271 (1999).
10. L. Ma, B. Gong, T. Tran, M.S. Wainwright, *Catal. Today* **63**, 499 (2000).
11. A. D. Tomsett, H. E. Curry-Hyde, M. S. Wainwright, D. J. Young and A. J. Bridgewater, *Appl. Catal.* **33**, 119 (1987).
12. A. D. Tomsett, D. J. Young, M. R. Stammbach and M. S. Wainwright, *J. Mat. Sci.* **25**, 4106 (1990).
13. L. Ma and M. S. Wainwright, Chemical Industries (Dekker), **89** (*Catal. Org. React.*), 225 (2002).

4. Hydrogenation Catalysis in a Nylon Recycle Process

Alan M. Allgeier, Theodore A. Koch, and Sourav K. Sengupta

INVISTA™ Nylon Intermediates
Experimental Station, Wilmington, DE 19880-0302

sourav.k.sengupta-1@invista.com

Abstract

A process for the hydrogenation of adiponitrile and 6-aminocapronitrile to hexamethylenediamine in streams of depolymerized Nylon-6,6 or a blend of Nylon-6 and Nylon-6,6 has been described. Semi-batch and continuous hydrogenation reactions of depolymerized (ammonolysis) products were performed to study the efficacy of Raney® Ni 2400 and Raney® Co 2724 catalysts. The study showed signs of deactivation of Raney® Ni 2400 even in the presence of caustic, whereas little or no deactivation of Raney® Co 2724 was observed for the hydrogenation of the ammonolysis product. The hydrogenation products from the continuous run using Raney® Co 2724 were subsequently distilled and the recycled hexamethylenediamine (HMD) monomer was polymerized with adipic acid. The properties of the polymer prepared from recycled HMD were found to be identical to that obtained from virgin HMD.

Introduction

Recycling of waste carpets has become an increasingly important problem in recent years. Carpet waste generally includes both industrial carpet waste and post-consumer carpet waste. Currently, landfill capacity is nearing maximum utilization. In the United States, a minimum of 25% recycle content is required in government carpet installations, and customer mills are also asking for fibers with a high recycle content. A critical problem of recycling Nylon-6, and -6,6 is the capability of converting the polyamide in carpet waste to monomeric components, which can subsequently be reused as pure spinnable Nylon. Cost-effective recovery processes in which the monomers are available for reuse without further conversion reactions are especially important.

A Nylon recycle based on depolymerization via ammonolysis has been shown to be a technically feasible route for the recovery of high purity Nylon intermediates (1-6). In the ammonolysis process, the secondary amides of the Nylon (-6 and / or –6,6) fibers/polymers react with ammonia to break the Nylon chain and form a primary amide and an amine (Figure 1). The primary amide can subsequently be dehydrated to form a nitrile group. The net result is that the ammonolysis product is predominantly a mixture of four major components.

Hexamethylenediamine (HMD) and adiponitrile (ADN) are formed from Nylon-6,6, while 6-aminocapronitrile (ACN) and caprolactam (CL) are formed from Nylon-6. The ammonolysis product, which also contains many minor byproduct components, is fractionated by distillation with the HMD, ACN, ADN, and CL in one fraction. This fraction is subsequently hydrogenated to form HMD. Caprolactam remains intact during the hydrogenation reaction.

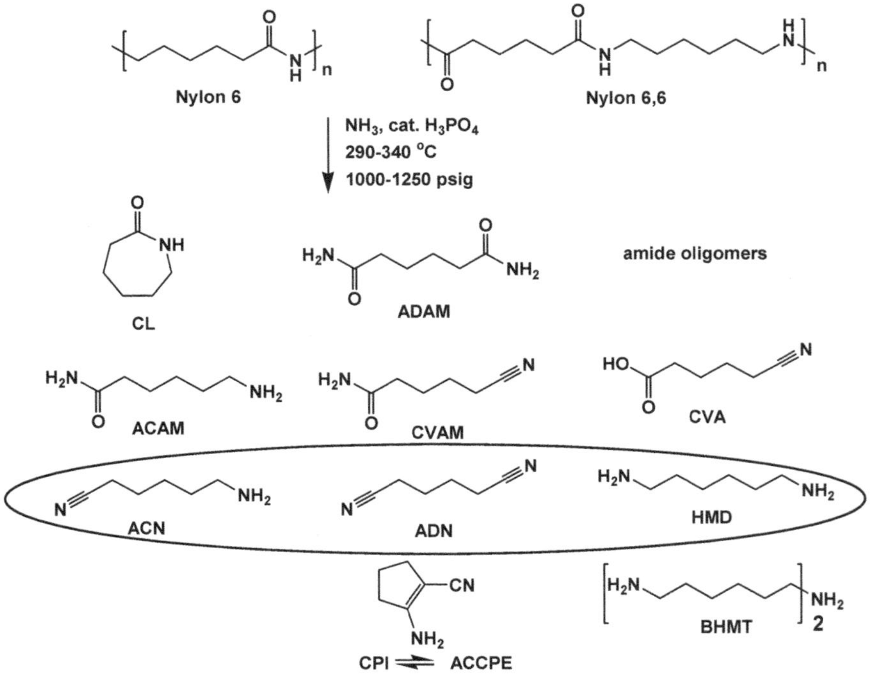

Figure 1 Ammonolysis of Nylon-6 and Nylon-6,6 leads to a variety of products. The ultimate products of ammonolysis and dehydration are ideally ACN, ADN and HMD, which can be converted to "Recycle HMD" via hydrogenation.

The crude recycled HMD and CL can be separated by distillation and subsequently refined to produce monomer grade HMD and CL, respectively. The caprolactam can be directly polymerized to Nylon-6 or it can be recycled back to the ammonolysis reactor, where it can be converted into ACN.

Since the hydrogenation step of the Nylon ammonolysis process had not been studied extensively, it was deemed necessary to investigate the feasibility of a relatively low-pressure (less than 1000 psig) and low-temperature (less than 100°C) process for the hydrogenation of depolymerized Nylon-6,6 and/or a blend of Nylon-6 and -6,6 products.

Results and Discussion

Drawing heavily from prior experience in hydrogenation of nitriles (7-10) and of ADN to ACN and/or HMD (11), in particular, we decided to restrict the scope of this investigation to Raney® Ni 2400 and Raney® Co 2724 catalysts. The hydrogenation reactions were initially carried out in a semi-batch reactor, followed by continuous stirred tank reactor to study the activity, selectivity, and life of the catalyst.

Semi-batch screening A limited number of catalyst scouting experiments were carried out in the semi-batch reactor using crude ammonolysis product before and after carbon dioxide and ammonia removal. The semi-batch reaction studies of the hydrogenation of crude ammonolysis product were conducted in a 300cc pressurized, stirred tank reactor at 500 psig (35.5 MPa) and 75°C. Hydrogen consumption and gas chromatographic analysis were employed to follow the course of the reaction.

To evaluate the influence of residual carbon dioxide and ammonia, a depolymerized Nylon product was prepared with of the following molar composition: 10.8% HMD, 19.9% ACN, 9.5% ADN, 17.0% CL, 1.4% 6-aminocaproamide (ACAM), 2.2% 5-cyanovaleramide (CVAM) 0.1% bis(hexamethylene)triamine (BHMT) and 39.1% other products. Among the "other products" at least thirty four separate compounds have been identified by GC/MS. The largest concentration products are amide dimers, i.e. intermediate products from the ammonolysis process and some do bear nitrile groups capable of hydrogenation. This depolymerized Nylon product (150 g), from which carbon dioxide and ammonia had not been stripped, was charged to the reactor with Raney® Co 2724 (4.2 g). Operating at 500 psig (3.55 MPa) and 75°C, the reaction showed an initial hydrogen uptake rate of 5.25 psi (36.2 kPa)/minute. After 340 minutes the reaction had consumed 490 psig (3.48 kPa) hydrogen (Figure 2); a sample was removed from the reactor for analysis. It comprised 34% HMD, 19% CL and by-products. No ACN or ADN was detected (detection limit 100 ppm).

A similar reaction was conducted with Raney® Ni 2400 and 0.4 g 50% NaOH(aq), instead of Raney® Co. The reaction showed an initial hydrogen uptake rate of 3.6 psi (24.8 kPa)/min (Figure 2). Even after 526 minutes, the reaction had only consumed 385 psig (2.76 MPa); analysis showed 27% HMD, 5% ACN, 20% CL, and by-products. The insufficient hydrogen uptake and the shape of the hydrogen uptake profile suggested catalyst deactivation might have been occurring. In the absence of sodium hydroxide deactivation was more rapid.

For comparison to the above reactions, depolymerized Nylon was prepared and stripped of CO_2 and NH_3 in a standard distillation. The depolymerized Nylon had the following molar composition; 12.9% HMD, 21.7% ACN, 11.6%

ADN, 19.4% CL, 1.3% ACAM, 2.2% CVAM and 30.9% others and was hydrogenated with Raney® Co 2724 under identical conditions to the above. The reaction showed an initial hydrogen uptake rate of 7.8 psi (53.8 kPa)/minute. After 240 minutes, the reaction had consumed 525 psig (3.72 MPa); a sample was removed from the reactor for analysis. It comprised 39% HMD, 18% CL, and by-products. The reaction showed no evidence of catalyst deactivation. While the rate of hydrogen consumption was detectably larger in this experiment than in the Raney® Co experiment with CO_2 and NH_3, the differences are not sufficiently large to infer a mechanistic difference.

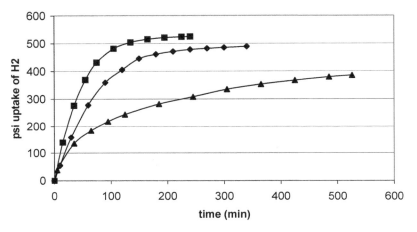

Figure 2 Hydrogenation of crude ammonolysis product. All reactions were conducted with 4.2 g catalyst, 150 g of feed, 75 °C and 500 psig. ♦) Raney® Co, CO_2 and NH_3 not removed; ■) Raney® Co, CO_2 and NH_3 were removed; ▲) Raney® Ni, CO_2 and NH_3 not removed.

The product mixtures from the Raney® Co experiments comprised 34-39% HMD, 18-19% caprolactam and other undesirable byproducts. Poor accounting of analytical samples was observed for the crude ammonolysis product and the products of hydrogenation (72-75%), some of the unaccounted material is water (5-10%), which cannot be detected by the FID detector. Other sources may include remaining high boilers (oligomers), which are not transported through the GC column. It should be noted that any residual amide dimers or higher order oligomers could be isolated by distillation and fed back to the ammonolysis reactor to increase the overall process yield. The product mixture from the Raney® Ni reaction included 5% unreacted 6-aminocapronitrile (ACN), indicating that the reaction did not proceed to high conversion. A model synthetic reaction mixture composed of 50% HMD and 50% ACN did proceed to high conversion in a hydrogenation reaction conducted with the Raney® Ni/caustic system. Deactivation was not observed in the batch experiment, which generated a product of 95% HMD and 99% accounting. To better understand the mechanism of Raney® Ni deactivation, ongoing work is aimed at elucidating the differences between the Ni and Co catalysts through controlled reactivity studies and examination of the catalyst surface via ESCA (12).

   Continuous hydrogenation of crude ammonolysis product using Raney® Co 2724 catalyst. Based on the results obtained from the semi-batch reactions, the hydrogenation of ammonolysis products was investigated in a continuous stirred tank reactor with Raney® Co 2724, to evaluate the productivity and lifetime of the catalyst. The feed to the reactor, obtained by reacting a mixture of Nylon-6,6 and Nylon-6 with ammonia, comprised 8.5% HMD, 23.6% ACN, 8.5% ADN, 26.2% CL, 0.4% ACAM, 0.03% BHMT, and 6% Others. HMD (120 g), 50%(w/w) NaOH(aq) (0.6 ml, to initially condition the catalyst), and pre-activated Raney® Co 2724 slurry catalyst (10 g: 5 g of dry catalyst in 5 g water) were loaded into the stirred tank reactor, which subsequently operated at 500 psig (35.5 MPa) and 85-95°C. Crude ammonolysis product (depolymerized Nylon), which had been stripped of CO_2 and NH_3, was then continuously added to the reactor at the rate of 12 mL per hour. The hold up time of the mixture in the reactor was 10 hours. The product distribution as a function of gram ADN

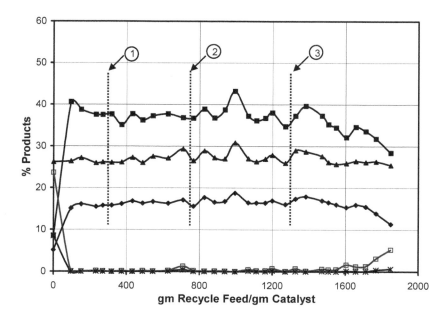

Figure 3 Hydrogenation of recycled Nylon-6 and Nylon-6,6 ammonolysis feed in the presence of 5 g of Raney® Co 2724 catalyst at a total pressure of 500 psig, and temperature of 85 to 90°C, at a feed flowrate of 12 ml/h. Hexamethylenediamine (■), caprolactam (▲), adiponitrile (*), 6-aminocapronitrile (□), and other byproducts (♦).

fed per gram of catalyst is presented in Figure 3. A few process changes were made during the course of the hydrogenation run including (1) increased the reaction temperature from 85 to 90°C at 318 g recycle feed/g of catalyst; (2) started adding ammonium hydroxide solution at the rate of 0.6 cc/h at 766 g recycle feed/g of catalyst; and (3) stopped feeding ammonium hydroxide solution and started adding water at the rate of 0.12 mL/h at 1326 g recycle feed/g of catalyst.

Addition of ammonium hydroxide and water were explored to evaluate their influence upon catalyst activity and selectivity. The data in this study suggest that there was little influence of ammonium hydroxide on reaction rate and selectivity. The data, however, were not sufficient to definitively define the role of these additives and investigation of these effects will be the subject of future exploration. Examination of Figure 3 may lead to the conclusion that water is actually harmful to the life of the catalyst but such a preliminary hypothesis is overly simplistic, acknowledging that the ammonium hydroxide additive comprises 70% water.

The hydrogenation products from the continuous run using Raney® Co 2724 were subsequently distilled and the product hexamethylenediamine monomer (i.e. "recycled HMD") was polymerized with adipic acid. The properties of the polymer prepared from recycled HMD were found to be identical to that obtained from virgin HMD, indicating that the continuous hydrogenation of ammonolysis product offers potential for the commercial production of recycled Nylon.

Conclusions

The technical feasibility of a relatively low-pressure (less than 1000 psig) and low-temperature (less than 100°C) process for the hydrogenation of depolymerized (ammonolysis) Nylon-6,6 and/or a blend of Nylon-6 and -6,6 products has been described. While Raney® Co 2724 showed little or no sign of deactivation during the semi-batch hydrogenation of the ammonolysis products, before and after CO_2 and NH_3 removal, Raney® Ni 2400 showed signs of deactivation even in the presence of caustic. Raney® Co 2724 proved to be an effective and robust catalyst in a continuous stirred tank reactor study.

Experimental Section

The depolymerized Nylon used in the hydrogenation process was obtained by the ammonolysis of a mixture of Nylon-6 and Nylon-6,6 (described elsewhere, see reference 2). Hydrogenation reactions were conducted in 300 cc stirred pressure vessels. For semi-batch reactions hydrogen was constantly replenished to the reactor from a 1L reservoir to maintain a reactor pressure of 500 psig and all of the reactions were conducted with the same operating parameters and protocol. In continuous stirred tank studies hydrogen flow was controlled using a Brooks mass flow controller and a backpressure regulator. The feed to the reactor was supplied from an Isco syringe pump.

The hydrogenation processes were performed at a relatively low temperature and pressure in the presence of promoted Raney® Ni 2400 and Raney® Co 2724 catalysts (13) in this study but any common nitrile hydrogenation catalysts (e.g. Fe, Ru, Rh, bulk or supported catalysts) could be used. The advantage of using a low temperature and pressure process is that it lowers the investment cost of an industrial process. Raney® Ni 2400 is promoted with Cr and Raney® Co 2724 is promoted with Ni and Cr. The particle sizes for both catalysts were in the range $25 - 55$ μm. The BET surface area of Raney® Ni 2400 and Raney® Co 2724 are 140 m^2/g and 76 m^2/g, respectively, and the active surface area of the Ni and Co catalysts are 52 and 18 m^2/g, respectively, based on CO chemisorption (Grace Davison Raney® Technical Manual, 4th Edition, 1996).

Acknowledgements

Peter Kassera is acknowledged for helpful discussions and leadership of the Nylon Ammonolysis program. Leon Scott is acknowledged for assistance in distillation work. Nancy Singletary is acknowledged for assistance in polymer preparation.

References

1. A. Lambert and G.H. Lang, GB Pat. 1,172,997 to ICI (1969).
2. R. J. McKinney, US Pat. 5,302,756 to DuPont (1994).
3. J. A. J. Hendrix, M. Booij, and Y. H. Frentzen, US Pat. 5,668,277 to DSM (1997).
4. G. Kalfas, *Polymer Reaction Engineering*, **6**, (*1*), 41-67 (1998).
5. H. P. Kasserra, Science and Technology of Polymers and Advanced Materials: Emerging Technologies and Business Opportunities, [Proceedings of the International Conference on Frontiers of Polymers and Advanced Materials], 4th, Cairo, Jan. 4-9, 1997, P. N. Prasad, Ed., Plenum, New York, (1998), p. 629 - 635.
6. S. B. Fergusson, Y. Yan, US Pat. 6,627,046 to DuPont (2003).
7. H. Greenfield, *Ind. Eng. Chem.*, **6**, 142 (1967).
8. S. N. Thomas-Pryor, T. A. Manz, Z. Liu, T. A. Koch, S. K. Sengupta and W. N. Delgass, Chemical Industries (Dekker), **75** (*Catal. Org. React.*), 195-206 (1998).
9. T. A. Johnson and D. P. Freyberger, Chemical Industries (Dekker), **82** (*Catal. Org. React.*), 201-227 (2001).
10. D. J. Ostgard, M. Berweiler, S. Röder, and P. Panster, Chemical Industries (Dekker), **89** (*Catal. Org. React.*), 273–294 (2003).
11. S. K. Sengupta, T. A. Koch, and K. R. Krause, US 5,900,511 to DuPont (1999).
12. A. M. Allgeier and M. W. Duch, *J. Catal.*, submitted for publication.
13. S. Montgomery, EP 212,986 to W.R. Grace (1994).

5. Hydrogenation of Dodecanenitrile Over Raney Type and Supported Group VIII Metal Catalysts. Effect of Metal on Selectivity

Sándor Göbölös, József L. Margitfalvi

*Chemical Research Center, Hungarian Academy of Sciences,
1025 Budapest, Pusztaszeri út 59-67 Hungary*

gobolos@chemres.hu

Abstract

Supported Co, Ni, Ru, Rh, Pd and Pt as well as Raney Ni and Co catalysts were used for the hydrogenation of dodecanenitrile to amines in stirred SS autoclaves both in cyclohexane and without a solvent. The reaction temperature and the hydrogen pressure were varied between 90-140 °C and 10-80 bar, respectively. Over Ni catalysts NH_3 and/or a base modifier suppressed the formation of secondary amine. High selectivity (93-98 %) to primary amine was obtained on Raney nickel, Ni/Al_2O_3 and Ru/Al_2O_3 catalysts at complete nitrile conversion. With respect to the effect of metal supported on alumina the selectivity of dodecylamine decreased in the order: Co~Ni~Ru>Rh>Pd>Pt. The difference between Group VIII metals in selectivity can be explained by the electronic properties of d-band of metals. High selectivity to primary amine was achieved on base modified Raney Ni even in the absence of NH_3.

Introduction

Catalytic hydrogenation of nitriles is an industrially important route for the manufacture of amines [1]. The amines have wide ranging applications such as solvents and intermediates of agrochemicals and pharmaceuticals. Fatty amines are used as components of different chemical products, e.g. emulsifiers, softeners, corrosion inhibitors. The hydrogenation of nitriles typically gives a mixture of primary, secondary and tertiary amines. This hydrogenation is usually carried out on heterogeneous catalysts under high pressure and temperatures of 55-150 °C. The catalysts most often used are Group VIII metals on Al_2O_3, SiO_2 or MgO, Raney nickel or Raney cobalt [2].

While the desired product of the hydrogenation of nitriles is often the primary amines, the proportion of primary/secondary/tertiary amines in the product is strongly affected by the nature of metal. In the hydrogenation of nitriles on Group VIII metals, the selectivity of primary amine decreases in the order: Co>Ni>Ru>Rh>Pd>Pt [1]. The difference between Group VIII metals in selectivity to primary amine is explained by the difference in the electronic properties of the metals [3]. Pd and Pt preferentially form tertiary amines while Cu and Rh are selective towards secondary amines and Co, Ni and Ru have high selectivity for primary amines [4]. For instance, in the hydrogenation of acetonitrile, the selectivity to primary amine decreased in the order of Ru>Ni>Rh>Pd>Pt. Ru is highly selective for primary amine and Pt for tertiary amine. The selectivity of the metals is related to their propensity for multiple bond formation. The metal known for the highest propensity to form multiple bonds has the highest selectivity to primary amine [5]. Furthermore, the inhibition of the sites responsible for condensation reactions resulting in secondary and tertiary amines can be affected by the addition NH_3 or base compounds, e.g. alkalis [1].

In this study Al_2O_3 supported Co, Ni, Ru, Rh, Pd and Pt, as well as Raney Ni and Co catalysts were used for the hydrogenation of dodecanenitrile (RCN) to amines in cyclohexane and without a solvent. The effect of metal, reaction conditions and modifiers on the selectivity was investigated.

Experimental Section

Catalysts - A commercial Raney nickel (RNi-C) and a laboratory Raney nickel (RNi-L) were used in this study. RNi-C was supplied in an aqueous suspension (pH < 10.5, Al < 7 wt %, particle size: 0.012-0.128 mm). Prior to the activity test, RNi-C catalyst (2 g wet, 1.4 g dry, aqueous suspension) was washed three times with ethanol (20 ml) and twice with cyclohexane (CH) (20 mL) in order to remove water from the catalyst. RCN was then exchanged for the cyclohexane and the catalyst sample was introduced into the reactor as a suspension in the substrate. RNi-L catalyst was prepared from a 50 % Ni-50 % Al alloy (0.045-0.1 mm in size) by treatment with NaOH which dissolved most of the Al. This catalyst was stored in passivated and dried form. Prior to the activity test, the catalyst (0.3 g) was treated in H_2 at 250 °C for 2 h and then introduced to the reactor under CH. Raney cobalt (RCo), a commercial product, was treated likewise. Alumina supported Ru, Rh, Pd and Pt catalysts (powder) containing 5 wt. % of metal were purchased from Engelhard in reduced form. Prior to the activity test, catalyst (1.5 g) was treated in H_2 at 250 °C for 2 h and then introduced to the reactor under solvent. 10 % Ni and 10 % Co/γ-Al_2O_3 (200 m^2/g) catalysts were prepared by incipient wetness impregnation using nitrate precursors. After drying the samples were calcined and reduced at 500 °C for 2 h and were then introduced to the reactor under CH.

Activity tests - Catalytic activity tests were carried out in 70 or 300 ml stainless steel stirred autoclaves (Parr Co.) at a stirring rate of 1000 rpm. Reaction conditions are given in the Tables.

Analysis of reaction products - Liquid reaction products were analyzed by GC using a capillary column CP-Sil-8CB (WCOT Fused Silica - stationary phase: 5% phenyl-methyl-polysiloxane, length: 50 m, ID: 0.32 mm, OD: 0.45 mm, film thickness: 0.25 μm). Liquid samples were diluted with ethanol in ratio sample/ethanol = 1/15. Hexadecane was used as an internal standard in the quantitative analysis.

Results and Discussion

Raney-nickel catalysts - The effect of NH_3 and base modifier on the activity and selectivity of RNi-C catalyst is shown in Table 1. The addition of NH_3 significantly decreased the pseudo first-order rate constants, the conversion of RCN and the selectivity to R_2NH. Upon increasing the reaction time (t) on

Table 1 Hydrogenation of RCN without solvent on RNi-C catalyst at T= 125 °C.

No	NH_3/RCN	t, h	C, %	RNH_2	SB	R_2NH	k, h^{-1}
1	0	1	99.7	83.5	0.1	16.4	5.06
2		2	99.7	77.1	0.1	21.8	
3		4	99.7	66.2	-	31.7	
4	0.30	1	78.5	92.8	6.2	0.6	2.61
5		2	89.8	93.1	6.1	0.7	
6		4	99.8	93.7	4.6	1.6	
7	0.86	1	30.9	94.1	5.6	0.2	1.44
8		2	42.4	94.9	4.8	0.3	
9		4	65.9	96.1	3.6	0.3	
10	0^a	1	96.5	98.9	0.3	0.8	1.75
11		2	98.6	98.8	0.5	0.7	
12		4	99.8	98.6	0.6	0.8	
13	0^b	1	38.5	89.3	10.7	0	0.453
14		2	92.1	90.3	10.7	0	
15		4	98.5	94.7	2.1	3.2	

Abbreviations: NH_3/RCN molar ratio; t=reaction time; C=conversion of nitrile; RNH_2, SB and R_2NH=selectivity of primary amine, Schiff-base and secondary amine in %, respectively. k=pseudo-first order rate constant determined from the C *vs. t* dependence at low conversions. *Reaction conditions*: $V_{reactor}$=300 ml; V_{RCN}=100 ml (0.45 mole); $m_{catalyst}$=1.4g; P_{H2}=10 bar was maintained during reaction; [a]instead of NH_3 a base modifier was used; [b]RCo catalyst was used.

unmodified catalyst the selectivity of R_2NH was enhanced at the expense of primary amine (RNH_2). In the presence of NH_3 the selectivity of RNH_2 slightly increased with the reaction time, while that of R_2NH increased due to the hydrogenation of Schiff-base (SB). It is noteworthy that high selectivity toward RNH_2 is achieved on base modified RNi-C catalyst even in the absence of NH_3. It is also interesting to note that in the absence of NH_3 the hydrogenation activity of RNi-C catalyst is significantly higher than that of the RCo catalyst; however, the latter is more selective to RNH_2 (Table 1: No 1-3 and No 13-16).

It is also important to note that the ln100/(100-C) *vs.* t plot indicated no linear correlation at higher conversions (not shown). In the absence of NH_3 there was a relative acceleration of the rate; whereas, in its presence, a retardation effect was observed. These results suggest that in the absence of NH_3 the strong adsorption of RCN on the catalyst inhibits the formation of SB in the condensation of imine with RNH_2. After the consumption of most of RCN, the condensation and reduction resulting in R_2NH can take place at a higher rate. The same effect was already observed by Pasek et al. [6]. However, NH_3 can compete not only with RNH_2 and H_2 but also with RCN at high conversion for the adsorption sites causing retardation. Therefore, high selectivity (93-98 %) to primary amine was obtained over Raney nickel catalysts at almost complete nitrile conversion.

The hydrogenation of RCN on RNi-L (in fact a skeletal catalyst) was performed in CH. The effect of reaction temperature and NH_3 on the activity and selectivity is given in Table 2. The addition of NH_3 at T = 90 °C resulted in slight decrease of activity and significant increase of RNH_2 selectivity. As expected the increase of the reaction temperature resulted in higher activity and $SB+R_2NH$ selectivity at the expense of RNH_2 selectivity. However, the

Table 2 Hydrogenation of RCN in cyclohexane over RNi-L catalyst.

No	Cat.	T, °C	t, h	C, %	RNH_2	SB	R_2NH	k, h^{-1}
1	RNi[a]	90	1	42.7	73.5	5.1	17.3	0.563
2			4	89.7	57.1	4.1	31.8	
3	RNi	90	1	37.5	84.7	10.9	1.6	0.464
4			4	86.9	92.8	5.2	1.2	
5	RNi	110	1	63.1	90.6	4.8	2.7	0.989
6			4	98.6	93.9	1.3	3.9	
7	RNi	125	1	69.9	87.3	3.8	3.6	1.28
8			4	99.9	83.4	0.3	14.8	

Abbreviations: Cat. = catalyst; otherwise see Table 1; *Reaction conditions*: $V_{reactor}$=70 ml; V_{CH}=V_{RCN}=10 ml (0.045 mole); $m_{catalyst}$=0.3 g; P_{H2}=80 bar (at RT) was not maintained during reaction NH_3/RCN=0.25; [a]without ammonia.

selectivity of SB is higher at lower temperatures. The selectivity to RNH_2 decreased with reaction time for the experiment performed without NH_3. The apparent activation energy of the hydrogenation of RCN on RNi-L catalyst was 30.5 kJ/mol, which is close to the value 46 kJ/mol measured in the liquid phase hydrogenation of acetonitrile on CoB amorphous alloy catalyt [7]. RNi-C is more active than RNi-L catalyst (compare Table 1: No 4 and 6 and Table 2: No 7 and 8).

Conversions obtained in the hydrogenation of RCN in CH on alumina supported Group VIII metals indicated the lower activity of these catalysts compared to the Raney type catalysts (see Table 3). Almost complete conversion (except Co/Al_2O_3) was achieved on M/Al_2O_3 catalyst at 140 °C for 9 h instead of

125 °C and 4 h for Raney catalysts. Due to lower reducibility and stronger interaction of Co- and Ni-oxides with alumina, 10 wt % metal was used. Despite the higher metal content of these catalysts they were less active than the alumina supported noble metals and their selectivity to RNH_2 was lower than that of Ru. The selectivity pattern on noble metals was in good agreement with literature data [1,4].

Table 3 Hydrogenation of RCN in cyclohexane over M/Al_2O_3 catalysts.

No	Catalyst, M	C, %	RNH_2	$SB+R_2NH$	Others
1	10Co	85.5	75.7	24.1	0.2
2	10Ni	98.9	82.5	16.7	0.8
3	5Ru	97.8	94.4	2.9	2.7
4	5Rh	99.7	72.7	19.8	7.5
5	5Pd	99.9	27.1	69.4	8.1
6	5Pt	99.9	9.6	81.4	9.0

Abbreviations: M=metal; see Table 1 and Table 2; Reaction conditions: $V_{reactor}$=100 ml; V_{CH}=100 ml; V_{RCN}=10 ml (0.045 mole); $m_{catalyst}$=1.5g; P_{H2}= 50 bar (at RT) was not maintained during reaction; T=140 °C; t=9 h; NH_3/RCN=1.

The selectivity of RNH_2 on M/Al_2O_3 and Raney catalysts decreased in the order: Co~Ni~Ru>Rh>Pd>Pt. This order corresponds to the opposite sequence of reducibility of metal-oxides [8] and standard reduction potentials of metal-ions [9]. The difference between Group VIII metals in selectivity to amines can probably been explained by the difference in the electronic properties of d-bands of metals [3]. It is interseting to note that the formation of secondary amine, i.e. the nucleophilic addition of primary amine on the intermediate imine can also take place on the Group VIII metal itself. Therefore, the properties of the metal d-band could affect the reactivity of the imine and its interaction with the amine. One could expect that an electron "enrichment" of the metal d-band will decrease the electron donation from the unsaturated -C=NH system, and the nucleophilic attack at the C atom by the amine [3]. Correlation between selectivity of metals in nitrile hydrogenation and their electronic properties will be published elsewhere.

Acknowledgements

SG and JLM wish to thank the Hungarian Scientific Research Fund (OTKA Grant) N° T-32065 and T 43570 for financial support, respectively. One of us (SG) is grateful for the financial support by the American Chemical Society-Petroleum Research Foundation (ACS-PRF).

References

1.   J. Volf and J. Pasek, in Catalytic Hydrogenation, Ed. L. Cerveny, Elsevier, Amsterdam, 1986, vol. 27, p. 105.

2. H. Greenfield, *Ind. Eng. Chem. Prod. Res. Dev.,* **6**, 142 (1967).
3. D. Tichit, F. Medina, R. Durand, C. Mateo, B. Coq, J. E. Sueiras and P. Salagre, in Heterogeneous Catalysis and Fine Chemicals IV, Eds. H. U. Blaser, A Baiker and R. Prins, Elsevier, Amsterdam, 1997, vol. 108, p. 297.
4. C. L. Thomas, Catalytic Processes and Proven Catalysts Academic Press, New York, 1970.
5. Y. Huang, and W. M. H. Sachtler, *J. Catal.,* **184**, 247 (1999).
6. J. Pasek, N. Kostova, and B. Dvorak, *Collect. Czech. Chem. Comm.,* **46**, 1011 (1981).
7. H. Li, Y. Wu, H. Luo, M. Wang, and Y. Xu, *J. Catal.,* **214**, 15 (2003).
8. D. Kulkarni, and I. E. Wachs, *Appl. Catal. A,* **237**, 121 (2002).
9. CRC Handbook of Chemistry and Physics, 70th Edition, Ed. R. C. Weast, D151-158, 1989-1990.

b. Anchored and Supported Hydrogenation Catalysts

6. Highly Efficient Metal Catalysts Supported on Activated Carbon Cloths

Alain Perrard and Pierre Gallezot

Institut de Recherches sur la Catalyse-CNRS,
2, av. Albert Einstein, 69626 Villeurbanne Cedex, France

gallezot@catalyse.cnrs.fr

Abstract

Activated carbon cloths (ACC) were oxidized and loaded with platinum-group metals (Ru, Pd, Pt) by cationic exchange or anionic adsorption. The ACC present a very narrow distribution of micropores that were not enlarged by subsequent treatments. The metal particles were homogeneously distributed inside the carbon fibers in the form of 2-3 nm particles. Ru/ACC catalysts exhibited a fair activity in glucose hydrogenation and gave higher yields in sorbitol (>99.5%) than Ru/C catalysts prepared from carbon powder or extrudates. The catalysts were also active in glucosone hydrogenation to fructose.

Introduction

Metal catalysts supported on activated carbon cloths (ACC) have been little studied in the past, particularly in liquid phase reactions (1-3). These catalysts present potentially a number of significant advantages with respect to conventional activated carbons in powder or granular form, e.g., high rate of mass transfer from the liquid phase, no need of decantation or filtration, and high flexibility to fit into any reactor geometry. The present work was intended to prepare and characterize the texture of well-dispersed ruthenium, palladium and platinum catalysts supported on activated carbon cloths. Two reactions of industrial importance, the hydrogenation of glucose to sorbitol and of glucosone to fructose were conducted on Ru/ACC and Pd/ACC catalysts, respectively, in an autoclave equipped with a special device holding the activated carbon cloth.

Results and Discussion

The BET surface of ACC, oxidized ACC and Pt/ACC were 1300, 680, and 580 m^2g^{-1}, respectively. Surprisingly, the distribution of pore radius in the three samples exhibited 4 sharp peaks centered at the same position at 0.37, 0.55, 0.75, 0.95 nm, respectively (Table 1). Therefore, neither the NaOCl oxidizing treatment, nor the metal loading modified the micropore size. However, the peak heights decreased in the series ACC >> ACC(oxidized) > Pt/ACC resulting in a decrease of the differential volumes dV/dr given in Table 1. Therefore, the micropores were not enlarged but their access was restricted because of the formation of carboxylic groups and of pore blockage by metal particles. However, new micropores centered at 1.1 and 1.3 nm were formed by the oxidizing treatment.

Table 1 Distribution of pore radius and pore volume.

micropore radius	dV/dr $(cm^3\ nm^{-1}g^{-1})$		
r (nm)	ACC	ACC (oxidized)	Pt/ACC[a]
0.37	3.1	1.3	0.8
0.55	2.1	0.6	0.5
0.75	4.3	1.3	0.5
0.95	3.7	2.7	1.7
1.10	0	0.8	0.8
1.30	0	0.3	0.5

[a] 10 wt% Pt.

Figure 1 gives a SEM image at low magnification showing the texture of the woven ACC and a TEM image of the ultramicrotome section of a carbon fiber of a 10 wt% Pt/ACC catalyst prepared by ion exchange. In spite of the high metal loading all the Pt-particles were uniformly distributed in the micropores in the form of 2-3 nm particles. Similar high metal dispersions were obtained with palladium and ruthenium catalysts.

The specific activity of two Ru/ACC catalysts in glucose hydrogenation and their selectivity to sorbitol are given in Table 2. The activities were 2 to 3 times higher than those measured on Ru/C catalysts in trickle bed reactor (4), but smaller than those measured on Ru-catalysts supported on activated carbon powder (5), all catalysts having the same metal dispersion. The most striking result is that the selectivity at near total conversion is higher than 99.5%, a value higher than that measured on Ru/C catalysts supported on carbon powders or extrudates. Because of the importance of high selectivity for the preparation of pure sorbitol used in pharmaceuticals, the Ru/ACC catalysts were patented for this application (6). The selectivity to sorbitol depends upon the occurrence of parallel or consecutive isomerization reactions such as the epimerization of sorbitol to mannitol. The high selectivity of ACC-supported catalysts is probably due to the lower chance of sorbitol epimerization than on conventional catalysts. Indeed, because the micropores in ACC open directly to the outer surface (7) one can expect a faster desorption of sorbitol than on conventional activated carbons where the molecules must pass through the macro and meso-pore systems.

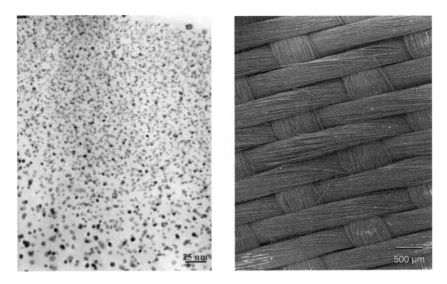

Figure 1 A) SEM view of ACC (left field) and B) TEM view of 10 wt%Pt/ACC (right field).

Table 2 Initial specific activity and selectivity to sorbitol of Ru-catalysts in glucose hydrogenation.

	1.9% Ru/ACC[a]	0.9% Ru/ACC[b]	5% Ru/C[c]
Activity $(mmol\ h^{-1}g_{Ru}^{-1})$	1800	2400	2500
Selectivity (%)	99.8 (98% conv.)	99.5 (99.7% conv.)	99.0 (99% conv.)

[a]Prepared by ion-exchange; [b]prepared by anionic adsorption; [c]ref. (5).

The catalytic hydrogenation of glucosone (keto-glucose) **1** into fructose **2** is a possible industrial route to obtain fructose from glucose which is first enzymatically oxidized to glucosone. One of the challenges is to minimize dehydroxylation reactions leading to unwanted deoxyhexoses. The hydrogenation of 10 wt% aqueous glucosone solution was conducted at 80°C under 80 bar H_2-pressure on 2.5 wt%Pd/ACC catalyst in the presence of small amounts of $NaHCO_3$ added to increase the pH of reaction medium. The catalyst activity was 500 mmol $h^{-1}g_{Pd}^{-1}$, but the fructose yield was only 70% because the pH decreased in the course of reaction.

Experimental Section

The ACC precursor was a woven rayon cloth which was dry cleaned, precarbonized by pyrolysis at 400°C, and carbonized at 1200°C under nitrogen. The final ash content was 0.1 wt%. The carbon cloth was then activated at 900°C under CO_2 to get a 50% burn-off. The ACC was oxidized in NaOCl solution to increase the number of surface carboxylic groups. Then the metal was loaded by cationic exchange of the protons of the carboxylic groups by $Ru(NH_3)_6^{3+}$, $Pd(NH_3)_4^{2+}$ or $Pt(NH_3)_4^{2+}$ cations in ammonia solutions. Metal loading was also carried out by anionic adsorption on the oxidized ACC, i.e., from metal chlorides in hydrochloric medium The NaOCl oxidation and metal loading procedure was described in a previous work (8). During the oxidation and metal loading steps the cloth was fixed on a cylindrical glass grid set around the inner wall of the treatment vessel equipped with a stirrer.

Nitrogen adsorption isotherms were measured with a sorbtometer Micromeretics Asap 2010 after water desorption at 130°C. The distribution of pore radius was obtained from the adsorption isotherms by the density functional theory. Electron microscopy study was carried out with a scanning electron microscope (SEM) HitachiS800, to image the texture of the fibers and with a transmission electron microscope (TEM) JEOL 2010 to detect and measure metal particle size. The distribution of particles inside the carbon fibers was determined from TEM views taken through ultramicrotome sections across the carbon fiber.

Hydrogenation reactions were carried out in a 300 mL stirred autoclave equipped with a cylindrical grid built from stainless steel, fitting around the autoclave walls. Rectangular pieces of ACC were cut to dimension and fixed around the grid as shown in Figure 2. This setup allows a stable attachment of the ACC at high stirring speed and makes easy the recycling experiments.

Glucose hydrogenation to sorbitol was carried out on 40 wt% aqueous solution at 100°C, under 80 bar H_2-pressure, in the presence of Ru/ACC catalysts prepared by cationic exchange and anionic impregnation. The hydrogenation of 10 wt% aqueous glucosone solution was conducted at 80°C under 80 bar H_2-pressure on 2.5 wt%Pd/ACC catalyst in the presence of small amounts of $NaHCO_3$ added to increase the pH of reaction medium. Reaction kinetics was followed by HPLC and GC analysis of the reaction medium at different time intervals.

Figure 2 Experimental setup for liquid phase reaction with ACC catalysts (the grid supporting the ACC is partially lifted out of the autoclave).

Acknowledgements

Carbon Industries, a subsidiary of Messier-Bugatti Co., is acknowledged for the supply of activated carbon cloths.

References

1. Yu. Matatov-Meytal and M. Sheintuch, *Appl. Catal. A: General*, **231**, 1 (2002).
2. M.C. Macias Perez, C. Salinas Martinez de Lecea, and A. Linares Solano, *Appl. Catal. A: General*, **151**, 461 (1997).
3. Yu. Shindler, Yu. Matatov-Meytal, and M. Sheintuch, *Ind. Eng. Chem. Res.*, **40**, 3301 (2001).
4. P. Gallezot, N. Nicolaus, G. Flèche, P. Fuertes, and A. Perrard, *J. Catal.*, **180**, 51 (1998).
5. P. Gallezot and A. Perrard, unpublished results.
6. P. Parmentier and A. Perrard, EP 989557285, WO 9802506, to Messier-Bugatti (1998).
7. I. Mochida, Y. Korai, M. Shirahama, S. Kawano, T. Hada, Y. Seo, M. Yoshikawa, and A. Yasutake, *Carbon*, **38**, 227 (2000).
8. P. Gallezot, A. Giroir-Fendler, D. Richard, Chemical Industries (Dekker), **47** (*Catal. Org. React.*), 1-17 (1992).

7.   Anchored Wilkinson's Catalyst

Clementina Reyes, Setrak Tanielyan and Robert Augustine

Center for Applied Catalysis, Seton Hall University, South Orange, NJ 07079

augustro@shu.edu

Abstract

Wilkinson's catalyst, $Rh(Ph_3P)_3Cl$, was the first viable homogeneous hydrogenation catalyst. Only a few years after its discovery, several reports were published on the adaptation of this catalyst using a triphenyl phosphine substituted polystyrene as an insoluble ligand. This approach to heterogenizing this classic catalyst has not met with any significant success. We have been successful, however, in anchoring the Wilkinson's catalyst to a phosphotungstic acid modified alumina using our Anchored Homogeneous Catalyst technology.

We report here a number of examples of the use of this anchored catalyst for the hydrogenation of different substrates at moderate to high substrate/catalyst ratios along with a direct comparison of these results with those obtained using the homogeneous Wilkinson under the same conditions. Also presented will be some examples of the use of the anchored catalyst in long term continuous reactions. Reaction rates, selectivities and the extent of metal loss will be presented where appropriate.

Introduction

Only a few years after the discovery of the Wilkinson catalyst, $Rh(Ph_3P)_3Cl$, (1) attempts were being made to attach the active component to an insoluble support material (2,3). It was considered that by converting this homogeneous catalyst into a heterogeneous species one would be able to combine the activity and selectivity of the homogeneous catalyst with the ease of separation inherent to a heterogeneous species and, thus, have the 'best of both worlds.' The earliest reports of work in this area involved the modification of a polystyrene to introduce diphenylphosphine ligands to which the rhodium was then attached. This approach of using a solid heterogeneous ligand is still the most commonly employed, albeit with some modification (4,5). However, these so-called 'tethered' catalysts can be prone to loss of metal during the reaction, are frequently not as active or selective as the homogeneous analog, and frequently lose their activity on attempted re-use (6,7). Some recent publications (8,9) describe the preparation and use of a tethered Wilkinson for both hydrogenations and hydrogen peroxide decomposition. In one, very high turnover numbers are reported for the hydrogenation of cyclohexene with no reported metal loss as determined by uv absorption of the reaction mixture, (8) but there is no mention in the other of the extent of metal loss during the reactions nor of any high TON reactions (9).

Several years ago we described a new approach to the preparation of 'heterogenized homogeneous' catalysts (10-13). It involved the use of a heteropoly acid (HPA) as an anchoring agent to attach a homogeneous catalyst to a variety of support materials giving what has since been termed 'Anchored Homogeneous Catalysts' (AHC) (14-18). This method has been used to prepare an Anchored Wilkinson catalyst which was compared to a commercially available polymer supported Wilkinson with respect to activity and metal loss in a variety of solvents (16). In every instance, the anchored species was superior to the polymer supported material. These reactions, however, were run at low substrate/catalyst ratios (turnover numbers-TON's), usually between 50 and 100. While these low TON reactions made it possible to obtain a great deal of information concerning the anchored Wilkinson and other AHC's, in order to establish that these catalysts could be used in commercial applications it was necessary to use them in reactions at much higher TON and, also, to make direct comparisons with the corresponding homogeneous catalyst under the same reaction conditions.

Results and Discussion

The anchored Wilkinson catalyst (AHC-Wilk) was used to promote the hydrogenation of 1-hexene at a substrate/catalyst ratio of 10,000 (Turnover number – TON). These reactions were run in 10% toluene/ethanol at 35°C and 50 psig of hydrogen. The catalyst was pre-hydrogenated in this solvent for two hours under reaction conditions. The catalyst could be re-used several times simply by removing the reaction mixture by a cannula after the hydrogenation was complete and then adding additional solvent and substrate and continuing the hydrogenation. During this procedure the reactor is kept at a positive pressure of hydrogen or zero-grade argon. The hydrogen uptake curves for three uses of the same catalyst sample with a 10,000 TON per use are shown in Figure 1a. The amount of Rh in each of these reaction mixtures was below the level of detection (0.5 ppm). The rates shown are the initial rates of the reaction.

This should be compared with the hydrogen uptake curve for the corresponding 10,000 TON homogeneously catalyzed reactions shown in Figure 1b. In this case the catalyst could not be re-used after removal of the first reaction mixture. Instead, additional 1-hexene was added to the reaction mixture at the end of the first hydrogenation. The second batch of substrate was hydrogenated at about half the rate as the first. We have observed this apparent deactivation of homogeneous catalysts during large TON hydrogenations several times with other catalysts. Toluene was added to the reaction mixture on the assumption that it would stabilize the catalyst. It has recently been found,

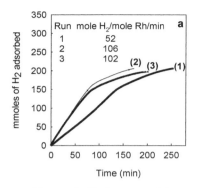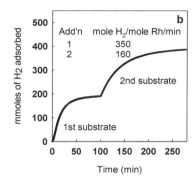

Figure 1 Hydrogen uptake curves for 1-hexene hydrogenations run at 35°C and 50 psig of hydrogen in 10% toluene/EtOH with a stirring rate of 1700 rpm. a) AHC-Wilk catalyst; b) Homogeneous Wilkinson's catalyst.

However, that toluene was not needed. The same type of results, only with a faster rate, have been obtained with an absolute EtOH solvent.

Contrary to the observations made previously with low TON reactions, here the homogeneous catalyst was more active than the anchored species. Once the homogeneous catalyst had been used for the high TON hydrogenation, though, it appeared to have been somewhat deactivated. The anchored catalyst, however, is capable of re-use several times. In this instance there was an increase in activity after the first use, even though the catalyst had been pre-hydrogenated.

Table 1 Effect of stirring rate on the initial rate of hydrogen uptake (mmoles H_2/mmole Rh/min) in the hydrogenation of 1-hexene over AHC-Wilk.

Run	1200 rpm	1700 rpm	2000 rpm
1	49	52	64
2	80	106	122
3	80	102	120

One of the more obvious reasons for the slower reaction rates observed with the anchored catalyst as compared to the homogeneous catalysts is the fact that hydrogen and substrate diffusion are more important in heterogeneously catalyzed reactions than in those promoted by homogeneous catalysts. In Table 1 are listed the hydrogenation rates for 10,000 TON hydrogenations of 1-hexene over AHC-Wilk at different stirring rates. It is obvious from these data that the agitation rate was not sufficient to overcome the diffusion control of these heterogeneously catalyzed reactions. With the reactor system used, 2000 rpm is the maximum value reasonably attainable.

Another interesting comparison of the homogeneous Wilkinson catalyst with AHC-Wilk is in the high TON hydrogenation of cyclohexene in 10%

toluene/ethanol at 35°C and 50 psig. Figure 3 shows the hydrogen uptake curves for three uses of the same AHC-Wilk for the hydrogenation of cyclohexene each at a 12,350 TON. These reactions all went to completion at a rate of 55-65 mmole H_2/mmole Rh/min. The rhodium levels in the reaction

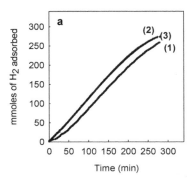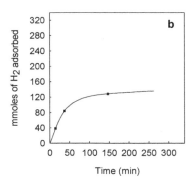

Figure 2 Hydrogen uptake curves for cyclohexene hydrogenations run at 35°C and 50 psig of hydrogen in 10% toluene/EtOH with a stirring rate of 1700 rpm. a) AHC-Wilk catalyst; b) Homogeneous Wilkinson's catalyst.

mixtures were also below the level of detection. While the corresponding homogeneously catalyzed hydrogenation proceeded with an initial rate of 130 mmole H_2/mmole Rh/min, the reaction stopped at 50-60% conversion (Figure 4). This later reaction was repeated five times with different samples of cyclohexene and fresh catalysts with the same result obtained each time.

Experimental Section

The AHC Wilkinson catalyst $(Rh(Ph_3P)_3/PTA/Al_2O_3)$ (AHC-Wilk) was prepared using the general procedure described previously (10,12). The catalyst contained 0.5% Rh which corresponds to a 1:1 Rh:PTA ratio and about a 4.5% load of the anchored complex. The hydrogenations were run using the low pressure apparatus previously described (19) under the conditions listed in the discussion.

Acknowledgements

Financial support of this work from Johnson Matthey is gratefully acknowledged.

References

1.   J. A. Osborn, F. H. Jardine, J. F. Young and G. Wilkinson, *J. Chem. Soc., A*, 1711 (1966).
2.   J. P. Collman, L. S. Hegedus, M. P. Cooke, J. R. Norton, G. Dolcetti, D. N. Marquardt, *J. Amer. Chem. Soc.*, **94**, 1789 (1972).

3. W. Dumont, J. C. Poulin, T. P. Daud, H. B. Kagan, *J. Amer. Chem. Soc.,* **95**, 8295 (1973).
4. I. F. J. Vankelecom, P. A. Jacobs, in D. E. DeVos, I. F. J. Vankelecom, P. A. Jacobs (Eds.) Chiral Catalyst Immobilization and Recycling; Wiley-VCH: New York, NY, 2000, pp 19-80.
5. D. J. Bayston, M. E. C. Polywka, in D. E. DeVos, I. F. J. Vankelecom, P. A. Jacobs (Eds.) Chiral Catalyst Immobilization and Recycling; Wiley-VCH: New York, NY, 2000, pp 211-234.
6. F. R. Hartley, Supported Metal Complexes; Reidel: Dordrecht, 1985.
7. W. Keim, B. Dreissen-Holscher, B. in G. Ertl, H. Knozinger, J. Weitkamp, (Eds.) Handbook of Heterogeneous Catalysis; Wiley-VCH: New York, NY, 1977, pp. 231-240.
8. M. E. Quiroga, E. A. Cagnola, D. A. Liprandi and P. C. L'Argentiere, *J. Mol. Catal. A:Chem.,* **149**, 147 (1999).
9. S.-G. Shyu, S.-W. Cheng and D.-L. Tzou, *Chem. Commun.,* 2337 (1999).
10. S. K. Tanielyan, R. L. Augustine, US Patents 6,005,148 (1999) and 6,025,295 (2000), to Seton Hall University.
11. S. K. Tanielyan, R. L. Augustine, R. L. PCT Int. Appl., WO-9828074: *Chem. Abstr.,* **129**, 109217 (1998).
12. R. Augustine, S. Tanielyan, S. Anderson and H. Yang, *Chem. Commun.,* 1257 (1999).
13. S. K. Tanielyan, R. L. Augustine, Chemical Industries (Dekker), **75** *(Catal. Org. React.),* 101 (1998).
14. R. L. Augustine, S. K. Tanielyan, S. Anderson, H. Yang, Y. Gao, Chemical Industries (Dekker), **82** *(Catal. Org. React.),* 497 (2000).
15. P. Goel, S. Anderson, J. Nair, C. Reyes, G. Gao, S. Tanielyan, R. Augustine, Chemical Industries (Dekker), **89** *(Catal. Org. React.),* 523 (2003).
16. C. Reyes, Y. Gao, A. Zsigmond, P. Goel, N. Mahata, S. K. Tanielyan and R. L. Augustine, Chemical Industries (Dekker), **82** *(Catal. Org. React.),* 627 (2000).
17. S. Anderson, H. Yang, S. K. Tanielyan and R. L. Augustine, Chemical Industries (Dekker), **82** *(Catal. Org. React.),* 557 (2000).
18. R. L. Augustine, S. K. Tanielyan, N. Mahata, Y. Gao, A. Zsigmond and H. Yang, *Appl. Catal. A*, in Press.
19. R. L. Augustine, S. K. Tanielyan, Chemical Industries (Dekker), **89** *(Catal. Org. React.)*, 73 (2003).

8. Carbon Supported Anchored Homogeneous Catalysts

Pradeep Goel, Clementina Reyes, Setrak Tanielyan, and Robert Augustine

Center for Applied Catalysis, Seton Hall University, South Orange, NJ 07079

augustro@shu.edu

Abstract

In almost all of the previous descriptions of our Anchored Homogeneous Catalysts an alumina was the support used. However, the technique of using a heteropoly acid as the anchoring agent does not limit itself to only alumina as the solid support for these species. Some brief mention has been made of the use of carbon, clay, silica and lanthana as supports but nothing extensive was done with these materials. We report here some examples of the use of carbon supported anchored homogeneous catalysts for the hydrogenation of prochiral substrates at moderate to high substrate/catalyst ratios along with a comparison of these results with those obtained using the corresponding homogeneous catalysts under the same conditions. Particular attention will be paid to the use of the anchored chiral Rh(DuPhos) and Rh(BoPhoz) species in the hydrogenation of dimethyl itaconate. Reaction rates, product enantioselectivities and the extent of metal loss will be presented where appropriate.

Introduction

Our Anchored Homogeneous Catalysts (AHC's) are composed of a support material, a catalytically active organometallic complex and a heteropoly acid (HPA) used to anchor the complex to the support. The most common HPA used in our work is phosphotungstic acid (PTA). Previous studies on the use of these AHC's have been concerned with studying the effect which different reaction variables had on the activity, selectivity and stability of these catalysts (1-9). These reactions were typically run at relatively low substrate/catalyst ratios (turnover numbers-TON's), usually between 50 and 100. While these low TON reactions made it possible to obtain a great deal of information concerning the AHC's, in order to establish that these catalysts could be used in commercial applications it was necessary to apply them to reactions at much higher TON's and, also, to make direct comparisons with the corresponding homogeneous catalyst under the same reaction conditions. Almost all of these reactions used catalysts supported on alumina, but some mention was made of the use of other support materials as well (1-4).

We report here the use of the carbon supported anchored catalysts, Rh(COD)(Me-DuPhos)/PTA/C (AHC-1) and Rh(COD)(BoPhoz)/PTA/C (AHC-2) (10) for the enantioselective hydrogenation of dimethyl itaconate (DMIT).

Results and Discussion

Rh(COD)(Me-DuPhos)/PTA/C (AHC-1)

In Table 1 are listed the TOF's and product ee's observed with both AHC-1 and the homogeneous species, Rh(COD)(Me-DuPhos)$^+$BF$_4^-$ (3) for the hydrogenation of DMIT in MeOH solution at 50°C and 50 psig of hydrogen. The hydrogenations were run at several different TON's with both catalysts. The homogeneous catalyst was more active than the anchored material but the product ee's were the same in both reactions. However, addition of more DMIT to the homogeneously catalyzed reaction mixture after the first amount had been hydrogenated resulted in complete deactivation of the catalyst, regardless of the TON. The anchored catalyst, though, could be re-used several times.

Table 1 Comparison of reaction rate and selectivity between AHC **1** and the homogeneous analog, **3**, in the hydrogenation of DMIT[a].

TON	mL MeOH	mL DMIT	[DMIT] M	AHC **1** TOF hr^{-1}	AHC **1** %ee	Homo **3** TOF hr^{-1}	Homo **3** %ee
1,000	15	1.4	0.7	4,000	96	16,500	96
5,000	15	7.0	2.3	5,700	96	28,000	96
10,000	15	14.0	3.4	6,600[a]	96	20,500	96

[a] Reactions run at 50°C and 50 psig.
[b] 0.9 ppm Rh in the reaction solution.

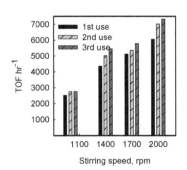

Figure 1 Effect of stirring speed on the rate of hydrogenation of DMIT over AHC-2. 5000 TON each reaction, ee's = 95-97%. Reactions run at 50° and 50 psig.

One of the more obvious reasons for the slower reaction rates observed with the AHC as compared to the homogeneous catalysts is the fact that hydrogen and substrate diffusion are more important in heterogeneously catalyzed reactions than in those promoted by homogeneous catalysts. Figure 1 depicts the relationship between the stirring rate and reaction TOF for the hydrogenation of DMIT over AHC-1. All hydrogenations were run at 5,000 TON with a separate batch of the catalyst used at each stirring speed. These catalysts were each re-used three times. It is obvious from these data that the agitation rate was not sufficient to overcome the diffusion control of these reactions. Another factor in the use of the high stirring speeds with our reactors is that at speeds above about 1700 rpm there is a tendency to deposit a ring of catalyst on the top of the reaction medium away from contact with the substrate.

Even though the catalysts were pre-hydrogenated the reaction rate increased with successive use but the product ee's remained constant at 96-97%. This increase in reaction rate with catalyst re-use was not found with higher TON hydrogenations as shown by the data listed in Table 2. Better results were obtained by running the hydrogenations at 75psig and 50°C with the reaction data also shown in Table 2 for three 20,000 TON hydrogenations, three 50,000 TON hydrogenations, a repeat of the 50,000 TON reaction with only two uses of the catalyst and a single 100,000 TON hydrogenation. The amount of Rh found in these TON reaction mixtures was below the level of detection (<1 ppm).

Table 2 Reaction data obtained on hydrogenation of DMIT over AHC-1 with multiple re-use of the catalyst at different TON's.

TON	Use #	TOF hr^{-1}	%ee
10,000[a]	1	6000	96
	2	6000	96
	3	6000	96
	4	6000	96
20,000[b]	1	5500	96
	2	4500	96
	3	4000	96
50,000[b]	1	6500	96
	2	5000	96
	3	3500	96
50,000[b]	1	6500	96
	2	5000	96
100,000[b]	1	6500	96

[a] Hydrogenations run at 50°C and 50 psig.
[b] Hydrogenations run at 75°C and 50 psig.

Rh(COD)(BoPhoz)/PTA/C (AHC-**2**)

The Rh(COD)(BoPhoz)/PTA/C (AHC-**2**) was prepared and used to promote the hydrogenation of DMIT. Data for 5,000 TON and 10,000 TON reactions are given in Table 3. In the first set of reactions, the four hydrogenations had identical rates (TOF hr^{-1}). With the 10,000 TON reactions the first three runs had the same reaction rate while the fourth and fifth showed about a 20% decrease, probably because the catalyst was exposed to hydrogen too long after run number three was completed. These reactions were all run using 3% toluene in methanol. The toluene was used as a co-solvent in these reactions in order to stabilize the catalyst for repeated re-use. Less than 1 ppm of Rh was detected in each of these reaction mixtures.

Table 3 Reaction data obtained on hydrogenation of DMIT over AHC-**2** at different TON's.

TON	Use #	TOF hr^{-1}	%ee
5,000[a]	1	5000	90
	2	5000	90
	3	5000	90
	4	5000	90
10,000[a]	1	7500	90
	2	7500	90
	3	7500	90
	4	6000	90
	5	6000	90
50,000[b]	1	18,500	90
50,000[c]	1	20,000	89
142,000[c]	1	12,500	88

[a] Hydrogenations run at 40°C and 75 psig in 3% toluene in MeOH.
[b] Hydrogenations run at 40°C and 75 psig in MeOH.
[c] Hydrogenations run at 50°C and 75 psig in MeOH.

However, the presence of toluene, even in a small amount, decreased the rate of hydrogenation for these reactions. Some very high TON reactions were run using AHC **2** for a single use without the presence of toluene. Table 3 also lists reaction data for a 50,000 TON hydrogenation run at 40°C and 75 psig. for a 50,000 TON reaction run at 50° and 75 psig and a 142,000 TON hydrogenation of DMIT run at 50°C and 75 psig. Less than 0.5 ppm of Rh was found in these reaction solutions.

Experimental Section

The carbon supported AHC catalysts, (Rh(COD)(Me-Duphos)/PTA/C (AHC-**1**) and Rh(COD)(BoPhoz)/PTA/C, (AHC-**2**) were prepared using the general procedure described previously (1-3). The catalysts contained about 0.9% Rh which corresponds to about an 8.5% load of the anchored complex. The hydrogenations were run using the low pressure apparatus previously described (11) under the conditions listed in the discussion.

Acknowledgements

Financial support of this work from Johnson Matthey is gratefully acknowledged.

References

1. S. K. Tanielyan, R. L. Augustine, US Patents 6,005,148 (1999) and 6,025,295 (2000), to Seton Hall University.
2. S. K. Tanielyan, R. L. Augustine, PCT Int. Appl., WO-9828074: *Chem. Abstr.*, **129**. 109217 (1998).
3. R. Augustine, S. Tanielyan, S. Anderson and H. Yang, *Chem. Commun.*, 1257 (1999).
4. S. K. Tanielyan, R. L. Augustine, Chemical Industries (Dekker), **75** *(Catal. Org. React.)*, 101 (1998).
5. R. L. Augustine, S. K. Tanielyan, S. Anderson, H. Yang, Y. Gao, Chemical Industries (Dekker), **82** *(Catal. Org. React.)*, 497 (2000).
6. P. Goel, S. Anderson, J. Nair, C. Reyes, G. Gao, S. Tanielyan, R. Augustine Chemical Industries (Dekker), **89** *(Catal. Org. React.)*, 523 (2003).
7. C. Reyes, Y. Gao, A. Zsigmond, P. Goel, N. Mahata, S. K. Tanielyan, and R. L. Augustine, Chemical Industries (Dekker), **82** *(Catal. Org. React.)*, 627 (2000).
8. S. Anderson, H. Yang, S. K. Tanielyan, and R. L. Augustine, Chemical Industries (Dekker), **82** *(Catal. Org. React.)*, 557 (2000).
9. R. L. Augustine, S. K. Tanielyan, N. Mahata, Y. Gao, A. Zsigmond, and H. Yang, *Appl. Catal. A*, **256**, 69 (2003).
10. The BoPhoz complex, $Rh(COD)(Me\text{-}BoPhoz)^+BF_4^-$, has been described: N. W. Boaz, S. D. Debenham, E. B. Mackenzie, and S. E. Large *Org. Lett.*, **4** (2002) 2421.
11. R. L. Augustine and S. K. Tanielyan, Chemical Industries (Dekker), **89** *(Catal. Org. React.)*, 73 (2003).

9.   Catalytic Hydrogenation of Cinnamaldehyde on Pt/Carbon Catalysts: The Effects of Metal Location, and Dispersion on Activity and Selectivity

Nabin K. Nag

Engelhard Corporation, Process Technology Group, 23800 Mercantile Road, Beachwood, OH 44122

nabin.nag@engelhard.com

Abstract

In continuation of a previous work (1), catalytic hydrogenation of cinnamaldehyde has been studied in slurry phase using a high-pressure autoclave. A series of carbon powder (CP)-supported Pt catalysts with widely varying Pt dispersion and Pt location on the support has been used in the study. The purpose has been to find out how the location of the metal on the support and its dispersion affect the two parallel reaction paths, namely the hydrogenation of the $C=O$ and $C=C$ bonds.

It is found that the hydrogenation of the $C=O$ bond in cinnamaldehyde is relatively more facilitated (as compared with the conjugated $C=C$ bond) on catalysts where Pt remains more concentrated near the pore mouths of the carbon particles (the highly edge-coated (HEC) catalysts). On the contrary, the $C=C$ hydrogenation selectivity is higher with highly dispersed (HDC) catalysts where the metal remains well dispersed throughout the whole porous region of the support carbon.

Introduction

In an earlier paper (1) the results of hydrogenation of cinnamaldehyde on various metals including Pt, Pd, Ru, and some bimetallic combinations of them, supported on aluminas, were reported. The selectivities for the hydrogenation of $C=O$ and $C=C$ bonds were found to depend on factors including the type of alumina, the method of catalyst preparation and the location of the metals on the support surface. In continuation of that work, a different catalyst system, namely Pt supported on activated carbon powders, has been used to study the chemoselectivity. From an organic reactions mechanism point of view this reaction provides an opportunity for studying the various factors that control the selectivities for the hydrogenation of two parallel reactions involving the conjugated $C=C$ and $C=O$ bonds. The locations of these two bonds in cinnamaldehyde are such that the $C=O$ bond may attach to an active catalyst site without facing any steric hindrance; it may even penetrate the pore mouths leaving the bulk of the molecule outside the pores. However, the C=C bond, being flanked by the terminal C=O group and the aromatic ring, is subject to steric hindrance. The steric hindrance is expected to be high especially in situations where the active sites are buried within the porous structure of the support (2).

Two distinct types of catalysts, differing in Pt dispersion and its location on the porous carbon support, have been used for the study: In one Pt has considerably higher concentration on the edges of the carbon particle (Figure 1) than within the pores. In the second series, Pt remains very well dispersed throughout the whole porous regions (Figure 2). The importance of metal location in the catalyst on the chemoselectivity is demonstrated in this work. Blackmond et al. (2) have shown the effect of steric hindrance on the selectivities for the saturation of C=O and C=C bonds during the hydrogenation of α,β unsaturated aldehydes on metal catalysts supported on zeolites. The results obtained in the present investigation are in good agreement with their work.

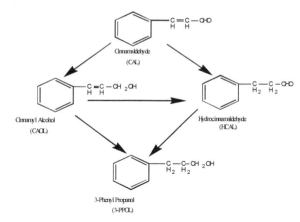

Scheme 1 Reaction Network for Cinnamaldehyde

Results and Discussion

In keeping with the earlier work (1), the activity of the catalysts is reported as the reciprocal of the time (hour) taken for 50% conversion of cinnamaldehyde. The selectivity for a particular product is defined as:

Selectivity for product A =   (Number of moles of a product A) X 100
                                                Total number of moles of reactant converted

Typical products distribution, and selectivity profiles were similar to what was reported in the earlier publication (1). The reaction network is shown in Scheme 1. The catalyst properties and test data are given in Table 1.

Table 1 Catalyst properties, and performance data.

Catalyst	% Pt Dispersion	Pt:C* (ESCA)	Activity** (1/h)	Selectivity for CAOL	HCAL	3-PPOL
1. CP-1/HDC	60	0.0021	1.30	10.4	23.5	63.1
2. CP-1/HEC	14	0.023	0.67	44.1	27.8	28.1
3. CP-2/HDC	60	0.0029	0.57	23.1	37.4	39.1
4. CP-2/HEC	18	0.037	0.57	47.1	29.6	23.1
5. CP-3/HDC	60	0.0022	0.83	16.4	30.4	53.2
6. CP-3/HEC	13	0.012	1.33	24.1	19.7	56.2

Note: CP-1/HDC and CP-1/HEC denote highly dispersed and highly edge-coated Pt respectively on CP-1. CAOL, HCAL and 3-PPOL denote cinnamyl alcohol, hydrocinnamaldehyde and 3-phenyl propanol respectively.
* Atomic ratio as measured by ESCA.
**Activity: Reciprocal of time (h) for 50% conversion.

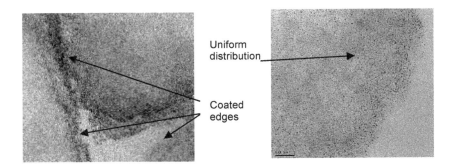

Uniform distribution

Coated edges

Figure 1 Typical **HEC** catalyst with High Pt concentration at the edges.

Figure 2 Typical **HDC** catalyst with uniform Pt distribution.

From the results given in Table 1, several key observations are made:
- All HEC catalysts have low Pt dispersion and high Pt to C ratios (ESCA); and there is more Pt on the edges of the support. These catalysts are more selective towards the hydrogenation of the C=O bond, as compared with the corresponding HDC catalysts.
- All the HDC catalysts have higher Pt dispersion and lower Pt to C atomic ratios (ESCA). These attributes are indicative of Pt being uniformly distributed throughout the porous region of the support. These catalysts are in general more selective towards the hydrogenation of the C=C bond.
- No correlation is found between catalyst activity and the mode of disposition of Pt on the carbon support.
- High or low dispersion, and the concomitant high and low Pt:C (ESCA) values are in accord with the TEM pictures shown above.

Although no general trend between catalyst activity and metal location (HDC vs. HEC) is found, the location of Pt on the support shows considerable effect on the selectivities for the hydrogenation of the C=O and C=C bonds. As mentioned above the selectivity is considerably higher for the C=C bond than the C=O bond on all the HDC catalysts, with the exception of the one based on CP-1 (where the selectivities are comparable). However, on the HEC catalysts, without exception, the selectivity for the hydrogenation of the C=O bond is substantially higher as compared with the C=C bond. This reversal is due to two factors: One, the higher concentration of Pt on the edges (pore mouths) of the carbon particles, and two, the terminal location of the C=O group (opposite to the benzene ring) in the cinnamaldehyde molecule. In a competition with the C=C bond, the C=O bond gets relatively easier access to the Pt sites at the edges on the carbon particles before the reactant molecule enters the pores to reach the active sites within the pores. One can imagine a molecule being adsorbed on the edge sites through the terminal C=O bond while the bulk of the molecule remaining outside the carbon particles (a dangling situation). This advantage in selective adsorption of the C=O bond on the edge sites in the HEC catalysts leads to higher selectivity for the hydrogenation of the C=O bond.

From the present work a model of how this catalytic hydrogenation process takes place may be presented. When the active sites are all within the pores and well dispersed throughout the whole porous region of the support (like in the HDC catalysts), a reactant molecule needs to enter the pores before any hydrogenation reaction can take place. In fact, in this situation, the C=C bond hydrogenation gets preference, as the data reveal. However, when there is an excess concentration of the metal on the edges of the support or in the pore mouths (like in the HEC catalysts), the C=O bond gets the preference owing to its strategic location in the cinnamaldehyde molecule. Due to its terminal location in the molecule the C=O bond gets preferentially adsorbed on the edge sites even before the reactant molecule enters the pores. This leads to higher hydrogenation selectivity for the C=O bond. The C=C bond simply cannot compete with the C=O bond outside the pores because steric hindrance, arising due to its being flanked by the C=O and the aromatic ring, impedes its adsorption and subsequent reaction outside the pores (although some reaction outside the pores on the edge sites cannot be totally ruled out). In summary, by manipulating the location of the catalytically active metal in a porous support, the chemoselectivity for the hydrogenation of α,β unsaturated aldehydes may be altered substantially.

Experimental Section

Three carbon powders from three different commercial manufacturers were used to make six catalysts. For each carbon, two types of catalysts, namely HEC and HDC, were prepared using slurry-phase preparation methods. For the HDC catalysts a chloroplatinic acid solution containing the requisite amount of Pt (to generate a nominal 1.5 wt% Pt/CP) was added to an alkaline CP slurry. This was followed by reduction with sodium formate at elevated temperatures. After reduction, the solid was separated, washed, and stored. The HEC catalysts were made by following a similar method with some variations in the hydrolysis and reduction steps (Engelhard proprietary method). Catalysts were tested as follows: A 2% (wt/vol) methanol solution of cinnamaldehyde (Aldrich) was charged into the reactor (autoclave from Autoclave Engineering), and the air was replaced by nitrogen purges. After further purging (3 times) with hydrogen, the pressure was raised 30 psig, and the temperature was raised to 100°C. After this, the final pressure of hydrogen was adjusted to 100 psig, and this point was taken as the '*zero*' time of the run. Samples were collected at various intervals and analyzed by gas chromatographic method (see ref. 1 for further details).

The details of ESCA measurements have been given in an earlier publication (1). A dynamic pulse chemisorption method was applied for measuring Pt dispersion taking 1:1 as the CO:Pt stoichiometry. A Jeol 2010 machine (200kV) with a LaB6 filament was used for generating the TEM images which were collected by a Gatan 2K CCD camera. A microtome technique was used to prepare the samples.

Acknowledgements

Thanks are due to Nancy Brungard, and George Munzing of Engelhard for the ESCA and TEM works.

References

1. N. K. Nag, Chemical Industries (Dekker) **89**, (*Catal. Org. React.*), 415 (2002). (A comprehensive literature review may be found in this reference.)
2. D. Blackmond, R. Oukaci, B. Blanc, and J. Gallezot., *J. Catal.,* **131**, 401 (1991).

10. Para-substituted Aniline Hydrogenation Over Rhodium Catalysts: Metal Crystallite Size and Catalyst Pore Size Effects

Kenneth J. Hindle, S. David Jackson, and Geoff Webb

Department of Chemistry, The University, Glasgow G12 8QQ, Scotland

Abstract

The hydrogenation of para-substituted anilines over rhodium catalysts has been investigated. An antipathetic metal crystallite size effect was observed for the hydrogenation of *p*-toluidine suggesting that terrace sites favour the reaction. Limited evidence was found for catalyst deactivation by the product amines. Catalysts with pore diameters less than 13.2 nm showed evidence of diffusion control on the rate of reaction but not the cis:trans ratio of the product.

Introduction

Alicyclic amines are used as pesticides, plasticizers, explosives, inhibitors of metal corrosion and sweetening agents as well as having uses in the pharmaceuticals industry. Aniline hydrogenation has been studied in the literature with the main reaction products cyclohexylamine, dicyclohexylamine, *N*-phenylcyclohexylamine, diphenylamine, ammonia, benzene, cyclohexane, cyclohexanol and cyclohexanone [1-9]. The products formed depend on the catalyst used, reaction temperature, solvent and whether the reaction is performed in gas or liquid phase. For example high temperature, gas-phase aniline hydrogenation over Rh/Al_2O_3 produced cyclohexylamine and dicyclohexylamine as the main products [1].

However few studies have been performed on substituted anilines [9, 10]. Freifelder and Stone [9] found that an alkyl substituent on the aromatic ring had very little effect on substituted-aniline hydrogenation, when using ruthenium dioxide as the catalyst. Ranade and Prins [10] investigated the hydrogenation of o-toluidine using (S)-proline as a chiral auxiliary with the aim of achieving diastereoselective hydrogenation.

The hydrogenation of 4-alkyl substituted anilines results in the production to two possible isomeric forms, cis and trans, and both of these isomers have two possible conformations in the chair configuration of the cyclohexyl ring. However experimentally it has been found that one cis and one trans conformer dominate. The dominant isomers are shown in Figure 1.

Thermodynamic data are not available to allow a calculation of the equilibrium ratio of the isomers, but a theoretical analysis of the energies [11] has confirmed that the trans isomer is the most stable.

This paper examines the hydrogenation of aniline, *p*-toluidine, and 4-*tert*-butylaniline over a series of 2.5 % Rh/SiO$_2$ catalysts, comparing reaction rates and product selectivities. Further studies concentrated on examining support particle size and average metal crystallite size effects on *p*-toluidine hydrogenation and the support pore size effects on 4-*tert*-butylaniline hydrogenation.

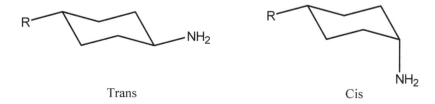

Trans   Cis

Figure 1 Trans and cis isomers of 4-alkyl substituted aniline.

Experimental Section

The catalysts used throughout the research were 2.5 % Rh/SiO$_2$ catalysts prepared by incipient wetness. Grace Catalysts supplied the catalyst supports and the catalysts were prepared by Johnson Matthey.

Hydrogenation experiments were performed using a Buchi autoclave stirred tank reactor. The reaction took place in a 500 cm^3 glass vessel surrounded by an oil heating jacket with a maximum operating temperature limit of 473 K. The temperature was measured in the liquid slurry with an accuracy of ± 1 K. The maximum operating pressure of the vessel was 6 barg and a magnetically driven stirrer performed the mixing. A Buchi pressflow gas controller controlled the pressure of the reactor, the flow of H$_2$ or N$_2$ and measured the H$_2$ consumed in the reaction.

Pre-reduced catalyst (typically 2 g) was added with 300 ml of 2,2,4-trimethylpentane (Lancaster Synthesis Ltd) to the reactor vessel. An *in-situ* reduction of the catalyst was performed by sparging H$_2$ (280 cm^3 min^{-1}) through this mixture for 30 minutes while stirring at 800 rpm. During this time the vessel was heated to 328 K. The stirrer was then turned off.

Aniline (0.0196 moles, Sigma-Aldrich), *p*-toluidine (0.0196 moles, Sigma-Aldrich) or 4-*tert*-butylaniline (0.0196 moles, Sigma-Aldrich), was dissolved in 50 ml of 2,2,4-trimethylpentane in a conical flask (the flask was heated gently to 323 K to aid dissolution of *p*-toluidine).

Immediately after the reduction process, the reactant and solvent solution was injected into the reactor vessel (giving a concentration of 0.0560 mol l^{-1}) and stirred at 800 rpm for 5 s to allow mixing. The stirrer was turned off, the vessel purged with N_2 twice at a flow rate of 280 cm^3 min^{-1}, before being pressurised with N_2 to 1 barg. A sample, 2.5 cm^3, was taken. The vessel was de-pressurised, before being purged with H_2 twice at a flow rate of 280 cm^3 min^{-1}, then pressurised with H_2 to 2 barg. Once the vessel was pressurised the stirrer speed was set to 1000 rpm. Samples of 2.5 ml were taken at time intervals and analysed using G.C.

Results

A variety of catalysts were used to examine the effect of changing pore size, metal crystallite size, and catalyst particle size. The catalyst characterisation is reported in Table 1.

Table 1 Characterisation of 2.5 % Rh/silica catalysts.

Catalyst	Dispersion (%)	Metal crystallite size (nm)	Metal Surface Area ($m^2 g^{-1}$)	Pore diameter (nm)	PSD (50) (microns)
M1081	71	1.6	7.8	2.3	10.7
M1272	60	1.8	6.6	6.1	10.0
M1079	94	1.2	10.3	10.6	9.5
M1273	37	3.0	4.1	13.2	10.2
M1074	43	2.6	4.7	13.2	24.0
M1038	41	2.7	4.5	13.2	49.9
M1035	32	3.5	3.5	13.2	96.3
M1078	50	2.2	5.5	13.2	241.0

The hydrogenation of toluene, aniline, *p*-toluidine, and 4-*tert*-butylaniline was examined over catalyst M1273. The reaction profile for the reactions is shown in Figure 2. From this it can be seen that the order of reactivity is aniline > toluene > *p*-toluidine > 4-*tert*-butylaniline. The hydrogenation products were methylcyclohexane from toluene, cyclohexylamine from aniline, 4-methyl-cyclohexylamine (4-MCYA) from *p*-toluidine, and 4-*tert*-butylcyclohexylamine (4-tBuCYA) from 4-*tert*-butylaniline. At 50 % conversion the cis:trans ratio of 4-MCYA was 2, while tBuCYA it was 1.6.

The effect of support particle size and metal crystallite size was investigated for *p*-toluidine hydrogenation using catalysts M1074, M1038, M1035, and M1078. The rate of reaction was found to be the same for each catalyst at 1 mmol min^{-1}. However this does not take into account the different metal surface areas. Hence a Turnover Frequency (TOF) was calculated for each catalyst by dividing the rate of reaction (moles min^{-1} g^{-1})

by the moles of active Rh per gram of catalyst (moles g⁻¹). A plot of TOF versus catalyst particle size showed no trend (not shown). However a plot of TO vs metal crystallite size revealed a straight-line relationship (Figure 3). The absence of an effect due to catalyst particle size was confirmed by crushing a sample and showing that the TOF was identical.

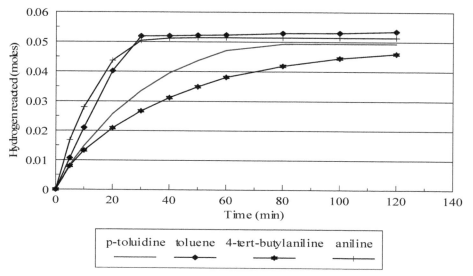

Figure 2 Hydrogenation of aniline, toluene, p-toluidine, 4-tert-butylaniline, over M1273. Temperature 338 K and 2 barg hydrogen pressure.

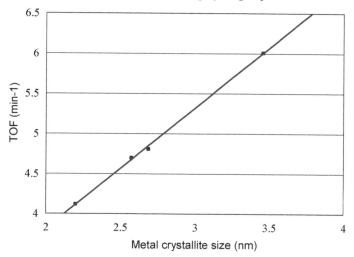

Figure 3 Relationship between TOF and metal crystallite size.

The effect of catalyst pore size on the hydrogenation of 4-tert-butylaniline was examined using catalysts M1081, M1272, M1079, and M1273. The effect on the rate is shown in Table 2.

Table 2 Effect of pore size on initial hydrogenation rate.

Catalyst	M1081	M1272	M1079	M1273
Pore Size (nm)	2.3	6.1	10.6	13.2
Rate (mmol.min^{-1})	0.2	0.7	0.9	0.9

The TOF was calculated for the 4-*tert*-butylaniline hydrogenation over each of the four catalysts. Using the relationship between metal crystallite size and TOF shown in Figure 3, the TOF for each catalyst was normalised to 1 nm and back calculated to a rate expression to remove the metal crystallite size effect from the rate. With the rate expressed in this way the effect of the pore size is clearly seen (Figure 4).

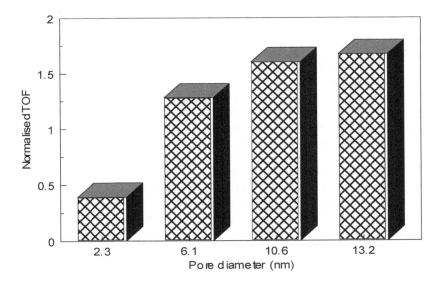

Figure 4 Effect of pore size on TOF.

In Figure 5 the variation of the 4-tBuCYA cis:trans ratio is plotted against the conversion for each of the catalysts.

Discussion

The literature on the hydrogenation of aniline and substituted anilines suggests that the amine functionality can act as a poison [3, 4, 12 and 13] especially over platinum catalysts. The more basic the nitrogen the more

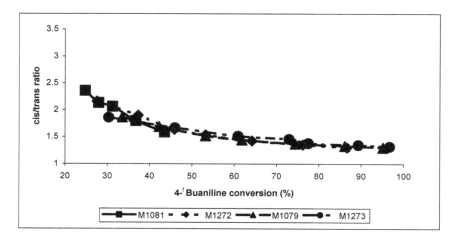

Figure 5 Variation in 4-tBuCYA cis:trans ratio with conversion and catalyst.

effective the amine is as a poison, so the aliphatic amine products are more deleterious than the aromatic amines. However there is no literature concerning the hydrogenation over rhodium catalysts. Therefore one of the initial experiments was to examine whether there was obvious deactivation, by comparing the hydrogenation of aniline with that of toluene. From Figure 2 it can be seen that the initial rate of aniline hydrogenation is faster than that of toluene, suggesting that the aromatic amine causes no serious deactivation. However in the later stages of the reaction the instantaneous rate of toluene hydrogenation was faster than the rate of aniline hydrogenation. This change over in the later stages may well be linked to a slight deactivation due to the cyclohexylamine product. This would be in keeping with the concept of the aliphatic amine being a more effective poison. Even so, there was no gross deactivation as had been observed with platinum systems [3, 4, 12 and 13].

The effect on the rate of hydrogenation of changing the para substituent in substituted anilines was investigated using aniline, *p*-toluidine, and 4-*tert*-butylaniline over catalyst M1273 (Figure 2). The order of reactivity was found to be aniline > *p*-toluidine > 4-*tert*-butylaniline. As expected as the para substituent was changed from -H to -CH$_3$ to -C(CH$_3$)$_3$, the rate decreased. This is almost certainly due to steric effects rather than poisoning as the pK$_a$ of the respective aliphatic amines will be similar. Further steric effects can be seen in the cis/trans ratios observed between cis and trans 4-MCYA (2:1) and cis and trans 4-tBuCYA (1.6:1). The thermodynamically favoured product is, as expected, the trans isomer [11] whereas experimentally in both systems it is the cis isomer that is found to be the major product. The formation of the cis isomer as the primary product and subsequent isomerisation is a common feature in heterogeneous catalysis [14, 15]. However with the larger *tert*-butyl group the thermodynamic driving force for isomerisation to the trans isomer is greater with the 4-tBuCYA.

The metal crystallite size effect found with *p*-toluidine hydrogenation was a surprise. Figure 3 reveals that as the crystallite size increases the TOF increases. This behaviour suggests that the plane face/terrace sites rather than edge or corner sites favour the reaction. Indeed the increase in the number of C^3_9 atoms [16] with crystallite size is similar to the activity increase observed in the *p*-toluidine hydrogenation. Examination of the literature reveals no other comparable studies for substituted anilines. However in the literature of benzene and toluene hydrogenation some support for such behaviour can be found. Benzene hydrogenation over Rh/alumina [17] and Ni/silica [18] show a complex crystallite size effect. Toluene hydrogenation over platinum catalysts [19] was shown to display an antipathetic crystallite size effect (i.e. increasing TOF with increasing crystallite size), as was toluene hydrogenation over iridium catalysts [20]. These results suggest that for ring hydrogenation, the ring prefers a flat, parallel adsorption mode rather than an edge-on adsorption.

The effect of the support pore size on 4-*tert*-butylaniline hydrogenation is reported in Table 2 and as a TOF in Figure 4. When the pore diameter is only 2.3 nm there is a significant effect on the rate. This is to be expected, as at this pore diameter there will be significant diffusional effects limiting access to active sites within the pores. However it can be seen that there is also an effect when the pore diameter is 6.1 nm. The effect is much less marked but quite clearly defined. With 4-*tert*-butylaniline hydrogenation there is also the formation of the cis and trans-4tBuCYA. Figure 5 shows the cis:trans ratio against conversion for each of the catalysts. It can be seen that there is a common line upon which all the catalysts lie. Hence the pore size does not affect the cis:trans ratio. This is slightly surprising as if there is a diffusional constraint on the transfer of product from reactive site to the bulk liquor, then we might expect a larger extent of isomerisation towards the thermodynamic equilibrium with catalysts having a smaller pore size.

Conclusions

The hydrogenation of *p*-substituted anilines has been studied over a series of Rh/silica catalysts. Limited evidence was found for any deactivation compared to the significant deactivation reported in the literature for platinum catalysts [3, 4, 13]. Increasing the size of the *para*-substituent reduced the rate of reaction. An antipathetic metal crystallite size effect was observed for the hydrogenation of *p*-toluidine, indicating that the hydrogenation was more favoured on terrace sites. The principal product was the cis-isomer with a slow isomerisation to the trans-isomer taking place during the course of the reaction. Finally a pore size effect was observed with pore diameters of 10 nm and less showing evidence of diffusional constraint on the rate of reaction.

Acknowledgements

The authors acknowledge the support of the EPSRC and the following companies, Johnson Matthey, Grace Davison, Robinson Brothers, BP-Amoco and Accelerys, as part of the Innovative Manufacturing Initiative (IMI).

References

1.   V. Vishwanathan and S. Narayanan, *J. Chem. Soc. Commun.*, 78 (1990).
2.   D. V. Sokolskii, A. Ualikhanova, and A. E. Temirbulatova, *React. Kinet. Catal. Lett.*, **20**, 35 (1982).
3.   H. Greenfield, *J. Org. Chem.*, **29**, 3082 (1964).
4.   E. L. Pitara, B. N'Zemba, J. Barbier, F. Barbot, and Miginiac, *J. Mol. Catal. A*, **106**, 235 (1996).
5.   B. T. Mailyubaev and A. Ualikhanova, *Russian Journal of Applied Chemistry*, **66**, 1739 (1994).
6.   S. Narayanan, R. Unnikrishnan, and V. Vishwanathan, *Appl. Catal. A*, **129**, 9 (1995).
7.   S. Narayanan and R. Unnikrishnan, *J. Chem. Soc. Faraday Trans.*, **93**, 2009 (1997).
8.   G. Mink and L. Horváth, *React. Kinet. Catal. Lett.*, **65**, 59 (1998).
9.   M. Freifelder and G. R. Stone, *J. Org. Chem.*, **27**, 3568 (1962).
10.   V. S. Ranade and R. Prins, *J. Catal.*, **185**, 479 (1999).
11.   D. J. Willock, Personal Communication.
12.   J. M. Devereux, K. R. Payne, and E. R. A. Peeling, *J. Chem. Soc.*, 2845 (1957).
13.   E. B. Maxted and M. S. Briggs, *J. Chem. Soc.*, 3844 (1957).
14.   S. Siegel, *Adv. in Catal.*, **16**, 123 (1966).
15.   P. N. Rylander, in Hydrogenation Methods, Academic Press, London, 1985.
16.   M. Che and C. O. Bennett, *Adv. Catal.*, **36**, 55 (1989).
17.   S. Fuentes and F. Figueras, *J. Catal.*, **61**, 433 (1980).
18.   J. W. E. Coenen, R. Z. C. van Meerten, and H. Th. Rijten, *Proc. 5th Int. Congr. Catal.*, **1**, 671 (1972).
19.   B. D. Chandler, A. B. Schabel, and L. H. Pignolet, *J. Phys. Chem. B*, **105**, 149 (2001).
20.   Z. Xu, F.-S. Xiao, S. K. Purnell, O. Alexeev, S. Kawi, S. E. Deutsch, and B. C. Gates, *Nature*, **372**, 346 (1994).

11.   Reductive Alkylation During Nitrobenzene Hydrogenation Over Copper/Silica

David S. Anderson, S. David Jackson,[*] David Lennon, and Geoff Webb

Department of Chemistry, The University, Glasgow G12 8QQ, Scotland

Abstract

The reductive alkylation reaction occurring during the hydrogenation of nitrobenzene in a 1-hexanol solvent is not between the aniline formed and the 1-hexanol but is between 1-hexanol and a surface species retained by the catalyst. This surface species is not formed during the aniline 1-hexanol reaction. There is also inhibition of the reaction between aniline and 1-hexanol.

Introduction

The reduction of nitrobenzene to aniline is a major industrial process at the heart of the production of polyurethanes, and it is also often used as a marker reaction to compare activities of catalysts [1,2]. It can be performed over a variety of catalysts and in a variety of solvents. As well as its main use in polyurethanes, aniline is used in a wide range of industries such as dyes, agrochemicals, by further reaction and functionalisation. Reductive alkylation is one such way of functionalising aromatic amines [3, 4]. The reaction usually takes place between an amine and a ketone, aldehyde or alcohol. However it is possible to reductively alkylate direct from the nitro precursor to the amine and in this way remove a processing step. In this study we examined the reductive alkylation of nitrobenzene and aniline by 1-hexanol.

Experimental Section

The catalyst used throughout this study was prepared by impregnation. To a slurry of silica (M5 Cab-O-Sil, Surface Area 200 m^2g^{-1}) in water sufficient copper nitrate solution was added to give a loading of 8.6 % w/w Cu. The resulting suspension was dried at 353 K until a free flowing powder was obtained.

Copper metal surface area was determined by nitrous oxide decomposition. A sample of catalyst (0.2 g) was reduced by heating to 563 K under a flow of 10 % H_2/N_2 (50 cm^3min^{-1}) at a heating rate of 3 deg.min^{-1}. The catalyst was then held at this temperature for 1 h before the gas flow was switched to helium. After 0.5 h the catalyst was cooled in to 333 K and a flow of 5 %N_2O/He (50 cm^3min^{-1}) passed over the sample for 0.25 h to surface oxidise the copper. At the end of this period the flow was switched to 10 % H_2/N_2 (50 cm^3min^{-1}) and the sample heated at a heating rate of 3 deg.min^{-1}. The hydrogen up-take was quantified, from this a surface area could be determined [5, 6]. The metal surface area calculated was 5.47 m^2g^{-1}, giving a dispersion of 9.5% and a metal particle size of 10.7 nm.

The reductive alkylation reactions were performed in a 1 atm stirred tank reactor with hydrogen sparged (200 cm^3min^{-1}) through the solution. The catalyst (0.5 g) was reduced *in situ,* by heating under a flow of 10 % H_2/N_2 at 3 $Kmin^{-1}$ from ambient to 563 K and holding at this temperature for 1 h. During this procedure the catalyst was supported on a glass sinter and the gas flow was through the catalyst bed. After 1 h in 10 % H_2/N_2 the gas flow was switched to helium and the temperature held for another 0.5 h. The catalyst was then cooled to room temperature and the reactor inverted to transfer the catalyst into the stirred tank reactor under a flow of helium. The reaction mix of 80 cm^3 of 0.1 M nitrobenzene (or aniline) in 1-hexanol was added and the system heated to the reaction temperature (383 K). The reaction was initiated by switching the gas flow from helium to hydrogen. The reaction solution was stirred using a magnetic stirrer and the absence of diffusion control was determined by a variety of means [7]. Samples (0.3 cm^3) were removed from the reactor at regular intervals and were analysed by gas chromatography.

Table 1 Nitrobenzene hydrogenation in 1-hexanol at 383 K.

Time min	% NB[a] conversion	% AN[a] yield	% HBA[a] yield	% HA[a] yield
0	0.0	0.0	0.0	0.0
60	10.5	9.2	1.1	0.0
180	32.0	24.6	1.3	0.0
300	53.4	37.2	1.1	0.0
420	77.2	49.7	2.0	0.6
480	85.5	54.9	2.7	0.7
600	100	57.7	1.7	0.6
660	100	58.8	1.0	1.3
750	100	56.6	1.4	2.5
840	100	57.7	1.5	3.6
930	100	57.7	1.8	4.8
1020	100	58.8	2.1	5.9
1111	100	57.7	1.9	6.9
1200	100	54.9	1.2	8.0
1290	100	58.8	1.4	8.9
1380	100	57.1	2.2	10.2
1440	100	56.0	4.4	10.8

[a]NB, nitrobenzene; AN, aniline; HBA, N-hexylenebenzenamine; HA, N-hexylaniline

Results

A mixture of aniline, nitrobenzene, and 1-hexanol was left for 5.5 h at 406 K in the absence of a catalyst to confirm that any reaction was a function of the catalyst and not due to a stoichiometric reaction. No reaction was observed.

Aniline and 1-hexanol were left over a reduced catalyst at 383 K in a helium atmosphere. After 0.3 h N-hexylenebenzenamine (HBA) was detected, no N-hexylaniline (HA) was observed.

The results from nitrobenzene hydrogenation over Cu/silica are shown in Table 1. It can be seen that N-hexylaniline is only formed significantly once all the nitrobenzene has been consumed. No other by-products were detected.

Table 2 Reaction of Aniline with 1-hexanol at 383 K.

Time min	% AN conversion	% HBA selectivity	% HA selectivity	% HBA yield	% HA yield
0	0.0	0.0	0.0	0.0	0.0
60	7.8	8.3	42.5	0.6	3.3
120	9.7	6.4	51.7	0.6	5.0
184	12.2	4.9	56.0	0.6	6.8
260	14.1	4.8	61.7	0.7	8.7
300	16.0	4.4	64.2	0.7	10.3
360	16.6	3.8	71.6	0.6	11.9
480	19.1	2.9	73.3	0.6	14.0
651	21.6	2.8	82.3	0.6	17.8
740	24.8	2.8	81.6	0.7	20.2
840	29.8	2.5	72.8	0.7	21.7
948	32.3	2.2	73.7	0.7	23.8
1080	31.7	2.5	81.0	0.8	25.6

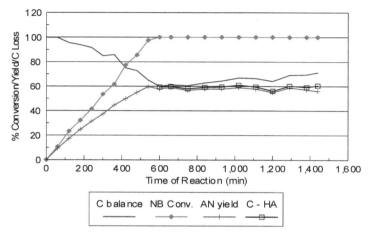

Figure 1 Reaction of nitrobenzene to form aniline. The C-HA term displays the carbon balance assuming all the HA is formed from a surface species.

Discussion

It is clear from Tables 1 and 2 that HA is formed in the nitrobenzene hydrogenation system and the direct aniline reaction. However the rate of production of HA is significantly faster (x2.5) from the aniline reaction compared with the nitrobenzene reaction. Examination of the production of HA in the nitrobenzene system shows that HA is only produced once all the nitrobenzene has reacted, suggesting that the reaction is one between aniline and 1-hexanol. However closer examination of the data reveals that this is not the case. The rate of HA formation (1.7 μmol.min^{-1}.g^{-1}) determined once all the nitrobenzene has reacted to aniline, is significantly slower than that found in the direct aniline/1-hexanol reaction (4.7 μmol.min^{-1}.g^{-1}). Also there is no obvious reduction in the amount of aniline present in solution. When a mass balance is taken in both systems we find that a mass balance in the aniline system is a steady 95 % throughout the time of the reaction. However the mass balance for the nitrobenzene system shows a fall as the hydrogenation of nitrobenzene proceeds (Figure 1). Once all the nitrobenzene has reacted HA production increases, however the yield of aniline remains constant. Whereas the mass balance starts to slowly increase. Therefore it appears that the 1-hexanol is not reacting with aniline but with a surface species. In Figure 1 the effect on the mass balance if the production of HA is subtracted (C – HA) is shown. It can be seen that the value remains constant indicating that it is the surface species that is reacting and not the aniline. Also the rate of HA formation is much slower compared to the rate obtained with aniline (4.7 c.f. 1.7 μmol.min^{-1}.g^{-1}), indicating that a species inhibits aniline reaction. Therefore by carrying out the reductive alkylation during the hydrogenation of nitrobenzene we have accessed a separate route to HA using a surface species that is much less reactive than aniline and that there is an inhibition of direct aniline alkylation.

The loss of 30 % of the starting reactant is significant and this amount of material is in considerable excess over the potential number of surface sites available on the copper surface. However the same is not true for the silica support. The hydroxyl density of silica can vary from 2.7 OH nm^{-2} to 15.9 OH nm^{-2} [8 – 11] with an average value around 6 OH nm^{-2}. For a silica with a surface are of 200 m^{2}g^{-1}, a hydroxyl concentration of 6 OH nm^{-2} translates into 1.2×10^{21} OH groups on the surface of the sample used in the experiments. The amount of nitrobenzene at the start of the reaction was 4.82×10^{21} molecules, 30 % of this is 1.45×10^{21}, which compares well with the number of available hydroxyl groups. The reaction between a surface hydroxyl and aniline is similar to that between aniline and 1-hexanol:

$$\equiv SiOH + Ph\text{-}NH_2 \rightarrow \equiv Si\text{-}NH\text{-}Ph + H_2O.$$

Hence the 30 % that are "lost" could be on the support and slowly react with 1-hexanol to form HA. However when aniline is reacted there is no significant loss of material, which suggests that aniline cannot interact directly with the surface hydroxyls. This suggests that the interaction between aniline and the support hydroxyls is not as simple as shown above, rather it is more likely that the reaction operates via a spillover mechanism involving an intermediate in the nitrobenzene hydrogenation sequence rather than aniline. The alkylation reaction between aniline and 1-hexanol takes place on the metal function, therefore the reaction with the "missing" aniline associated with the support will be slow as it requires a reverse spillover and a diffusion across the support surface.

Finally we must address the difference in rate between the in situ alkylation of aniline at the end of nitrobenzene hydrogenation and that found for the direct reaction between aniline and 1-hexanol. One significant difference between the systems is the presence of water. Two moles of water are produced for every mole of nitrobenzene converted so the final water concentration will be in excess of the aniline concentration. In the direct reaction no water was present. Water is a known poison for a range of catalytic reactions [8 – 11] and we have shown in our own laboratories that water will reduce the rate of nitrobenzene hydrogenation [12]. Therefore it is likely that the deactivation of the aniline reaction after nitrobenzene hydrogenation is due to inhibition by water.

These results indicate that the one-pot reductive alkylation using nitro-precursors rather than the amine may be more complex than previously thought.

Acknowledgements

The authors would like to thank the referees for their helpful suggestions.

References

1. K. Weissermel and H.-J. Arpe, Industrial Organic Chemistry, VCH, Weinheim, 1997.
2. M. Arai, A. Obata, and Y. Nishiyama, *React. Kinet. Catal. Lett.*, **61**, 275 (1997).
3. P. Rylander, Hydrogenation Methods, Academic Press, London, 1990.
4. M. P. Reynolds, and H. Greenfield, Chemical Industries (Dekker), **68** (*Catal. Org. React.*), 343, (1996); V. L. Mylroie, L. Valente, L. Fiorella, and M. A. Osypian, Chemical Industries (Dekker), **68** (*Catal. Org. React.*), 301, (1996) and references therein.
5. G. C. Bond and S. N. Namijo, *J. Catal.,* **118**, 507 (1989).
6. K. J. Sorensen and N. W. Cant, *Catal. Lett.,* **33**, 117 (1995).
7. D. S. Anderson, Ph.D. Thesis, University of Glasgow, 1999.
8. G. Centi, S. Perathoner, Z. S. Rak, *Applied Catalysis B*, **41**, 143 (2003).
9. J. Li, X. Zhan, Y. Zhang, G. Jacobs, T. Das, B. H. Davis, *Applied Catalysis A*, **228**, 203 (2002).
10. J. P. Claverie and R. Soula, *Progress in Polymer Science*, Volume **28**, 619 (2003).
11. M. E. Harlin, A. O. I. Krause, B. Heinrich, C. Pham-Huu, M. J. Ledoux, *Applied Catalysis A*, **185**, 311 (1999).
12. E. A. Gelder and S. D. Jackson, unpublished results.

12. Hydrogenolysis of Butyl Acetate to Butanol over Naphtha Reforming Type Catalysts in Conventional and High Throughput Slurry Phase Reactors

S. Göbölös, M. Hegedüs, I. Borbáth, and J. L. Margitfalvi

Chemical Research Center, Hungarian Academy of Sciences, 1025 Budapest, Pusztaszeri út 59-67 Hungary; gobolos@chemres.hu

Abstract

Butyl acetate was hydrogenolyzed to butanol over different $RePt/Al_2O_3$ and Re modified $SnPt/Al_2O_3$ naphtha reforming type catalysts both in a conventional autoclave and a high throughput (HT) slurry phase reactor (AMTEC-SPR16). Catalytic tests performed in a single autoclave revealed that Re is a crucial ingredient for high conversion (95 %) and selectivity (90 %) at 235 °C and 6 MPa. The addition of Re to the $SnPt/Al_2O_3$ catalyst significantly improved its activity. The specific activity showed a maximum at a Re content of 0.8 wt. %. In the HT activity tests, the reaction order for butyl acetate and the apparent activation energy was close to zero and 115 kJ/mole, respectively. TPR of RePt catalysts indicated the presence of the oxide precursors of separate Pt clusters, RePt alloy phase and Re ions stabilized by the support. It is proposed that in the activation of the carbonyl group "M_1^{n+}-M_2^0" ensembles (M_1 = Re, Sn; M_2 = Pt, Re, RePt or SnPt alloy) are involved.

Introduction

The catalytic hydrogenolysis of esters to two alcohols was first described by Folkers and Adkins in 1931 and has since assumed considerable industrial importance [1]. Conventionally the reaction is carried out in the liquid phase using a copper chromite catalyst at 10 MPa H_2 pressure and ~ 250 °C [2,3]. Due to the concern for the toxicity of Cr nowadays, $CuO-ZnO/Al_2O_3$ catalysts are used [4]. Supported SnRu [5], or SnRh [6] catalysts with high metal loading have gained much scientific interest in the transformation of less reactive carbonyl compounds such as amides, esters and acids to the corresponding products. It has been reported that in Cu- [7] and Ru-based [8] catalysts the mutual presence of both ionic and metallic species is required to activate the carbonyl group and the hydrogen, respectively.

Only scare data is available in the literature on the application of rhenium containing mono- or bimetallic catalysts in the hydrogenolysis of esters to alcohols. Decades ago Broadbent and co-workers studied the hydrogenation of organic carbonyl compounds (aldehydes, ketones, esters, anhydrides, acids, amides, etc.) on different reduced bulk rhenium oxides [9,10]. Recently a few publications have reported the transformation of different carbonyl compounds over Re containing catalysts [11,12]. In the bimetallic catalysts, the Re was combined with Pd [13] or Rh [14]. In addition, supported SnPt catalysts have been found to be active and selective in the hydrogenation of unsaturated aldehydes to the corresponding alcohols [15]. It is noteworthy that both Re and Sn are frequently used components of commercial naphtha reforming catalysts [16]. However, there is no data in the literature on the use of reforming type SnPt, RePt or ReSnPt catalysts for the hydrogenolysis of esters to alcohols. Despite this lack of data, we assumed that bimetallic naphtha reforming catalysts containing Pt and Re or Sn promoters could also be applied for the hydrogenation of carbonyl compounds. It has been proved that in the reforming catalysts at least part of the Sn or Re is in ionic form [17,18]. Therefore, one might expect that ionic Sn or Re species as Lewis acid sites in naphtha reforming catalysts can also activate the $C=O$ group of carbonyl compounds.

In this study butyl acetate (AcOBu) was hydrogenolysed to butanol over alumina supported Pt, Re, RePt and Re modified SnPt naphtha reforming catalysts both in a conventional autoclave and a high throughput (HT) slurry phase reactor system (AMTEC SPR 16). The oxide precursors of catalysts were characterized by Temperature-Programmed Reduction (TPR). The aim of this work was to study the role and efficiency of Sn and Re in the activation of the carbonyl group of esters.

Experimental Section

Catalysts - Alumina supported 0.5 wt. % Pt (0.5Pt) and 0.5 wt. % Re (0.5Re) catalysts in the form of beads (d=1 mm) were commercial products. γ-Alumina supported (GFSC-200RP, extrudates, d=1 mm, length=3-4 mm, SA=200 m^2/g) 1.5 wt. % Re catalyst (1.5Re) was prepared by incipient wetness impregnation using aqueous solution of NH_4ReO_4. The impregnation was followed by drying at 105 °C for 24 h. SnPt/Al_2O_3 commercial reforming catalyst (beads, d=1 mm) containing 0.3 wt. % Sn, 0.3 wt. % Pt and 1 wt. % Cl will be dubbed as 0.3Sn0.3Pt. This catalyst was modified with different amounts of Re ranging from 0.1 to 1.5 wt. %. Re was introduced to SnPt catalyst by incipient wetness impregnation using NH_4ReO_4. The impregnated catalysts were dried at both RT and 105 °C for 24 h.

Five RePt/Al_2O_3 catalysts, three with low metal loading ~ 0.2 - 0.3 wt. % (see Table 2) and two with 1 wt. % for both Re and Pt, were commercial products and were designated 0.3Re0.3Pt or 1Re1Pt. The chlorine content of these catalysts was 0.9 wt. %. Commercial catalysts were obtained either in reduced (red.) or calcined (calc.) form and used in their pre-formulated shape, i.e. beads (b) (d = 1 mm) or extrudates (e) (d = 1 mm, length = 3-4 mm).

Temperature-programmed reduction - Prior to TPR the catalysts were treated in different atmospheres as given in Figures 1-3 and Table 1. In the TPR experiments, the catalyst (0.08 g) was heated to 800 °C at a rate of 10 °C/min in a flowing 5% H_2-Ar mixture (W = 40 mL/min).

Activity test - Prior to the activity test catalyst (2 or 4 g) was pre-treated in O_2, N_2 and H_2 at 400 °C for 1, 0.5 and 2 h, respectively. The first series of liquid phase hydrogenolyses of AcOBu was carried out in a 300 mL stirred SS autoclave (Parr Co.). The reaction conditions were: AcOBu (50 ml, 0.38 mol); P_{H2} = 6 MPa (at RT); reaction temperature = 235 °C; reaction time = 24 h; and stirring rate (s.r.) = 600 rpm. Further experiments were performed in a slurry phase parallel reactor system SPR16 (Advanced Micro Technologies AMTEC, Chemnitz, Germany). Prior to the activity test 1Re1Pt(e,red.) catalyst (0.1 g) was *in situ* reduced in H_2 at 250 °C and 1.0 MPa for 1 h. The reaction order for AcOBu and apparent activation energy were determined in 16 parallel 15 ml reactors operated simultaneously. The reaction conditions were as follows: the solvent (heptane); AcOBu; (0.38, 0.57, 0.76 and 0.97 mole/L); P_{H2} = 6 MPa (maintained during test); reaction temperature = 215, 220, 225 and 230 °C; reaction time = 3.6 h; and, s.r. = 1000 rpm.

Analysis of reaction products - Liquid reaction products were analyzed by gas chromatography using a capillary column (type: WCOT Fused Silica, stationary phase: 5% phenyl-methyl-polysiloxane: length - 50 m; ID - 0.32 mm, OD - 0.45 mm; film thickness - 0.25 μm).

Calculation of kinetic parameters - In the experiments carried out in the single autoclave the H_2 pressure was not maintained and the consumption of H_2 controlled the conversion of AcOBu, which could be described by pseudo-first order rate constant. In the activity tests performed in SPR16 the conversion of AcOBu increased linearly up to ca. 50 % with reaction time. Initial reaction rates were calculated from AcOBu conversion vs. reaction time dependence, the initial concentration of substrate and the amount of catalyst or the amount of promoters in 1 g of catalyst.

Results and Discussion

Temperature-programmed reduction - Figure 1 shows the TPR curves of alumina supported Pt, Re and PtRe catalysts (Table 1 – No 1-3). In the TPR curve of Pt catalyst the low temperature peak at 230 °C can be attributed to the reduction of PtO_2, whereas the peak at ~ 350 °C can be assigned to platinum species interacting more strongly with the support [19,20]. The maximum of the TPR curve of Re/Al_2O_3 catalyst was observed at ~ 450 °C which is similar to that obtained in the case of rhenium catalyst prepared by using NH_4ReO_4 [19]. The amount of H_2 consumed in TPR indicated incomplete reduction of Re. As for the 1Re1Pt(e,red.) catalyst the H_2 consumption was normalized for 0.5 wt. % metal content of both Re and Pt. In this catalyst Pt facilitated the reduction of Re suggesting that Pt was near the rhenium [19]. The shape of the TPR curve and the characteristic temperatures of the peaks indicated simultaneous reduction of platinum and rhenium at low temperature ($\sim 300\ °C$) and thus the formation of bimetallic species [21].

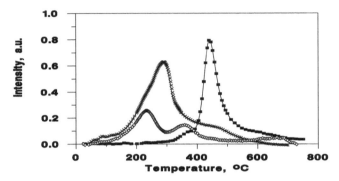

Figure 1 TPR of alumina supported Pt, Re and Pt-Re catalysts. Pre-reduced samples were calcined in air at 400 °C: (X)-1Re1Pt; (v)-0.5Re; (M)-0.5Pt.

Figure 2 shows the influence of pre-treatment on the TPR of 1Re1Pt(e,calc.) catalyst (Table 1, No. 6-8). Treatment at 250 °C in Ar resulted in a single symmetric peak at about 280 °C in good agreement with previous findings. It has been found that on hydrated catalyst the reduction of Re is catalyzed by Pt, resulting in simultaneous reduction of both metals at ~300 °C [21]. As seen in Figure 2 the shape of the TPR curves of catalysts treated in Ar at 250 °C and oxidized in air at 400 and 500 °C is completely different. The oxidation at higher temperatures resulted in more complicated TPR spectra with broad overlapping bands. This can be an indication on the segregation of Re and Pt oxide phases during oxidation treatment. The high temperature part of the spectra indicates the presence of separate oxidized Re species in the catalyst.

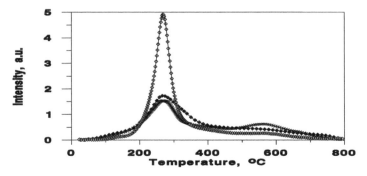

Figure 2 Effect of pre-treatment on TPR of 1Re1Pt(e,calc.) catalyst. Sample

was treated prior to TPR: (M) - in Ar at 250 °C; (Λ) - in air at 400 °C; (X) - in air at 500 °C.

Figure 3 shows that the higher the temperature of oxidation the higher the amount of the hydrogen consumed in high temperature part of TPR for the reduction of separate rhenium oxide species (Table 1, No 4-5). The low temperature peak between 200 and 300 °C indicates the presence of the oxide precursor of the bimetallic phase. All these results indicated that the lower the temperature of drying or calcination the more intimate the contact between Re and Pt, the higher the reducibility of rhenium and the smaller the amount of consumed hydrogen in the high temperature part of TPR spectra.

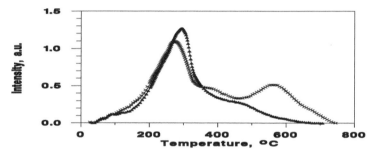

Figure 3 Effect of calcination temperature on TPR of 1Re1Pt(e,red.) catalyst. Sample was calcined in air prior to TPR: (σ) - at 400°C; (X) - at 500 °C.

The degree of reduction of oxide phases determined from TPR is given in Table 1. PtO_2 and Re_2O_7 phases were supposed to be present in the catalyst samples treated differently prior to TPR. In the TPR measurement PtO_2 can be fully reduced, whereas only 86 % of the Re_2O_7 is reduced in 0.5Re(b) catalyst. In the 1Re1Pt(e,red.) and 1Re1Pt(e,calc.) catalysts the reduction of both phases was practically complete. In the case of sample dried in argon at 250 °C only 79 % of the rhenium oxide phase was reduced. This may suggest that the oxidation degree of Re in the precursor state was lower than +7. The most interesting result was that 1Re1Pt(e,red.) catalyst oxidized and then reduced at 400 °C can be further reduced in the TPR experiment (Table 1, No 5). The hydrogen uptake indicated that in this catalyst after reduction at 400 °C for 2 hours half of the rhenium was in the form of Re^{4+}. This indicates that the treatment used prior to activity test results in a catalyst containing also ionic rhenium species. In 0.3Sn0.3Pt(b) catalyst 35 % of SnO_2 was reduced to Sn^0 and 65 % to SnO.

Table 1 Reduction degree of Pt and Re (or Sn) in the different catalysts.

No	Catalyst/Treatment	PtO_2 reduction, %	Re_2O_7 reduction , %
1	0.5Pt(b)/O400	100	-
2	0.5Re(b)/O400	-	86
3	1Re1Pt(e,red.)/O400	100	96
4	1Re1Pt(e,red.)/O500	100	100
5	1Re1Pt(e,red.)/O400H400	100	50 (as for ReO_2)
6	1Re1Pt(e,calc.)/Ar250	100	79
7	1Re1Pt(e,calc.)/O400	100	100
8	1Re1Pt(e,calc.)/O500	100	100
9	0.3Sn0.3Pt(b)O400	100	68 (as for SnO_2)

Treatments in Ar, O_2 (O) or H_2 (H) were carried out at 250, 400 and 500 °C.

Catalytic activity - Pseudo first order rate constant and reaction rates listed in Table 2 indicate that, due to the absence of ionic promoter species the activity of Pt/Al$_2$O$_3$ catalyst is very low. The addition of rhenium to the Pt/Al$_2$O$_3$ catalyst increased the hydrogenolysis activity at least by a factor of 20. The activity of the different types of Re-Pt catalysts with low metal content (c.a. 0.3 wt. %) is in the same order of magnitude. However, the 1Re1Pt catalyst is about three times more active than the catalyst with lower metal content. The most active catalyst is 1.5Re due probably to the presence of both metallic and ionic Re. There is a good correlation between the logarithmic plot of conversion vs. reaction time used for the calculation of k. The continuous decrease of the specific reaction rate related to the amount of Pt in the bimetallic catalysts can probably attributed to the decrease of the dispersion of platinum. The specific reaction rate related to the amount of Re in the catalysts is in the same range except for the 0.5Re catalyst in which most of the Re is in ionic state (see TPR curve in Figure 1).

Table 2 Hydrogenolysis of AcOBu over Pt, Re and RePt catalysts.

No	Catalyst	k, h^{-1}	R^2	R^0	$r^0{}_{Pt}$	$r^0{}_{Re}$	C, %
1	0.5Pt	0.0021	0.883	0.221	8.62	-	4.9
2	0.5Re	0.0219	0.989	2.33	-	88.4	40.9
3	1.5Re	0.1149	0.991	12.36	-	156.3	93.7
4	0.35Re0.17Pt	0.0297	0.991	3.05	350.0	165.4	51.0
5	0.34Re0.21Pt	0.0306	0.992	3.04	282.4	169.7	52.0
6	0.3Re0.3Pt	0.0273	0.995	2.82	183.4	178.4	48.1
7	1Re1Pt(red.)	0.0964	0.973	7.73	150.8	146.7	58.3[a]

Abbreviations: k = pseudo-first order rate constant; R^2 = correlation coefficient; r^0 = initial reaction rate in mmole/g$_{catalyst}$xh; $r^0{}_{Pt}$ and $r^0{}_{Re}$ = reaction rate related to the amount of Pt and Re, respectively; C = conversion of AcOBu at t=24 h; selectivity to BuOH was above 90 %; [a]2 g catalyst was used.

The activity data obtained on SnPt and ReSnPt catalysts are given in Table 3. The comparison of the data given in the first row of Table 2 and Table 3 indicate that SnPt catalyst is significantly more active than monometallic Pt catalyst. Upon increasing the amount of Re added to the Sn-Pt/Al$_2$O$_3$ catalyst the activity of the trimetallic catalysts monotonously increased. The selectivity to butanol showed up a maximum in the range of Re content of 0.5-0.75 wt. %.

Table 3 Hydrogenolysis of AcOBu on Re modified 0.3Sn0.3Pt(b,red.) catalyst.

No	Re, wt. %	k, h^{-1}	R^2	R^0	Conv., %	Sel., %
1	0	0.0216	0.983	2.24	40.5	88.8
2	0.10	0.0269	0.968	2.75	47.6	90.0
3	0.14	0.0314	0.995	3.32	52.9	90.8
4	0.25	0.0405	0.994	4.11	62.2	92.1
5	0.50	0.0667	0.984	6.45	79.8	93.9
6	0.75	0.0889	0.997	8.94	88.2	92.3
7	1.00	0.1013	0.996	10.35	91.2	91.5
8	1.20	0.1058	0.997	11.26	92.1	91.5
9	1.50	0.1116	0.986	11.43	93.1	91.4

Abbreviations: k = pseudo-first order rate constant; R^2 = correlation coefficient; r^0 = initial reaction rate in mmole/g$_{catalyst}$ h; Conv. = conversion of AcOBu at 24 h reaction time; Sel. = selectivity to BuOH.

Besides butanol and ethanol ethyl acetate, ethyl butyrate and butyl butyrate was detected in the reaction mixture.

The activity results clearly indicate that both Sn and Re significantly improves the activity of platinum in the hydrogenolysis of AcOBu. As seen in Figure 4 the specific activity (r$_{Sn+Re}$, mol/mol$_{promoter}$xh) of trimetallic catalysts has a maximum in the function of promoter concentration (Sn+Re) at ~ 70 mmol/g$_{catalyst}$ (0.75 wt. % Re).

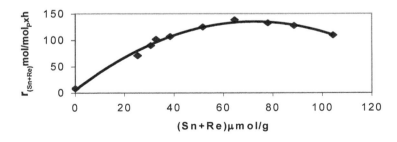

Figure 4 Correlation between the specific reaction rate measured on Re-modified 0.3Sn0.3Pt(b) catalysts and the concentration of promoters (Sn+Re).

Figure 5 shows the specific reaction rate has a maximum at Re/(Re+Sn+Pt) concentration ratio of 0.5. The trimetallic catalysts at concentration ratio of 0.4-0.6 are almost twice as active than the 0.3Sn0.3Pt(b) catalyst and the

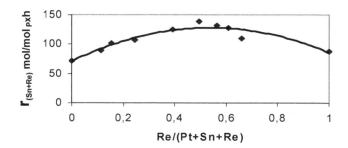

Figure 5 Correlation between the specific reaction rate measured on Re-modified 0.3Sn0.3Pt(b) catalysts and the concentration ratio of Re to total metal content.

monometallic 1.5Re(e). Activity data shown in Figure 5 clearly indicate the promoting effect of Re in the hydrogenolysis of AcOBu. It can be assumed that Re (and/or Sn) probably in its ionic form was involved in the activation of the C=O group of butyl acetate. Literature data indicate that in reforming type SnPt and RePt catalysts part of the promoter is in ionic state [17,18]. In addition our TPR studies (see Table 1) confirmed that in RePt catalyst oxidized and then reduced at 400 °C half of the rhenium was in the form of Re^{4+}. Therefore it is suggested that atomic closeness of positively charged Re or Sn species and platinum or bimetallic nanocluster, i.e. the formation of "metal ion-metal nanocluster" ensembles is required to achieve high catalytic activity.

Preliminary results on AcOBu conversion and reaction rates obtained on 1Re1Pt(e,red.) catalyst in the SPR16 parallel reactor are listed in Table 4. The reaction order in AcOBu was slightly negative but close to zero (poor correlation coefficient was obtained at lower temperatures, see Table 5). The activation energy only slightly depended on the concentration of AcOBu. Both reaction order and activation energy obtained was in good agreement with those reported in the literature for gas phase hydrogenolysis of methyl- and ethyl acetate [22].

Table 4 Hydrogenolysis of AcOBu on 1Re1Pt(e,red.) catalyst in SPR16 reactor. Effect of temperature and concentration on conversion and reaction rate.

T, °C	C = 0.38 M $C\%/r^0$	c = 0.57 M $C\%/r^0$	c = 0.76 M $C\%/r^0$	c = 0.97 M $C\%/r^0$
215	20.8/1.968	15.4/1.954	11.5/1.941	8.7/1.834-
220	31.9/2.691	21.0/2.661	15.7/2.655	12.8/2.699
225	44.5/3.758	28.3/3.589	20.5/3.467	16.4/3.458
230	54.4/4.594	35.3/4.467	26.0/4.385	20.5/4.315

Abbreviations: c = initial concentration of AcOBu in mole/L; C = conversion of AcOBu at t=3.6 h; r^0 = initial reaction rate in mmole/$g_{catalyst}$ h.

Table 5 Hydrogenolysis of AcOBu on 1Re1Pt(e,red) catalyst in SPR16 reactor. Reaction order for AcOBu and apparent activation energy.

T, °C	N	R^2	c, M	Ea, kJ/M	R^2
215	-0.067	0.754	0.38	118.0	0.991
220	-0.001	0.002	0.57	114.0	0.995
225	-0.096	0.956	0.76	113.5	0.999
230	-0.068	0.999	0.97	113.5	0.985

Abbreviations: c = initial concentration of AcOBu in mole/L; n = reaction order for AcOBu; E_a = apparent activation energy in kJ/mole.

The above results clearly indicated that both Sn and Re probably in their ionic form (see TPR results) significantly improved the activity of platinum for the hydrogenolysis of AcOBu. Therefore, it is suggested that atomic closeness of positively charged Re or Sn species and metal nanoclusters, i.e. the formation of "M_1^{n+}-M_2^0" ensembles (M_1 = Re, Sn; M_2 = Pt, Re, RePt or SnPt alloy) is required to achieve high catalytic activity in the hydrogenolysis of esters.

Acknowledgements

SG and JLM wish to thank the Hungarian Scientific Research Fund (OTKA Grant) N^o T-32065 and T 43570 for financial support, respectively. SG thanks the Am. Chem. Soc. Petroleum Research Foundation for financial support.

References

1. K. Folkers, and H. Adkins, *J. Am. Chem. Soc,* **54**, 1145 (1931).
2. H. Adkins, *Organic Reactions,* **8**, 1 (1954).
3. B. Miya, US Pat. 4,252,689, to Kao Soap Co. Limited. (1981).
4. M. Schneider, K. Kochloefl, G. Maletz, Ger. Offen. DE 4142899 A1, to Sued-Chemie AG, Germany (1993).
5. Y. Pouilloux, F. Autin, C. Guimon, and J. Barrault, *J. Catal.,* **176**, 215 (1998).
6. R. Snappe, and J. P. Bournonville, Ger. Offen 3,217,429 (1981).
7. R. Hubault, M. Dage, and J. P. Bonnelle, *Appl. Catal.,* **22**, 231 (1986).
8. V. M Deshpande, K. Ramnarayan, and C. S. Narasimhan, *J. Catal.,* **121**, 174 (1990).
9. H. S. Broadbent, *Ann. N. Y. Acad. Sci.,* 145, 58-71 (1967).
10. H. S. Broadbent, V. L. Mylroie, W. R. Dixon, *Ann. N.Y. Acad. Sci.,* 172, 194-207 (1970).
11. C. Hirosawa, N. Wakasa, and T. Fuchikami, *Tetrahedron Lett.,* **37**, 6749 (1996).
12. K. Tahara, E. Nagahara, Y. Itoi, S. Nishiyama, S. Tsuruya, and M. Masai, *React. Kinet. Catal. Lett.,* **59**, 15 (1996).
13. I. D. Dobson, EP 0 286 280 A1, to BP Chemicals Limited (1988).
14. T. Fuchikami, T. Ga, N. Wakasa, and K, Tawara, Jpn. Kokai Tokkyo Koho, JP 09,132,541, to (1997).

15. V. Ponec, *J. Mol. Catal. A*, **133,** 222 (1998).
16. J. H. Sinfelt, in Catalysis Science and Technology J. R. Anderson and M. Boudart, Eds., Springer-Verlag, Berlin-Heidelberg, 1981, vol.1, p. 287-300.
17. J. L. Margitfalvi, I. Borbáth, M. Hegedüs, S. Göbölös, and F. Lónyi, *React. Kinet. Catal. Lett.,* **68,** 133 (1999).
18. M. F.I. Johnson, and V. M. Leroy, *J. Catal.,* **35,** 434 (1974).
19. A. S. Fung, M. R. McDevitt, P. A. Tooley, M. J. Kelley, D. C. Köningsberger, and B. C. Gates, *J. Catal.*, **140,** 190 (1993).
20. R. Prestvik, K. Moljord, K. Grande, and A. Holmen, *J. Catal.*, **174,** 119 (1998).
21. M. Rønning, T. Gjervan, R. Prestvik, D. G. Nicholson, and A. Holmen, A., *J. Catal.,* **204,** 292 (2001).
22. A. K. Agarwal, N. W. Cant, M. S. Wainwright, and D. L. Trimm, *J. Mol. Catal.,* **43,** 79 (1987).

13.   New Ways for the Full Hydrogenation of Quinoline in Mild Conditions

M. Campanati,[a] **S. Franceschini,**[b] **O. Piccolo,**[c] **and A. Vaccari**[b*]

[a]*Endura SpA, Viale Pietramellara 5, 40121 Bologna, Italy*
[b]*Dipartimento di Chimica Industriale e dei Materiali, Università di Bologna, INSTM-UdR Bologna Viale del Risorgimento 4, 40136 Bologna, Italy*
[c]*Consultant, via Bornò 5, 23896 Sirtori LC, Italy*

*Fax: +39-051-2093680; vacange@ms.fci.unibo.it

Abstract

The liquid-phase hydrogenation of quinoline (Q) under mild reaction conditions (T = 373 K, P_{H2} = 2.0 MPa) was investigated using a commercial Rh/Al$_2$O$_3$ catalyst (5 wt% Rh) in the presence of different amines. With aliphatic amines, the best yield in decahydroquinoline (DHQ) was obtained by co-feeding cyclopropylamine (30%), while with aromatic amines, the yield reached 49% with 2-ethylaniline. Key factors favoring full hydrogenation of Q to DHQ seem to be the steric hindrance of the amine and its electronic interaction with the active sites of the catalyst. Finally, an alternative strategy to obtain the full hydrogenation was to tailor the acidity of the catalyst support, when a 58% yield of DHQ was attained using another commercial Rh/Al$_2$O$_3$ (5 wt% Rh) catalyst with a higher surface acidity.

Introduction

The hydrogenation of quinoline (Q) and its derivatives is of considerable industrial interest for production of petrochemicals, fine chemicals, agrochemicals and pharmaceuticals (1-4). Transition or noble metals supported on different solids have been tested in liquid-phase hydrogenation of Q to decahydroquinoline (DHQ). This reaction occurs in two steps (5) (Figure 1): (i) formation of 5,6,7,8-tetrahydroquinoline (bz-THQ) and mainly 1,2,3,4-tetrahydroquinoline (py-THQ) as intermediates, and (ii) their hydrogenation to DHQ. The reaction conditions for the first step include a temperature of about 373 K, a H$_2$ pressure of 0.1-7.0 MPa, and the use of an alcohol solvent (6-8). The conversion of both bz-THQ and py-THQ to DHQ requires longer reaction times and more drastic reaction conditions, typically a temperature of 448-553 K, a H$_2$ pressure of 11.0-21.0 MPa, and the use of a strong Brönsted acid as solvent, which prevents catalyst poisoning by the basic nitrogen atom of the substrate, intermediate(s) or product (9-12). The use of strong Brönsted acids implies the use of expensive enameled reactors, in order to avoid corrosion phenomena and product contamination.

The role of the reaction parameters in mild hydrogenation of Q has been investigated previously using different noble metal-based commercial catalysts (13); Rh systems were superior, although in all cases only partial hydrogenation to py-THQ was observed. This deactivation might be due to adsorption on the active sites of reaction intermediate(s) (a, b or c) formed in the first hydrogenation step Q → py-THQ (Figure 1). One way to remove this deactivation was the use of highly dispersed metal catalysts, with very small crystallites that hamper the strong adsorption of reaction intermediate(s), although a higher reaction temperature (473 K) was required (14). Another way to reduce catalyst deactivation was by the addition of a Lewis base to the reaction mixture; this favored complete hydrogenation probably by competitive adsorption on the catalyst surface (13).

The aim of this work was to investigate more thoroughly this latter strategy, by studying the role of the properties of different added aliphatic or aromatic amines. Furthermore, we studied the possibility of eliminating the deactivation by modifying the surface acidity of the support, through testing a different commercial catalyst with the same Rh-content.

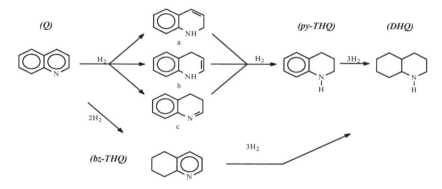

Figure 1 Proposed reaction pathways for the hydrogenation of quinoline (a = 1,2-dihydroquinoline, b = 1,4-dihydroquinoline, c = 3,4-dihydroquinoline).

Experimental Section

The commercial 5 wt % Rh supported on γ-Al$_2$O$_3$ catalysts were supplied in the reduced form by Engelhard (Rh/Al$_2$O$_3$-1) and Johnson Matthey (Rh/Al$_2$O$_3$-2), and were used without further treatment. Quinoline (Q), 1,2,3,4-tetrahydroquinoline (py-THQ), decahydroquinoline (DHQ), aniline (AN), benzylamine (BzAM), cyclopropylamine (CprAM), dipropylamine (DprAM), 2-ethylaniline (2-ETAN), 3-ethylaniline (3-ETAN), 2-ethylpyridine (2-EtP), isopropylamine (iPrAM), N,N-diethylcyclohexylamine (N,N-DetCeAM), N-ethylaniline (N-ETAN), propylamine (PrAM), tert-butylamine (tBuAM), tripropylamine (TprAM), 1-octanol and 2-propanol were purchased from

Aldrich Chemicals (purity ≥ 98 %), and were used as received. The pK_a values of the amines were calculated using the Advanced Chemistry Development (ACD) Software Solaris V4 program (15). X-ray diffraction powder patterns were collected using a Philips PW 1050/81 diffractometer (40 kV, 25 mA), equipped with a PW 1710 unit, and Cu $K\alpha$ radiation ($\lambda = 0.15418$ nm). BET surface area and pore volume values were determined by physical adsorption of N_2 at 77 K, using a Carlo Erba Sorptomatic 1900. H_2-chemisorption measurements were performed with a Micromeritc Autochem 2910. Before each adsorption (physical or chemical) measurement, the samples were degassed for 12 h at 423 K and 1.33×10^{-3} Pa. The catalyst surface acidity was determined by temperature-programmed desorption (TPD) tests of ammonia, using a ThermoQuest TPD/R/O 1100.

The catalytic tests were carried out using a 300 mL stainless-steel Parr reactor, equipped with a magnetic stirrer and digital oven-controller. The standard conditions for the experiments were: 0.30 g of catalyst, 41 mmoles of Q, and 41 mmoles of aliphatic or aromatic amine in 2-propanol to a total volume of 100.0 mL, $P_{H2} = 2.0$ MPa and T = 373 K. Before being heated, the autoclave was purged three times with He at 0.3 MPa. When the system reached the reaction temperature, the H_2 was introduced at the set pressure and the reaction started. 2.0 mL were periodically sampled, 100.0 μL of 1-octanol added as internal standard, and the mixture diluted to 10.0 mL with 2-propanol. Quantitative analyses were carried out using a Perkin Elmer Autosystem XL gas chromatograph, equipped with a PE-5 column (30 m x 0.25 mm, film thickness 0.25 μm) and a FID detector. The response factors were obtained using standard solutions of pure compounds in 2-propanol; for the non-commercially available bz-THQ, the factor used was that for py-THQ. The yields reported are referred to the starting quinoline. Qualitative analyses were carried out using a GC-MS Hewlett-Packard GCD 1800A system, equipped with a HP-5 column (30 m x 0.25 mm, film thickness 0.25 μm) and mass spectrometer detector.

Results and Discussion

1. Addition of aliphatic or aromatic amines as co-catalyst

The behavior of the different amines depends on at least four factors: basicity, nucleophilicity, steric hindrance and solvation. In the literature (16), 126 aliphatic and aromatic amines have been classified by a statistical analysis of the data for the following parameters: molar mass (mm), refractive index (n_D), density (d), boiling point (bp), molar volume, and pK_a. On such a premise, a Cartesian co-ordinate graph places the amines in four quadrants (16). In our preliminary tests, amines representative of each quadrant have been investigated, and chosen by consideration of their toxicity, commercial availability and price (Table 1).

Table 1 Amines co-fed in the hydrogenation of quinoline, their quadrant classification (16) and pK_a values (15); n.a. = not available.

Amine	Quadrant	pK_a
N,N-DetCeAM	1	10.69
2-EtP	2	5.89
BzAM	2	9.40
PrAM	3	10.66
CprAM	3	9.10
tBuAM	4	10.69
DprAM	4	10.94
TprAM	4	9.99
iPrAM	4	10.68
AN	n.a.	4.61
2-ETAN	n.a.	4.36
3-ETAN	n.a	4.73
N-ETAN	n.a.	5.12

In previous use of the Engelhard catalyst Rh/Al$_2$O$_3$-1 (13,14), only partial hydrogenation of Q to py-THQ was observed (Table 2), and only the addition of N,N-diethylcyclohexylamine (N,N-DetCeAM) (first quadrant) was investigated. Compared to the test in the absence of the amine (Table 2), there was a slight decrease in the conversion rate of Q and yield of py-THQ, and more significant amounts of the partially hydrogenated intermediate bz-THQ (by hydrogenation of the carbocyclic ring) and the fully hydrogenated DHQ. Thus, the N, N-DetCeAM not only modified the interaction of the deactivating intermediate(s) (Figure 1, a-c) to favor complete hydrogenation to DHQ, but also affected the distribution of intermediates; the heterocyclic ring of the Q has been electronically stabilized, and the hydrogenation of the carbocyclic ring favored (17).

The study has now been extended to two amines of the second quadrant: 2-ethylpyridine (2-EtP) and benzylamine (BzAM), both containing an aromatic ring that might favor a stronger interaction with the active sites on the catalyst surface, and thus increase the competition with deactivating intermediate(s). The increased yields of both the unusual intermediate bz-THQ and DHQ were particularly evident for BzAM, while 2-EtP showed behavior similar to that of the first sector amine N,N-DetCeAM. The different behaviors might be justified by considering the higher basicity of BzAM versus 2-EtP (pK_a = 9.40 and 5.89, respectively), which should favor, together with the presence of an aromatic ring, competition with the deactivating intermediate(s) (Figure 1, a-c).

Next analyzed was the behavior of the third quadrant amines, propylamine (PrAM) and cyclopropylamine (CprAM). Surprisingly, PrAM not only did not favor the formation of DHQ, but also decreased the conversion rate of Q in comparison to data observed in the absence of amine, suggesting an effective

Table 2 Catalytic activity for the hydrogenation of quinoline in the presence of aliphatic or aromatic amines (0.3 g of the 5 wt% Rh/Al$_2$O$_3$-1 catalyst; T = 373 K; P$_{H2}$ = 2.0 MPa; reaction time (where not indicated) = 6 h; 41 mmoles of quinoline (82 in the test in absence of amine) and 41 mmoles of amine in 100.0 mL of 2-propanol).

Amine added	Sector	Quinoline (conv. %) (1 h)	(6 h)	Py-THQ (yield %)	Bz-THQ (yield %)	DHQ (yield %)
None	--	80	99	99	<1	0
N,NDetCeAM	1	70	99	82	9	8
2-EtP	2	97	100	87	7	5
BzAM	2	94	100	61	15	24
PrAM	3	63	94	90	4	0
CprAM	3	80	100	59	11	30
tBuAM	4	94	97	68	<1	29
DprAM	4	97	98	79	<1	18
TprAM	4	97	98	73	0	25
iPrAM	4	95	96	76	1	19

competition with Q for adsorption on the catalytically active sites. On the contrary, CprAM did not modify the conversion rate of Q, and improved significantly the yield of both bz-THQ and particularly DHQ. Noteworthy is that PrAM has a higher basicity (pK$_a$ = 10.66, the same as for N,N-DetCeAM, see Table 1) than CprAM (pK$_a$ = 9.10). Furthermore, the pK$_a$ value for CprAM is slightly lower than that of BzAM, notwithstanding the higher DHQ yield obtained with CprAM. Thus another key parameter to be considered is the steric hindrance of the amine, which could hinder its adsorption on the catalyst surface, while allowing for competition with the deactivating intermediate(s).

Finally, four amines of the fourth quadrant were investigated, tert-butylamine (tBuAM), dipropylamine (DprAM), tripropylamine (TprAM) and isopropylamine (iPrAM). All these amines increased significantly the rate of Q conversion (see particularly the 1 h data in Table 2) together with increased DHQ yields, while there was negligible formation of intermediates hydrogenated on the carbocyclic ring (bz-THQ) (in contrast to the observations with the other amines). As these amines are all aliphatic with similar basicities (9.99 < pK$_a$ < 10.94), the data likely shed light on the specific role of steric hindrance. In fact, the DHQ yields increase with increasing steric hindrance of the amines.

This preliminary screening gives the following activity order for hydrogenation of Q to DHQ, likely providing evidence for the capacity of the amine to inhibit deactivation by the intermediate(s) on the catalyst active sites:

CprAM ≥ tBuAM > TprAM ≥ BzAM > iPrAM ≥ DprAM >> N,N-DetCeAM > 2-EtP >> PrAM (no formation of DHQ)

The basicity of the co-fed amine seems to be an important factor favoring complete hydrogenation of Q, as the more basic amine (through the N lone-pair) likely interacts increasingly with the catalyst sites, and competes with the deactivating intermediate(s). However, the amine may adsorb so strongly (as in the case of PrAM) to slow down significantly the rate of Q hydrogenation. A key factor controlling the effect of the amine is thus their steric hindrance that may inhibit irreversible adsorption of the amine, and favor competition with the deactivating intermediate(s); this would consequently allow for complete hydrogenation to DHQ.

Table 3   Catalytic activity for the hydrogenation of quinoline in the presence of aromatic amines (0.3 g of the 5 wt% Rh/Al$_2$O$_3$-1 catalyst; T = 373 K; P$_{H2}$ = 2.0 MPa; reaction time (where not indicated) = 6 h; 41 mmoles of quinoline and 41 mmoles of amine in 100.0 mL of 2-propanol).

Amine	Quinoline (1h)	(conv. %) (6 h)	Py-THQ (yield %)	Bz-THQ (yield %)	DHQ (yield %)
AN	95	97	74	<1	23
2-ETAN	100	100	50	1	49
3-ETAN	100	100	52	1	47
N-ETAN	99	100	54	1	45

In the previous analysis for the second quadrant amines, there was evidence that the presence of an aromatic ring (BzAM) increased competition with the deactivating intermediate(s) and significantly the amount of DHQ obtained. The study was thus extended to other aromatic amines: aniline (AN), 2-ethylaniline (2-ETAN), 3-ethylaniline (3-ETAN) and N-ethylaniline (N-ETAN). These amines are not classified in the literature analysis of amine properties (16), although aniline and pyridine were studied by statistical analysis of their solvent properties and classified in the same sector (16). By analogy, we hypothesize that these model aromatic amines should be classified in the second sector. Thus, they may aid in an understanding of the specific role of the aromatic ring and the effect of an alkyl substituent.

All these amines increased significantly the rate of Q conversion, while the final DHQ yield obtained by co-feeding AN (Table 3) was very similar to that observed with BzAM (Table 2); with AN, the formation of bz-THQ was almost negligible, evidencing a lower stabilizing effect on the heterocyclic ring of Q. The presence of an alkyl side-chain improved significantly the DHQ yields that reached the highest values observed (Table 3). As this positive alkyl chain effect was essentially independent of the alkyl position (either on the carbocyclic ring or on the N-atom), both electron and steric effects have to be considered. While the presence of a electron-donating substituent improves the basicity of the amine, thus favoring its interaction with the catalyst active sites, the steric hindrance of an alkyl chain will inhibit strong adsorption of the amine. Thus, key factors to favor complete hydrogenation of Q to DHQ are considered to be the steric hindrance of the amine and its electronic interaction with catalyst active sites, which may favor desorption of the deactivating intermediate(s) formed in the first step of the Q hydrogenation.

2. Surface acidity of the catalyst

In our previous paper (13), it has been shown that under mild catalysis conditions, the addition of excess acetic acid did not inhibit deactivation due to irreversible adsorption on the active sites of a reaction intermediate(s) formed in the step Q → py-THQ. As the addition of a stronger Brønsted acid likely requires the use of an expensive, enameled reactor, and the co-feeding of an aliphatic or aromatic amine requires a recycling step to be scheduled (with an increase of the production costs), the possibility of achieving hydrogenation of Q to DHQ by tailoring the surface acidity of the support was investigated. A different commercial catalyst with the same Rh-content (Rh/Al$_2$O$_3$-2 by Johnson Matthey) was tested.

Table 4 Characterization data for the two commercial, 5 wt% Rh supported on γ-Al$_2$O$_3$ catalysts.

Catalyst	BET surface area (m^2/g)	Metal dispersion (%)	Metal crystallite size (nm)	Surface acidity (μmoles of NH$_3$/g$_{Cat}$)
Rh/Al$_2$O$_3$-1	139	17.8	6.2	280
Rh/Al$_2$O$_3$-2	129	20.6	5.3	490

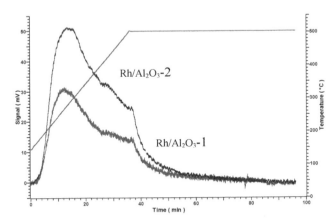

Figure 2 TPD profiles for the two commercial, 5 wt% Rh supported on γ-Al$_2$O$_3$ catalysts.

The XRD powder patterns of the two catalysts were not significantly different, both showing mainly the pattern of the support (γ-Al$_2$O$_3$) together with the presence of metallic Rh (2θ ~ 41°). Also the main chemical-physical characterization data were analogous for the two catalysts (Table 4), with a similar BET surface area and only a slightly higher metal dispersion for the Rh/Al$_2$O$_3$-2 catalyst. On the contrary, there was a significantly higher surface acidity for this latter catalyst in the TPD tests of NH$_3$ (Figure 2); the test showed similar trends for both catalysts, with the presence of at least two types of acid sites (maximum desorption peaks at 543 and 688 K), but with about twice the amount of NH$_3$ being desorbed from Rh/Al$_2$O$_3$-2.

These differences in surface acidity are reflected dramatically in the catalytic activities (Figure 3a and b); the Rh/Al$_2$O$_3$-2 system shows an increase in formation of the intermediate hydrogenated on the carbocyclic ring (bz-THQ), and particularly in the DHQ yield that reached the highest observed value of 58% after 6 h of reaction. It is noteworthy that the amounts of both partially hydrogenated intermediates (bz-THQ and py-THQ) decrease with time, with corresponding increased amounts of DHQ, implying that a higher surface acidity possibly inhibits irreversible adsorption of the deactivating intermediate(s), probably by formation of the corresponding ion(s) and a change in the nature of the adsorption (12, 18). Furthermore, it must be noted that use of the Rh/Al$_2$O$_3$-2 catalyst, with the addition of an equimolar amount of 2-EtP, did not modify the catalytic activity (unlike for the Rh/Al$_2$O$_3$-1 catalyst system); this is in agreement with the previously assumed competitive-adsorption mechanism between the amine (aliphatic or aromatic) and the deactivating intermediate(s).

Conclusions

The main problem in the liquid-phase hydrogenation of Q, using commercial alumina-supported Rh catalysts under mild conditions, is the inhibition of full hydrogenation to DHQ by a reaction intermediate(s) formed in the first hydrogenation step. Thus, acetic acid or other acid solvents have to be used for the hydrogenation when amines are present as products or reagents (19). Surprisingly, we observed that this deactivation may be removed by co-feeding equimolar amounts of suitable aliphatic or particularly aromatic amines.

The results obtained with different amines cannot be explained merely on the effects of amine basicity. Thus, to obtain complete hydrogenation of Q to DHQ, the basicity has to be tailored by other factors such as the steric hindrance of the amine and its electronic interaction with the catalyst active sites; this seems to be favored by the presence of an electron-rich aromatic ring. Of note, the positive effect of substituted aromatic amines, with a 49% DHQ yield being obtained for ethylanilines, is independent of the substituent position of the alkyl group.

Considering the possible increase of production costs due to the recycling

a)
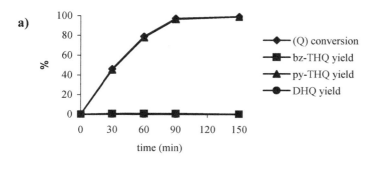

b)
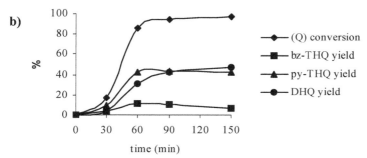

Figure 3 Activities in the hydrogenation of quinoline for the commercial catalysts (a) Rh/Al_2O_3-1 and (b) Rh/Al_2O_3-2 (0.3 g of catalyst; T = 373 K; P_{H2} = 2.0 MPa; 82 mmoles of quinoline in 100.0 mL of 2-propanol).

of the amine, an alternative strategy to achieve complete hydrogenation of Q is to tailor the acidity of the catalyst support; a 58% yield in DHQ, after 6 h, was obtained using a commercial catalyst (5 wt% Rh) with a relatively high surface acidity.

Acknowledgements

Financial support from the Ministero per l' Istruzione, l' Università e la Ricerca (MIUR, Rome) and the American Chemical Society – Petroleum Research Foundation (ACS-PRF) is gratefully acknowledged. Thanks are due to Engelhard and Johnson Matthey for providing the commercial catalysts, and to Prof. B. R. James (University of British Columbia) for his careful editorial work.

References

1. R. T. Shuman, P. L. Ornstein, J. W. Paschal, and P. D.Gesellchem, *J. Org. Chem.,* **55,** 738 (1990).

2. J. M. Schaus, D. L. Huser, and R. D. Titus, *Synth. Commun.*, **20**, 355 (1990).
3. G. Perot, *Catal. Today,* **10**, 447 (1991).
4. P. Boyssou, C. Le Goff, and J. Chenault, *J. Heterocycl. Chem.*, **29**, 895 (1992).
5. C. W. Curtis and D. R. Cahela, *Energy Fuels,* **3**, 168 (1989).
6. H. Adkins and H. R. Bilica, *J. Am. Chem. Soc.,* **70**, 695 (1948).
7. J. E. Shaw and P. R. Stapp, *J. Heterocycl. Chem.,* **24**, 1477 (1987).
8. F. Santangelo, C. Casagrande, G. Miragoli, and V. Vecchietti, *Eur. J. Med. Chem.,* **20**, 877 (1994).
9. S. Tsushima and J. Sudzuki, *J. Chem. Soc. Jpn.,* **64**, 1295 (1943).
10. M. Freifelder, *Adv. Catal.,* **14**, 203 (1963).
11. R. L. Augustine, Catalytic Hydrogenation, Marcel Dekker, New York, 1965, p. 104-123.
12. P. N. Rylander, Hydrogenation Methods, Academic Press, London, 1990, p. 133-147.
13. M. Campanati, A. Vaccari, and O. Piccolo, *J. Mol. Catal. A: Chemical,* **179**, 287 (2002).
14. M. Campanati, M. Casagrande, I. Fagiolino, M. Lenarda, L. Storaro, M. Battagliarin, and A. Vaccari, *J. Mol. Catal. A: Chemical,* **184**, 267 (2002).
15. Advanced Chemistry Development (ACD) Software Solaris V4 (1994-2004): www.scifinder.scholar.com.
16. R. Carlson, Design and Optimization in Organic Synthesis, Elsevier, Amsterdam, 1992, p. 374, 379 and 408.
17. M. Campanati, S. Franceschini, A. Vaccari, and O. Piccolo, Chemical Industries (Dekker), **89**, (*Catal. Org. React.*), 441-452 (2003).
18. R.M. Skomoroski and A. Schriesheim, *J. Phys. Chem.,* **65**, 1340 (1961).
19. M. Freifelder, *J. Org. Chem.,* **26**, 1835 (1961).

14. Some Recent Developments in Heterogeneous Hydrogenation Catalysts for Organic Synthesis

Jianping J. P. Chen and Charles R. Penquite

Engelhard Corporation, 23800 Mercantile Road, Beachwood, OH 44122

Jp.chen@engelhard.com

Abstract

This paper discusses typical properties of powdered precious metal catalysts related to catalytic performance in several organic syntheses. Some new developments and trends in precious metal catalysts, such as new generations of palladium (DeLink™), platinum and rhodium heterogeneous catalysts including immobilized homogeneous catalysts (LiganNet™) for hydrogenation in organic synthesis are discussed.

Introduction

An important factor in developing fine chemicals and pharmaceuticals, agrochemicals, fragrances and flavors, food additives, and dyes and pigments, is choosing the reaction route to the final product. Catalytic routes have proven to be one of the most effective ways in simplifying the reaction routes to these compounds by increasing product selectivity and reducing waste and hazardous materials handling. Organic synthetic chemists and engineers seek more active, selective and environmentally benign catalysts for their synthesis applications. To meet increasing market demand, catalyst manufacturers are closely working with synthesis chemists and engineers to select and develop better catalysts enabling a wide range of different synthetic transformations. A number of new precious metal catalysts, such as Pd/C, Pt/C and immobilized Rh catalyst, have been successfully commercialized in recent years. Examples of recent developments and trends will be illustrated using several widely practiced reactions, such as debenzylation, nitro group reduction, and chemo-selective and stereo-selective hydrogenations.

DeLink™ – a New Generation of Pd/C Catalysts

Activated carbon supported palladium catalysts have been widely used in organic chemical synthesis (1). A comprehensive review surveys the research and development work on preparation of supported palladium catalysts covering the 1990-2000 period (2). The proven catalysts are: 5-10% Pd on carbon. Sometimes high palladium catalysts, such as Pearlman catalyst (20%Pd(OH)$_2$/C), are preferred for difficult debenzylation reactions. There has not been a significant breakthrough on commercially produced Pd/C catalysts until DeLink™ catalysts were introduced. In the following sections, we will focus on the effects of catalyst preparation and discuss the properties of DeLink™ catalysts.

Superior Activity In the synthesis of fine chemicals, especially in pharmaceutical applications, abundant examples of synthesized molecules containing multiple functional groups are known. Some of the reactive functional groups need to be temporarily protected during synthesis by introducing protecting groups. Using benzyl protection groups is one of the most common methodologies in complex organic synthesis. At the end of synthesis, protected functional groups can be deprotected by hydrogenolysis of the benzyl group. Rylander (3), Freifelder (4) and Seif *et al.* (5) have reported that an activated carbon supported palladium catalyst is the most commonly used catalyst for debenzylations.

Our earlier work (6-8) shows that using debenzylation of 4-benzyloxy phenol (C$_6$H$_5$CH$_2$OC$_6$H$_4$OH, 4-BP) to hydroquinone (C$_6$H$_4$(OH)$_2$) as a model reaction, one can evaluate the relative activity of different catalysts. As shown in Figure 1 and Table 1, the catalytic performance results demonstrate that DeLink™ 3%Pd/CPS4 (carbon powder support) has activity equal to current commercially supplied 5%Pd/CPS1 and CPS2 catalysts. Using the same technology, DeLink™ 5% and 10%Pd/CPS4, also are developed in this work. These new catalysts have superior catalytic activity. CPS1 and CPS2 supported catalysts are the regular commercial catalysts and CPS4 supported ones are the new DeLink™ catalysts. Table 1 summarizes the key properties and reaction rate constants on the catalysts used in this work. This test shows that catalysts with the same palladium metal loading have similar CO chemisorption or metal dispersion, but different activity.

The hydrogen uptake curves (Figure 1) show that the reaction completes in less than 5000 seconds for DeLink™ 5%Pd/CPS4 as compared to about 10,000 seconds for standard catalysts 5%Pd/CPS1 and 5%Pd/CPS2. Figure 1 also shows the reaction is complete in less than 1500 seconds for DeLink™ 10%Pd/CPS4, but it took up to 4000 seconds to complete the reaction on standard 10%Pd/CPS1 and 10%Pd/CPS2 catalysts. The calculated rate constants of 5%Pd/CPS4 are about twice those of 5%Pd/CPS1 and 5%Pd/CPS2 on both catalyst weight and metal basis (Table 1). DeLink™ 3%Pd/CPS4 catalyst that is 40% lower than the current commercial catalysts, 5%Pd/CPS2 and 5%Pd/CPS3 has slightly faster hydrogen uptake rate.

Assuming zero order kinetics, the reaction rate constants can be calculated from the slope of the hydrogen uptake curve. Table 1 shows that the first three catalysts have similar rate constants on catalyst weight basis, from 5.6×10^{-3} to

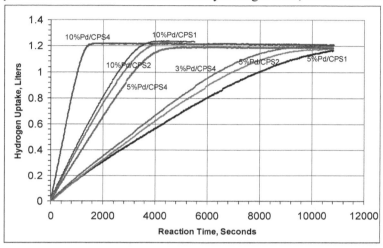

Figure 1 Hydrogen uptake curves of 3%Pd/CPS4 and 5%Pd and 10%Pd on CPS1, CPS2 and CPS4. The reaction conditions are: 10 g 4-(benzyloxy) phenol in 100 methanol, hydrogen pressure 1.1 bar, agitation rate 200 rpm, temperature 35°C, catalyst loading 3wt%.

Table 1 Key Properties and Activity of the DeLink™ (Pd/CPS4) vs. Classical Catalysts (Pd/CPS1 and CPS2).

Catalyst Description	CO Chemisorption ml CO/g cat.	Rate Constant moles/min/g cat X 100[a]	Catalyst PSD Span[c]	Filtration Time[b]
3%Pd/CPS4	2.19	0.65	2.4	5'40"
5%Pd/CPS1	3.74	0.56	5.2	11'33"
5%Pd/CPS2	3.27	0.62	3.5	5'59"
5%Pd/CPS4	3.77	1.12	2.6	5'20"
10%Pd/CPS1	5.11	1.57	5.6	10'30"
10%Pd/CPS2	5.43	1.49	3.5	5'49"
10%Pd/CPS4	6.05	3.72	2.5	5'23"

[a]Reaction conditions: 3% catalyst loading, 10 g 4-BP in 100 ml methanol, 35°C, 2000rpm, 1.1 bar H_2.
[b]Filtration time represented the time for filtration of 350 ml of debenzylation product of 10% 4-BP in methanol and 1.2g of catalyst over 55 mm diameter #42 Whatman filter paper under 24" Hg vacuum.
[c]PSD: particle size distribution; Span =[$D(v,0.9) - D(v,0.1)$]/ $D(v,0.5)$ (9).

6.5×10^{-3} mol.min^{-1}.g cat^{-1}, although 3%Pd/CPS4 has 40% less palladium. The activity test results show that rate constants of 5%Pd/CPS4 are about twice those of 5%Pd/CPS1 and 5%Pd/CPS2 and the reaction rate constants of 10%Pd/CPS4 are more than twice those of 10%Pd/CPS1 and 10%Pd/CPS2.

Table 1 shows that increasing Pd loading levels from 3% to 5% and 10% on CPS4 results in greater increases in catalyst weight basis reaction rate constants than is seen with traditional supports. For the same CPS, 10%Pd catalyst has more than double the rate constants of 5%Pd catalyst and more than five times the rate constants of 3%Pd catalyst. This indicates that the turn over frequency numbers of the reaction on higher metal loading catalyst are higher although they have lower Pd metal dispersion. One of the possible reasons is the synergistic effect of adsorption of substrate and hydrogen over the larger palladium metal crystallites on higher metal loading catalysts that can facilitate faster reaction.

The new DeLink™ 3%Pd/CPS4 catalyst has been tested for debenzylation of N-phenylbenzylamine also. Under the same reaction conditions, 3%Pd/CPS4 has approximately the same activity as the current commercially produced 5%Pd/CPS2 in this N-debenzylation reaction on catalyst weight basis.

Improved Filtration Rate Filterability is an important powder catalyst physical property. Sometimes, it can become more important than the catalyst activity depending on the chemical process. When a simple reaction requires less reaction time, a slow filtration operation can slow down the whole process. From a practical point of view, an ideal catalyst not only should have good activity, but also it should have good filtration. From catalyst development point of view, one should consider the relationship between catalyst particle size and its distribution with its catalytic activity and filterability. Smaller catalyst particle size will have better activity but will generally result in slower filtration rate. A narrower particle size distribution with proper particle size will provide a better filtration rate and maintain good activity.

Table 1 shows that catalysts prepared on the same type of support have about the same particle size distribution (PSD). Table 1 also shows that the newly developed CPS4 supported catalysts have the smallest span of particle size distribution, therefore, it has the fastest filtration rate. Filtration rate is measured by measuring the filtration time of 350 ml of 4-benzyloxyphenol debenzylation products.

Figure 2 shows the particle size distribution of the three catalysts. The distribution curves show that 10%Pd/CPS4 has the narrowest distribution. The particle size distribution and filtration rates of the three catalysts are summarized in Table 1. The span represents the width of the distribution, which is independent of the median size. The definition of the span of the particle size is shown in the footnotes of Table 1. $D(v,0.5)$ is median size of the distribution and $D(v,.1)$ and $D(v,0.9)$ are particle sizes with cumulative volume percentile distribution of 10% and 90% respectively, and v in the expressions refers to volume distribution.

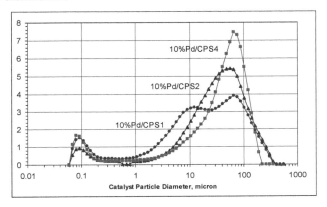

Figure 2 Particle size distribution of 10%Pd on CPS1, CPS2 and CPS4.

The span of particle size distribution (PSD) is an indication of the broadness of catalyst particle size distribution. Generally speaking, under the same filtration conditions and similar average particle size, a catalyst with larger span of PSD will have slower filtration. Table 1 shows that CPS4 supported catalysts have the fastest filtration rates followed by CPS2 then CPS1 supported catalysts; This can be explained by the PSD span and the distribution curves. For CPS1 type catalyst, the span is in the range of 5.2 to 5.6; for CPS2 type catalysts, the span is about 3.5; and for CPS4 type catalysts, the span is in the range of 2.4 to 2.6. The differences in the span of PSD of these three types of catalyst are shown in Figure 2 also. Among these three catalysts, 10%Pd/CPS1 has the broadest tri-modal distribution; 10%Pd/CPS2 has a narrower bi-modal distribution; and 10%Pd/CPS4 has the narrowest bi-modal distribution. 5%Pd catalysts on these CPS have similar PSD.

Improved Selectivity   Hydrogenolysis of benzyl protecting groups in the presence of an aromatic halogen, that is, selective debenzylation versus dehalogenation has received little attention (10). In an earlier report, H_3PO_3 was used as a catalyst inhibitor to increase the selective debenzylation of a tertiary amine in the presence of aromatic halogen (11). Recently, Wang and Chen were able to accomplish high selective debenzylation by using DeLink™

10%Pd/CPS4 catalyst without using inhibitor (12). As shown in Figure 3, the reaction requires selective debenzylation versus dehalogenation of the

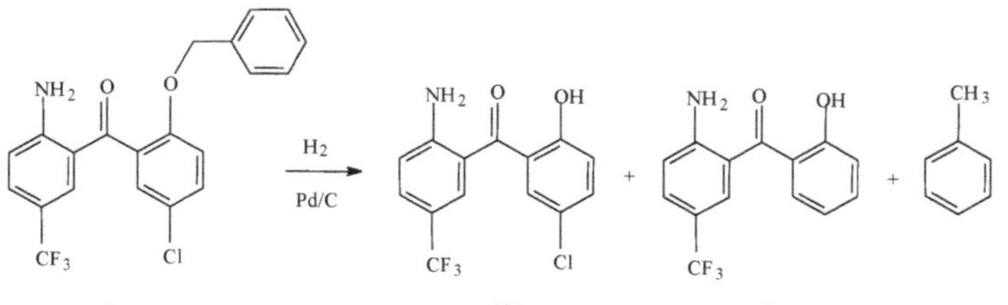

Figure 3 Debenzylation of a pharmaceutical intermediate containing benzyl ether and aromatic chlorine.

intermediate (S). In the earlier development stage, catalytic transfer hydrogenolysis of (S) was studied using either formic acid, ammonium formate or sodium hypophosphite as hydrogen donors. The debenzylation proceeded well but along with about 5% of dechlorinated by-product (B). The dehalogenation proceeds quickly under basic and neutral conditions but slowly in the presence of acid (13). Study on different types of catalyst show that edge coated, unreduced Pd/C (ESCAT 147) gives the highest activity, followed by edge coated, reduced (ESCAT 142) and uniform, reduced (ESCAT 168) under the same reaction conditions. However, all of these catalysts made more than 1% de-chlorinated by-product. Further study demonstrated that with DeLink™ 10%Pd/CPS4 catalyst and proper control of the reaction conditions, selectivity for the desired product (A) can be as high as >99% and the undesired by-product, dechlorinated compound (B), can be controlled to <1%. The reactant and product concentration profiles reveal that only 0.8% dechlorinated by-product was formed even after 19 hr reaction (12). Kinetics analysis and mechanism study establish the model that dehalogenation occurs primarily from consecutive hydrogenolysis if not exclusively from the desired product (A).

Strong Poison Resistance   Hydrogenation or hydrogenolysis of an organic molecule containing sulfur is a challenging task due to the poisoning effect of sulfur. Hydro-debromination of 4-bromothioanisole ($BrC_6H_4SCH_3$, 4-BTA) to thioanisole by catalytic transfer hydrogenation is a good model reaction to evaluate the catalyst poison resistance. As shown in Figure 4, DeLink™ 10%Pd/CPS4 is the most resistant to sulfur poisoning. The reaction was complete in about 21 hours. Other catalysts only gave about 50 to 70% conversion at the same reaction time.

Other Supported Palladium Catalysts

Approved for use in the U.S. in the early 1990s, Sertraline (Zoloft™) is the most prescribed drug of its class. In the original synthesis, the carbonyl group of the

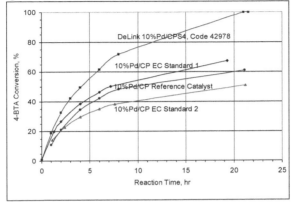

Figure 4 Hydrogenolysis of 4-BTA on 10%Pd/CP catalysts: 62°C, 500 RPM, catalyst loading 25% of substrate, NH₄COOH/4-BTA=6, methanol solvent.

dichlorophenyltetralone is converted to an imine using methylamine in tetrahydrofuran or toluene. This step uses TiCl₄ as a dehydrating agent to drive the reaction equilibrium toward the imine. The imine product is isolated and then hydrogenated to form amine isomers (*cis:trans*, 6:1) using hydrogen and a Pd/C catalyst in tetrahydrofuran. The mixture of amine isomers is crystallized to selectively isolate the racemic *cis* hydrochloride salt, which is resolved *via* D-mandelic acid in ethanol to obtain the desired (S,S)-*cis* isomer. In the final step, sertraline mandelate is converted to the hydrochloride salt in ethyl acetate (14,15).

As shown in Figure 5, the new process, developed by Pfizer manufacturing chemists, the first three reaction steps are carried out without isolating the intermediates. The team chose to use the more environmentally benign ethanol as the only solvent. TiCl₄ is no longer needed to drive the first step since the imine has poor solubility in ethanol and precipitates from the reaction mixture. More importantly, switching catalyst from Pd/C to Pd/CaCO₃ results in significant improvement in *cis* amine selectivity (*cis:trans* = 18:1). In total, the overall yield has been nearly doubled to 37% and raw materials for the synthesis have been cut by 60, 45, and 20%, respectively, for methylamine, dichlorophenyltetralone, and D-mandelic acid and eliminated the use of TiO₂, HCl, and NaOH.

Changing catalyst support from carbon to calcium carbonate leads to dramatic improvement of the *cis/tran* ratio from 6:1 to 18:1, that is the *cis* selectivity increases from 85.7% to 94.7%. The reason for better selectivity on CaCO3 supported catalyst is attributed to its lower surface area leading to lower hydrogenation activity, but more selective to the desired product. The successful commercialization of the new route for sertraline synthesis demonstrates that for a stereoselective hydrogenation reaction, improve product selectivity can be achieved by proper selection of catalyst support.

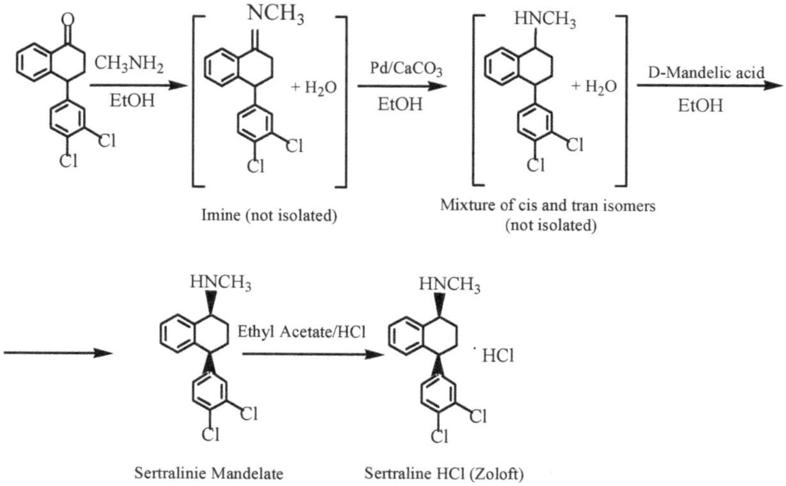
NCH3
HNCH3
CH3NH2
EtOH
+ H2O
Pd/CaCO3
EtOH
+ H2O
D-Mandelic acid
EtOH
Cl
Cl
Imine (not isolated)
Mixture of cis and tran isomers
(not isolated)
HNCH3
HNCH3
Ethyl Acetate/HCl
· HCl
Sertralinie Mandelate
Sertraline HCl (Zoloft)

Figure 5 Postapproval route for Pfizer's antidepression drug required fewer solvents and reagents and gave higher product yield (14).

Pt/C Powder Catalyst

Pt/C powder catalyst is one of the most common catalysts for fine chemical synthesis. Recently, Engelhard introduced (16,17) a new class of Pt/C powder catalysts designed for slurry phase hydrogenation of large molecules, such as: aromatic nitro group hydrogenation, halonitro aromatic hydrogenation, para-aminophenol synthesis. Compared to existing Pt/C catalysts, these new catalysts offer better Pt accessibility to large molecules and demonstrate higher activity in certain structure-sensitive reactions. This improved, more sharply edge coated, Pt/C catalyst minimizes diffusional interference with reactant and metal crystallite contacting. When more metal crystallites are on the outside of the pores of the carrier material, *i.e.* when there is more edge coating, there is greater accessibility to large molecules.

This new preparation method also deposits larger metal crystallites on the support. Thus the dispersion of the metal crystallites for these new catalysts is significantly lower. As a result, CO adsorption values vary from one third to one half of the values obtained for standard Pt/C catalysts. This can be beneficial for structure-sensitive reactions. For example, the lab test results show that these newly developed, low-dispersed, highly edge-coated Pt/C catalysts deliver better selectivity in halo-nitroaromatic selective hydrogenation reactions. In the hydrogenation of 1-chloro-2-nitrobenzene, the use of this new edge-coated, low-dispersed catalyst leads to 25% less dehalogenation compared to standard Pt/C catalysts (16,17).

This technology has broad applicability. For instance, using the same carbon support, test results show that a new Pt/C catalyst with edge-metal location and low dispersion resulted in 36% more activity than ESCAT 20 in a standard nitrobenzene (SNB) test (Figure 6). Using the same technology with a different carbon support yielded a catalyst with 57% more activity than ESCAT 20 in SNB test (16,17).

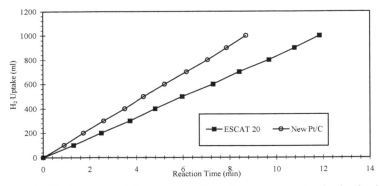

Figure 6 Hydrogen uptake vs. reaction time under standard nitrobenzene hydrogenation test conditions: 5 mg Pt, temperature = 29°C.

Supported Rh Homogeneous Catalyst: Tethering Technology
One of the advantages of homogeneous catalysts is the ability to perform chemical reactions with high chemo-, regio-, and enantio-selectivity. However, homogeneous catalysts have not been widely used in industrial scale due to the difficulties in catalyst separation after the reaction and other reasons. Moreover, the costs for the use of homogeneous catalysts are high. Therefore, it would be beneficial if the homogeneous catalyst could be anchored on a support and re-used. This enables ease of catalyst separation and catalyst reuse while maintaining the good selectivity of the homogeneous catalyst.

There are many reports on anchoring homogeneous catalysts, some of workers encountered metal leaching or lost activity upon re-use (18-22). Recently, Tanielyan and Augustine (23) invented a new way to prepare supported homogeneous catalysts. This method allows preformed homogeneous catalysts that consist of a cationic metal ion, a neutral mono- or bidentate ligand and an anionic counter ion to be anchored on a support. As shown in Figure 7, phosphotungstic acid (PTA) is used as an anchoring agent between the support material and the homogeneous catalyst.

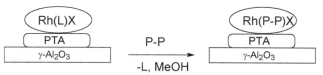

Figure 7 Schematic representation of tethering technology.

Using this technique, Brandts et al. (24-26) have successfully anchored two homogeneous catalysts, that is [(*R,R*)-(Me-DuPHOS)Rh(COD)]BF$_4$ and the non-chiral [Rh(DiPFc)(COD)BF$_4$].

Recently, Engelhard introduced LigandNet^{TM}, immobilized Rh(COD)$_2$BF$_4$ (26). LiganNet^{TM} can be used to prepare and test the immobilized version of the equivalent homogeneous Rh-catalyst. This material can be modified with a simple ligand exchange reaction, in which one 1,5-cyclooctadiene ligand (COD) is replaced by a chiral ligand, to give the desired homogeneous catalyst. LigandNet^{TM} modified with (*S*)-BINAP, (*R,R*)-MeDuPHOS, (*R,R*)-DiPAMP, (*S*)-DIOP has been successfully applied in chiral hydrogenation reactions.

Complex γ-Al$_2$O$_3$/PTA/[(*R,R*)-(Me-DuPHOS)Rh(COD)]$^+$ (**1**) was prepared and tested in the hydrogenation of the prochiral substrate methyl-2-acetamidoacrylate (MAA). After full conversion, the products were separated from the catalyst and analyzed for Rh and W content and product selectivity. The catalyst was re-used three times. Analytical results show no rhodium leaching is observed. Complex **1** maintains its activity and selectivity in each successive run. The first three runs show tungsten (W) leaching but after that no more W is detectable. The leached W comes from the excess of PTA on alumina. The selectivity of both tethered and non-tethered forms gave the product in 94% ee.

With tethering technology, immobilized Rh(DiPFc) catalyst, a very selective hydrogenation catalyst was developed by Engelhard (26) and Chirotech (27). Rh(DiPFc)(COD)BF$_4$ is one of most interesting homogeneous catalysts due to its chemo-selectivity and sulfur tolerant behavior. The anchored complex γ-

$Al_2O_3/PTA/[Rh(DiPFc)(COD)]^+$ (**2**) was used in the reduction of thiophene-2-carboxaldehyde (TCHO) (24). All reactions were carried out in an *i*-PrOH/H_2O solvent mixture to avoid acetal formation. After full conversion the products were separated from the catalyst and analyzed for Rh and W content. The results show that this anchored homogeneous catalyst maintains its activity through 4 consecutive runs (24). The only product observed is the corresponding thiophenic alcohol in quantitative yield. Conventional heterogeneous catalysts (such as Pd/C) gave only limited activity due to sulfur poisoning and do not give this type of chemo-selectivity. However, about 3% total Rh was leached in the first two runs.

Brandts et al. pointed out that (24, 25) although many advantages can be given for this tethering technology, it also has limitations. Not all solvents can be used because some solvents (e.g. MeOH) cause certain anchored complexes to leach from the support, and certain charged functional groups cause leaching, e.g. quaternary ammonium moieties and carboxylate salts. Once the anchored catalyst is activated with hydrogen one has to work under strict oxygen-free conditions, similar to the corresponding activated homogeneous catalysts. In conclusion, tethering technique has several advantages over the corresponding homogeneous catalysts. Multiple re-uses of it are possible, and the product is less contaminated and sometimes is more active and selective.

Acknowledgements

Helpful discussions with Drs. Jim Brandts and Hans Donkervoort are appreciated. The authors thank Mr. Jim Kell, Mr. Mike Baran, and Mrs. Eileen Davis for their help and contributions.

References

1. H.-U. Blaser, A. Indolese, A. Schnyder, H. Steiner, and M. Studer, *J. Mol. Catal. A: Chemical,* **173**, 3 (2001).
2. M. L. Toebes, J. A. van Dillen, and K. P. de Jong, *J. Mol. Catal. A: Chemical* **173**, 75 (2001).
3. P. N. Rylander, Catalytic Hydrogenation over Platinum Metals, Academic Press, New York, 1967, pp. 450.
4. M. Freifelder, Practical Catalytic Hydrogenation, Technique and Applications, Wiley-Interscience, New York, 1971, p. 398.
5. L. S. Seif, K. M. Partyka, and J. E. Hengeveld, Chemical Industries (Dekker), **40** *(Catal. Org. React.),* 1990, p. 197.
6. J. P. Chen, D. S. Thakur, A. F. Wiese, G. T. White, and C. R. Penquite, Chemical Industries (Dekker), **89** *(Catal. Org. React.),* 313 (2002).
7. J. P. Chen and C. R. Penquite, *Specialty Chemical Magazine,* 22(9), p. 22, October (2002).

8. J. P. Chen, A. F. Wiese, and C. R. Penquite, CCN Enabling Catalytic Synthesis, Engelhard Corporation, October (2002).
9. Mastersizer Reference Manual, *Software Version 2.1 and later, Chapter 2, Understanding Particle Sizing,* p. 25, Malvern Instrument Ltd.
10. M. Studer and H.-U., Blaser, *J. Mol. Catal. A: Chemical,* **112,** 437 (1996).
11. M. G. Scaros, et al., Chemical Industries (Dekker), **62** *(Catal. Org. React.),* 457 (1995).
12. S. Y Wang, and J. P. Chen et al., *"Selective Debenzylation in the Presence of Aromatic Chlorine on Pd/C Catalysts: Effects of Catalyst Types and Reaction Kinetics",* paper presented at 20th Organic Reactions Catalysis Society Meeting, March 21-25, 2004, Hilton Head Island, SC, USA.
13. J. Li and S.Y. Wang et al., *Tetrahedron Lett.,* **44**(21), 4041 (2003).
14. A. M. Rouhi, Green Chemistry for Pharma, *C&EN,* April 22, 2002.
15. S. K. Ritter, Green Challenge, *C&EN,* July 1, 2001.
16. J. G. Donkervoort, private communication.
17. *Introducing Engelhard's New, Low-Dispersed, Edge-Coated, Platinum-on-Carbon Powder Catalysts,* Engelhard New Product Bulletin, 2001.
18. D. J. Bayston, J. L. Fraser, M. R. Ashton, A. D. Baker, M. E. C. Polwka, and E. Moses; *J. Org. Chem.,* 9, 7313 (1998).
19. C. Langham, D. Bethell, D. F. Lee, P. McMorn, P. C. Bulman Page, D. J. Willock, C. Sly, F. E. Hancock, F. King, and G. J. Hutchings, *Appl. Cat. A.,* **182,** 85 (1999).
20. R. Margalef-Català, P. Salagre, E. Fernández and C. Claver, *Cat. Letters,* **60,** 121 (1999).
21. F. Gelman, D. Avnir, H. Schumann and J. Blum; *J. Mol. Cat. A.,* **146,** 123 (1999).
22. H. Gao and R. J. Angelici, *Organometallics,* **18,** 989 (1999).
23. (a) S. K. Tanielyan and R. L. Augustine, US Pat. 6,025,295 to Seton Hall University. (b) R. Augustine, S. Tanielyan, S. Anderson, H. Yang *Chem. Commun.,* 1257 (1999).
24. J. A. M. Brandts, J. G. Donkervoort, C. Ansems, P. H. Berben, A. Gerlach and M. J. Burk, Chemical Industries (Dekker), **82** *(Catal. Org. React.),* 573 (2000).
25. J. A. M. Brandts and P. H. Berben,. *Org. Proc. Res. Devel., 7, 393,* (2003).
26. Engelhard Technical Datasheet, *LiganNet* Rh(COD)$_2$BF$_4$, 2003.
27. M. J. Burk, A. Gerlach, and D. Semmeril, *J. Org. Chem.,* **65,** 8933, (2000)

c. Hydrogenation of Interesting Substrates and Renewable Sources

15. Hydrogenation of Nitrile Butadiene Rubber Catalyzed by [Ir(COD)py(PCy₃)]PF₆

Jianzhong Hu, Neil T. McManus, Garry L. Rempel*

Department of Chemical Engineering, University of Waterloo,
Waterloo, Ontario, Canada N2L 3G1

Abstract

The application of [Ir(COD)(PCy₃)(py)]PF₆, ((COD) 1,4 cyclooctadiene, (PCy₃) tricyclohexylphosphine, (py) pyridine) "Crabtree's catalyst"[1] for the hydrogenation of nitrile butadiene rubber (NBR) has been investigated for the first time. The catalyst was efficient for selective hydrogenation of olefinic groups in NBR. A preliminary kinetic study was used to explore the possible catalytic pathway. The process was found to be first order with respect to [Ir] at relatively low hydrogen pressures. As pressure was increased the process showed first to zero order behavior with respect to [Ir], implying that at low [Ir] the active complex is mononuclear, but as hydrogen pressure is increased dimerization to a less active binuclear species may occur.

Introduction

The residual carbon-carbon double bond in nitrile butadiene rubber (NBR) can be catalytically hydrogenated to yield its tougher and more stable derivative, hydrogenated nitrile butadiene rubber (HNBR).[2] This class of specialty elastomer was developed to expand the range of operating environments possible for nitrile butadiene rubber NBR in environments that expose the rubber to chemical and thermal attack.

A number of different catalysts have been shown to be effective for the hydrogenation of NBR. The rhodium complexes: RhCl(PPh₃) and RhH(PPh₃)₄ are known to exhibit excellent hydrogenation activity and selectivity with respect to crosslinking that makes them the basis for industrial HNBR production.[3] Other catalyst systems based on osmium[4] and ruthenium[5] complexes have been investigated and show good hydrogenation activity but promote polymer cross-linking which makes them less attractive systems because of the detrimental effect on HNBR processibility.[6]

---

*Author for correspondence.

The efficiency of Crabtree's catalyst as a catalyst for small molecule hydrogenation has been known for many years. Unlike many homogeneous hydrogenation catalysts, Crabtree's catalyst is able to reduce hindered olefins at favourable rates.[7] It has never been reported as a catalyst for the hydrogenation of rubber except for its use in the hydrogenation of bulk PBD.[8] This paper describes the first use of Crabtree's catalyst in the hydrogenation NBR. Kinetic data are presented and analyzed to understand the underlying chemistry.

Experimental Section

Crabtree's catalyst was obtained from Strem Chemicals. The NBR grade used throughout was Krynac 38.50 (NBR containing 38% acrylonitrile units) from Bayer Rubber Inc. Solvents employed (chlorobenzene, acetone and 2-butanone) were used as received from Fisher Chemical Co. The purity of H_2 and D_2 (O_2 free) from Linde-Union Carbide was reported to be 99.99%.

Apparatus and Procedures

In the kinetic study, data were obtained from the amount of hydrogen consumed by the reaction over time using an automated gas uptake control system. The apparatus is designed to maintain isothermal and isobaric conditions while monitoring H_2 consumption. Procedures used were as described previously.[3-5]

Kinetic data were obtained at various reaction conditions and substrate concentrations.

Product Characterization

Spectra of solution cast films of the hydrogenated NBR were recorded on a Nicolet 520 FT-IR spectrophotometer. The final degree of olefin conversion was confirmed by infrared analysis.[9]

HNBR Viscosity

The viscosities of products in solution (0.5 Weight %/volume) were measured by dilute solution viscometry using a Cannon Ubellohde viscometer at 35 °C.

Results and discussion

In the experiments carried out, the rate of hydrogenation was first order with respect to [C=C] from 30 to 90% conversion. Pseudo first order rate constants (k') were determined for experiments over a range of conditions in order to measure the effect of different reaction parameters. The maximum hydrogenation rate constant recorded in this study was an order of magnitude less than the rate of H_2 mass transfer[10] and so gas uptake measurement reflected the inherent chemically controlled kinetics of the system.

Effect of Catalyst Concentration

Two sets of experiments were performed to determine the effect of catalyst concentration [Ir] on the rate of hydrogenation. These were carried out at hydrogen pressures of 200 psi and 350 psi, (see Figure 1). The data obtained at 200 psi showed a first order dependence on the catalyst level over the chosen [Ir] range (see Figure 1). This suggests a mononuclear species was probably active for this range of conditions. In contrast, reactions at a hydrogen pressure of 350 psi, over a wider range of catalyst concentrations, demonstrated an apparent first to zero dependence with respect to the catalyst concentration. Dimerization or trimerization of the coordinatively unsaturated (catalytically active species) in the presence of hydrogen to produce inactive iridium-hydride complexes is a well known side reaction for Crabtree's catalyst[7] and this could explain the trend, as its likelihood would be greater at higher catalyst and hydrogen concentrations.

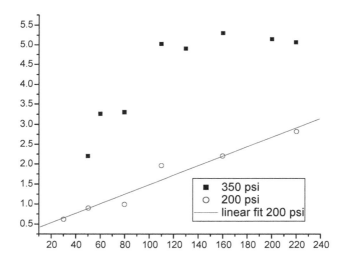

Figure 1 Effect of [Ir], at two reaction pressures; [RCN] = 172 mM, T = 130 °C.

The magnitudes of the rate constants for the iridium catalyst were close to those obtained for rhodium [3] and osmium [5] based catalyst systems at similar conditions. However, the unusual dependence on catalyst concentration affects its general utility in comparison to other homogeneous catalysts for the hydrogenation of NBR.

Effect of Hydrogen Pressure

The results shown in Figure 2 demonstrated that the rate of hydrogenation is pseudo-first order with respect to the hydrogen pressure. The first order rate dependence implies that primarily a single reaction pathway is probably involved in the reaction of the polymer unsaturation with hydrogen. If more than one process were involved, the relative contribution of each pathway should change with varying hydrogen pressure, and thus the dependence might deviate from first order behavior.

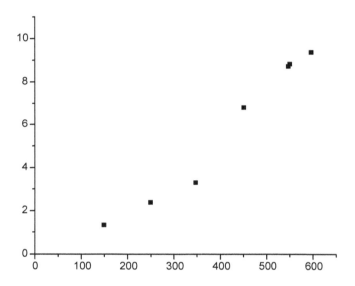

Figure 2 Influence of [H$_2$] on hydrogenation rates; [Ir] = 80 μM, [RCN] = 169 mM, T = 130 °C.

The relationship between hydrogen pressure (P$_{H2}$) and kinetics of reaction has been found to vary for homogeneous catalyst systems based on the different platinum metals.   The RhCl(PPh$_3$)$_3$ system exhibits a first to zero order dependence as the P$_{H2}$ is increased.[3] Martin et al. observed a linear relationship between k$'$ and P$_{H2}$ for the RuHCl(CO)(PCy$_3$)$_2$ system.[5] The OsHCl(CO)(PCy$_3$)$_2$ system, demonstrated a distinct second to zero order dependence of k$'$ on the P$_{H2}$,[4] implying that a truly unique mechanism underlies the observed catalytic process. The interaction between P$_{H2}$ and catalyst concentration, indicated by the observation that high P$_{H2}$ and high [Ir] leads to zero order behavior with respect to [Ir] is the main feature that distinguishes this system.

Effect of Nitrile Concentration.

A series of experiments were carried out at varying polymer concentrations to examine the effect of varying nitrile concentration (the nitrile concentration range was 86-258 mM). The relationship between polymer (nitrile) concentration and reaction rate for this system is shown in Figure 3. It shows that increasing the NBR concentration, (increase in the concentration of nitrile functionality) inhibits the catalytic activity of the catalyst i.e. the observed rate constants showed an inverse dependence on nitrile concentration.

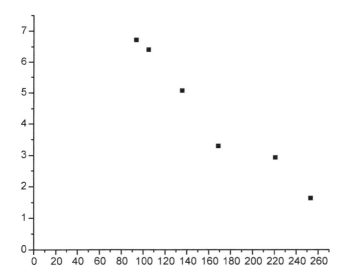

Figure 3 Influence of [RCN] on the hydrogenation rate; [Ir] = 80 μM, P_{H2} = 350 psi, T = 130 °C.

Rhodium,[3] osmium [4] and ruthenium [5] based catalyst systems are affected by nitrile in a similar way. This arises from the relatively high affinity of complexes of these metals towards nitrile group coordination.[11] The resulting equilibrium between free catalyst and catalyst with bound nitrile reduces the effective catalyst concentration and hence reaction rate for a given set of conditions.

Effect of Temperature

A series of experiments was carried out over the temperature range of 120-140°C, where [Ir] (80 μM), [C=C] (275 mM) and P_{H2} (350 psi) remained constant. An Arrhenius plot of the data is illustrated in Figure 4. The linear trend in ln k' versus 1/T yields an apparent activation energy of 68.6 kJ/mole. The value of the activation energy indicates that the reaction is under chemical control and the diffusion of the reactants is not a rate-determining factor under these conditions.

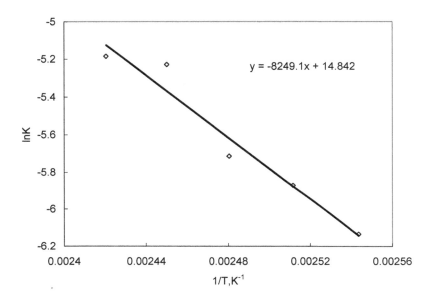

Figure 4 Arrhenius plot (120-140 °C); [Ir] = 80 μm, [RCN] = 172 mM, P_{H2} = 350 psi, Krynac 38.50.

Hydrogen Kinetic Isotopic Effect

If C-H bond formation or breaking is involved in the rate determining step of the catalytic cycle the replacement of hydrogen by deuterium raises the relative activation energy, and thus will reduce the reaction rate all other reaction parameters being equal. Thus the magnitude of the ratio $k'H_2/k'D_2$ is a simple indicator for the cleavage of a bond to hydrogen in the rate-determining step.

Two trials employing D_2 were carried out for this system. The results demonstrate the existence of an average kinetic isotopic effect, $k'H_2/k'D_2 = 1.27$. This suggests the involvement of C-H bond formation in the rate-determining step of the mechanism.

Product Viscosity

The product viscosity of HNBR produced using Crabtree's catalyst was studied in relation to how it varies with catalyst concentration. The results are illustrated in Figure 5. In comparison to results obtained for other catalyst systems the absolute values compare unfavorably with results obtained using Rh catalyst technology. The viscosities obtained are significantly higher.[7] However the results compare favorable with what was observed using Ru and Os based catalysts. Results with these systems led to high viscosities and the viscosity increased with catalyst concentration. In the case of the Ir catalysts the viscosity numbers are fairly high but there is no significant increase in viscosity with catalyst concentration.

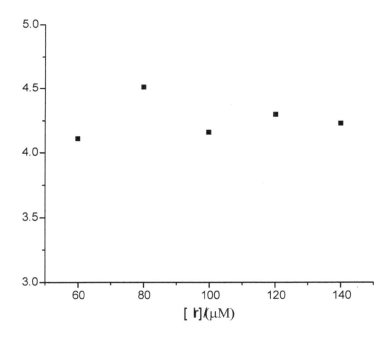

Figure 5 Relative viscosity of HNBR with respect to [Ir].

Proposed Mechanism and Rate Law

The overall study showed that the rate of reaction has a first order dependence on both hydrogen concentration and total [Ir] (up to certain limiting hydrogen and catalyst concentrations) and an inverse dependence on the nitrile concentration. The observed kinetic dependence of the pseudo first order rate constant (k') for the hydrogenation of C=C in NBR may be summarized by the expression show in Equation (1).

$$k' = \frac{k[H_2][Ir]}{a+b[CN]} \quad (1)$$

Using the 16-18-electron rule and the kinetic dependencies described, a reaction mechanism has been proposed and is shown in Figure 5. The mechanism suggests that the active catalytic species is the disubstituted cation $[Ir(py)(PCy_3)]^+$, which is formed rapidly from the catalyst precursor after the COD ligand is hydrogenated to cyclooctane and removed. The active centre then presumably reacts with H_2 to generate a hydride containing catalytic centre. The substrate double bond coordinates to the resulting Ir-hydride intermediate. Hydride transfer to the resulting coordinated olefin complex occurs next, to form an Ir-alkyl complex. The alkyl complex is then cleaved by hydrogen to form hydrogenated polymer and regenerate the cationic active species. It is also assumed that the active species is in equilibrium with an 18 electron, 6 coordinated inactive complex with nitrile ligands, to explain the inverse dependence on nitrile concentration.

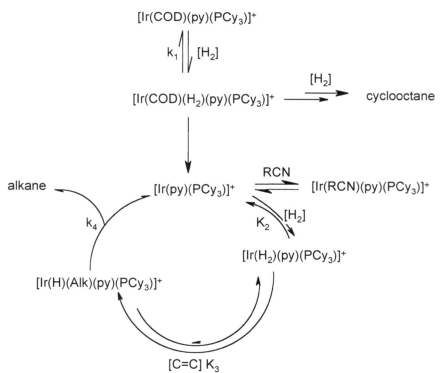

Figure 6 A catalytic pathway of NBR hydrogenation with Crabtree's catalyst.

According to the proposed mechanism, the hydrogenation of olefin by iridium catalyst should conform to the following rate expression,

$$-\frac{d[C=C]}{dt} = k_4[Ir(H_2)(Alk)(py)(PCy_3)] \tag{2}$$

Application of a steady state assumption to each of the equilibria leading to the formation of $[Ir(H_2)(Alk)(py)(PCy_3)]$ provides a means of relating the concentration of every species to the rate determining step. Due to the removal of cyclooctene or cyclooctane pushing the reaction to the direction of the generation of active species, the formation of the active metal fragment is rapid, the concentration of the dihydrido-(cod) Ir complex in solution will be zero. The following equilibria for the system can be formulated.

$$K_3 = \frac{[Ir(H_2)(Alk)(py)(PCy_3)]}{[Ir(H_2)(py)(PCy_3)][C=C]} \tag{3}$$

$$K_2 = \frac{[Ir(H_2)(py)(PCy_3)]}{[Ir(py)(PCy_3)][H_2]} \tag{4}$$

$$K_{CN} = \frac{[Ir(RCN)(py)(PCy_3)]}{[Ir(py)(PCy_3)][RCN]} \tag{5}$$

Calculation of the mass balance on the total amount of Ir charged to the system, $[Ir]_T$, yields the concentration of the active center as a function of the total amount of Ir charged to the system.

$$[Ir]_T = [Ir(H)(Alk)(py)(PCy_3)] + [(Ir(H_2)(py)(PCy_3)] + [Ir(RCN)(py)(PCy_3)] + [Ir(py)(PCy_3)] \tag{6}$$

Manipulating expressions (2) to (6) to solve for $[Ir(H)(alk)(py)(PCy_3)]$ gives the general rate expression:

$$-\frac{d[C=C]}{dt} = \frac{K_2K_3k_4[Ir]_T[C=C][H_2]}{1+K_2K_3[C=C][H_2]+K_2[H_2]+K_{CN}[RCN]} \tag{7}$$

The functional form of this rate expression is consistent with the behavior of the iridium system observed throughout the kinetic investigations. The coordination of nitrile to iridium is anticipated to produce more than a simple inhibitory effect. Being the dominant equilibrium in the mechanism, nitrile coordination may produce the observed first order dependence of the reaction rate with respect to hydrogen. Given $K_{CN}[RCN]$ is the predominant term in the denominator, the rate expression may be reduced to the form of (8) which is first order with respect to both olefin and $[H_2]$.

$$-\frac{d[C=C]}{dt} = k_4[Ir(H)(Alk)(py)(PCy_3)] = \frac{\alpha[Ir]_T[C=C][H_2]}{\beta + K_{CN}[RCN]} \quad (8)$$

This mechanism can only be regarded as of limited utility since it does not take into account the zero order dependence on catalyst that is observed under some conditions. More investigation is needed to expand the understanding of the system over a wider range of conditions using a rigorous statistical design to try and determine the extent of interactions between the different reaction parameters.

Conclusions

Crabtree's catalyst is an efficient catalyst precursor for the selective hydrogenation of olefin resident within nitrile butadiene rubber (NBR). Its activity is favorably comparable to those of other catalyst systems used for this process. Under the conditions studied the process is essentially first order with respect to [Ir] and hydrogen pressure, implying that the active complex is mononuclear. Nitrile reduces the catalyst activity, by coordination to the metal center. At higher reaction pressures a tendency towards zero order behavior with respect to catalyst concentration was noted. This indicated the likelihood of further complexity in the system which can lead to possible formation of a multinuclear complex that causes loss of catalyst activity.

Acknowledgement

We gratefully acknowledge the support of the Natural Sciences and Engineering Research Council of Canada.

References

1. R. H. Crabtree and W. D. Mark, *J. Org. Chem.*, **51**, 2655 (1986).

2. K. Hashimoto, N. Watanabe and A. Yoshioka, *Rubber World*, **190,** 32 (1984).

3. J. S. Parent, N. T. McManus and G. L. Rempel, *Ind. Eng. Chem. Res.,* **35**, 4417 (1996).

4. J. S. Parent, N. T. McManus and G. L. Rempel, *Ind. Eng. Chem. Res.*, **37** 4253 (1998).

5. P. Martin, N. T. McManus and G. L Rempel, *J. Mol. Catal.*, **126**, 116 (1997).

6. J. S. Parent, N. T. McManus and G. L. Rempel, *J. App. Polym. Sci.*, **79**, 1618 (2001).

7. J.W. Suggs and S. D. Cox, *Tetrahedron Lett.*, **22**, 303 (1981).

8. L. R. Gilliom and K. G. Honnell, *Macromolecules*, **25**, 6066 (1992).

9. A. J. Marhsall and I. R. Jobe, *Rubber Chem. and Tech.*, **63**, 244 (1990).

10. J. S. Parent, *Ph.D. Thesis*, University of Waterloo, (1996).

11. J. S. Parent, N. T. McManus, G. L. Rempel, W. P. Power, and T. B. Marder, *J. Mol. Catal.*, **135**, 285 (1998).

16. Catalytic Hydrogenation and Hydrogenolysis of Lignin Model Compounds Using *In-Situ*, Water-Soluble Ru/P(CH₂OH)₃ Systems

Maria B. Ezhova,[a] Andy Z. Lu,[a] Brian R. James,[a]* and Thomas Q. Hu[b]*

[a]*Department of Chemistry, University of British Columbia, 2036 Main Mall, Vancouver, B.C., V6T 1Z1*
[b]*Vancouver Laboratory, Pulp and Paper Research Institute of Canada, 3800 Wesbrook Mall, Vancouver, B.C., V6S 2L9*

brj@chem.ubc.ca, Thu@paprican.ca

Abstract

Chemical modification of lignin without degradation of its polymeric structure could lower the production cost of higher quality paper, particularly with regard to stabilization toward light-induced yellowing, where the presence of so-called α-carbonyls and α-hydroxyl groups is of key importance. In order to explore which chromophore groups can be modified, hydrogenation of aromatic lignin model compounds has been tested, using as catalysts *in-situ*, water-soluble ruthenium species containing tris(hydroxymethyl)phosphine, $P(CH_2OH)_3$. Such systems are active for hydrogenation of activated alkenes, carbonyl groups and, of more interest, hydrogenolysis of some aromatic alcohols. Catalyst activity depends on the phosphine/Ru ratio, the nature of the substrate (particularly substituents of the aromatic ring), and the solvents used. Thus, hydrogenation of a $>C=O$ functionality can give either $>CH(OH)$ or $>CH_2$ depending on the nature of the arene fragment.

Introduction

The pulp and paper industry is a multibillion-dollar business. Production of high quality printing paper involves the use of low-yield, lignin-free, chemical pulp, while employment of less expensive, high-yield, lignin-rich, mechanical pulp leads to lower quality paper (1). As a sequel to our efforts to find an improved industrial process for the bleaching and brightness stabilizing of lignin-rich mechanical wood pulps, we are developing methods for catalytic hydrogenation of such pulps using H_2O-soluble metal complexes (1-6). Hydrogenation using heterogeneous and homogeneous catalysts has been used previously to convert lignin to commercially valuable chemicals and for investigation of lignin structures, but usually at severe conditions of temperature and pressure, with accompanying significant hydrogenolysis/cleavage of substituent side-chains (7, 8). Studies by our group, using milder conditions, have shown that Ru complexes containing H_2O-soluble, sulfonated phosphines (2), and Ru-arene (2) or *n*-trioctylamine (4) systems, can catalyze the hydrogenation of alkenyl and carbonyl bonds, and aromatics, in lignin model compounds (LMCs), and in milled wood lignin without degradation of its polymeric structure. We then demonstrated that use of *in situ* Ru/tris(hydroxymethyl)phosphine (THP) [Ru/P(CH$_2$OH)$_3$] hydrogenation catalysts led to a significant bleaching of mechanical pulps (1). In order to understand which chromophoric groups can be hydrogenated with Ru/THP/H_2 catalysts, we have now studied such systems for the hydrogenation of LMCs. In this paper, results on hydrogenation of C=O and C=C bonds, and the hydrogenolysis of alcohols in such model compounds, are presented. More generally, reports on transition metal complexes containing THP or other hydroxymethyl functionalized phosphines have been limited (9-15), and their application in biphasic or aqueous catalysis has been rare (10-13).

Experimental Section

Solid P(CH$_2$OH)$_3$ (THP, > 90% purity) and [P(CH$_2$OH)$_4$]Cl (THPC, 80% aq. solution) were purchased from Strem Chemicals and Aldrich, respectively; THP was also prepared in similar purity from THPC by a literature procedure (16). Lignin model substrates **1**, **3**, **5**, **6**, **10** and **13** (see eqs. 2-6, below), were used as received from Aldrich. Carbinols **11b** and **11c** (eq. 5) were prepared from a reported borohydride reduction of **10b** and **10c**, respectively (17). Solutions of *in situ* catalyst precursors were prepared via a ligand exchange reaction between RuCl$_2$(PPh$_3$)$_3$ and THP, ideally exemplified by eq. 1 (but see Results and Discussion): a selected amount of THP dissolved in deoxygenated H_2O (25 mL) was added dropwise via a cannula to a brown solution of RuCl$_2$(PPh$_3$)$_3$ (18) (785 mg, 0.82 mmol) in deoxygenated CH$_2$Cl$_2$ (25 mL) in a Schlenk tube, and the resulting mixture stirred at room temperature (r.t., ~20 °C) for 2 h. The yellow-brown aqueous layer was then decanted from the almost colourless ($n > 3$) or light brown ($n \leq 3$) CH$_2$Cl$_2$, which was washed with H_2O (3 x 15 mL); the water fractions were combined, and the volume typically made up to 100 mL by additional H_2O. This solution (containing ~8.2 mM Ru) was filtered in air, and stored in a dark glass flask; solutions with THP/Ru ratios of 0.5 - 6 were prepared.

$$RuCl_2(PPh_3)_3 + n\ THP \longrightarrow n\ PPh_3 + RuCl_2(THP)_n(PPh_3)_{3-n}\ ;\ n = 1\text{-}3 \qquad (1)$$

Catalytic hydrogenations were performed generally under 500 psig of H_2 at 25 or 80 °C in H_2O or 1:1 H_2O/EtOH media, typically with a total volume of 5 mL; the substrates were essentially insoluble in water, but dissolved readily in the mixed medium. H_2O/NaOH and H_2O/buffer imply, respectively, use of a 1:1 mixture of the Ru catalyst solution with either 5.0 mM aq. NaOH or a pH 10 buffer solution (made from 44 mM NaHCO$_3$ and 17.6 mM NaOH (19)). Catalyst concentrations were either 4.1 or 2.05 mM, while the substrate (S) concentration was always 100 mM. The reaction mixture was placed into a glass sleeve equipped with a magnetic stir bar in air. The sleeve was placed in a steel autoclave that was then charged at r.t. with H_2 (pressures throughout are given in psig). The reaction mixture was stirred at the required temperature for selected times, after which the H_2 was released and the products extracted with 4:1 ethyl acetate/acetone (3 x 10 mL); analysis was done by GC (HP-17; 100 °C for 2 min, 10 °C/min to 220 °C, and 3 min at 220 °C) and/or ^{1}H NMR. Hydrogenated products were identified either by GC-MS or GC co-injection of the commercially available or isolated and fully characterized material. NMR spectra at r.t. were recorded on a Bruker AV300 spectrometer, with residual protons of deuterated solvents (^{1}H, relative to external $SiMe_4$) and external $P(OMe)_3$ being used as references; ^{31}P data are reported relative to 85% aq. H_3PO_4. UV-Vis spectra were recorded on an HP 8452A Diode Array spectrophotometer.

Results and Discussion

Nature of the Ru/THP species. Our attempts to characterize the precursor catalyst species have been unsuccessful, as the systems are complicated. Eq. 1 shows a hypothetical, idealized situation with increasing substitution of PPh_3 by THP. *In-situ* ^{31}P{^{1}H} NMR data on the extracted aqueous (D_2O) layer from the synthesis at 3:1 THP/Ru with [Ru] $\sim 10^{-2}$ M do show complete consumption of the added THP (δ_P –23) and generation of overlapping singlets and multiplets in the δ_P 18.3-16.3 region; free THP is first seen at a THP:Ru ratio of ~4, implying that 3-4 THP ligands are involved in the chemistry. Of note, another group (14) has isolated the complex $Ru(THP)_2[PH(CH_2OH)_2]_2Cl_2$ (**A**) from the exchange reaction shown in eq. 1, in a 2-phase, H_2O/CH_2Cl_2 system. **A** was structurally characterized; 2 of the THP groups have eliminated formaldehyde to give the product, which shows the expected 2 triplet, A_2B_2 ^{31}P{^{1}H} pattern at δ_P 13.5 and 9.7 (14). Unfortunately, experimental details were not presented (14), but we note that our signal for the monooxide of THP ($\delta_P = 50$), formed when solutions with THP:Ru > 4 are handled in air, is 5 ppm less than that given in ref. 14, so species **A** may well be present in our systems. Mixtures of species are certainly present depending on the THP:Ru ratio, but we were unable to form solutions containing just a single species even at high THP:Ru ratios. Further, solutions of the species are somewhat photo-sensitive, and are not stable over 24 h in daylight in the presence or absence of air. Nevertheless, Beer's Law was obeyed in the UV-Vis for freshly made solutions up to $\sim 10^{-2}$ M [Ru] in air: λ_{max} 466 nm ($\varepsilon = 140$ M^{-1} cm^{-1}) at THP/Ru = 3, and λ_{max} 448 ($\varepsilon = 95$) at THP/Ru = 10. Our attempts to isolate characterizable complexes were unsuccessful. Complex **A** has also been isolated from an EtOH solution of $RuCl_3$ and THP at r.t. (14), while refluxing of the same solution has yielded the structurally characterized, dimeric complex $[Ru(\mu-PCH_2OH)\{\mu-P,O-P(CH_2O)(CH_2OH)_2\}Cl_2]_2$, which contains bridged phosphido groups and bridged monoalkoxides derived from THP (20). The chemistry of Ru/THP solutions is clearly complex.

Surprisingly, although the composition of species in the Ru solutions used for catalysis slowly changed with time, their catalytic activities did not. For example, hydrogenation of 3,4-dimethoxyacetophenone (**10c**) to the alcohol (**11c**) (eq. 5) in H_2O/buffer with THP/Ru = 3, S/Ru = 25, H_2 = 500, after 22 h at 80 °C gave conversions between 68-71% when using fresh or 7- or 14-day old catalyst solutions. Plausible explanations are that under H_2: (i) the various catalyst precursor species present are converted to the same active hydride species or, less likely, (ii) under H_2, a mixture of hydrides of comparable activity are formed.   In the absence of Ru, no hydrogenation or other reaction was observed; this point is worth mentioning in view of the recently reported metal-free, phosphine-catalyzed hydration and hydromethoxylation of activated olefins (21). That conversions increase with increasing [Ru], and vary with the THP/Ru ratio, rule out that the hydrogenations are mass transfer limited.

Hydrogenation of lignin model compounds. Derivatives of 3-methoxystyrene (**1a** and **1b**), isoeugenol (**3a**) and 3,4-dimethoxypropenylbenzene (**3b**), eugenol (**5a**), benzaldehyde (**6a**), 3-methoxybenzaldehydes (**6b** and **6c**), 3-methoxy-acetophenones (**10a** and **10c**), acetovanillone (**10b**), 3-methoxycarbinols (**11b** and **11c**), and 4-hydroxy-3-methoxycinnamaldehyde (**13**) were chosen as lignin model compounds for testing the *in-situ,* water-soluble Ru/THP catalyst systems.

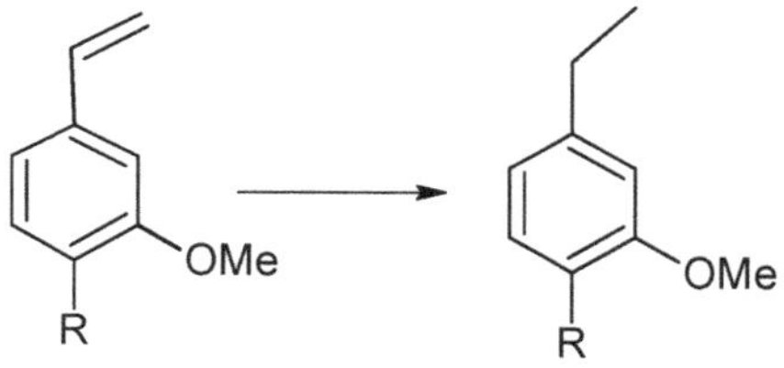

(2)

(1)   **(2)**
(**a**) R = OH; (**b**) R = OMe

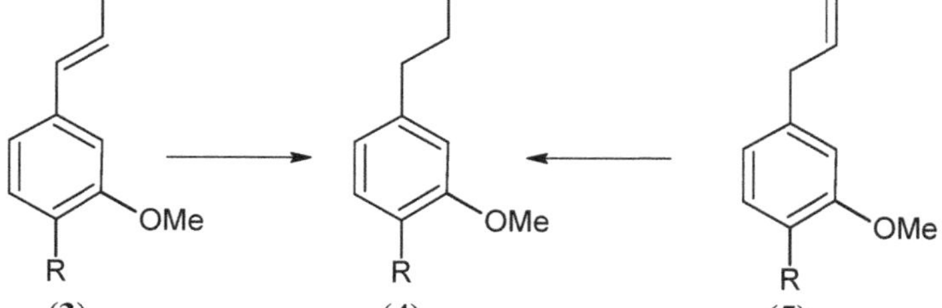

(3)

(3)   **(4)**   **(5)**
(**a**) R = OH; (**b**) R = OMe

$$(4)$$

(6)   **(7)**   **(8)**   **(9)**

(a) R = R' = H; (b) R = OH, R' = OMe; (c) R = R' = OMe

$$(5)$$

(10)   **(11)**   **(2)**   **(12)**

(a) R = H; (b) R = OH; (c) R = OMe

$$(6)$$

(13)   **(14)**   **(15)**   **(4a)**

The activated C=C bond in substituted styrenes (**1**) or propenyl arenes (**3**) is hydrogenated to give **2** and **4**, respectively (eqs. 2, 3), with the less substituted alkenyl moiety in **1** being more easily reduced. For example, hydrogenation of **1a**, **1b**, **3a** and **3b** for 22 h at our "standard" conditions (H$_2$O/buffer, THP/Ru = 3, S/Ru = 25, H$_2$ = 500, at 80 °C) gives respective conversions of 100, 100, 59 and 31%. Consistent with this commonly observed trend (22, 23), the terminal C=C bond in eugenol (**5a**), less hindered than that in its isomer **3a**, shows 88% conversion under the same conditions (eq. 3). Of interest, when an OH (vs. an OMe) group is *para* to the alkenyl group, hydrogenation is more effective as seen by the data above for **3a** vs. **3b**, and for **1a** vs. **1b** (50 and 15%, respectively, under the same conditions but after 0.5 h reaction time). Whether this results from electronic factors for OH (or O⁻) vs. OMe, or simply increased solubility of the OH-containing substrate, is uncertain. The alkene-containing substrates can also be hydrogenated at 25 °C.

The catalyzed hydrogenation of an aldehyde- vs. a ketone-carbonyl is invariably faster because of steric effects (23), and the data for **6** vs. **10** are in line with this (eqs. 4 and 5). Thus, conversions of **6a-c** after 0.5 h at standard conditions are 86, 47, and 97%, respectively, while corresponding values for **10a-c** after 4 h are 78, 36 and 49%, respectively. Indeed, the aldehydes can be reduced at 25 °C under otherwise identical conditions (**6b** gives 38% conversion after 4 h, and **6c** gives 99% after 15 h). The above reactivity trend for the ketones **10a-c** shows that the hydrogenation rates depend on the substituent *para* to the carbonyl functionality and increase in the order H > OMe > OH. For the aldehyde susbtrates, the more limited data (substrate **6** with R = H and R' = OMe was not available) suggest a similar *para*-substituent effect (at least OMe > OH). Note that this is the reverse trend to that observed for reduction of the activated C=C systems described above.

Of more significant interest, the hydrogenation products of **6** or **10** (under the standard conditions) depend strongly on the *para*-substituent of the aromatic ring of the substrate. Thus, the presence of a *p*-OH leads to formation of the fully reduced products **8b** or **2a**, while a *p*-H or -OMe gives solely the alcohols (**7a,c** or **11a,c**; eqs. 4 and 5); with H_2O/EtOH as solvent, the corresponding ethers **9** or **12** are formed. As expected from these findings, hydrogenation of the carbinol (**11b**) containing a *p*-OH gave 100% yield (after 22 h) of the ethyl derivative **2b**, while there was no hydrogenation of the *p*-OMe substrate (**11c**). The net hydrogenolysis of the *p*-OH substituted alcohol (**11b**) could be a direct step, rather than proceeding via dehydration to an alkene such as **1a**, followed by hydrogenation; at least, no alkene intermediate is seen by NMR analysis of the extracted products, although the rate of hydrogenation of the C=C bond is greater than that of hydrogenolysis of the alcohols (e.g., see the data above for the reduction of **1a** vs. **10b**). Further, there is no evidence for the Ru-catalyzed dehydration of the alcohol **11b** in the absence of H_2. Worth noting is that alcohols such as **11b** or **11c** when passed through the analytical GC column (a polysilane material) are partially dehydrated to the styrene product, and thus it is essential that product analysis using the alcohols or the precursor aldehydes and ketones as substrates is done by NMR.

The cinnamaldehyde substrate (**13**) with the standard conditions for 22 h shows 70% conversion to a mixture of mainly (65%) the saturated alcohol (**15**), with 10% unsaturated alcohol (**14**), 15% hydrogenolysis product (**4a**), and 10% of other trace materials (eq. 6). Product profiles as a function of time have not been investigated, but formation of **4a** again likely derives from a direct hydrogenolysis of **15**.

Hydrogenolysis of an aldehyde or ketone carbonyl to $>CH_2$ is an important organic transformation, and classical procedures such as the Clemmenson and Wolff-Kishner reactions have limitations (24, 25); heterogeneous catalytic systems and several two-step procedures are also known (1, 24, 26). Our observation of this conversion in what is essentially a 2-phase medium (aqueous/substrate), with the catalyst remaining in the aqueous phase, is highly significant. More detailed studies are on-going on the use of larger quantities of neat substrate, and on the scope of the reaction (including the potential for the commercially important, catalyzed hydrogenolysis of the carbonyl group in fatty acids and esters to give alcohols (27)). We are unaware of other homogeneous systems that catalyze the observed hydrogenolysis process. The closely related (supposed homogeneous) catalyzed H_2-hydrogenation of aromatic ketones and aldehydes using a Rh(I)-diene/β-cyclodextrin system at r.t. in THF (24) involves colloidal Rh that eventually deposits as metal (1). Studies on the re-cycling of the Ru/THP system are in progress; a 2nd use of the recovered catalyst solution (for substrate **10b**) revealed a 20% loss in activity.

Influence of THP/Ru ratio and solvent systems. Many empirical studies were carried out on variation of conversions with the THP:Ru ratio, defined as R, which was varied from 0.5 to 6.0. Invariably, in the H_2O/buffer "standard" conditions (and other solvent systems – see below), conversions for any selected reaction time decreased when $R > 3$, but this was not usually the optimum ratio. For the ketone **10b**, the maximum conversion was at $R = 3$, but for ketone **10c** and the alkene substrates such as **1b** and **3a**, R was closer to 1; for **6c**, the aldehyde substrate, optimum conversion was at $R \sim 2$. The unknown nature of the catalytic species present in solution makes any discussion of these data meaningless.

As mentioned in the Experimental Section, the substrates are essentially insoluble in water, at least at r.t.; presumably, solubility increases with temperature, but this was not investigated. We did test some systems in H_2O/EtOH in which the substrates were soluble, but conversions at selected reaction times were usually lower than under the standard buffered aqueous conditions; e.g. for 3,4-dimethoxybenzaldehyde (**6c**), acetovanillone (**10b**), and 3,4-dimethoxyacetophenone (**10c**), conversions were decreased by factors of 6 to 8. However, the aq. EtOH systems were not buffered; addition of a base such as Et_3N (Et_3N:Ru = 3) in this medium, however, did increase the conversion rates of these substrates by factors of 2-4, but this possibly results from enhanced formation of Ru-hydride(s) via heterolytic cleavage of H_2 (22, 23, 28). In contrast, hydrogenation of isoeugenol (**3a**) was ~4 times more effective in H_2O/EtOH than in H_2O/buffer, and addition of the Et_3N marginally decreased conversions. Of note, the yield of the hydrogenolysis product **4a** from reduction of the cinnamaldehyde (eq. 6) in the H_2O/EtOH system with added Et_3N was increased to 25% of the products, with a corresponding decrease in the saturated alcohol (**15**).

Hydrogenations were also tested in the industrially attractive medium of unbuffered, "pure" H_2O under otherwise our standard conditions; conversions in the 20-60% over 24 h were attained for various substrates, with the higher values for the alkenyl substrates, and the lower conversions for the aldehyde and ketone substrates. The H₂O/NaOH medium could be used instead of H₂O/buffer, with resulting relatively minor variations in conversions. Similar effects were found on varying the ionic strength of the H₂O/buffer media by changing the concentration of buffer reagents by a factor of 4 (see Experimental Section); e.g. the mentioned conversion of 49% for **10c** was increased to 65% on increasing the buffer concentration by 20, while the 36% noted for **10b** was correspondingly decreased to 20%. Single-phase H₂O/dioxane or 2-phase H₂O/Cl(CH₂)₂Cl systems were completely inactive.

Conclusions

In situ, water-soluble Ru/P(CH₂OH)₃ species are effective catalysts for hydrogenation of olefinic and aldehyde/ketone carbonyl groups in a range of lignin model compounds. Of significance, alcohol functionalities also undergo hydrogenolysis, i.e. >CH(OH) → >CH₂; this is an important transformation that is rarely catalyzed by homogeneous catalytic systems. Although the nature of the air-stable catalyst(s) is poorly understood, they operate in 2-phase systems, the substrate being essentially insoluble in water.

Acknowledgements

We thank the Natural Sciences and Engineering Research Council of Canada for financial support via Mechanical and Chemimechanical Pulps Network and Strategic grants, and Colonial Metals Inc. for loans of RuCl₃ ·3 H₂O.

References

1. T. Q. Hu and B. R. James, in Chemical Modification, Properties and Usage of Lignin (ed. T. Q. Hu), Kluwer Academic/Plenum, New York, 2002, p. 247, and refs. therein; T. Q. Hu, G. Cairns, and B. R. James, Proc. 11ᵗʰ Intern. Symp. Wood Pulping Chem., Nice, France, 2001, Vol.1 p. 219, and refs. therein.

2. T. Y. H. Wong, R. Pratt, C. S. Leong, B. R. James, and T. Q.Hu, Chemical Industries (Dekker), **82**, (*Catal. Org. React.*), 255 (2000).

3. T. Q. Hu, B. R. James, and Y. Wang, *J. Pulp Paper Sci.*, **25(9)**, 312 (1999).

4. B. R. James, Y. Wang, C. S. Alexander, and T. Q. Hu, Chemical Industries (Dekker), **75**, (*Catal. Org. React.*), 233 (1998).

5. T. Q. Hu, B. R. James, and C-L. Lee, *J. Pulp Paper Sci.*, **23(5)**, J200 (1997).

6. B. R. James, Y. Wang, and T. Q. Hu, Chemical Industries (Dekker), **68**, (*Catal. Org. React.*), 423 (1996); T. Q. Hu, B. R. James, and C.-L. Lee, *J. Pulp Paper Sci.,*. **23(4)**, J153 (1997).

7. B. F. Hrutfiord, Reduction and Hydrogenolysis in Lignins – Occurrence, Formation, Structure and Reactions (ed. K. V. Sarkanan), Interscience, New York, 1971, p. 487; I. A. Pearl, The Chemistry of Lignin, Marcel Dekker, New York, 1967, p. 339.
8. S. W. Eachus and C. W. Dence, *Holzforschung*, **29(2)**, 41 (1975).
9. R. Schibli, K.V. Katti, W.A. Volkert, and C.L. Barnes, *Inorg. Chem.*, **40**, 2359 (2001).
10. A. Fukuoka, W. Kosugi, F. Morishita, M. Hirano, L. McCaffrey, W. Henderson, and A. Komiya, *J. Chem. Soc., Chem. Commun.*, 489 (1999).
11. J. Holz, D. Heller, R. Stürmer, and A. Börner, *Tetrahedron Lett.*, **40**, 7059 (1999)
12. K. V. Katti, H. Gali, C. J. Smith, and D. E. Berning, *Acc. Chem. Res.*, **32**, 9 (1999), and refs. therein.
13. B. Drießen-Hölscher and J. Heinen, *J. Organometal. Chem.*, **570**, 141 (1998).
14. L. Higham, A. K. Powell, M. K. Whittlesey, S. Wocadlo, and P. T. Wood, *J. Chem. Soc., Chem. Commun.*, 1107 (1998).
15. K. N. Harrison, P. A. T. Hoye, A. G. Orpen, P. G. Pringle, and M. B. Smith, *J. Chem. Soc., Chem. Commun.*, 1096 (1989).
16. J. W. Elliss, K. N. Harrison, P. A. T. Hoye, A. G. Orpen, P. G. Pringle, and M. B. Smith, *Inorg. Chem.*, **31**, 3026 (1992).
17. T. Q. Hu, G. Leary, and D. Wang, *Holzforschung*, **53(1)**, 43 (1999).
18. P. S. Hallman, T. A. Stephenson, and G. Wilkinson, *Inorg. Synth.*, **12**, 237 (1970).
19. Handbook of Chemistry and Physics, 50th ed. R.C. West, The Chemical Rubber Co., Cleveland, OH, D-102 (1969).
20. A. Romeroso, T. Campos-Malpartida, M. Serrano-Ruiz, and M Peruzzini, Proc. Dalton Discussion 6, Organometallic Chemistry and Catalysis, York, UK, 2003, Poster 15.
21. I. C. Stewart, R. G. Bergman, and F. D. Toste, *J. Am. Chem. Soc.*, **125**, 8698 (2003).
22. B. R. James, in Comprehensive Organometallic Chemistry (eds. G. Wilkinson, F. G. A. Stone, E. A. Abel), Pergamon Press, Oxford, 1982, Vol. 8, Ch. 51.
23. P. A. Chaloner, M. A. Esteruelas, F. Joö, and L. Oro, Homogeneous Hydrogenation, Kluwer Academic Publishers, Dordrecht, 1994.
24. H. A. Zahalka and H. Alper, *Organometallics*, **5**, 1909 (1986).
25. J. March, Advanced Organic Chemistry, 2nd ed., McGraw-Hill, New York, 1977, p.1119.
26. W. F. Maier, K. Bergmann, W. Bleicher, and P.v. R. Schleyer, *Tetrahedron Lett.*, **22**, 4227 (1981).
27. L. Ma and M. S. Wainwright, Chemical Industries (Dekker), **89**, (*Catal. Org. React.*), 225 (2003), and refs. therein; J. Ladebeck and T. Regula, Chemical Industries (Dekker), **82**, (*Catal. Org. React.*), 403 (2001).
28. S. N. Gamage, R. H. Morris, S. J. Rettig, and B. R. James, *J. Organometal. Chem.*, **309**, C59 (1986).

17. Catalytic Preparation of Pyrrolidones from Renewable Resources

John G. Frye, Jr., Alan H. Zacher, Todd A.Werpy, and Yong Wang

*Physical Sciences Laboratory, Pacific Northwest National Laboratory,
Richland, WA 99352*

john.frye@PNL.gov

Abstract

Use of renewable resources for production of valuable chemical commodities is becoming a topic of great national interest and importance. This objective fits well with the U.S. DOE's objective of promoting the industrial bio-refinery concept in which a wide array of valuable chemical, fuel, food, nutraceuticals, and animal feed products all result from the integrated processing of grains, oil seeds, and other bio-mass materials. The bio-refinery thus serves to enhance the overall utility and profitability of the agriculture industry as well as helping to reduce the USA's dependence on petroleum.

Pyrrolidones fit well into the bio-refinery concept since they may be produced in a scheme beginning with the fermentation of a portion of the bio-refinery's sugar product into succinate. Pyrrolidones are a class of industrially important chemicals with a variety of uses including polymer intermediates, cleaners, and "green solvents" which can replace hazardous chlorinated compounds.

Battelle has developed an efficient process for the thermo-catalytic conversion of succinate into pyrrolidones, especially N-methyl-2-pyrrolidone. The process uses both novel Rh based catalysts and novel aqueous process conditions and results in high selectivities and yields of pyrrolidone compounds. The process also includes novel methodology for enhancing yields by recycling and converting non-useful side products of the catalysis into additional pyrrolidone. The process has been demonstrated in both batch and continuous reactors. Additionally, stability of the unique Rh-based catalyst has been demonstrated.

Introduction

Recent advances in biotechnology have resulted in improved fermentation processes for the production of useful chemical intermediates. One such process was developed a few years ago during the U.S. DOE sponsored CRADA (Cooperative Research and Development Agreement) with Applied Carbochemicals (ACC). Under this CRADA a consortium of four national labs including Argonne National Laboratory (ANL), Oak Ridge National Laboratory (ORNL), National Renewable Energy Laboratory (NREL), and Pacific

Northwest National Laboratory (PNNL), and industry partner ACC collaborated to develop an overall process for the conversion of glucose to a slate of commodity chemicals (1). PNNL was tasked with identifying appropriate catalysts and process conditions for the conversion of succinate to products such as gamma-butyrolactone (GBL), 1,4-butanediol, (BDO), tetrahydrofuran, (THF), and pyrrolidones (also known as pyrrolidinones). Catalytic routes and process conditions were initially explored using reagent quality succinate feedstocks, then later with actual fermentation derived feedstocks. Fermentation derived feedstocks, even though they had been processed through a series of clean up steps, performed poorly compared to reagent quality feed. Conversion rates for fermentation derived succinic acid were roughly an order of magnitude lower than for pristine succinic acid feeds. Residual proteinaceous materials that had not been removed during clean up steps were thought to be the most likely cause of catalyst fouling. Project funding, at that point in time, did not allow a thorough investigation into the cause of the catalyst fouling. Battelle's corporate organization (BCO) funded a follow up project to develop technology that would allow fermentation derived diammonium succinate to be converted to N-methyl-2-pyrrolidone (NMP). During the CRADA work, it was observed that when methanol is added to an aqueous diammonium succinate hydrogenation, the product mixture contained both 2-pyrrolidone (2-P) and N-methyl-2-pyrrolidone (NMP). The combined yield of 2-P and NMP was nearly ninety mole percent with an NMP : 2P mole ratio of ~ 2 : 1. N-methylsuccinimide (NMS) was identified as an important intermediate that is formed during the reaction. One of the objectives of the BCO funded effort was to determine how NMS compared to a mixture of diammonium succinate and methanol for the production of NMP. Published physical properties suggested that NMS may be easily purified by distillation or sublimation. This suggested that if NMS could be synthesized from an impure fermentation broth, then NMS might easily be separated from potential catalyst poisons, allowing better downstream conversion of NMS to NMP (2,3).

Experimental Section

Catalyst Preparation

Catalysts were prepared by impregnation of metal precursor solutions onto a commercially available granular coconut shell carbon support (20X50 mesh Engelhard # 712A-5-1589-1, C.T.C. # = ~ 95%), using the method of incipient wetness. Following impregnation, catalysts were dried, then reduced with hydrogen in the reactor vessels prior to use. Typical reductions were carried out at 120°C for 4 hours, using a 20% H_2 in N_2 reduction mixture. During catalyst screening under the ACC CRADA, Rh/C catalysts afforded the highest yields of 2-P from the hydrogenation of aqueous diammonium succinate solutions (1). Therefore, all of the catalyst compositions used in this study were based on Rh. The main catalyst compositions prepared for testing during this study are as follows:

A: 5%Rh/C (uniform metal distribution)
B: 2.5%Rh2.5%Re/C (uniform metal distribution)

C: 2.5%Rh2.5%Zr/C (pseudo-eggshell metal distribution)

Additional details on the catalyst preparations can also be found in the patents (2,3). Chemicals used in the catalyst preparations were obtained from the following sources:

$Rh(NO_3)_3$ stock solution (10.37% Rh by wt.) – Engelhard Corp.
Perrhenic Acid stock solution (52.091% Re by wt.) – Engelhard Corp.
Zirconyl Nitrate, hydrate – Aldrich Chemical Co.
Nitric Acid, 70%, A.C.S. Reagent Grade – Aldrich Chemical Co.
D.I. Water, from Milli-Q Plus high purity water system

Preparation of Aqueous Feedstock Solutions

Feedstock Solution #	1	2	3
Feeds mole ratio NH_3/Succinate/MeOH/H_2O	2:1:2:20.2	2:1:1:22	1:1:1:23
Wgt.% Composition			
SuccinicAcid	20.35	20.35	20.35
NH_3	5.88	5.87	2.93
Methanol	11.09	5.52	5.52
Water	62.69	68.27	71.20

Feedstock Solution #	4	5	6	7
Feeds mole ratio Succinimide/MeOH/H2O	1:25.9	1:1:24.9	1:2:23.2	1:2:16.7
Wgt.% Composition				
N-methyl succinimide	19.49	None	None	None
Succinimide	None	17.07	17.07	21.34
Methanol	None	5.52	11.04	13.85
Water	80.51	77.41	71.88	64.85

Additional details on the feedstock preparations can be found in the patents (2,3). All chemicals used in the preparation of the above feedstock solutions were reagent grade materials obtained from the following sources:

Succinic Acid, 99+% - Aldrich Chemical Co.
Succinimide, 98% - Aldrich Chemical Co.
N-Methylsuccinimide, 99% - Aldrich Chemical Co.
Ammonium Hydroxide, Certified A.C.S. PLUS – Fisher Scientific Co.
Methanol, Optima Grade – Fisher Scientific Co.
D.I. Water, from Milli-Q Plus high purity water system

Catalytic Testing for NMP Production – Batch Reactor

Catalytic testing was conducted in a 450 ml Parr Hasteloy C autoclave reactor equipped with magnetic drive stirring and a gas-dispersing stirrer, and was operated in a batch mode. Generally, ~ 3.75g of unreduced catalyst was placed in the empty reactor vessel, for reduction prior to adding feed solution. After catalyst reduction, the reactor headspace would be evacuated and the aqueous feedstock solution (typically ~ 150 cc of soln.) added through the reactor diptube port. The reactor was then pressurized with H_2 to 1000 psig and heated to reaction temperature. Upon reaching the desired temperature, the H2 inlet to the reactor is opened to a pressure regulated H_2 supply (usually ~ 1900 psig), allowing constant pressure operation over the course of the reaction. Samples of product solution are removed periodically to monitor the course of the reaction.   Product solutions were analyzed by liquid and gas chromatographic methods.   Gaseous products were sampled only at the completion of the runs and analyzed by G.C.

Non-Catalytic Testing for NMS Synthesis – Batch Reactor

Non-catalytic NMS synthesis runs were also conducted in a 450 ml Parr Hasteloy C autoclave reactor equipped with magnetic drive stirring and a gas-dispersing stirrer, and was operated in a batch mode. Feedstock solutions (usually ~ 150 cc of solution) were placed in the empty reactor vessels and purged with N_2 to remove air, pressurized with ~ 500 psig of N_2 pressure, then heated up to temperature. Upon reaching the desired temperature, the N_2 inlet was opened allowing the system to operate at constant pressure (usually either 1900 psig or 2200 psig). Samples of product solution were removed periodically to monitor the course of the reaction. Product solutions were analyzed by liquid and gas chromatographic methods. Gaseous products were sampled only at the completion of the runs and analyzed by G.C.

Catalytic Testing for NMP Synthesis – Continuous Flow Reactor

One catalyst composition, pre-selected from batch reactor screening runs, was prepared and tested in a small trickle bed reactor.   In this reactor configuration both gas and liquid feeds are introduced into the reactor in a co-current downflow manner.   The reactor is an Autoclave Engineers 316 SS medium pressure reactor tube (0.750" O.D. X 0.098" wall thickness with coned and threaded ends). An internal thermowell tube runs from the bottom end up through the entire heated length of the reactor tube, so that the reactor's temperature profile can be probed. The reactor tube is jacketed by a 2" diameter X 18" long heat exchanger through which hot oil is circulated to provide uniform heating. Then catalyst is generally packed within the middle 14" of the 18" heated zone. Bed temperatures can generally be maintained within +/- ~ 0.5 °C. The oil was heated and circulated by a Julabo closed system heated pump. Liquid feeds were supplied to the reactor by an ISCO high-pressure syringe pump, equipped with a syringe heating jacket and heated liquid delivery lines. Feed gases were metered into the reactor using Brooks high-pressure mass flow controllers. A dome-loaded back pressure regulator was used to maintain the reactor at the desired system pressure. Gaseous and liquid products are separated in chilled collection vessels, where liquid products were periodically drawn off and analyzed. Gas products were measured with a wet test meter, and samples were periodically withdrawn and analyzed by gas chromatography.

For this series of runs, the reactor body was packed with 40 ml of 2.5%Rh-2.5%Re on 20X50 mesh 95% CTC carbon with 10 ml of quartz chips above and below the bed. The catalyst was first reduced at 120C for 16 hours with 4% hydrogen in argon at atmospheric pressure. After reduction, the gas was switched to hydrogen flowing at 18 SLPH (standard liters per hour), the pressure increased to 13 MPa, and the bed temperature raised to 200 °C. After equilibrating the temperature and gas flow rate, the liquid feed was added via high-pressure ISCO liquid pump at 100 ml/hr. From previous trickle bed experience, a total of 3 or more bed volumes of feed were allowed to pass through the reactor before it was considered equilibrated. Upon equilibration, approximately 20 to 30 ml of product was collected in a sample vial before the reactor conditions were adjusted and allowed to equilibrate again. During each sample period, a gas sample was taken from the low-pressure side of the back-pressure regulator for analysis. The liquid products were analyzed by GC and HPLC. To idle the reactor following a series of run conditions, the catalyst was first washed with approximately 3 bed volumes of deionized water and the bed temperature lowered to 100 °C. The reactor was held at pressure under reducing conditions with a low flow of hydrogen between runs.

Analytical Procedures
Sample analysis was performed using a combination of Gas Chromatography (GC) and High Pressure Liquid Chromatography (HPLC).

GC Analysis
GC analysis for methanol, 1-propanol, 1-butanol, pyrrolidine, N-methylpyrrolidine, 2-pyrrolidinone, N-methyl-2-pyrrolidinone, gamma-butrolactone, dimethylsuccinate, and N-butyl-2-pyrrolidinone was performed with a Hewlett-Packard Model 5890 Gas Chromatograph equipped with a 30-meter, 0.53 mm I.D., 0.50-micron film, Nukol capillary column (Supelco, Bellefonte, PA) and a flame ionization detector (FID).

HPLC Analysis
HPLC analysis for succinic acid, succinamic acid, succinamide, succinimide, N-methylsuccinimide, butyric acid, and propionic acid was performed with a Waters Model 515 HPLC pump equipped with a Waters Model 2410 Refractive Index Detector. Separations were performed with an Aminex HPX-87H 300mm column (Bio-Rad Laboratories, Hercules, CA) operated at 35°C, and using 0.005 M H_2SO_4 elluent.

Catalytic Preparation of Pyrrolidiones

Results and Discussion

Table 1 shows the results for a series of NMP synthesis runs that were run in the batch mode to investigate the effects of using different feedstocks, temperatures, and catalysts. The data shows that use of aqueous ammonium succinate/methanol feedstock mixtures at 265°C produces significant amounts of NMP.

Table 1 Synthesis of NMP – batch reactor results.

Catalyst	Run Temp. (°C)	Feedstock Solution	% Conv. @ Max. Yield	Time (hrs.) @ Max. Yield	Max. Yield (NMP + 2P)	NMP / 2P (Mole Ratio)
5%Rh/C	265	1	98.5	21.0	83.3	1.58
2.5%Rh/2.5%Zr/C	265	1	97.5	21.3	95.2	2.31
2.5%Rh/2.5%Re/C	265	1	90.6	8.0	90.0	1.15
5%Rh/C	265	2	95.5	5.0	80.7	0.55
2.5%Rh/2.5%Re/C	200	1	93.5	24.0	47.2	0.21
5%Rh/C	265	3	89.6	7.0	70.9	0.86
2.5%Rh/2.5%Zr/C	200	3	80.4	23.5	24.8	0.16
5%Rh/C	265	4	95.6	8.0	74.1	2.15
2.5%Rh/2.5%Re/C	265	4	88.7	5.0	53.7	5.28
2.5%Rh/2.5%Zr/C	200	4	93.5	24.0	81.1	38.1
2.5%Rh/2.5%Re/C	200	4	94.8	8.0	88.8	67.3

*All tests at 1900 psig (13.2 MPa)

N-methylsuccinimide is observed growing in as an intermediate early in the reaction sequence and is subsequently converted to pyrrolidone products. However, a significant amount of the succinate inventory has also been converted to other species, such as succinimide, succinamic acid, succindiamide, and methyl substituted analogues of these. As a consequence, NMP yields less than ~ 67% are observed for those feeds. The figure below shows a simplification of the equilibrium that exists in the aqueous phase between the various ammonia containing succinate derivatives (N-methyl derivatives and potential polymeric species have not been included in the figure to conserve space). Feeds containing a higher stoichiometric amount of methanol (MeOH:succinate = 2:1 instead of 1:1) generally gave greater amounts of NMP. However, when aqueous ammonium succinate/methanol feedstock mixtures are hydrogenated at 200°C, very little intermediate formation of N-methylsuccinimide is observed, and much lower selectivity to NMP is also observed.

Where aqueous NMS solutions are used as the feedstock, higher NMP : 2P product ratios are generally observed. However, when aqueous NMS feed is run at 265°C, free methanol is observed in the earliest product samples. Loss of the methyl group from NMS via high temperature hydrolysis apparently becomes significant at this temperature. With aqueous NMS as the feedstock, and when the reaction temperature is lowered to 200°C, NMP : 2P molar product ratios as high ~ 67 : 1 were observed for the more active 2.5%Rh2.5%Re/C catalyst. At the same temperature, even the less active 2.5%Rh2.5%Zr/C catalyst displayed a respectable NMP : 2P product ratio of ~ 38 : 1.

Table 2 Synthesis of NMS – batch reactor results.

Temp. (°C)	Pressure (MPa)	Feedstock Solution	Time (hrs.) @ Max. Yield	Max. Yield of NMS (%)	% Conversion @ Max. NMS Yield
265	13	1	3.0	66.9	100
300	13	1	3.5	83.3	100
265	13	5	4.0	43.2	75.3
300	13	6	2.5	82.3	97.8
300	15	6	3.0	79.0	99.1
300	13	7	0.5	87.5	89.0

Table 2 shows the results for a series of non-catalytic NMS synthesis runs that were conducted in the batch mode to test the effect of different feedstock compositions and temperatures. As this was a preliminary study, only a few scoping experiments were conducted. As can be seen from the table, operation at 300°C appears to result in higher NMS yields. The third entry in the table was the only feedstock run that did not have a stoichiometric excess of methanol. It gave the lowest NMS yield and had the slowest rate of NMS formation. The other runs had a twofold stoichiometric excess of methanol. The 2nd, 4th, and 6th entries in the table were run under very similar conditions and appeared to give very similar NMS yields. The final entry in the table used a feedstock with the same molar feed ratios as the two previous runs listed in the table. However, the feedstock solution for the last run was prepared using ~ 25% higher succinimide concentration (lower water concentration). The higher concentration feedstock resulted in the highest NMS yield (~87.5%) of the conditions tested and in the shortest amount of time.

Table 3 shows the results for a series of trickle bed reactor runs that were done with a 2.5%Rh2.5%Re/Engelhard 95% CTC 20X50 mesh carbon catalyst composition. This catalyst displayed the highest NMP yield in previous batch reactor testing (~ 87% NMP yield after 8 hours). The feedstock solution used in these runs is the same as the solution #4 described above. A feed rate of 100 ml/hr. was used. This was based on extrapolation of the conversion rate observed in the batch reactor runs (grams NMS converted per gram of catalyst per hour). However, the respective residence times in the two reactors are quite different. Where the residence time in the batch reactor is 8 hours, the residence time in the flow reactor is only about 24 minutes. Under the initial set of conditions, the NMS conversion level in the flow reactor (~ 93.6% NMS conversion) compares very favorably to that which was seen in the batch reactor (~ 94.8% NMS conversion). The NMP : 2P molar ratio for the flow reactor product solution is also very high (~ 79 : 1). However, the NMP+2P combined yield is low by comparison with batch reactor results (~ 52.6% compared to ~ 87%), and the flow results had a low overall mass balance. CHN analysis of the feed and product solutions indicated that carbon inputs to and outputs from the reactor were nearly identical, but much of the carbon in the flow reactor product solutions could not be accounted for in known compounds, and no significant extra peaks were observed in HPLC analyses. The presence of an oligomeric type product material was suspected. This general phenomenon has also been observed by others (4). Process conditions were changed to try to maximize NMP + 2P yields. NMP selectivity was improved by increasing the reactor temperature from 200°C to 250°C and also by reducing the hydrogen pressure in the flow reactor.

To test the hypothesis of oligomeric products formation, a composite product solution was made up from the products collected from the runs shown in columns 2 and 3 in Table 3. A portion of this composite product solution was sampled and analyzed and is shown in Table 4 under the column heading marked "pre-hydrolysis". A second portion of the composite product solution

Table 3 Synthesis of NMP – continuous flow reactor.

Feedstock Solution	4	4	4	4	4	4	4
H₂ flow (slph)	18.2	18.2	14.4	8.3	8.3	18.2	18.2
H₂ / NMS	4.3	4.3	4.6	4.2	4.2	4.6	4.6
psig	1900	1900	1900	1900	1900	1900	1200
Temp. (°C)	200	180	180	200	220	250	250
feed (ml/hr)	100	100	80	50	50	100	100
conversion %	93.6	78.0	85.3	98.7	99.4	97.8	88.2
NMP selec.	55.5	50.0	52.4	57.1	60.0	64.1	75.7
2P selec.	0.7	0.0	0.0	1.1	1.8	0.0	0.6

*Catalyst Charge was ~ 40ml (18.31 g) of a 2.5%Rh2.5%Re/Engelhard 95% CTC granular carbon support (20X50 mesh particle size)

was placed in a 300 cc Parr autoclave with no hydrogenation catalyst being present and heated under nitrogen pressure to 250°C for 15.5 hours (overnight under essentially hydrolytic conditions). After cooling, the hydrolyzed product mixture was removed from the autoclave and analyzed. The analysis results are shown in Table 4 under the heading marked "post-hydrolysis". The observed concentration of NMP in the "post-hydrolysis" product solution is ~ 62% higher than was observed in the "pre-hydrolysis" product solution compared to the rest of the components

Table 4 Hydrolysis of continuous flow reactor product solution.

Product Component	Pre-Hydrolysis	Post Hydrolysis
% Succinimide	0.05	0.00
% N-Methylsuccinimide	3.09	2.98
% Methanol	0.15	0.13
% Gamma-Butyrolactone	0.15	0.04
% N-Methyl-2-Pyrrolidinone	6.42	10.42
% N-Butyl-2-Pyrrolidone	0.02	0.03
% 2-Pyrrolidinone	0.00	0.00

respectively. This would suggest that the oligomeric material present in the "pre-hydrolysis" product solution was already a hydrogenated specie and upon hydrolysis, and subsequent ring closure, yields additional quantities of NMP product. Although the exact structure(s) are not known, this observation strongly suggests that recycle of by-products in a commercial process will improve overall NMP yield.   The figure shown below suggests a reaction pathway thought to be operative in the aqueous phase hydrogenation of NMS and also a possible structure of the oligomeric material present in the "pre-hydrolysis" product solution.

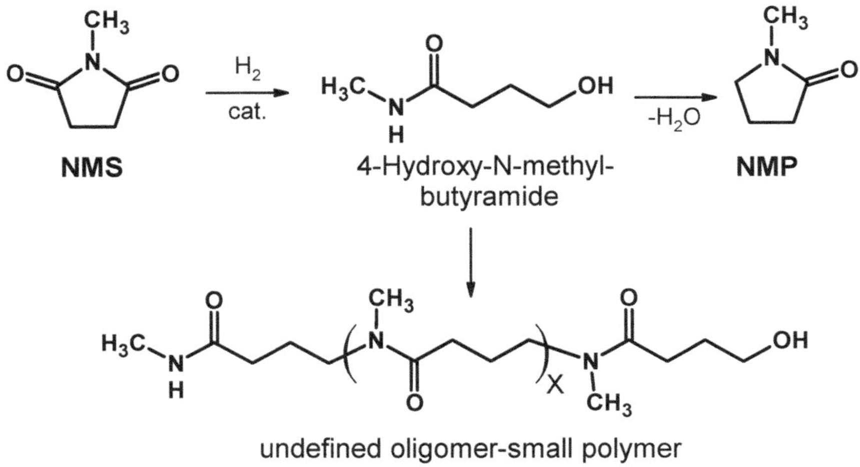

NMS   4-Hydroxy-N-methyl-   **NMP**
                  butyramide

undefined oligomer-small polymer

Acknowledgements

We would like to thank Battelle Memorial Institute for supporting this work and also to the U.S. DOE(OIT) under whose sponsorship this work was begun. We would also like to acknowledge Mr.Todd Hart who conducted most of the batch reactor screening runs and also Mr. Mark Butcher who performed the product analyses. Our deepest appreciation goes to Dr. James F. White for his assistance in the preparation of this paper.

References

1. Nghiem et al., An Integrated Process for the Production of Chemicals from Biologically Derived Succinic Acid, A.C.S. Symposium Series, 2001, vol.748, pp. 160-173.
2. Todd Werpy, John G. Frye, Jr., Yong Wang, and Alan H. Zacher, US Pat. 6,603,021 to Battelle Memorial Institute (2003).
3. Todd Werpy, John G. Frye, Jr., Yong Wang, and Alan H. Zacher, US Pat. 6,632,951 to Battelle Memorial Institute (2003).
4. Kou-Chang Liu and Paul D. Taylor, US Pat. 4,824,967 to GAF Corp. (1989).

18.   Hydrogenation of Glutamic Acid to Value Added Products

Johnathan E. Holladay and Todd A. Werpy

Pacific Northwest National Laboratory, Richland, WA 99352

john.holladay@pnl.gov

Abstract

Glutamic acid (Glu) provides a platform to numerous compounds through thermochemical approaches such as hydrogenation, cyclization, decarboxylation and deamination. In this paper preliminary results of the hydrogenation of Glu are discussed with an emphasis on controlling the selectivity of carbonyl reduction. Under thermal conditions Glu cyclizes to give 5-oxopyrrolidine-2-carboxylic acid (Pyroglu). At low temperatures (70 °C) Glu can be hydrogenated to give 2-aminopentane-1,5-diol (glutaminol) (1). At elevated temperatures (150 °C) and in the absence of added acid 5-hydroxymethyl-2-pyrrolidinone (5HMP) is the main product. While at low pH, normally used for hydrogenation of amino acids, the pyrrolidinone carbonyl is also reduced resulting in pyrrolidin-2-ylmethanol (prolinol). A mechanism for the various products is suggested. The results show that hydrogenation of Glu has characteristics unique from those of other amino acids, and that paradigms in the literature do not apply for Glu.

Introduction

Technology to convert biomass to chemical building blocks provides an opportunity to displace fossil fuels and produce novel building blocks while increasing the economic viability of bio-refineries (see Figure 1), an important goal of the United States Department of Energy (DOE). The strategy for the development of products from biomass needs to be two fold: (1) identify opportunities for direct displacement of petrochemical derived products and (2) develop novel building blocks that contain functionality that cannot easily or cost effectively be derived from petrochemical sources. An example of the former are products produced from fermentation derived succinate, including: 1,4-butanediol, THF, γ-butyrolactone, or N-methyl-2-pyrrolidinone. For the bio-based processes to compete against petroleum-based processes in this example, succinic acid must be produced at a cost competitive with petroleum derived maleic anhydride.

Amino acids, such as Glu, derived from fermentation provide an excellent example of the latter. Glu can be converted to numerous monomeric intermediates that have unique functionality not provided by petrochemical intermediates, including the five-carbon motif and chirality. Glu can also be directly made into polymers. Recently Ichimaru Pharcos announced the "volume production" of γ-polyglutamic acid for use in cosmetics, paints, water purification and biodegradable plastics (2). In this paper, we report on our preliminary studies on the thermochemical conversion of Glu to products and intermediates.

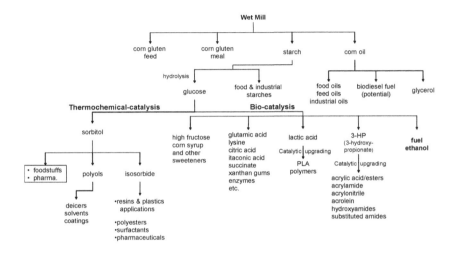

Figure 1   A biorefinery produces fuels and chemicals from biomass; one embodiment today is the corn wet mill.

Background

Although catalytic hydrogenation has been part of the chemical landscape for many years, there has been relatively little work directed toward the hydrogenation of amino acids (3,4) and even less directed toward the mechanism of hydrogenation (5,6). On a practical level the pioneering work of Adkin's group (3) in the 1930s and 1940s demonstrated that carboxylic acid esters can be converted to alcohols at mild temperatures (25–150 °C) and high hydrogen pressures (>15 MPa). Important to Adkin's work is the demonstration that carboxylic reduction can be done while maintaining a high degree of stereochemical purity. The catalyst systems of the day, nickel, copper or chromium oxides, worked well with esters but not with free acids. Development of ruthenium (7) and rhenium (8) catalysts in the 1950s solved this problem.

There has been a substantial body of work directed at hydrogenation of dicarboxylic acids (9,10), particularly maleic and succinic acids (11), in light of the development of 1,4-butanediol and tetrahydrofuran from butane-based maleic anhydride in the 1970s. The simple C_5 dicarboxylic acid, glutaric acid, and its alkyl esters have been the subjects of numerous studies to produce 1,5-pentanediol as a product. Catalysts used include copper chromite, (12,13), palladium, and cobalt (14). High selectivity (>95%) and high conversion (>98%) are achieved over the $CuCrO_x$ catalysts. Conversion over homogeneous ruthenium catalysts is also reported (15,16). In general, achieving high selectivity to the diol product is easier with glutaric and adipic acids than with succinic acid, because of the thermodynamic favorability of succinic acid cyclization during reaction which detracts from the linear diol formation.

Direct hydrogenation of amino acids to amino alcohols was first examined by Adkins et al. (3) via the esters, and recently studied by Antons and Beitzke in patents (4). Using Ru/C catalysts at high pressures (>14 MPa) and mild temperatures (70-150 °C), Antons demonstrated the conversion of carboxylic acids and amino acids with retention of optical activity in the product alcohols. High yields (>80%) and high enantiomeric purity (>97% in many cases) were achieved. Broadbent et al. had demonstrated earlier that under certain conditions hydrogenation of amino acids can be accompanied by deamination (8).

When we first contemplated thermochemical products available from Glu, a search of the literature revealed no studies expressly directed at hydrogenation to a specific product. Indeed, the major role that Glu plays in hydrogenation reactions is to act as an enantioselectivity enhancer (17,18). Glu (or a number of other optically active amino acids) is added to solutions containing Raney nickel, supported nickel, palladium, or ruthenium catalysts and forms stereoselective complexes on the catalyst surface, leading to enantioselective hydrogenation of keto-groups to optically active alcohols. Under the reaction conditions used, no hydrogenation of Glu takes place.

Since the inception of our work Jere, Miller and Jackson have published kinetic and stereochemical data on the hydrogenation of alanine (19). Important in their analysis is the observation that amino acids must be in their protonated form to undergo facile hydrogenation since reduction of carboxylate anions is significantly more endothermic than protonated acids (19). Control of pH is important for two reasons: at neutral pH amino acids exist as zwitterions and the resultant hydrogenation products are basic. For these reasons a full equivalent of phosphoric acid (or similar acid) is required to obtain high yields.

$R\diagdown\diagup CO_2H$		$R\diagdown\diagup CO_2^-$		$R\diagdown\diagup CO_2^-$
$\overset{+}{N}H_3$	$\rightleftharpoons$	$\overset{+}{N}H_3$	$\rightleftharpoons$	NH_2
Protonated predominates		Zwitterion predominates		Carboxylate predominates
pH < 2		pH ~ 5 - 9		pH > 10

Jere et al. also reported on the stereoretentive C-H bond activation in aqueous phase catalytic hydrogenation of alanine (19). They demonstrated that hydrogenation of the carboxylic functionality is a stereo retentive process. Racemization occurs through a distinct process.

Also since the inception of our work Antons *et al.* reported on the reduction of several amino acids including Glu and Pyroglu (1). This study employed extremely high catalyst loadings of unsupported ruthenium and rhenium. The resultant products, glutaminol from Glu and 5HMP from Pyroglu, were achieved in fair yields (58 – 65%) with 98% enantiomeric excess—i.e. high optical purity (see

Figure **2**). However, the work clearly shows the need for catalyst improvement and leaves open the opportunity to identify other novel products. With these results in mind we can now turn to our preliminary work in this area.

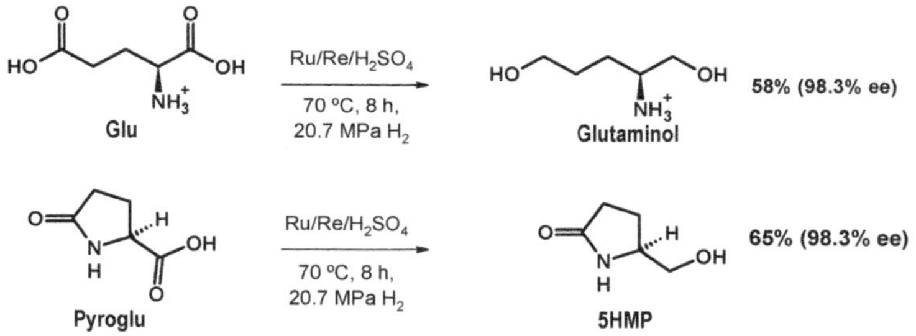

Figure 2 Hydrogenation of Glu and Pyroglu (ref. 1).

Experimental Section

Glutamic acid (Glu), monosodium glutamate, Pyroglu and 5HMP were purchased from commercial venders. Hydrogenation reactions were done in a 300 mL Parr autoclave using a continuous hydrogen gas feed at pressures and temperatures indicated. Reactant solutions contained 0.22 M substrate, and 0 – 0.29 M phosphoric acid in deionized water. Catalysts used were various precious metals on carbon supports as provided by Engelhard and Degussa or prepared in house and generally added at 1% loading based on total feed. Analysis was done either on a Waters HPLC with RID detector or an Agilent HPLC with UV detection using various columns. ^{13}C NMR spectra were taken on a VXR-300 MHz spectrometer at 75 MHz. MALDI Mass spectral data was obtained on a Perceptive Biosystems MALDI Mass Spectrometer.

Results and Discussion

Early in our work we discovered that at elevated temperatures Glu has a strong propensity to cyclize, even when dilute in neutral or acidic aqueous media. Although the reaction is thermal and does not require a catalyst the cyclization rate appeared to be enhanced in the presence of metal on carbon catalysts. Ring opening is possible but requires base (20). The propensity for cyclization which forms substituted pyrrolidinones provides additional possible products. Furthermore, it provides a potential method for selectively protecting the pendant carboxylic acid allowing one to selectively reduce the amino acid. Cyclization is slow at temperatures below 100 °C and was not noted in the Antons' work. However, for hydrogenations at 150 °C, essentially identical results were obtained whether starting with Glu or Pyroglu since the rate for cyclization was competitive with the rate for hydrogenation. Pyroglu is water soluble and hence easier to handle than Glu (Glu water solubility = 0.864 wt% at 25 °C, Merck Index).

The cyclization of Glu to Pyroglu changes the pH requirement since the amide nitrogen is not as basic as an amino group. In fact, as we shall see, the addition of acid promoted further chemistry that was unanticipated. This will be illustrated by the discussion of two key hydrogenation experiments. In these experiments Ru on carbon support was used. Pd, Pt and Rh were also tested but not as effective. The hydrogen pressure was maintained at 13.7 MPa. Lower pressures resulted in slower rates. The reactions were done in aqueous solution using 0.22 M Pyroglu as starting material, although essentially identical results were obtained with Glu. In **Experiment I** 0.29 M phosphoric acid was added. In **Experiment II** no additional acid was employed. The results of the two experiments are shown in Figure 3.

In **Experiment I** (presence of phosphoric acid) the starting material, Pyroglu, was rapidly consumed within two hours. 5HMP was initially formed and subsequently consumed. During the course of the reaction no other products were observed by the HPLC analytical method employed. By comparison, in **Experiment II** (absence of phosphoric acid), 5HMP was formed but not consumed, at least at the rate observed in **Experiment I** (Figure 3b). But, surprisingly, the reaction rate in **Experiment II** slowed after one hour, well before the starting amino acid was fully consumed. No further reaction was noted after 2 hours. Control experiments verified that the catalyst was still viable, i.e. that it had not been poisoned, and that it could be re-used to produce identical results. Analysis indicated that the pH of the reaction solution in **Experiment II** rose sharply from the initial pH of 3 to a final pH of 9. The increase of pH would be accompanied by ionization of Pyroglu. Thus we can account for the reaction inhibition over time. 5HMP, a neutral compound, could not account for the rise in pH. The possibility that ammonia was being released during the reaction, as observed by Broadbent et al. (8), was considered; however, analysis demonstrated this not to be the case. Incidentally, analysis of the reaction off-gas showed no significant formation of methane or other over-hydrogenation products in either experiment. Total organic carbon analysis showed that essentially no carbon was lost from solution. Whatever the product(s) of Experiment I or the basic compound(s) of Experiment II were, they were not being detecting by our HPLC method.

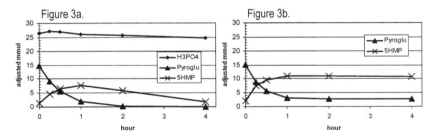

Figure 3   Hydrogenation of Pyroglu (similar results were obtained when starting with Glu).

^{13}C NMR was helpful in determining what had occurred in the two experiments. Reaction solutions from **Experiment I** and **II** were concentrated and examined by NMR without any further modification. The spectra are shown below (see Figure 4). ^{13}C NMR data showed that a single major product was formed in **Experiment I**. The unanticipated product was identified as prolinol (the minor peaks in the spectrum correspond to a small amount of 5HMP). Conversely, in **Experiment II** (no additional acid), the spectrum indicated a more complex mixture of compounds with the major product being 5HMP. The spectrum also contained resonances assigned to Pyroglu and prolinol, as minor components (21). ^{13}C NMR assignments were confirmed by comparisons with spectra of known compounds. MALDI mass spec data verified the presence of compounds that had the molecular weights of those identified by NMR. In addition it appears that very minor amounts of other cyclic amines were formed since molecular weights consistent with 2-methylpyrrolidine and pyrrolidine were also observed.

Although we did not anticipate the hydrogenation of the lactam carbonyl, this transformation is known in the literature with the earliest examples coming from the pioneering work of Adkins et al. using Cu-Cr oxides (22).

Hydrogenation of amides normally result in the formation of alcohols, whereas lactams give cyclic amines. The work presented in this paper is the first example that we are aware in which a lactam was hydrogenated by a ruthenium catalyst. To verify that acid promoted the lactam carbonyl hydrogenation, two

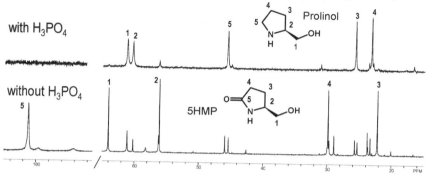

Figure 4 ^{13}C NMR data from concentrating the solutions from **Experiments I** and **II**.

controls were run under the conditions of **Experiments I** and **II** using 5HMP as the starting material. In the absence of phosphoric acid (neutral pH) 5HMP remained essentially unreacted (about 5% conversion). The small amount of reaction products consisted of hydrogenolysis products, 5-methyl-2-pyrrolidinone and 2-pyrrolidinone, rather than prolinol. Conversely, when the reaction was repeated in the presence of phosphoric acid, the substrate was quickly converted to prolinol in high yields (reaction time about 3 h).

A proposed mechanism that accounts for all the observations of **Experiment I** and **II** is shown in Figure 5. Under the reaction conditions employed Glu rapidly cyclizes to Pyroglu. Facile hydrogenation of Pyroglu produces 5HMP. Under neutral conditions 5HMP is stable and the lactam carbonyl is not appreciably hydrogenated. Under low pH conditions further reduction of the lactam carbonyl occurs in a mechanism consistent with that shown in Figure 5 (hydrogenation followed by anchimeric assisted water loss and a second hydrogenation) to afford prolinol. Prolinol is a basic compound (pK$_a$ ~ 11.3). In the presence of phosphoric acid, prolinol forms a phosphate salt, the amino acid remains protonated and goes to a full conversion. In the absence of phosphoric acid, Pyroglu itself is sufficiently acidic to promote a small amount of lactam hydrogenation. Prolinol forms a salt with the amino acid and the reaction halts before the Pyroglu is fully converted. Our initial struggles at identifying prolinol can be attributed to our HPLC method, which was not amenable to prolinol (as verified later by subjecting prolinol standards to the method). In **Experiment I** the final selectivity of 5HMP to prolinol was 1:98. In **Experiment II** the selectivity was 6:1 (~85% 5HMP). Due to the

limited resources for this exploratory work, determination of stereo-integrity of
the reaction was not completed and will have to be reported later.

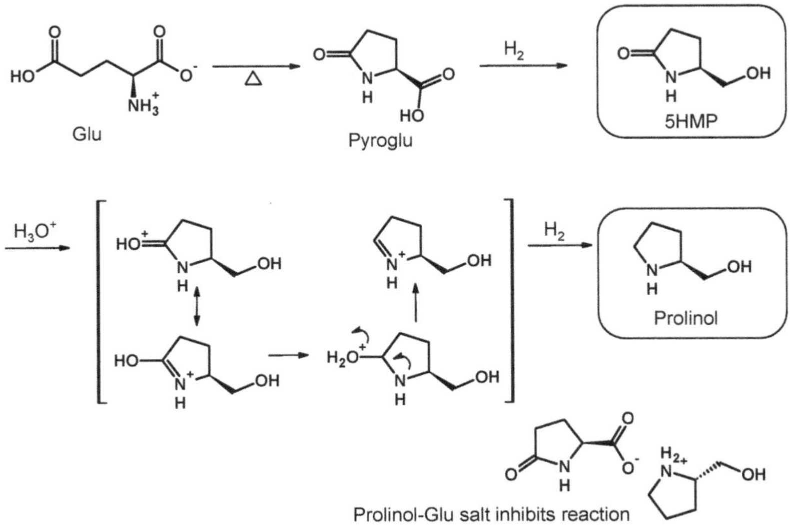

Figure 5 Proposed reaction mechanism, and products observed.

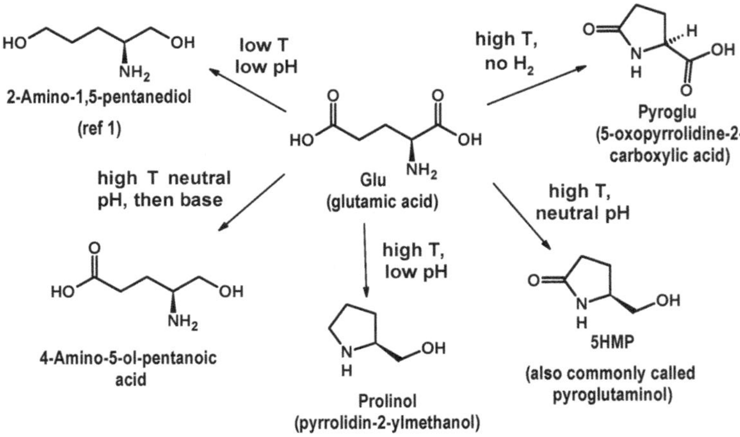

Figure 6 Products demonstrated from Glu.

In conclusion we have presented our initial findings of exploratory research
aimed at converting Glu to various products, including substituted
pyrrolidinones. The chemistry highlights the dual role of acid in amino acid

hydrogenation. Stoichiometric amounts of acid are generally required for high conversion of amino acids, initially to protonate the zwitterionic amino acid and then to neutralize the basic product. The caveat for Glu is that cyclization results in a lactam nitrogen that is significantly less basic than amines. In this case, the addition of acid surprisingly promotes the subsequent hydrogenation of the less reactive lactam carbonyl. By careful control of reaction conditions we were able to obtain 5HMP or prolinol selectively. Based on an observation that Pyroglu can be ring opened using base (20) we contemplate that 5HMP might also be subsequently ring opened to form 4-amino-5-hydroxy pentanoic acid. The various products are shown in Figure 6.

Acknowledgement

This work was funded under a Laboratory Directed Research and Development (LDRD) grant administered by Pacific Northwest National Laboratory (PNNL). PNNL is operated by Battelle for the US Department of Energy. We acknowledge the help of April Getty with ^{13}C NMR and Catherine Petersen with MALDI Mass Spectra. Keith Peterson assisted in reactor support and Tom Wietsma in analytical support. Finally we appreciate the help of Jim White and Dennis Miller for helpful discussions.

References

1. A. Antons, A. S. Tilling, and E. Wolters, US Pat. 6,310,254, to Bayer Aktiengesellschaft (2001).
2. Japan Chemical Week, April 24, 2003.
3. (a) E. Bowden and H. Adkins, *J. Am. Chem. Soc.*, **56**, 689 (1934). (b) H. Adkins and H. R. Billica, *J. Am. Chem. Soc.*, **70**, 3118 (1948). (c) H. Adkins and H. R. Billica, *J. Am. Chem. Soc.*, **70**, 3121 (1948).
4. S. Antons and B. Beitzke, U.S. Pat. 5,536,879, to Bayer Aktiengesellschaft (1996); S. Antons U.S. Pat. 5,731,479, to Bayer Aktiengesellschaft (1998).
5. W. Rachmady and M.A. Vannice, *J. Catal.*, **192**, 322 (2000).
6. M. A. N. Santiago, M. A. Sanchez-Castillo, R. D. Cortright, and J. A. Dumesic, *J. Catal.*, **193**, 16 (2000).
7. J. E. Carnahan, T. A. Ford, W. F. Gresham, W. E. Grigsby, and G. F Hager, *J. Am. Chem. Soc.*, **77**, 3766 (1955).
8. H. S. Broadbent, G. C. Campbell, W. J. Bartley, and J. H. Johnson, *J. Org. Chem.*, **24**, 1847 (1959).
9. K. Tahara, H. Tsuji, H. Kimura, T. Okazaki, Y. Itoi, S. Nishiyama, S. Tsuruya, and M. Masai, *Catalysis Today,* **28**, 267 (1996).
10. M. Toba, S. Tanaka, S. Niwa, F. Mizukami, Z. Koppany, L. Guczi, K.-Y. Cheah, and T.-S. Tang, *Appl. Catalysis A; General,* **189**, 243 (1999).
11. T. Turek and D. L. Trimm, *Catal. Rev. Sci. Eng.*, **36**, 645 (1994).
12. H. Nagahara, M. Ono and K. Nakagawa, JP Pat. 01085937 A2 19890330 to Asahi Chemical Industry Co. (1989).

13. A. Corry, UK Pat. GB 2169896 A1 19860723, to Imperial Chemical Industries (1986).
14. I. Iliuta, M. Bulearca, and L. Lazar, *Romanian Rev. Chim. (Bucharest)*, **46(8)**, 725 (1995).
15. F. Mesich, J. Bedford, and E. Dougherty, German Pat. DE 2131696 19711230 to Celanese Corp. (1971).
16. M. Bianchi, G. Menchi, F. Francalanci, F. Piacenti, U. Matteoli, P. Frediani, and C. Botteghi, *J. Organometal. Chem.*, **188**, 109 (1980).
17. G. Smith and M. Musoiu, *J. Catal.*, **60**, 184 (1979).
18. T. Osawa, T. Harada, and T. Akira, *J.Catal.*, **121**, 7 **(1990)**.
19. F. T. Jere, D. J. Miller, and J. E. Jackson, *Org. Lett.*, **5**, 527 (2003).
20. In a hydrogenation experiment starting with Pyroglu at 150 °C using two equivalents of base, Glu was formed as the sole product.
21. The ^{13}C NMR spectrum for **Experiment II** appeared to consist of two sets of resonances that corresponded to Pyroglu. We speculate this is due to a Pyroglu-prolinol salt adduct.
22. J. C. Sauer and H. Adkins, *J. Am. Chem. Soc.*, **60**, 402 (1938). B. Wojcik and H. Adkins, *J. Am. Chem. Soc.*, **56**, 247 (1934).

19.   Catalytic Hydrogenolysis of 5-Carbon Sugar Alcohols

Alan H. Zacher,[a] **John G. Frye, Jr.,**[a] **Todd A. Werpy,**[a] **and Dennis J. Miller**[b]

[a]*Pacific Northwest National Laboratory, Richland WA 99352*
[b]*Department of Chemical Engineering, Michigan State University, East Lansing, MI 48824*

PNNL-SA-39754

Abstract

PNNL, in cooperation with the USDOE and CRADA partners, National Corn Growers Association and Archer Daniels Midland, has evaluated a new class of catalysts based on nickel and rhenium with effective performance for highly selective, high conversion hydrogenolysis of five–carbon sugar alcohols to industrially useful glycols. The Ni/Re catalyst appears to exhibit preferential reductive cleavage of the carbon-carbon bonds of secondary carbons over primary carbons of the 5-carbon sugar alcohols tested. In addition, the catalyst has demonstrated the ability to produce glycerol and 1,2-propylene glycol in a controllable ratio. The rhenium containing catalysts are found to have higher activity and better selectivity to desired glycols than previously reported catalysts. A continuous flow reactor lifetime test of over 1500 hours also demonstrated the requisite high stability for an industrially attractive process.

Introduction

Use of renewable resources for production of valuable chemical commodities is becoming a topic of great national interest and importance. Catalytic hydrogenolysis of sugar alcohols to glycols fits this goal. However, previous demonstrations of the chemistry mainly has been focused on six carbon sugar alcohols [1,2,3,4,5]. Five carbon sugars and their alcohols, e.g. xylitol and arabitol, while merely mentioned in many of these patents, are an attractive alternative to six carbon sugar alcohols as they are potentially inexpensive materials available from controlled hydrolysis of very cheap corn fiber and other agricultural wastes.

Catalytic hydrogenolysis of arabitol or xylitol under basic aqueous conditions forms industrially useful glycols, generally a mixture of 1,2-propylene glycol (PG), glycerol, and ethylene glycol (EG). While this chemistry has been previously demonstrated [1,2,3,4,5], the catalysts and processes discussed in previous literature have not exhibited the high degree of activity and selectivity which would make production of these glycols from sugars or sugar alcohols economically attractive.

PNNL under a Cooperative Research and Development Agreement (CRADA) with the US Department of Energy Office of Energy Efficiency and Renewable Energy, the National Corn Growers Association (NCGA), and Archer Daniels Midland Company (ADM) recently examined the production of five carbon sugar alcohols from hemicellulose in corn gluten feed material [6]. The hydrolysis conditions hemicellulose is converted to xylose and arabinose and various water soluble oligomers. The sugars are then hydrogenated using well established processes to form xylitol and arabitol [7]. This work served as a launch pad for our investigation of 5-carbon sugar alcohols as a feedstock.

The scope of this work was to determine a catalyst preparation that would demonstrate the best yield to value added products to verify and complete this portion of the overall flowsheet from corn fiber. The scope did not include mechanistic studies, which will be completed at a later date if resources allow.

We evaluated a number of potential catalysts and conditions using xylitol as a model compound in a batch reactor. A catalyst was selected from this initial screening and examined in a continuous trickle-bed reactor to develop operating conditions. Finally, as resources allowed, the catalyst was evaluated in a trickle bed reactor to gain a concept of potential catalyst lifetime.

The use of 5-carbon sugar alcohols to produce glycols presents specific advantages over the use of 6-carbon sugar alcohols, such as sorbitol. Current and historical technology has demonstrated that internal C-C hydrogenolysis in basic aqueous solution is favored over primary C-C bond scission. This is based on the demonstrated product composition favoring 2 and 3-carbon hydrogenolysis products versus a theoretical product composition generated by the probability of equally likely C-C scission which would result in a product stream more resembling equimolar yields of 1, 2, 3, 4, and 5 carbon hydrogenolysis products [1,2,3,4,5]. With xylitol, this results in excellent selectivity as every internal C-C hydrogenolysis can result in a molecule of EG and a 3-carbon polyol (glycerol or PG) avoiding the production of erythritol or threitol as low value intermediates. In this work, a catalyst composition was developed that resulted in a satisfactory yield of mixed polyols.

Experimental Section

Catalyst Preparation

Initial catalysts were prepared by impregnation of metal salts on carbon supports using the method of insipient wetness. The supports were obtained from Calgon Carbon Corporation and consisted of a highly activated coconut-shell activated carbon with a Carbon Tetrachloride number (CTC) of 120%. Following impregnation, the catalysts were vacuum dried. Reduction of chloride containing catalysts was performed in a tube furnace with 4% H_2 in an inert gas. Reduction of the non-chloride containing catalysts and pre-reduced chloride catalysts was performed in the autoclave under 20% H_2 in an inert gas for 4 hours immediately prior to the introduction of the feedstock.

Catalysts used in the trickle bed reactor were supplied by Engelhard Chemical Catalyst Group prepared industrially to match in-house prepared catalysts used in our initial screening.

From previous experience with similar reactions, we chose the following catalysts for screening: 2.5% ruthenium (Ru), 2.5% ruthenium / 0.5% rhenium (Ru/Re), 2.5% ruthenium / 2.5% rhenium (Ru/Re+), 2.5% nickel / 2.5% rhenium (Ni/Re), 2.5% rhodium / 2.5% rhenium (Rh/Re), and 2.5% palladium / 2.5% rhenium (Pd/Re).

Catalyst Screening with Xylitol – Batch Reactor

Batch catalyst testing was performed in a 300-cc stainless steel Parr Autoclave. Xylitol was used for the model compound testing due to relative availability. Generally the batch reactor was loaded with 2.5g of dry catalyst and the catalyst was reduced in 20% H_2 in inert gas. Following reduction the reactor headspace was evacuated, covered with N_2, evacuated again, and then 105g of aqueous liquid feedstock consisting of 20wt% reagent grade xylitol and 1wt% NaOH was vacuum drawn into the batch reactor. The headspace was exchanged with H_2 and pressurized to 3400kPa prior to heating. Once the reactor was raised to operating temperature, the H_2 valve was opened whilst regulated to the desired operating pressure in semi-batch configuration, allowing the reactor to uptake H_2 and maintain the pressure. Catalysts were tested at 160°C to 240°C in 20° increments under 8300kPa pressure. Samples were taken upon reaching reaction temperature (t=0) and at t=1, 2, 3, and 4 hours after reaching temperature.

Comparison of Xylitol and Arabitol – Batch Reactor

The C-5 sugar alcohols produced from the hydrolysis of hemicellulose are both xylitol and arabitol [6]. Equivalence testing was performed with Ni/Re catalyst in the batch reactor to verify similar performance between xylitol and arabitol feedstocks. The operating conditions were 200°C and 8300kPa H_2 using the procedure outlined in section Catalyst Screening section.

Xylitol Testing in Trickle Bed Reactor

Continuous testing was performed in a 40cc Autoclave Engineering (AE) downflow trickle bed reactor with cocurrent H_2 flow. The bed was packed with 13g of Ni/Re on granular activated carbon occupying 35ml. The feed was 20% xylitol with 1% NaOH in water. The catalyst was reduced in situ with overnight H_2 flow at 280°C. Conditions included: liquid feed 25 to 350 ml/min, temperature of 140°C to 220°C, pressure of 5,500 to 11,000kPa, 5:1 H_2 to xylitol molar ratio, and base loading of 0.1% to 1% NaOH. The reactor was operated for a total of 240 hours. Lifetime testing was performed in the same reactor using this same catalyst under a related aqueous basic hydrogenation process for 1700 hours.

Analytical Procedures

Products were analyzed via Waters Model 515 HPLC Pump fitted with a Waters model 2410 refractive index detector. Separations was performed via an Aminex HP-87H 300mm column at 65°C using 0.005M H_2SO_4 as the mobile phase. Compounds calibrated for this work included xylitol, arabitol, erythritol, threitol, PG, EG, glycerol, lactate, 1-propanol, 2-propanol, ethanol, methanol, and the butanetriol isomers. Any compounds not visible by RID were not quantified in this work.

Results and Discussion

Catalyst Screening

Comparison of the catalysts in Figure 1 demonstrates that the Ni/Re catalyst yielded the highest activity of the catalysts under the conditions tested. The Rh/Re catalyst also showed excellent conversion at 240°C as did the Ru catalyst which is also known to be an excellent hydrogenolysis catalyst for conversion of sorbitol to these products [4,5]. The Ru/Re and Ru/Re+ catalysts also demonstrated high conversion, especially at high temperatures.

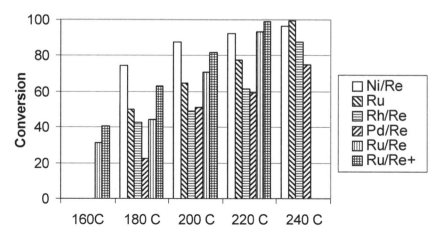

Figure 1 Comparison of Xylitol Conversion versus Temperature, 300cc Batch Reactor, 8,300kPa H_2, 4-Hour Sample.

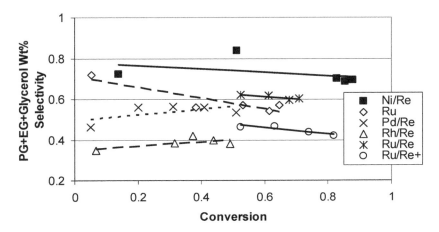

Figure 2 Conversion versus EG+PG+Glycerol Wt% Selectivity at 200°C, 300cc Batch Reactor, 8,300kPa H₂.

However, the product polyols mix selectivity is more critical than high feed conversion to overall process economics. The Ni/Re catalyst had the highest selectivity to PG, EG, and glycerol as shown in Figure 2. "Wt% Selectivity" is defined as weight of designated product measured divided by the weight of xylitol consumed. Overall, the lowest selectivity to the desired products was produced by the Rh/Re and the Ru/Re+ catalysts. While these two catalysts had very high xylitol conversions, they produced products other than PG, EG or glycerol. At temperatures above 200°C some of the desired polyol products were unstable on these catalyst and overall product selectivites degraded rapidly as conversion increases. Based on high conversion and desired yield, the Ni/Re catalyst was chosen for continuous testing.

Validity of Model Compound

Arabitol testing was performed to ensure that it behaved similarly to xylitol in this reaction to validate xylitol as a model compound. Testing was performed using Ni/Re catalyst from the initial batch screening. Shown in Table 1, the results from the two sugar alcohols were nearly equivalent. This gave some confidence that xylitol should be a valid model compound in the absence of actual hemicellulose derived feedstock.

Table 1 Comparison of Xylitol and Arabitol feed 4-Hour Sample on 2.5% Nickel / 2.5% Rhenium catalyst, 200°C, 300cc Batch Reactor, 8,300kPa H₂

Feed	Conv.	PG	Glycerol	EG	lactate	EtOH	PrOH	Other
Xylitol	88%	22%	16%	31%	9%	2%	1%	4%
Arabitol	86%	21%	13%	31%	11%	0%	0%	3%

Continuous Catalyst Evaluation

Selected conditions and results are shown in Table 2 that are representative of the catalyst performance. Continuous testing of the Ni/Re catalyst compared favorably with the baseline data generated for this catalyst in the batch reactor screening. At 200°C, the overall activity of the catalyst appeared slightly higher in the continuous reactor, achieving 94% conversion at a weight hourly space velocity of 2.5hr^{-1} (g xylitol/g catalyst/h) and 200°C compared to 88% conversion at an equivalent exposure in the batch reactor of 2.1hr^{-1} (g xylitol/g catalyst/h) achieved at the 4 hour sample at 200°C.

Table 2 Selected Conditions from Continuous Xylitol Testing, 13g of Ni/Re catalyst, aqueous 20.4% Xylitol 1% NaOH Feed, 35cc catalyst bed, 8,300kPa H$_2$, 5:1 H$_2$ to Xylitol molar ratio.

Temperature (°C)	140	160	170	170	180	180	200	200	200
Flow (cc/h)	25	50	25	50	25	50	50	150	200
WHSV (hr^{-1})	0.4	0.8	0.4	0.8	0.4	0.8	0.8	2.5	3.3
Conversion (%)	20.0	44.9	92.7	74.5	98.4	94.5	98.3	93.7	85.1
Product pH	12.6	12.5	12.5	12.1	12.1	11.7	n/m	n/m	n/m
Wt% Selectivities (%) PG	10.2	16.8	26.1	21.9	30.3	28.1	30.8	30.8	29.8
EG	24.5	26.7	28.2	28.0	28.6	29.2	27.9	29.7	35.2
Glycerol	22.3	23.2	13.6	16.9	8.6	11.3	7.4	8.2	9.7
Lactate	32.7	20.9	6.4	11.0	3.5	5.0	3.7	3.8	4.3
C4-itol*	1.9	2.6	1.8	2.1	1.7	1.6	2.1	1.7	1.7
Butanediol	0	0	2.1	1.0	2.5	2.5	2.2	3.0	2.5
Methanol	0	0	1.0	0	1.9	0.8	0	1.4	1.8
Ethanol	0	1.0	3.2	2.4	3.9	3.1	3.4	3.0	2.7
Propanols	0	1.0	3.1	2.0	3.1	2.8	2.3	2.3	2.2
Butanetriol	1.1	0	0.4	0.7	0.3	0.4	0.3	0.2	0.3

* "C4-itol" is defined as combined erythritol and threitol.

It is observed that more lactate was formed at conditions below 160°C and more PG was formed above 160°C. It is known that hydrogenation of lactate salts to PG is normally very difficult demonstrating that lactate is not an intermediate to PG in this basic reaction [8]. This indicates that study of the reaction pathway will be necessary to control and favor the obviously complex reaction that forms the desired PG product. However, the impact to the overall hemicellulose hydrolysis of the less desired lactate product can be mitigated, as the sodium lactate could be ion exchanged into the acid form and easily hydrogenated with high selectivity to PG over an appropriate catalyst thereby greatly improving the overall selectivity to PG, currently the highest value product from the reaction [8].

Batch and continuous data both indicated little or no change in the combined glycol product selectivity occurred on the Ni/Re catalyst as xylitol conversion increased while operating between 160°C and 200°C, 8,300kPa H_2, and weight hourly space velocities between 0.4 and 3.3 (hr^{-1}). Figure 3 shows that the selectivity remained mostly constant for EG and the combined glycol product even while the individual selectivities of PG and glycerol are changing as shown in Figure 4.

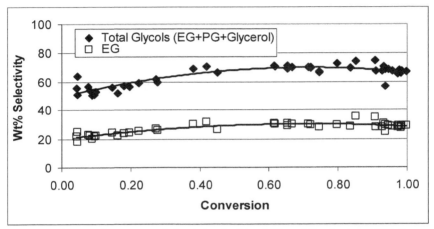

Figure 3 Conversion versus Selectivity for Total Glycol Product and EG from Continuous Xylitol Testing, 13g of Ni/Re catalyst, aqueous 20.4% Xylitol 1% NaOH Feed, 35cc catalyst bed, 8,300kPa H_2, 5:1 H_2 to Xylitol molar ratio.

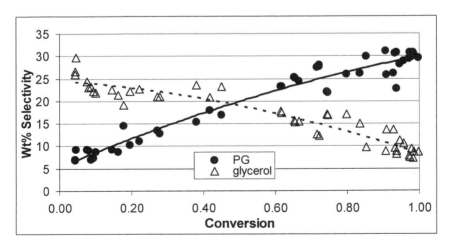

Figure 4 Conversion versus Weight Selectivity for Glyerol and PG from Continuous Xylitol Testing, 13g of Ni/Re catalyst, aqueous 20.4% Xylitol 1% NaOH Feed, 35cc catalyst bed, 8,300kPa H_2, 5:1 H_2 to Xylitol molar ratio.

The conversion versus selectivity plot in Figure 4 demonstrates that the process can generate a controlled ratio of PG and glycerol over the course of the reaction. By terminating the reaction at a given point, the ratio of PG and glycerol will be a predictable value. In the event of a change in market forces this property allows the ability to produce either C3 glycol to match industrial demands.

Long-Term Catalyst Stability

The Ni/Re on carbon catalyst was also evaluated in a 1700 hour continuous reactor test to determine the stability of the catalyst. This test was performed with a different model compound than xylitol. Shown in Figure 5, the results from the lifetime test of the Ni/Re catalyst operated at constant process conditions sampled intermittently for 1700 hours. This shows that for a similar aqueous hydrogenation reaction deliberately operated to near completion, the catalyst retained its activity and product selectivity even in the face of multiple feed and H_2 interruptions. We feel that this data readily suggests that the Ni/Re catalyst will retain its activity for xylitol hydrogenolysis.

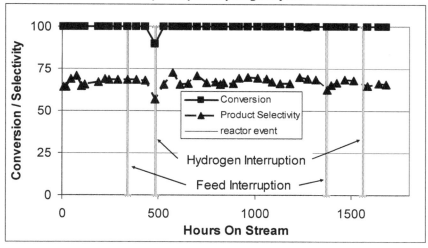

Figure 5 Conversion and Product Selectivity for Ni/Re Catalyst over 1700 hours, aqueous hydrogenation condition (unspecified).

Acknowledgements

This research was supported by the U.S. Department of Energy Office of Energy Efficiency and Renewable Energy in a Cooperative Research and Development Agreement with the National Corn Growers Association (NCGA) and the Archer Daniels Midland Company (ADM). Special thanks to Dr. James White of Engelhard for supplying catalyst and research consultation. Thanks to

Gary Neuenschwander, Keith Peterson, Mark Butcher, Eric Alderson, and Danielle Muzatko for excellent technical and analytical support.

References

1. A.W. Larchar, U.S. Patent 1,963,997 (1934).
2. H.S. Rothrock, U.S. Patent 2,004,135 (1935).
3. F. Conradin, G. Bertossa, and J. Giesen. U.S. Patent 2,852,570 (1958).
4. G. Gubitosa and B. Casale, U.S. Patent 5,326,912 (1994).
5. S. P. Chopade, D. J. Miller, J. E. Jackson, T. A. Werpy, J. G. Frye Jr., and A. H. Zacher, U.S. Patent 6,291,725 (2001).
6. K. Beery, C. Abbas, T. Werpy, and R. Orth, Oral Presentation 5-01, 24th Symposium on Biotechnology for Fuels and Chemicals, (2002).
7. D. C. Elliot, T. A. Werpy, Y. Wang, and J. G. Frye Jr., U.S. Patent 6,235,797 (2001).
8. Z. Zhang, J. E. Jackson, and D. J. Miller, *Applied Catalysis A: General*, **219**, 89-98 (2001).

II. Symposium on Novel Concepts and Approaches to Catalysis of Organic Reactions

a. Featured Technology Area: Combinatorial and Parallel Methods in Catalyst Design, Optimization and Utilization

20. Combinatorial Approaches for New Catalyst Discovery: The First Homogeneous Catalysts Derived from Combinatorial Technologies for a Variety of Cross-Coupling Reactions

George Y. Li

CombiPhos Catalysts, Inc., P.O. Box 220, Princeton, NJ 08542-0220

George.Y.Li@combiphos.com

Abstract

Palladium chlorides possessing phosphinous acid ligands (**POPd, POPd1** and **POPd2**) were found to be remarkably active and efficient catalysts in the presence of bases for a variety of cross-coupling reactions of aryl chlorides with arylboronic acids, olefins, amines, and thiols. ^{1}H and ^{31}P NMR studies argue that these phosphinous acid ligands in the complexes can be deprotonated to yield electron-rich anionic species, which is anticipated to accelerate the rate determining oxidative addition of aryl chlorides in the catalytic cycle. Herein, we report the first examples of these isolated air-stable palladium complexes served as efficient catalyst precursors for C-C, C-N, and C-S bond-forming reactions of aryl chlorides.

Over 100 pharmaceutical companies in the world have been applying these DuPont's air-stable coupling catalysts for the drug discovery and processes since 2001.

Introduction

Combinatorial chemistry has been developed and applied by the pharmaceutical industry to rapidly synthesize and screen thousands of compounds for the drug discovery and optimization in recent years. Combinatorial approaches have also led to the discovery of more efficient superconducting, photoluminescent, magnetoresistive, dielectric, and polymeric materials, and the development of heterogeneous catalysts. However, the combinatorial searches for homogeneous catalysts are often obstructed by inefficient synthetic and screening methods. In general, combinatorial synthesis of homogeneous ligands using solid-supported methods to facilitate the discovery process of new catalysts has been less successful. Phosphines of the formula R'R''R'''P have been used as powerful ligands in homogeneous catalysis for many years. However, these phosphines with formal valence three are notoriously susceptible to oxidation/deterioration, and difficult to store and handle. Thus, despite some applications in metal-catalyzed organic transformations using these compounds as ligands, the discovery and development of efficient, air- and moisture-stable ligands/catalysts are very important for industrial applications.

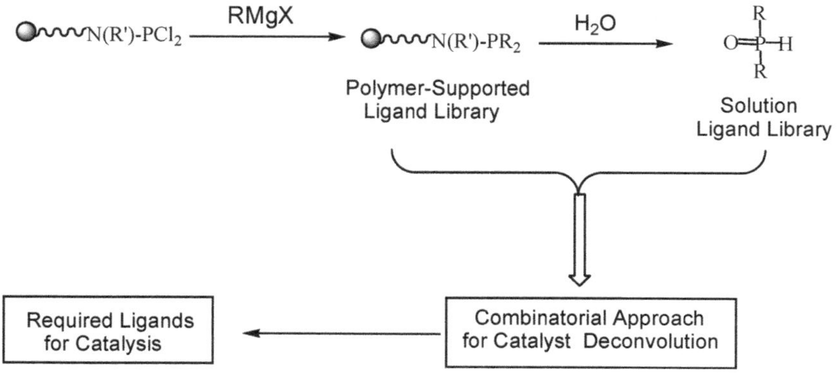

Scheme 1. Combinatorial Approaches for Catalyst Discovery

We successfully developed combinatorial approaches for the discovery of industrial ligands and/or catalysts [Scheme 1].[1] Air-stable phosphine oxides with formal valence five in the presence of transition metals undergo tautomerization to the less stable phosphinous acids, which subsequently and

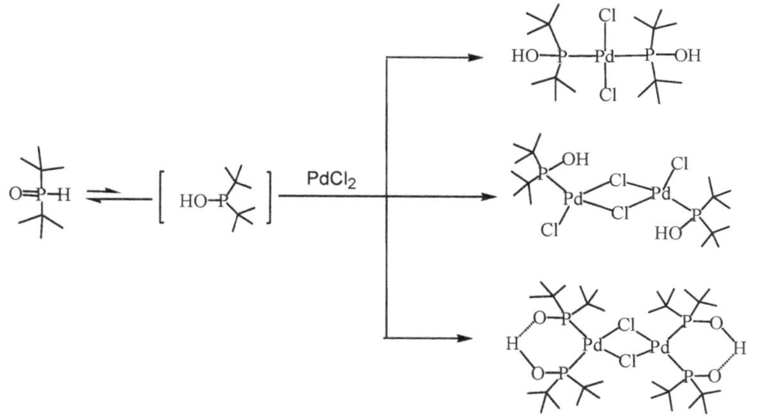

Scheme 2. Air-stable coupling catalysts derived from combinatorial approaches

expectedly coordinated to the metal centers through phosphorus atoms to form metal-phosphinous acid compounds [Scheme 2].[2,3]

The generated palladium chlorides possessing phosphinous acid ligands were found to be remarkably active and efficient catalysts in the presence of bases for a variety of cross-coupling reactions of aryl halides with arylboronic acids,[4] olefins,[5] amines,[4a] thiols,[4a,5a] Grignard reagent,[6] and tin reagent,[7] Herein, we summarize some examples of these isolated air-stable palladium complexes served as efficient catalyst precursors for C-C, C-N, and C-S bond-forming reactions of aryl halides. Initial observations regarding scope and mechanism are also discussed.

Results and Discussion

The **POPd**, **POPd1** and **POPd2** catalysts were prepared according to our reported procedures.[1,2] Well-characterized examples of this type of Pd(II) complexes are rare. **POPd**, **POPd1** and **POPd2** were isolated and characterized by 1-D, 2-D $^1H/^{13}C$, and ^{31}P NMR, as well as elemental analysis. The structures of **POPd1**, **POPd2** were confirmed by X-ray crystallography. As expected, the core structures of the **POPd1** and **POPd2** are P-Pd coordination with phosphinous acid ligands (RR'P-OH).

In contrast to recently reported highly active Pd/ArP(*t*-Bu)$_2$ and Pd/P(*t*-Bu)$_3$ systems for cross-coupling reaction,[8] a number of aspects of these air-stable **POPd**, **POPd1** and **POPd2** complexes are noteworthy and offer informative parallels and contrasts to the Pd/phosphine catalysts which are generated *in-situ* for cross-coupling of aryl chlorides. (Note: isolated phosphine complex Pd(P(*t*-Bu)$_3$)$_2$ does not appear to be the active catalyst for many cross-coupling processes.[8b]) These isolated, air- and moisture-stable **POPd**, **POPd1** and **POPd2** catalysts are all capable of catalyzing coupling reactions of a variety of aryl chlorides to yield the corresponding coupled products in high isolated yields. It can be seen that arylboronic acids (Table 1, entries 1-2),[4] amines (Table 1, entries 3-4),[5a] thiols ((Table 1, entries 5-8)[4a,5a] olefins (Table 1, entries 9-10),[5] Grignard reagent (Table 1, entries 11-12),[6] zinc reagent (Table 1, entries 13-14),[4a] and tin reagent (Table 1, entry 15)[7] can all be effected to yield a variety of coupled products. The present processes are effective for the coupling reactions of both electron-rich (Table 1, entries 1, 2, 11, and 12) and sterically hindered (Table 1, entries 1, 2, 12, and 15) aryl chlorides.

In regard to comparisons of catalytic effects of **POPd**, **POPd1** and **POPd2**, Table 1 offers data comparing these catalysts to those[8a] which are generated *in-situ*. It can be seen that the transformations are better or equally effective under comparable conditions of catalyst, concentration, and temperature. Interestingly, and in marked contrast to the typical catalytic systems employing excess free ligands to prevent catalysts from degradation in the catalytic processes, the present results (Table 1, entries 1 and 2) demonstrate that the air-stable Pd-complexes (**POPd**) for Suzuki cross-couplings of aryl chlorides are the same efficiency as those catalytic systems with excess ligands. More importantly, from the standpoint of organic synthesis, the direct use of **POPd**, **POPd1** and **POPd2** as catalysts for cross-couplings of aryl chlorides would be more practical owing to the difficulties in generating *in situ* catalysts and in handling extremely air-sensitive phosphine ligands. Entries 5-8 of Table 1 reveal that the organic sulfur compounds are tolerated by the present complexes as demonstrated by 1-methyl-1-propanethiol, 1-hexanethiol, and thiophenol.

With regard to a plausible mechanism, ^{31}P NMR studies of the reaction of **POPd2** with CsF in CD_3OD solution argue that in the presence of bases, the chloro-bridged **POPd2** is subject to cleavage giving mononuclear species,[9] and yielding an electron-rich phosphine-containing anionic complex,[10, 11] which is anticipated to accelerate the rate-determining oxidative addition of aryl chlorides in the catalytic cycle.[12] Further evidence for this pathway derives from the reaction of **POPd** with Et_3N in CH_2Cl_2, in which the **POPd** complexes are deprotonated to yield anionic dimer **POPd1** (Eq 2).

POPd (2 Isomers)

POPd1

(2)

In summary, these results demonstrate that air-stable **POPd**, **POPd1** and **POPd2** complexes can be directly employed to mediate the rate-limiting oxidative addition of unactivated aryl chlorides in the presence of bases, and that such processes can be incorporated into efficient catalytic cycles for a variety of cross-coupling reactions. Noteworthy are the efficiency for unactivated aryl chlorides; simplicity of use, low cost, air- and moisture-stability, and ready accessibility of these complexes. Additional applications of these air-stable palladium complexes for catalysis are currently under investigation.

Experimental Section

General procedure for Suzuki coupling reaction. A 100 mL of round-bottomed flask equipped with magnetic stir bar was charged with [(*t*-Bu)$_2$POH]$_2$PdCl$_2$ (POPd) (25 mg, 0.05 mmol), 2-chloroanisole (1.43 g, 10.0 mm), $C_6H_5B(OH)_2$ (1.83 g, 15.0 mm) and K_2CO_3 (4.15 g, 30.0 mmol) in 10 mL of THF. After the mixture was refluxed for 12 h, the reaction mixture was then cooled to room temperature, quenched with 50 mL of H_2O, and extracted with 300 mL of diethyl ether. The organic extracts were washed with H_2O (2 X 100 mL), brine (100 mL), and dried over $MgSO_4$, filtered, and the ether and THF removed from the filtrate by rotary evaporation. The resulting residue was chromatographed on silicon gel using 5% ethyl acetate/hexane as eluant. The eluate was concentrated by rotary evaporation followed by high vacuum to give 1.43 g (78 % yield) of 2-phenylanisole. It was >95% pure by ^{1}H NMR and GC/MS.

Table 1 Air-stable palladium-catalyzed cross-coupling reactions.

Entry	Halide	Coupling Partner	[Catalyst]	Base (Solvent)	Product	Yield (%) (Isolated)
1	Cl, OMe	⬡–B(OH)$_2$	POPd (0.5 mol%)	K$_2$CO$_3$ (THF)	OMe	78
2	Cl, OMe	⬡–B(OH)$_2$	POPd (0.5 mol%) (*t*-Bu)$_2$P(H)O (0.5 mol%)	K$_2$CO$_3$ (THF)	OMe	82
3	⬡–Br	indole (NH)	CombiPhos-Pd6[13] (5 mol %)	KOH (Toluene)		80
4	⬡–Cl	H$_2$N–⬡–	POPd (5.0 mol%)	NaO(*t*-Bu) (dioxane)	⬡–NH–⬡–	97
5	⬡–Cl	⋎–SH	POPd (5 mol%)	NaO(*t*-Bu) (DMSO)	⬡–S	70
6	⬡–Cl (N)	⋀⋀–SH	POPd1 (2.5 mol%)	NaO(*t*-Bu) (Toluene)	S⋀⋀	97
7	⬠–Cl	HSPh	POPd (3.0 mol%)	NaO(*t*-Bu) (Toluene)	⬠–SPh	88
8	⬠–Cl	⋀⋀–SH	POPd (3.0 mol%)	NaO(*t*-Bu) (Toluene)	⬠–S	97
9	⬡–Br	acrylate O-*t*-Bu	POPd2 (0.4 mol %)	K$_2$CO$_3$ (DMF)	Ph–CH=CH–CO$_2$*t*-Bu	64
10	Ac–⬡–Cl	acrylate O-*t*-Bu	POPd1 (1.5 % mol)	K$_2$CO$_3$ (DMF)	Ac–⬡–CH=CH–CO$_2$*t*-Bu	77
11	MeO–⬡–Cl	⬡–MgCl	POPd	THF	MeO–⬡–⬡	87
12	MeO, ⬡–Cl	⬡–MgCl	POPd2	THF	MeO	88
13	⬡–Cl	⬡–ZnCl	POPd (5 mol%)	THF/NMP		83
14	Ac–⬡–Cl	⬡–ZnCl	POPd2 (2 mol%)	THF/NMP	Ac–⬡–⬡	45
15	Cl, CN	⬡–SnMe$_3$	POPd (6 mol%)	Cy$_2$NMe (H$_2$O)	Ph, CN	62[7]

General procedure for C-N bond formation. A 50 mL of reactor was charged with 250.5 mg (0.50 mmol) of **POPd**, 1.27 g (10.0 mmol) of 4-chlorotoluene, 1.29 g (12.0 mmol) of 4-methylaniline and 1.35 g (14.0 mmol) of NaO(t-Bu) in 15 mL of dioxane. The resulting mixture was refluxed for 4 h before the reaction was cooled to room temperature and quenched with 50 mL of H_2O. The mixture was diluted with 300 mL of diethyl ether. The layers were separated, and organic layer was washed with H_2O (2 X 30 mL), brine (30 mL), and dried over $MgSO_4$, filtered, and the ether removed from the filtrate by rotary evaporation. The resulting residue was chromatographed on silica gel with hexane/ethyl acetate (50 : 1 volume ratio). The eluate was concentrated by rotary evaporation followed by high vacuum to give 1.92g (97% yield) of 4-methyl-N-(4-methylphenyl)benzenamine.

General procedure for C-S bond formation. A 50 mL of reactor was charged with 232.5 mg (0.25 mm) of **POPd1**, 1.36 g (12.0 mmol) of 3-chloropyridine, 1.18 g (10.0 mmol) of 1-hexanethiol and 1.92 g (20.0 mmol) of NaO-tBu in 15.0 mL of toluene. The resulting mixture was refluxed for 16 h before the mixture was cooled to room temperature and quenched with 100 mL of H_2O. The mixture was transferred to a separatory funnel, and extracted with EtOAc (2 X 200 mL). The layers were separated, and organic layer was washed with H2O (100 mL), brine (150 mL), and dried over $MgSO_4$, filtered, and the solvents removed from the filtrate by rotary evaporation. The final product was chromatographed on silica gel using ethyl acetate/hexane (5% volume ratio) as eluant. The eluate was concentrated by rotary evaporation to yield 1.90 g (97% yield) of 3-hexylthiopyridine.

General procedure for Heck coupling. A mixture of **POPd** (16.0 mg, 6 mol%), quinoline derivative (0.56 mmol), *tert*-butyl acrylate (356 mg, 2.8 mmol), and base (0.61 mmol) was stirred in 5 mL of anhydrous DMF at 135 °C for 24 h. The reaction mixture was allowed to cool to room temperature, quenched with water, and extracted with Et_2O. The combined organic layers were washed with water, dried over $MgSO_4$, and the solvents were removed under vacuum. The crude products were purified by flash chromatography on silica gel.

General procedure for the Negishi coupling. A solution of *o*-tolylmagnesium chloride (15 mL, 15 mmol, 1.0 M solution in THF) is treated with $ZnCl_2$ (32 mL, 16 mmol, 0.5 M solution in THF) dropwise over a period of 5 min. The resulting mixture was stirred at room temperature for 20 min before 1-methyl-2-pyrrolidinone (NMP, 22 mL) was added. After the mixture was stirred for 5 more min., 250.0 mg (0.5 mm) of **POPd**, 4-chloroacetophenone (1.55 g, 10 mmol) were added into the mixture above. The resulting mixture was refluxed for 2 h before it was cooled to room temperature, and aqueous HCl was added (1.0 M, 60 mL). The resulting mixture was extracted with Et_2O (4 X 100 mL), and the extracts were washed with H_2O (5 x 100 mL), and 100 mL of brine, then dried over $MgSO_4$, filtered and concentrated by rotary evaporation. The crude product was purified by column chromatography on silica gel (95 : 5-hexane : diethyl ether) to afford 0.95 g (45% yield) of required product.

General procedure for the reaction of aryl chlorides with Grignard reagents. A 50 mL of reactor was charged with 25.0 mg (0.05 mm) of **POPd**, 1.43 g (10.0 mm) of 4-chloroanisol and 10.0 mL of THF. The resulting mixture was stirred and then addition of 15 mL (15.0 mm, 1.0 M in THF) of o-tolylmagnesium chloride was made at room temperature over a period of 5 min. The resulting mixture was stirred at room temperature for 4 h before the reaction was quenched with 10.0 mL of H_2O, and the mixture was diluted with 300 mL of Et_2O. After separation of organic and aqueous phases, the organic phase was washed with 50 mL of H_2O, and 50 mL of brine, then dried over $MgSO_4$, filtered and concentrated by rotary evaporation. The crude product was purified by column chromatography on silica gel using t-butylmethylether/hexane (1% volume ratio) to afford 1.72 g (87% yield) of 4-(*o*-tolyl)anisole.

References

1. For combinatorial approaches to the synthesis of phosphine ligands, see:
   (a) G. Y. Li, P. J. Fagan, and P. L. Watson, *Angew. Chem. Int. Ed. Engl.*, **40**, 1106-1109 (2001).
   (b) G. Y. Li, (DuPont). US patent application, 60/197,031, Case Number: BC 1048 US PRV, (2000).
   (c) P. J. Fagan and G. Y. Li, PCT Int. Appl WO 2000021663 A2 20000420, 169 pp. APPLICATION: WO 1999-US23509 19991013.
2. G. Y. Li, *Angew. Chem. Int. Ed. Engl.*, **40**, 1513-1516 (2001).
3. For tautomerization of phosphine oxides and phosphinous acids, for an example see: D. E. C. Corbridge, *Phosphorus*, 5th Ed., Elsevier Science B. V., the Netherlands, 1990, Chapter 4.
4. (a) George Y. Li, *J. Org. Chem.*, **67**(11), 3643-3650 (2002).
   (b) W. Yang, Y. Wang, and J. R. Corte, *Org. Lett.*, **5**(17), 3131-3134 (2003).
5. (a) George Y. Li, G. Zheng, and A. F. Noonan, *J. Org. Chem.*, **66**(25), 8677-8681 (2001).
   (b) C. Wolf and R. Lerebours, *J. Org. Chem.*, **68**(18), 7077-7084 (2003).
6. G. Y. Li, *J. Organometal. Chem.* (Special Issue: 30 Years of the Cross-Coupling Reaction), **653**, 63-68 (2002).
7. C. Wolf and R. Lerebours, *J. Org. Chem.*, **68**(19), 7551-7554 (2003).
8. (a) A. F. Littke and G. C. Fu, *Angew. Chem. Int. Ed. Engl.*, **41**, 4176-4211 (2002).
   (b) A. F. Littke, C. Dai and G. C. Fu, *J. Am. Chem. Soc.*, **122**, 4020-4028 (2000).
9. F. A. Cotton and G. Wilkinson, *Advanced Inorganic Chemistry*, 5th Ed., John Wiley & Sons, New York, **1988**, pp. 921-922.

10. For an excellent paper on tri-coordinated anionic complexes $Pd^0L_2Cl^-$ as the real active catalysts, see: C. Amatore and A. Jutand, *Acc. Chem. Res.*, **33**, 314-321 (2000).

11. A reaction of POPd2 (100 mg, 0.147 mmol, ^{31}P NMR (CD_3OD): δ 148.0 (singlet) ppm) and CsF (50.0 mg, 0.329 mmol) in 1.0 mL of CD_3OD at room temperature for 5 min generates an insoluble mixture.

12. The low catalytic reactivity of aryl chlorides in cross-coupling reactions is usually attributed to their reluctance towards oxidative addition to Pd(0). For a discussion, see: V. V. Grushin and H. Alper, *Chem. Rev.*, **94**, 1047-1062 (1994), and reference therein.

13. CombiPhos-Pd6 is a mixture of POPd, POPd1, POPd2, and other coupling catalysts, which can carry out many cross-coupling reactions.

21. High Throughput Experiments in the Design and Optimization of Catalytic Packages for Direct Synthesis of Diphenylcarbonate

Grigorii L. Soloveichik, Kirill V. Shalyaev, Ben P. Patel, Yan Gao, and Eric J. Pressman

General Electric Global Research, One Research Circle, Niskayuna, NY 12309

soloveichik@crd.ge.com

Abstract

Diphenylcarbonate (DPC) is a key intermediate in the manufacture of LEXAN® polycarbonate resin by melt polymerization reaction of DPC and bisphenol A. One-step direct carbonylation of phenol to DPC is a potentially simpler route when compared with the commercial two-step process. Promising catalytic packages for one-step DPC synthesis consisting of a palladium catalyst, a lead co-catalyst and a halide (generally bromide) salt have been studied. Due to the complexity of DPC catalysis and the relatively high cost of conventional experiments, high throughput (HT) methodology was used both for design and optimization of lead based catalytic packages for DPC synthesis.

It was found that catalytic activity is affected by the order in which catalyst components are charged into the reaction, and by the Br:Pb ratio. Depending on the Br:Pb ratio, the precipitation of either $PbBr_2$ or polynuclear lead oxo(bromo)phenoxide species may occur. These compounds are capable of reducing Pd(II) to Pd(0) causing low reproducibility in HT experiments. The use of Pd bisphosphine complexes provided reproducible results even under heterogeneous conditions. Use of lead oxo(bromo)phenoxides in DPC synthesis instead of PbO yields higher Pd turnover numbers and selectivity. Use of an additional metal co-catalyst containing titanium or cerium to the palladium-lead-bromide system increases reaction rate. To decrease materials cost, expensive quaternary bromide salts were replaced with alkali metal bromides in combination with activating solvents. Finally, non-bromide containing catalytic packages for DPC synthesis have been developed. Results of HT experiments are in good correlation with those of batch experiments in 50 – 100 g scale.

Introduction

Diphenylcarbonate (DPC) is a key monomer for producing LEXAN® polycarbonate resin by melt polymerization reaction of DPC with Bisphenol A. Currently, DPC is produced by General Electric Company (GE) in Cartagena, Spain, using a two-step process (Eq. 1) The stoichiometric carbonylation of para-substituted phenols was discovered at GE in the mid 1970's. [1]. Catalytic reaction (Eq. 2) requires elevated pressure of carbon monoxide and oxygen, and a rather complex "catalytic package".

$$2 \text{ MeOH} + \text{CO} + 1/2 \text{ O}_2 \xrightarrow[- \text{H}_2\text{O}]{\text{CuCl}_2/\text{HCl}} \underset{\text{Me} \quad \text{Me}}{\overset{\text{O}}{\parallel}} \xrightarrow[- \text{MeOH}]{\text{PhOH}} \quad (1)$$

$$2 \quad \text{PhOH} + \text{CO} + 1/2 \text{ O}_2 \xrightarrow{\text{Pd catalyst}} \quad + \text{ H}_2\text{O} \quad (2)$$

The optimum catalytic package was shown to require, besides palladium, a co-catalyst (inorganic or organic compound capable of redox transformations) and a quaternary ammonium or phosphonium bromide salt. The presence of the co-catalyst is necessary because in its absence, the rate of palladium reoxidation by oxygen is very slow, and Pd turnover number (TON; mol DPC produced/mol Pd charged) is less then 10, which makes the process impractical. In 1981, Hallgren et al. [2] reported that Cu, Co and Mn catalyze Pd reoxidation with air oxygen increasing Pd TON by an order of magnitude. Later, Ce was used as an inorganic cocatalyst [3]. Addition of organic co-catalysts, for instance, hydroquinone [4] or p-benzoquinone [5], increased Pd TON. Mitsubishi patented the use of lead compounds as co-catalysts with noticeable improvement in catalyst productivity (up to 4100 Pd TON) [6]. In our previous described work [7], we screened a variety of two- and three-metal combinations as co-catalysts using high throughput (HT) methodology, and found that lead based binary and ternary combinations were consistently the most active (i.e. provided the highest Pd TON). In order to further examine these most promising commercially viable catalyst packages, we studied interactions between components in such systems, which led to the formulation of improved catalytic packages. HT methodology, described in detail elsewhere [7], was used for both the screening and optimization of catalytic packages, with the primary goal to replace one of the most expensive components, a quaternary bromide.

Results and Discussion

The classical Mitsubishi catalytic package for one-step DPC includes a palladium source, lead oxide, and a quaternary bromide dissolved in neat phenol [6]. This catalyst shows low reaction rate and a significant induction period. To improve catalyst performance, it was proposed to use additional redox co-catalysts. Small scale HT screening of one and two additional (to lead) metals as co-catalysts [7] at the same conditions found that only titanium, cerium, manganese, iron and zirconium showed a synergistic effect. Pd TON was maximal for the Pb-Ti pair (1068 vs. 412). Co-catalyst, consisting of two metals in addition to lead, allowed further increasing Pd TON [7]. However, the gain in the catalyst productivity was not sufficiently large to justify the increased complexity of the three-metal catalytic package. Large-scale runs showed that the binary Pb-Ti and Pb-Ce combinations had high reaction rate and no induction period. Therefore, these pairs along with lead only system were selected for optimization of DPC catalysts.

Interactions between catalyst components in lead based systems

It should be noted that all reaction mixtures using lead as a co-catalyst were heterogeneous. A variability study of the small-scale runs using this catalytic package showed that both block-to-block and across-the-block variability was very high in contrast to more conventional large-scale (60 g) runs. Such high variability made the search for new catalysts and especially their optimization practically impossible using the small-scale setup. As a result of the absence of stirring in the small-scale experiments, it was proposed that mass transport in heterogeneous mixtures was the cause of the observed variability. However, experiments with lead-cerium co-catalyst (vide infra) showed the main factor controlling catalytic activity was not heterogeneity but duration of the catalyst preparation (Table 1). One can see that Pd TON when using Pd(acac)$_2$ as catalyst decreases from 2300 – 2500 to 800 – 900 when catalyst components were maintained at 70 °C. Due to the sometimes lengthy period required for total reaction mixture formulation, catalyst components must be stable to storage, particularly in phenolic solution. Thus, aging studies focusing on the interactions between catalyst components were initiated.

Table 1 Time dependent performance in direct carbonylation of phenol using lead based catalysts with varying palladium sources (0.25 mM Pd, 12 eq. PbO, 5.6 eq. Ce(acac)$_3$, 400 eq. TBAB).

Pd stock age,[a] hr	Mixture age,[a] hr	Pd(acac)$_2$		Pd(dppb)Cl$_2$	
		Pd TON (av)	RSD	Pd TON (av)	RSD
4	4	801	9.1%	1926	3.7%
4	2	891	9.6%	2036	6.5%
4	0	2510	6.7%	2132	2.0%
2	2	841	4.6%	2000	5.1%
2	0	2330	7.4%	1879	7.3%
0	0	2368	2.0%	2018	29.8%
Average		1623		1998	
RSD		52.7%		4.4%	

[a] Solution of Pd compound in PhOH or reaction mixture were aged at 70 °C

Lead oxide is soluble in molten phenol (45-60°C) at a level of 20 wt%. Evaporation of phenol from such a solution yields lead (II) phenoxide. Reactions between lead oxide and quaternary bromide in phenol depend on temperature;

the QBr:PbO ratio and are described by Scheme 1. Below 140 °C, the precipitate formed in this reaction contains phases of $PbBr_2$ and $Pb(OH)Br$ (minor component). $Pd(acac)_2$ is unstable in phenolic solution and slowly decomposes to Pd black. The decomposition rate is higher in the presence of a bromide salt. Addition of $Pd(acac)_2$ solution to either filtrate or to solid $PbBr_2$ causes immediate formation of Pd black. Addition of $Pd(acac)_2$ to a fresh solution of $PbBr_2$ in PhOH (above 140 °C) yields an orange homogeneous solution of the Pd(II) salt. EXAFS spectrum of Pd black precipitated from such solutions is slightly different from the spectrum of Pd metal, and shows formation of Pd clusters containing Pd-Br bonds (Pd:Br = 6:1) and with Pd··Pd distance 2.81 Å (2.76 Å in Pd metal). Thus, reduction of Pd(II) and precipitation of palladium metal explains the high variability observed in experiments with no stirring.

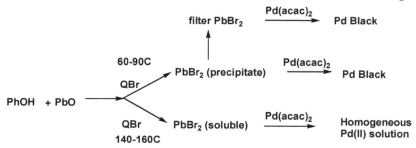

Scheme 1 Reactions between main components of lead based catalytic package.

To prevent palladium from precipitating, high temperature may be employed, although this would be impractical. A more attractive alternative involves use of Pd complexes containing stabilizing ligands. It is well known that phosphine ligands stabilize both Pd(II) and Pd(0) complexes. Presynthesized (isolated) Pd phosphine complexes compared with the same complexes prepared *in situ* would be expected to show less variability Screening of different phosphine ligands for this purpose revealed that bisphosphines were the best (Table 2). Phosphines with electron withdrawing groups are less active, probably due to lower stability of Pd complexes. It was shown by NMR that phosphine ligands were ultimately oxidized to phosphine oxides, but remain attached to palladium (as phosphines) during the active stage. Thus, the bis(diphenylphosphino)butane complex $Pd(dppb)Cl_2$ provided nearly an order of magnitude reduction in variability despite different aging conditions (relative standard deviation decreased from 53 % to less than 5 %, Table 1) and was used in further HT experiments as a lead source.

Table 2 Catalysts performance for different palladium phosphine complexes with lead based co-catalyst (0.25 mM Pd, 50 eq. PbO, 400 eq. TEAB).

Catalyst	Ligand	L:Pd	Pd TON (avg)	RSD*
$Pd(acac)_2$	-	-	3325	28%
$(MeCN)_2PdCl_2$	-	-	1289	42%
$[P(4\text{-}ClC_6H_4)_3]_2PdCl_2$	-	-	2028	10%
$(MeCN)_2PdCl_2$	$P(4\text{-}ClC_6H_4)_3$	2	1539	23%
$(PPh_3)_2PdCl_2$	-	-	1803	5%
$(MeCN)_2PdCl_2$	PPh_3	2	1475	54%
$(PMePh_2)_2PdCl_2$	-	-	1972	5%
$(MeCN)_2PdCl_2$	$PMePh_2$	2	2533	13%
$Pd(dppb)Cl_2$	-	-	3129	5%
$(MeCN)_2PdCl_2$	dppb	1	3089	11%

* - relative standard deviation

Lead co-catalysts

Interaction of lead oxide with bromide salts in phenol at the ratio Br: PbO less than 2 yields the precipitation of white sediments with the general formula $Pb_nO_m(OPh)_{(2-z)(n-m)}Br_{z(n-m)}$, where n = 4 – 10, m = 1 – 4, z = 0 – 1 and formation of (n-m) moles of water. Reaction takes place in the presence of a variety of bromide salts including quaternary and alkali metal bromides. In the latter case, the presence of a coordinating solvent, e.g. MeCN, is necessary. The composition of product lead bromophenoxides depends on the Br:PbO ratio, bromide salt and solvent. These complexes usually have Pb:O ratio of 4:1 or 3:1, and variable levels of bromide (including some bromide-free complexes) (Table 3).

Table 3 Analytical and EXAFS data for lead bromophenoxides.

Sample	% Br	EXAFS data		
	(calc)	Pb-O, Å	Pb-O, N	Pb-Pb, Å
$Pb_4O(OPh)_5Br$	5.7 (5.75)			
$Pb_6O_2(OPh)_6Br_2$	7.0 (8.02)	2.27	3.7	3.90
$Pb_6O_2(OPh)_7Br$	3.9 (3.98)	2.27	4.3	4.01
$Pb_6O_2(OPh)_{7.5}Br_{0.5}$	2.3 (1.98)	2.28	3.8	3.90
$Pb_6O_2(OPh)_8$	0.5 (0)	2.33	6.5	3.77
$Pb_{10}O_4(OPh)_{12}$	-	2.31	5.2	3.63

All compounds have large unit cells (from 3400 to 8350 Å^3) indicative of large, probably polynuclear molecules. This suggestion was confirmed by EXAFS found short Pb$\cdots$Pb non-bonding distances (Table 3). There are several published examples of lead alkoxide complexes. Complex Pb$_4$O(OSiMe$_3$)$_6$ has a tetrahedral metal skeleton with central μ$_4$ oxygen atom and μ$_2$ alkoxy groups with Pb-O = 2.28 Å and Pb-OR = 2.38 Å respectively [8]. Complex Pb$_6$O$_4$(O^{i}Pr)$_4$ has an octahedral metal skeleton (Pb$\cdots$Pb = 4.08 Å) with μ$_3$ oxygen atoms (Pb-O = 2.16 Å) and μ$_3$ alkoxy groups (Pb-OR = 2.41 Å) [9]. These distances are in good correlation with EXAFS data for synthesized lead bromophenoxides, taking into account that this method gives average values. One may suggest that the structure of Pb$_4$O(OPh)$_5$Br is close to the structure of Pb$_4$O(OSiMe$_3$)$_6$ [8], where one silyloxy group is replaced with bromide. In addition, the structure of hexanuclear Pb$_6$O$_2$(OPh)$_{(8-n)}$Br$_n$ likely corresponds in structure to the octahedral complex Pb6O4(OiPr)4. One of the complexes formed in the presence of NaBr/acetonitrile was crystallized from the reaction mixture as a solvate with acetonitile. Single crystal X-ray analysis found that the complex has the formula Pb$_{10}$O$_4$(OPh)$_{12}$·2MeCN, and a metal skeleton consisting of two lead octahedrons with a common edge bound by μ$_4$ and μ$_3$ oxygen atoms, as well as μ$_2$ and μ$_3$ phenoxy groups (Fig. 1). The complex has an average Pb-O distance 2.31Å, the same as determined by EXAFS. Interestingly, this decanuclear complex has three pairs of very short Pb$\cdots$Pb distances (Pb-Pb$_{(av.)}$ = 2.67 Å) that is close to the sum of covalent radii and much smaller than non-bonding Pb$\cdots$Pb distance (3.63 - 4.08 Å).

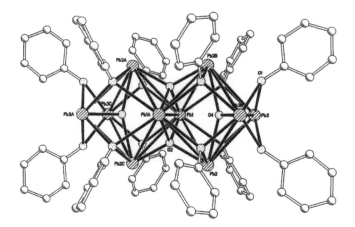

Figure 1 X-ray structure of the Pb$_{10}$O$_4$(OPh)$_{12}$·2MeCN complex (MeCN molecules and hydrogen atoms omitted).

Formation of polynuclear lead species with parameters close to isolated lead bromophenoxides during DPC synthesis was found by EXAFS of frozen active reaction mixtures (Pb-O = 2.34 Å, Pb···Pb = 3.83 Å). Noteworthy, in samples of final reaction mixtures, where catalyst was inactive, short Pb···Pb distances were absent. These polynuclear compounds have been tested as lead sources in large-scale runs (small scale reactions were inconclusive due to heterogeneity of reaction mixtures because these compounds are less soluble than PbO). It was found that the use of lead bromophenoxides instead of PbO increases both Pd TON (by 25-35%), and reaction selectivity (from 65 – 67 % to 75 – 84 %). Activity of different lead bromophenoxides was about the same (within experimental error) but the best selectivity was observed for complex $Pb_6O_2(OPh)_6Br_2$. Therefore, the gain in selectivity vs. loss due to additional preparation step should be analyzed for practical application.

Optimization of lead based DPC catalysts

Before optimization in HT reactions, we had to determine the extent of correlation between results in small (HT) and large-scale settings. Comparison of Pd TON for the same catalyst compositions (Pb-Ce co-catalyst) for both scales showed excellent correlation (R^2 = 0.986) that can be described by the following linear equation: Pd TON (large scale) = 1.65*[Pd TON (HT)] – 314. Therefore, the HT reaction setup allowed the running of multifactorial and multilevel DOEs in a single block containing multiple reaction formulations [7] To determine optimal composition of the lead based catalytic package, we ran response surface DOEs for formulations containing practical levels of NEt_4Br and NBu_4Br (TBAB). Results for both bromide salts were very similar (TBAB based formulations were ~15 % more active) and well described by a cubic model. Pd TON steadily grows with increases in lead concentrations and has a clear maximum at bromide level of 705 – 725 molar equivalents vs. Pd (Figure 2,a). Therefore, at some point, bromide starts to inhibit the reaction, possibly due to formation of coordinately saturated $[PdBr_4]^{2-}$ anion. Despite the absence of leveling of Pd TON with lead concentration, the practical limit of lead co-catalyst level is ~100 molar equivalents vs. Pd, due to the growing heterogeneity of the reaction mixture. Noteworthy, eventually almost all bromide is converted to bromophenols; after that, the reaction stops.

In two-metal co-catalyst Pd-Pb-bromide catalytic packages, Pd TON also increases with lead concentration, but decreases with concentration of a second co-catalyst (Figure 2,b). However, we found that addition of a small amount of titanium or cerium is justified because it improves reaction rate and eliminates the induction period characteristic of the lead only package. As a result of optimization, the best performing catalytic packages contain about 100 molar equivalents (vs. Pd) lead, 2 – 4 molar equivalents of a second co-catalyst (Ti or Ce), and 600 – 800 equivalents of a quaternary bromide.

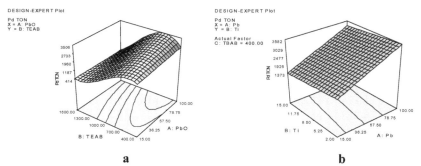

a   b

Figure 2 Pd TON response surface of for PbO-TBAB (a) and PbO-TiO(acac)₂-
TBAB (b) co-catalysts.

Catalytic packages containing NaBr

The use of large amounts of expensive quaternary bromide salts in lead
based catalytic DPC packages renders them cost prohibitive. Replacement of
those salts with the relatively inexpensive alkali metal bromides, however,
results in very low carbonylation activity. However, we found that alkali metal
bromides can be used in combination with activating solvents, such as nitriles,
ethers, and sulfones capable of solvating the metal cation (Table 4). The
catalyst performance (Pd TON and selectivity) in this case is comparable with
the performance of the same packages with quaternary bromide salts. Cost
analysis shows that systems with NaBr and acetonitrile or glymes are noticeably
less expensive that those with quaternary bromide salts, assuming reasonable
levels of recycle of the activating solvents.

Table 4 Catalysts performance of NaBr- activating solvent-based packages with
lead based co-catalyst (0.25 mM Pd, 50 eq. PbO).

2nd co-catalyst (eq)	NaBr, eq	Solvent (vol.%)	Pd TON	
			with solvent	no solvent
-	270	Triglyme (35)	1186	152
Ti (10)	270	Triglyme (35)	1249	472
-	650	MeCN (19.5)	2216	626
Ce (24)	240	NMP (35)	1339	540
-	270	Sulfolane (35)	1530	152
Ti (11)	270	Sulfolane (35)	1797	472
-	230	Tetraglyme (8.5)	4620 [a]	626 [a]
-	650	Tetraglyme (7.2)	6145 [a]	626 [a]
-	650	15-crown-5 (3.5)	5455 [a]	626 [a]

[a] – Large scale run

Non-bromide catalytic systems

Due to the high cost of quaternary ammonium bromide c in the catalytic package, we also studied the possibility of using salts of other halogens instead of bromides in lead based catalytic packages for DPC synthesis. HT screening showed that bimetallic co-catalysts containing lead and cerium, titanium or manganese demonstrated substantial activity with tetrabutylammonium chloride (TBAC) in the range of 800 – 1100 Pd TON (Figure 3) while lead co-catalyst alone showed low activity (400 – 500 Pd TON). In contrast to bromide based systems, the activity of chloride-based systems is not highly dependent on halide concentration. However, the increase of lead concentration from 12 to 50 equivalents vs. Pd increases Pd TON up to 2600 – 2800 with cerium co-catalyst.

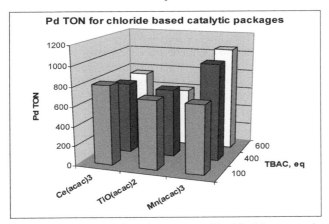

Figure 3 Performance of lead based systems with chloride co-catalyst.

Sodium iodide can also be used as a bromide replacement, yielding Pd TON of 1195. In this case, bisphosphine ligands are necessary to maintain high regioselectivity (DPC vs. phenylsalicilate) of carbonylation reaction (96 % vs. 72 % for Pd(acac)$_2$).

Conclusions

It has been shown that small-scale HT experiments can be successfully used for both screening and optimization of catalytic packages for DPC synthesis with lead co-catalyst, even in studies involving heterogeneous reaction mixtures. HT data are shown to be in good correlation with larger scale batch reactions. Depending on the Br:Pb ratio, PbBr$_2$ or polynuclear lead oxo(bromo)phenoxide species precipitate from lead based catalytic systems. These compounds accelerate the reducing Pd(II) to Pd(0), which causes low reproducibility in HT experiments. The use of Pd bisphosphine complexes reduces the rate of catalyst ageing and provides reproducible HT results even under heterogeneous conditions. Use of cerium or titanium compounds (in addition to lead compounds) does not increase Pd TON but accelerates the reaction and sometimes improve selectivity, possibly as a result of faster formation of polynuclear co-catalyst species. The bromide component of the catalytic package is more active than chloride or iodide analogs. However, we found conditions when these halogen salts show comparable activity. An important finding is the use of alkali metal bromides as an inexpensive bromide source in combination with activating solvents.

Experimental Section

HT and large scale experiments were performed as described in [7]. In HT runs, 116 2-ml GC vials (thin film, no stirring) were heated at 100° C for 3 h under 1500 psi of O_2/CO (10% O_2 in CO) The presented results are the average of at least two runs (three of four when RSD data presented). Large-scale (60 g) experiments were carried out in 450 mL Hastelloy C Parr reactor under the same conditions in the presence of 3A molecular sieves (35 g) s a desiccant.

Acknowledgements

Thanks to John Ofori, Bruce Johnson, Jay Spivack, Donald Whisenhunt, James Cawse, Emil Lobkovsky, Eric Williams, Timothy Chuck, David Smith, and Tracey Jordan for helpful discussions and experimental assistance.

References

1. A. J. Chalk, US Pat. 4,096,169 to General Electric (1978).
2. J. E. Hallgren, G. M. Lucas, and R. O. Matthew, *J. Organometal. Chem.,* **204**, 135 (1981).
3. M. Goyal, R. Nagahata, J. Sugiyama, M. Asai, M. Ueda, and K. Takeuchi, *Catal. Lett.,* **54**, 29 (1998).
4. M. Goyal, R. Nagahata, J. Sugiyama, M. Asai, M. Ueda, and K. Takeuchi, *J. Mol. Catal.,* **137**, 147 (1999).
5. A. Vavasori and L. Toniolo, *J. Mol. Catal.,* **139**, 109 (1999).
6. M. Takagi, H. Miyagi, Y. Ohgomori, and H. Iwane, US Pat. 5,498,789 to Mitsubishi Chemical Corporation (1996).
7. J. L. Spivack et al., *Applied Catalysis A,* **254**, 5 (2003).
8. C. Gaffney, P. G. Harrison, T. J. King, *Chem. Commun.,* 1251 (1980).
9. A. I. Yanovsky, N. Ya. Turova, E. P. Turevskaya, Yu. T. Struchkov, *Koord. Khim.,* **8**, 153 (1982).

22.   Rapid Optimization of Hydrogenation Process Variables by Combining Experimental Design and Parallel Reactors

Venu Arunajatesan, ShaRee L. McIntosh, Marisa Cruz, Roman F. Renneke, and Baoshu Chen

Degussa Corporation, 5150 Gilbertsville Hwy, Calvert City, KY, 42029

venu.arunajatesan@degussa.com

Abstract

In this paper we report a method for the rapid optimization of process conditions for a model reaction (selective hydrogenation of phenol to cyclohexanone over Pd/C in liquid phase), which in general could be applied to any reaction. Our goal is to optimize the process conditions that would maximize the yield of cyclohexanone within the reaction constraints. The variables examined include temperature (120 °C to 150 °C), pressure (15-30 bar), and NaOH concentration (410-820 ppm w/w). The liquid phase hydrogenation of phenol was conducted in a well-stirred autoclave at a constant hydrogen pressure in the absence of mass-transfer limitations. The number of experiments was minimized using the two-level factorial method. Further, conducting these experiments in two independent reactors in parallel shortened the optimization time. The activity and selectivity of the reaction was followed using hydrogen-uptake and product analysis by gas chromatography. A maximum cyclohexanone yield of 73% was obtained at 135 °C, and 22.5 bar with 615 ppm NaOH.

Introduction

Three-phase slurry reactors are commonly used in fine-chemical industries for the catalytic hydrogenation of organic substrates to a variety of products and intermediates (1-2). The most common types of catalysts are precious metals such as Pt and Pd supported on powdered carbon supports (3). The behavior of the gas-liquid-slurry reactors is affected by a complex interplay of multiple variables including the temperature, pressure, stirring rates, feed composition, etc. (1-2,4). Often these types of reactors are operated away from the optimal conditions due to the difficulty in identifying and optimizing the critical variables involved in the process. This not only leads to lost productivity but also increases the cost of down stream processing (purification), and pollution control (undesired by-products).

The conventional approach to optimize a heterogeneous catalytic process is based on heuristics, and trial and error, which could take a considerable amount of time. On the other hand, the use of combinatorial techniques, which has been applied successfully in homogeneous catalytic process, is expensive and not always suitable for identifying and testing heterogeneous catalysts. One of the simplest methods for optimization is the two-level factorial design (5-6). Here, the experiments are designed around a set of variables (temperature, pressure, etc.) that might affect the response (yield of the desired product) within the parameter space (120-150 °C, 15-30 bar, etc.). The capability of the equipment, cost, or prior knowledge of the reaction conditions could dictate the upper and lower limits of the parameter space. The minimum number of experiments needed to determine the effects of k parameters (main effects) and their interaction (interaction effects) is 2^k. Thus, for a three-parameter design, one needs $2^3 (=8)$ experiments. Adding multiple center points allows one to estimate the error and the curvature of the response surface.

The exothermic hydrogenation of phenol over a Pd/C catalyst is a series reaction that leads to the formation of cyclohexanone and cyclohexanol. Although the thermodynamic equilibrium favors the formation of cyclohexanol (ΔG = -211 kJ/mol) over cyclohexanone (ΔG = -145 kJ/mol), optimizing the variables could lead to higher yield of the ketone. In this paper, a two-level factorial design was applied to optimize the yield of cyclohexanone during the hydrogenation of phenol over a commercially available Pd/C catalyst (Degussa Corp.). The liquid-phase hydrogenation of phenol is an inherently safe industrial process currently in use for the production of cyclohexanone (7-11). More than 95% of the cyclohexanone produced is used in the manufacture of Nylon-6 and Nylon-6,6 through caprolactam (11). The total US production of caprolactam in 2003 was 840,000 MT/year in 1992 (12).

Experimental Section

The reactions were conducted in parallel using two of three identical high-pressure reactors (Parr Instrument Co.) in our lab. These 300 mL internally-baffled Hastalloy®C reactors are capable of operating up to 250 °C and 150 bar. The temperature, pressure, and stirring rates were monitored and independently controlled at the desired level by the SpecView32 data acquisition and control system. As the H_2 was consumed, the pressure in the reactor was maintained by a continuous flow of hydrogen from a high-pressure burette through a pressure regulator into the dead-ended reactor.

A typical experiment is conducted by loading 30 g phenol (0.318 mol), 60 g cyclohexane (0.714 mol), 100 µl of 50% NaOH (810 ppm w/w), and 0.15 g of a commercially available

Table 1 Characteristics of the 5% Pd/C.

Catalyst Characteristic	Value
BET Surface Area (m²/g)	818
Pore Volume (cc/g)	0.67
Dispersion (%)	19.1
Particle size: D_{50} (µm)	24.8

5% Pd/C (Degussa Corp.). The reactor was sealed, pressure tested, and the headspace was filled with H_2 (7 bar) before heating it to the operating temperature. Once the temperature reached within 10 °C of the operating temperature, the H_2 pressure was adjusted to 30 bar using the burette regulator and the stirring rate was set at 1000 rpm, and this was taken as time, t = 0 min. The consumption of hydrogen was then followed for the next 45 minutes at which point the reactor was rapidly cooled, and a liquid sample was taken out for analysis by gas chromatograph (Agilent, Inc.).

The reaction product was filtered to remove catalyst and analyzed in GC equipped with an HP5 (30 m X 0.32 mm X 0.25 μm) column. The temperature program used for analysis (31 °C – 35 min – 1 °C/min – 40 °C – 10 °C/min - 120 °C) ensured complete separation of the cyclohexanol, cyclohexanone, and phenol peaks. The conversion and selectivity were calculated directly from the area of each peak.

Three parameters that affect the catalyst activity and selectivity were chosen (8-9). They are temperature, pressure, and NaOH concentration. The upper and lower limits of the parameters (Table 2)

Table 2 Parameters and the range employed in the current work.

Factor	Parameter	Low	High
A	Temperature (°C)	120	150
B	Pressure (bar)	15	30
C	NaOH (ppm w/w)	410	820

were chosen based on the range observed in the literature. A two-level factorial design based on three variables, including two center points, leads to the ten experiments represented in Table 3. In this design, the experiments are conducted in the eight vertices of a cube to which center points are added to measure the experimental error.

Results and Discussion

Prior to conducting the DOE (design of experiments) described in Table 3, it was established that no reaction took place in the absence of a catalyst and that the reactions were conducted in the region where chemical kinetics controlled the reaction rate. The results indicated that operating the reactor at 1000 rpm was sufficient to minimize the external mass-transfer limitations. Pore diffusion limitations were expected to be minimal as the median catalyst particle size is <25 μm. Further, experiments conducted under identical conditions to ensure repeatability and reproducibility in the two reactors yielded results that were within ±5%.

Table 3. Conversion, selectivity, yield, and reaction rate at various conditions based on two-level factorial design.

Run	T (°C)	P (bar)	[NaOH] (ppm)	Conversion (%)	Selectivity (%)	Yield (%)	Rate (mol/g/h)
1	150	30	820	99	19	19	18.9
2	150	30	410	99	17	17	15.0
3	150	15	820	99	68	67	16.1
4	150	15	410	86	67	57	14.7
5	120	30	820	69	86	60	6.9
6	120	30	410	76	85	65	5.6
7	120	15	820	74	91	67	7.4
8	120	15	410	72	94	68	6.0
9	135	22.5	615	92	77	71	9.9
10	135	22.5	615	91	80	73	10.3

The DOE was completed in a random order and the results are tabulated in columns 5-8 of Table 3. The following observations can be readily made:

1. the reaction rate and conversion are higher at higher temperatures while the selectivity is typically lower (example, Run 1 *vs.* 5, 2 *vs.* 6, 4 *vs.* 8)
2. at a given temperature, the selectivity is higher at lower pressures (example, Run 1 *vs.* 3, 5 *vs.* 7)
3. at a given temperature and pressure, the concentration of NaOH does not affect the overall selectivity (example, Run 7 *vs.* 8)
4. the highest cyclohexanone yield was obtained at the center point (Run 9 and 10).

The relationship between the operating variables and the yield could be obtained using multiple regression. The model equation obtained from this regression can be used to predict the interplay between variables on the cyclohexanone yield. The regression coefficients for each parameter and their interaction are provided in Table 4. From this, the equation of the fitted model applicable to the parameter range examined can be written as,

Table 4 Regression coefficients for the model.

Parameter	Coefficient
Constant	-12.8
A: Temperature (°C)	0.67
B: Pressure (bar)	10.7
C: NaOH (ppm w/w)	-73.2×10^{-3}
AB	-86.7×10^{-3}
AC	0.73×10^{-3}
BC	-0.98×10^{-3}

$$\text{Yield} = -12.8 + 0.67*T + 10.7*P - 73.2 \times 10^{-3}*[\text{NaOH}] - 86.7 \times 10^{-3}*T*P + 0.73 \times 10^{-3}*T*[\text{NaOH}] - 0.98 \times 10^{-3}*P*[\text{NaOH}].$$

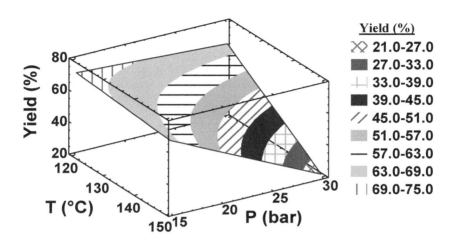

Figure 1 Predicted yield of cyclohexanone at various temperature and pressure ([NaOH] = 615 ppm w/w).

The R-squared value, which indicates how well the three chosen parameters account for the variability in the yield, was 84.2%. The analysis of variance indicates that only temperature and pressure (both P-value = 0.026) have significant impact at 90% confidence level. The P-value of 0.37 for [NaOH] indicates that, within the parameter space examined, the concentration of NaOH does not significantly affect the cyclohexanone yield. Based on the above equation, one can predict the cyclohexanone yield at any given condition within the parameter space chosen. Since [NaOH] does not have a significant effect on the yield, one can fix its value and plot the yield of cyclohexanone as a function of temperature and pressure (Figure 1).

Based on this analysis, a very high yield of cyclohexanone (69-75%) can be obtained at low temperature (120 °C) and pressure (15 bar); the experimental results of 67% and 68% are within error of the predicted values. In practice, although the yield is high, the reaction rates may not be high enough to be economically viable. Hence, it is imperative that other responses such as reaction rates are also considered during optimization. In this case, it is apparent from Table 3, that higher temperatures yield higher reaction rates. Hence an optimal combination of temperature (135-150 °C) and pressure (15 – 22.5 bar) where both the cyclohexanone yield and the reaction rates are high may be preferable. Despite the simplicity and elegance of this optimization method, one should be careful in extrapolating this regression model beyond its parameter space, as this could lead to large errors in the predicted yield. Either expanding the parameter space or using higher-order regression models can minimize this problem.

Conclusions

This study shows that the optimization of process conditions could be achieved rapidly by a judicious use of statistics and parallel reactors. A two-level factorial method with two center points was used to limit the total number of experiments to ten. Using two identical high-pressure reactors in parallel further shortened the time required to conduct these experiments. For the model reaction of phenol hydrogenation over a commercially available Pd/C, it was experimentally determined that the optimal yield was 73% at 135 °C, 22.5 bar, and 615 ppm w/w NaOH

References

1. J. D Super, Chemical Industries (Dekker), **82** (*Catal. Org. React.*), 35-49 (2001).
2. P. L. Mills and R. V. Chaudhari, *Catal. Today*, *37*, 376-404, (1997).
3. G. J. Hutchings, R. P. K. Wells and J. E. Bailie, *Sci. Prog., 82*, 233-250, (1999).
4. J. F. Jenck, in *Heterogeneous Catalysis and Fine Chemicals* (M. Guisnet *et al.,* ed.), Elsevier Science Publishers B. V., Amsterdam, 1991, p 1.
5. D. C. Montgomery, Design and Analysis of Experiments, 3rd Edn., John Wiley & Sons, Inc., New York, 1991, p 197.
6. C. D. Hendrix, *Chemtech,* 167, March (1979).
7. S. Narayanan, *Res. Ind., 34*, 296-300, (1989).
8. M. A. Guiterrez-Ortiz, A. Castano, M. P. Gonzalez-Marcos, J. I. Guiterrez-Ortiz and J. R. Gonzalez-Velasco, *Ind. Eng. Chem. Res., 33*, 2571-2577, (1994).
9. J. R. Gonzalez-Velasco, M. P. Gonzalez-Marcos, S. Arnaiz, J. I. Guiterrez-Ortiz and M. A. Guiterrez-Ortiz, *Ind. Eng. Chem. Res., 34*, 1031-1036, (1995).
10. J. F. VanPeppen and W. B. Fisher, US Pat. 4,164,515 to Allied Chemical Corp. (1979).
11. W. B. Fisher and J. F. VanPeppen, Cyclohexanol and Cyclohexanone in Kirk-Othmer Encyclopedia of Chemical Technology, John Wiley & Sons, Inc., New York, 1993.
12. Product Focus, *Chem. Week*, August 20, 2003, 26.

23. Design and Parallel Synthesis of New Oxidative Dehydrogenation Catalysts

Gadi Rothenberg,* E. A. (Bart) de Graaf, Jurriaan Beckers, and Alfred Bliek

Chemical Engineering Department, University of Amsterdam, Nieuwe Achtergracht 166, 1018 WV Amsterdam, The Netherlands.
gadi@science.uva.nl

Abstract

The pros and cons of oxidative dehydrogenation for alkene synthesis using doped cerianites as solid oxygen carriers are studied. The hydrogen oxidation properties of a set of ten doped cerianite catalysts ($Ce_{0.9}X_{0.1}O_y$, where X = Bi, In, La, Mo, Pb, Sn, V, W, Y, and Zr) are examined under cyclic redox conditions. X-ray diffraction, X-ray photoelectron spectroscopy, adsorption measurements, and temperature programmed reduction are used to try and clarify structure–activity relationships and the different dopant effects.

Introduction

Industrial dehydrogenation of alkanes to alkenes is a huge commercial process, as C_2–C_4 alkenes are the building blocks for the majority of polymers and numerous chemicals (1, 2). Unfortunately, the equilibrium favours the reactants and the forward reaction is energy-intensive (Scheme 1, left). This equilibrium shifts towards the products at higher temperatures, lower pressures, or when using larger alkanes (3). Another way to shift the equilibrium is to remove the hydrogen by-product, either by using a membrane (4, 5) or by chemical oxidation (Scheme 1, right). In the latter case, the heat of combustion can compensate for the endothermic dehydrogenation process, and the product is simultaneously purified.

Scheme 1

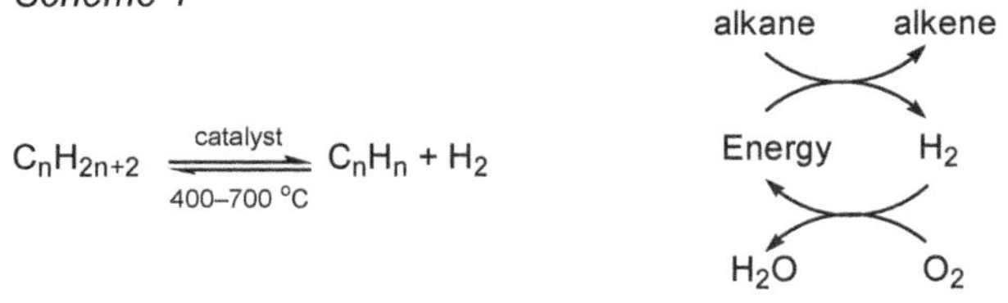

$$C_nH_{2n+2} \xrightleftharpoons[\text{400–700 °C}]{\text{catalyst}} C_nH_n + H_2$$

alkane → alkene

Energy → H_2

H_2O ← O_2

Hydrogen oxidation can be done *in situ*, using a bi-functional catalyst (6–11), or the two processes can be separated by feeding the dehydrogenation effluent to the hydrogen oxidation reactor (12, 13). The overall thermodynamics are the same, but the former approach has the advantage that the energy is delivered directly where it is consumed. The disadvantage is the explosion/runaway hazard – mixing oxygen, hydrogen, and hydrocarbons at high temperatures is, to put it mildly, unsafe.

One solution is to separate the reactants in space, using dense ion-conducting oxide membranes (14, 15). Another option, that we present here, is to separate the two processes in time, using solid oxygen carriers. Grasselli and co-workers (16, 17), and we (18) demonstrated selective oxidation of H_2 in the presence of alkanes at 500–700 °C using supported oxides of post-transition metals. The problem was that many metals melt at these temperatures, so when the metal oxide was reduced to metal$^{(0)}$ it liquified, causing sintering and deactivation.

To solve this problem, we used bimetallic doped cerianites. These materials are versatile oxygen exchangers. The redox cycle $Ce^{3+} \Leftrightarrow Ce^{4+} + e^-$ facilitates oxygen storage and release from the fluorite lattice. This makes them ideal for three-way catalysis and hydrocarbon fuel-cells (20, 21). The poor selectivity of pure CeO_2 in hydrogen oxidation can be improved by applying a shape-selective (22) or a chemo-selective (23) coating, or by doping (24). The doping method has an advantage: if the dopant is incorporated into the ceria lattice, it will not sinter at high temperatures. Furthermore, the redox chemistry of ceria is sensitive to small defects in the crystal lattice, so in principle a small amount of dopant could cause a significant change in the catalytic properties. In a preliminary communication, we reported the synthesis of ten bimetallic cerianites ($Ce_{0.9}M_{0.1}O_y$, where M = Bi, In, La, Mo, Pb, Sn, V, W, Y, and Zr, catalysts **1–10**, respectively, see Figure 1 **A**) and their application to hydrogen oxidation (25). Here, we study the hydrogen oxidation activity of these catalysts in hydrogen/argon and hydrogen/hydrocarbons mixtures, and discuss the influence of surface effects and crystal structure on the oxidation process.

Results and Discussion

In a typical reaction, a feed simulating the effluent at the end of the ethane dehydrogenation process (20% v/v C_2H_6, 20% C_2H_4, 5% H_2 and balance He) was flowed over the catalyst at 600 °C in a cyclic redox reaction set-up (Figure 2). The hydrogen in the feed reacts in each cycle with some of the lattice oxygen atoms, creating anion vacancies (Figure 1, **C**). These vacancies are then re-filled by an O_2 stream (Figure 1, **B**) that recharges the solid oxygen carrier.

Each reaction was repeated 19 times to examine the rechargeability and stability of the materials. Table 1 gives the key surface parameters, the average

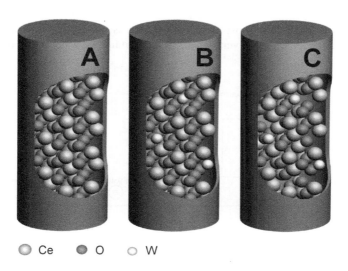

Ce   O   W

Figure 1 Computer-generated schematic drawn to scale showing the fluorite structure of pure CeO2 (**A**), and the vacancies present in the doped $Ce_{0.9}W_{0.1}O_y$ reservoirs when fully charged (**B**) and after releasing 30% of its oxygen atoms (**C**).

Table 1 Characteristics and performance of the oxygen reservoirs.

Entry	Catalyst /composition	Surface area /m^2g^{-1}	Pore Volume /mLg^{-1}	Pore Diameter /nm	Lattice Constant /nm	Bulk Conc. /mol%	Average Activity[a] /%	Selectivity[b] /%
1	**1** /$Ce_{0.9}Bi_{0.1}$	44.7	0.15	6.0	0.5418	1.7	33	15
2	**2** /$Ce_{0.9}In_{0.1}$	69.9	0.25	11.5	0.5403	3.8	35	24
3	**3** /$Ce_{0.9}La_{0.1}$	55.0	0.20	13.3	0.5416	1.3	44	34
4	**4** /$Ce_{0.9}Mo_{0.1}$	12.5	0.06	19.0	0.5424	n/a	11	18
5	**5** /$Ce_{0.9}Pb_{0.1}$	33.0	0.14	7.2	0.5410	4.5	29	11
6	**6** /$Ce_{0.9}Sn_{0.1}$	64.5	0.22	9.2	0.5405	2.3	16	75
7	**7** /$Ce_{0.9}V_{0.1}$	26.4	0.10	9.5	0.5421	n/a	28	10
8	**8** /$Ce_{0.9}W_{0.1}$	35.7	0.14	14.0	0.5416	n/a	23	97[c]
9	**9** /$Ce_{0.9}Y_{0.1}$	42.2	0.10	17.6	0.5411	4.7	22	<2
10	**10** /$Ce_{0.9}Zr_{0.1}$	71.2	0.23	9.4	0.5411	0.7	39	35

[a] Mol% fraction of lattice oxygen atoms available for oxidation, compared to the theoretical limit when all the cerium is reduced to Ce^{3+}. Values are the average of 19 repeated experiments, the distribution is ± 3%.
[b] mol% O_2 used to burn H_2 at 600 °C, determined by the amount of H_2O detected by GC and selective MS; values are the average of triplicate experiments.
[c] Six repeated experiments gave 97, 98, 97, 95, 98, and 97%.

activity (as % available oxygen), and the selectivity towards hydrogen oxidation for catalysts **1–10**. The most active combinations were **1**, **2**, **3**, **7**, and **10**, but CeW **8** was the real find: Over 97% of its oxygen uptake was used to oxidise

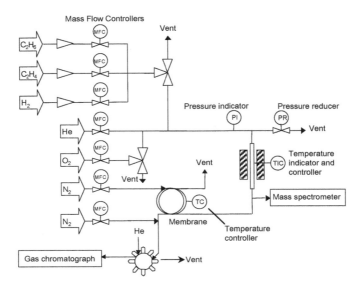

Figure 2 Reduced flowsheet of the experimental setup. The mass-flow controllers, the three-way valves in the oxidizing and reducing pipes, the multiple-way valve GC and the temperature of the reactor are all computer-controlled.

hydrogen. Samples with higher surface area exchanged more available oxygen, the only exception being the extremely low efficiency of $Ce_{0.9}Sn_{0.1}O_x$ (entry 6). Note that cerianites are known to retain their structure even under reducing atmospheres (26, 27).

Powder XR diffraction spectra confirm that all materials are single phase solid solutions with a cubic fluorite structure. Even when 10 mol% of the cations is substituted with dopant the original structure is retained. We used Kim's formula (28) and the corresponding ion radii (29) to estimate the concentration of dopant in the cerium oxide lattice. The calculated lattice parameters show that less dopant is present in the bulk than expected. As no other phases are present in the spectrum, we expect dopant-enriched crystal surfaces, and possibly some interstitial dopant cations. However, this kind of surface enrichment cannot be determined by XR diffraction owing to the lower ordering at the surface.

We then selected three samples for surface studies with X-ray photoelectron spectroscopy (XPS): $Ce_{0.9}Bi_{0.1}O_x$, $Ce_{0.9}Mo_{0.1}O_x$ and $Ce_{0.9}W_{0.1}O_x$. Bismuth ions fit well in the ceria fluorite lattice (30), while molybdenum and tungsten are uncommon cations in eight-fold coordination. The XPS results show that there is three times as much dopant on the surface compared to the overall composition (Table 2). The binding energy of the core levels also gives an indication for the metals' oxidation states. The cerium $3d_{5/2}$ peak is in agreement with literature values for all samples (31–33), though the binding energy of the Ce $3d_{5/2}$ for the molybdate sample is slightly higher than usual. The analysis of the oxidation state of ceria is difficult because the Ce 3d core level peak is subject to hybridization of the O 2p valence band with the 4f level. Five spin-orbit doublets can be found in the Ce 3d region (31). The oxidation state of bismuth is easier to determine. The binding energy of the Bi $4f_{7/2}$ at the surface is in good agreement with oxidic bismuth (34). The binding energy of Mo $3d_{5/2}$ at the surface of the $Ce_{0.9}Mo_{0.1}O_x$ is distinctively different from metallic Mo^0 (227.4–227.9 eV), Mo^{4+} (229.1–229.4 eV) and in fair agreement with Mo^{6+} (232.2–232.7 eV) (35–37). The binding energy of W $4f_{5/2}$ is 36.1 eV, in agreement with data for W^{6+} (35.5–36.4 eV) and distinctly different from metallic W^0 (31 eV) and W^{4+} (32.5 eV) (38, 39). Apparently the dopants are present at the surface as Bi^{3+}, Mo^{6+} and W^{6+}.

Table 2 Binding energies and surface composition as determined by XPS.

Catalyst/ Composition	BE Ce $3d_{5/2}$ (eV)		BE dopant (eV)	Ce:M ratio (XPS)[a]
1/ $Ce_{0.9}Bi_{0.1}$	882.1	Bi $4f_{7/2}$	158.9	3.5
4/ $Ce_{0.9}Mo_{0.1}$	882.7	Mo $3d_{5/2}$	232.7	2.5
8/ $Ce_{0.9}W_{0.1}$	882.1	W $4f_{7/2}$	36.1	3.6

[a] Calculated using the appropriate correction factors. The overall Ce:M ratio of each catalyst is 9:1.

Although the dopant dissolves in the ceria lattice, we cannot rule out the presence of an amorphous dopant-rich phase at the surface of the catalyst (even after severe calcining). XPS + XRD measurements show a dopant-lean bulk and a dopant-rich surface. The structural similarity of the different catalysts is supported by the surface area-pore volume relationship (Figure 3).

Doping the ceria lattice creates stress, easing the transport of oxygen atoms via interstitial sites. It also alters the oxygen vacancy mobility (40), the crystallite size (41), the acid/base properties (42), and the electronic affinity. All these will affect the catalytic properties. Consider, for example, the reduction with hydrogen shown in the temperature programmed reduction (TPR) profiles in Figure 5. All ten materials exhibit some hydrogen uptake below 600 °C. The $Ce_{0.9}Sn_{0.1}$ and $Ce_{0.9}Y_{0.1}$ samples have similar surface areas, but the former reduces at low temperatures and the latter only at high temperatures. Though the

TPR is suitable for characterization of the catalysts, it is not ideal for determining the oxygen storage capacity of the material (43).

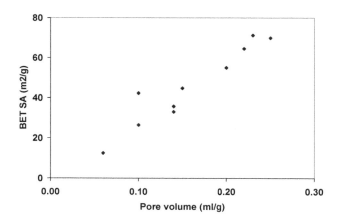

Figure 3 Total surface area of the doped ceria samples plotted versus pore volume as determined by nitrogen physisorption.

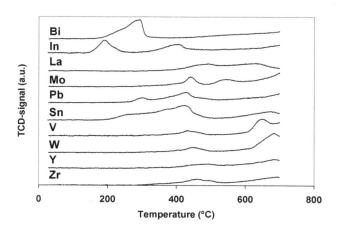

Figure 4 TPR patterns of the doped cerianites (ramp rate 300 °C/min, in 20 mL/min of 67 vol% H_2 in Ar).

The catalysts' surface structure and composition will determine the oxidation selectivity (44, 45). The surface of our catalysts contains lattice oxygen atoms, as well as metal ions (cerium or dopant) that can act as Lewis acids. For selective oxidation, the metal-oxygen bond has to be of intermediate strength (if the bond is too weak, the oxygen will react indiscriminately, and if the bond is too strong no reaction will occur) (46). The structure must be able to accommodate vacancies without collapsing. Moreover, the transfer of electrons, ion vacancies, and lattice oxygen atoms must be fast. Local dopant concentrations will facilitate the creation of (oxygen) vacancies (47). The surface of the catalyst is dynamic — the oxygen buffer is continuously drained and replenished. This ensures a dynamic concentration profile of oxygen species on the catalyst. The selectivity will change during the cyclic process with the oxygen concentration at the surface (48), because the requirements of the two oxidation reactions are different: one mole of lattice oxygen atoms is sufficient to oxidise one mole of H_2, but complete oxidation of ethylene requires six moles of oxygen atoms.

Depending on the electronegativity and the cation's oxidation state, the oxide can be either acidic or basic. Also, doping the ceria lattice does not automatically mean a substitution of a host cerium cation. Inomata and co-workers showed, in the case of $Ce_{(1-x)}Fe_xO_y$, that Fe^{3+} ions were located not only at Ce^{4+} sites, but also at interstitial sites (49). These interstitial ions bring about an increased barrier for electrons moving between Ce^{3+} and Ce^{4+}, and influence the redox potential and the ease of reducibility of the material.

Metal-doped cerianites can be used as selective and stable solid oxygen carriers for high-temperature hydrogen oxidation. The main advantage of these materials over the silica-supported metal oxides (18) is that the catalyst remains a solid, with no liquid phase forming in the reactor even at 600 °C. The stable oxygen uptake in repeated experiments is a further indication that these materials do not sinter during the reduction step. This compensates for the moderate surface area of these materials compared to silica and alumina. The XR diffraction and XPS studies show that there is considerable migration of dopant ions to the surface. While we do not have a conclusive mechanism for the oxidation reaction, the increased concentration of dopant at the surface supports the hypothesis that hydrogen is oxidized at dopant-enriched sites by lattice oxygen that migrates to the surface. Future work in our laboratory will include more extensive surface characterization and transient kinetic measurements to gain further understanding into this interesting system.

Experimental Section

X-ray photoelectron spectroscopy (XPS) was performed at 25 °C on a VG Escalab 200 spectrometer equipped with an Al Kα source and a hemispherical analyzer with a five-channel detector. Measurements were carried out at 20 eV pass energy. Traces of carbon were found on the samples, for which the charging was corrected using the C(1s) peak at 285.0 eV as a reference. XP spectra were fitted with CasaXPS (version 2.2.34). A linear background was applied and the d and f doublets were fitted using the appropriate area constraints. A detailed description of the catalyst synthesis and the XR diffraction analysis procedures was published (25).

Temperature programmed reduction (TPR) experiments. TPRs were performed for each material using a quartz reactor tube (4 mm i.d.), in which a 100 mg sample was mounted on loosely packed quartz wool. Samples were pre-dried overnight at 120 °C. The sample was heated at 5 °C /min up to 700 °C under 20 mL/min flow of a 2:1 mixture of H_2:Ar.

Hydrogen oxidation and catalyst regeneration. Each catalyst sample (50 mg) was mounted on quartz wool in a quartz tube reactor (3 mm inner diameter) and heated to 600 °C under He (50 mLmin^{-1}). The reactor temperature was maintained at 600 °C and the sample was subjected to a constant flow of 50 mLmin^{-1} of 1% v/v O_2 in He for 18 min. The reactor was then purged with pure He for 1 min, subjected to a reducing mixture of 5% H_2, 20% ethane and 20% ethylene (v/v in He) for 10 min, and purged again with pure He for 1 min. The product streams were analyzed by gas chromatography and selective mass spectrometry, enabling the separation and quantification of H_2, N_2, O_2, CO, CH_4, C_2H_4, C_2H_6 and CO_2, in this order. Oxygen exchange, in molkg^{-1}, was determined by subtracting the CO and CO_2 production from the total O_2 consumption in the oxidation step. This oxygen consumption was measured using a trace of argon in the oxygen flow, monitored at every time step by following the Ar and O_2 MS-signals. The amounts of CO and CO_2 produced were calculated using both GC- and MS-measurements.

Acknowledgements

We thank Dr. E. Eiser for the XRD measurements, M. Mittelmeijer-Hazeleger for surface characterisation, Dr. E. J. M. Hensen (TU Eindhoven) for the XPS measurements, and Dr. J. Louwen for discussions.

References

1.   D. Richards, *Chem. Market Rep.*, **259**, 7 (2001).
2.   J. Baker, *Eur. Chem. News*, 18 (1999).
3.   B. M. Weckhuysen, R. A. Schoonheydt, *Catalysis Today*, **51**, 223 (1999).
4.   Y. She, J. Han, Y. H. Ma, *Catalysis Today*, **67**, 43 (2001).
5.   Y. Yildirim, E. Gobina, R. Hughes, *J. Membrane Sci.*, **135**, 107 (1997).
6.   J. Mars, D. W. van Krevelen, *Chem. Eng. Sci.*, **3**, 41 (1953).

7. A. S. Bodke, D. A. Olschki, L. D. Schmidt, E. Ranzi, *Science*, **285**, 712 (1999).
8. M. Cherian, M. S. Rao, W. T. Yang, H. M. Jehng, A. M. Hirt, and G. Deo, *Appl. Catal., A Gen.*, **233**, 21 (2002).
9. P. Moriceau, B. Grzybowska, Y. Barbaux, G. Wrobel, C. Hecquet, *Appl. Catal. A Gen.*, **168**, 269 (1998).
10. M. Pedernera, M. J. Alfonso, M. Menéndez, J. Santamaría, *Chem. Eng. Sci.*, **57**, 2531 (2002).
11. Y. S. Choi, Y. K. Park, J. S. Chang, S. E. Park, A. K. Cheetham, *Catal. Lett.*, **69**, 93 (2000).
12. C. H. Lin, K. C. Lee, B. Z. Wan, *Appl. Catal. A Gen.*, **164**, 59 (1997).
13. R. K. Grasselli, D. L. Stern, J. G. Tsikoyiannis, *Appl. Catal. A Gen.*, **189**, 1 (1999).
14. C. Courson, B. Taouk, E. Bordes, *Catal. Lett*, **66**, 129 (2000).
15. A. Julbe, D. Farrusseng, D. Cot, C. Guizard, *Catalysis Today*, **67**, 139 (2001).
16. R. K. Grasselli, D. L. Stern, J. G. Tsikoyiannis, *Appl. Catal. A Gen.*, **189**, 9 (1999).
17. J. G. Tsikoyiannis, D. L. Stern, R. K. Grasselli, *J. Catal.*, **184**, 77 (1999).
18. L. M. van der Zande, E. A. de Graaf, G. Rothenberg, *Adv. Synth. Catal.*, **344**, 884 (2002).
19. J. Kašpar, P. Fornasiero, M. Graziani, *Catalysis Today*, **50**, 285 (1999).
20. S. D. Park, J. M. Vohs, R. J. Gorte, *Nature*, **404**, 265 (2000).
21. E. P. Murray, T. Tsai, S. A. Barnett, *Nature*, **400**, 649 (1999).
22. B. M. Maunders, S. R. Partington, WO Pat. 94/05608, to BP Chemicals Limited (1994).
23. E. A. de Graaf, A. Bliek, WO Pat. 02/055196, to Universiteit van Amsterdam (2002).
24. A. E. C. Palmqvist, M. Wirde, U. Gelius, M. Muhammed, *Nanostruct. Mater.*, **11**, 995 (1999).
25. G. Rothenberg, E. A. de Graaf, A. Bliek, *Angew. Chem. Int. Ed.*, **42**, 3365 (2003).
26. M. Ozawa, C.-K. Loong, *Catalysis Today*, **50**, 329 (1999).
27. S. Bernal, J. J. Calvino, M. A. Cauqui, J. M. Gatica, C. Larese, J. A. Pérez Omil, J. M. Pintado, *Catalysis Today*, **50**, 175 (1999).
28. D. J. Kim, *J. Am. Ceram. Soc.*, **72**, 1415 (1989).
29. R. D. Shannon, *Acta Cryst.*, **A32**, 751 (1976).
30. G. Li, Y. Mao, L. Li, S. Feng, M. Wang, X. Yao, *Chem. Mater.*, **11**, 1259 (1999).
31. A. Galtayries, R. Sporken, J. Riga, G. Blanchard, R. Caudano, *J. Electron Spectrosc. Relat. Phenom.*, **88-91**, 951 (1998).
32. D. Briggs, M. P. Seah, Practical Surface Analysis, John Wiley & Sons, New York, Vol. 1, 2nd edn. (1990).

33. V. I. Nefedov, D. Gati, B. F. Dzhurinskii, N. P. Sergushin, Y. V. Salyn, *Rus. J. Inorg. Chem.*, **20**, 2307 (1975).
34. W. E. Morgan, W. J. Stec, J. R. van Wazer, *Inorg. Chem.*, **12**, 953, (1973).
35. C. D. Wagner, W. M. Riggs, L. E. Davis, and J. F Moulder, Handbook of X-ray Photoelectron Spectroscopy, G. E. Muilenberg, Ed., Perkin Elmer Corporation, Eden Prairie, Minnesota, (1979).
36. I. Olefjord, B. Brox, U. Jelvestam, *J. Electrochem. Soc.*, **132**, 2854 (1985).
37. S. O. Grim, L. J. Matienzo, *Inorg. Chem.*, **14**, 1014 (1975).
38. S. Kumar, D. R. Chopra, G. C. Smith, *J. Vac. Sci. Technol. B.*, **10**, 1218 (1992).
39. G. E. McGuire, G. K. Schweitzer, T. A. Carlson, *Inorg. Chem.*, **12**, 2450 (1973).
40. P. Fornasiero, J. Kašpar, M. Graziani, *Appl. Catal. B Environ.*, **22**, L11 (1999).
41. C. de Leitenburg, A. Trovarelli, J. Llorca, F. Cavani, G. Bini, *Appl. Catal. A. Gen.*, **139**, 161 (1996).
42. M. G. Cutrufello, I. Ferino, V. Solinas, A. Primavera, A. Trovarelli, A. Auroux, C. Picciau, *Phys. Chem. Chem. Phys.*, **1**, 3369 (1999).
43. For zirconia-doped ceria it was shown that redox cycling can enhance the oxygen storage capacity, see F. Fally, V. Perrichon, H. Vidal, J. Kaspar, G. Blanco, J. M. Pintado, S. Bernal, G. Colon, M. Daturi, J.C. Lavalley, *Catalysis Today*, **59**, 373 (2000).
44. R. K. Grasselli, *Top. Catal.*, **21**, 79 (2002).
45. J. C. Védrine, *Top. Catal.*, **21**, 97 (2002).
46. J. M. Thomas, W. J. Thomas, Principles and Practice of Heterogeneous Catalysis, VCH, Weinheim, 1997, p. 29-30.
47. P. J. Gellings, H. J. M. Bouwmeester, *Catalysis Today*, **58**, 1 (2000).
48. P. Pantu, K. Kim, G. R. Gavalas, *Appl. Catal. A Gen.*, **193**, 203 (2000).
49. G. S. Li, R. L. Smith, H. Inomata, *J. Am. Chem. Soc.*, **123**, 11091 (2001).

24. Nanocluster-based Cross-coupling Catalysts: A High-Throughput Approach

Mehul B. Thathagar, Jurriaan Beckers, and Gadi Rothenberg*

van 't Hoff Institute for Molecular Sciences, University of Amsterdam, Nieuwe Achtergracht 166, 1018 WV Amsterdam, The Netherlands
e-mail gadi@science.uva.nl *Fax: +31 20 525 5604*

Abstract

Microwave irradiation is used to create 'hot spots' on metal clusters in solution, facilitating catalytic cycles in which these clusters participate. An 8×12 parallel screening system is built based on this concept, and tested using the Heck reaction as a case study. The spatial reproducibility of this system is examined and the pros and cons of monomode and multimode m/w setups are discussed.

Introduction

Carbon-carbon coupling reactions are very useful in organic synthesis (1). Petrochemistry gives us simple olefins and aromatics, and these can be coupled to construct a variety of agrochemical and pharmaceutical intermediates. Metal clusters feature in many C–C coupling mechanisms, *e.g.* in the Heck, Suzuki, and Ullmann-type reactions (2). Whether clusters are the actual catalysts for these reactions or not remains a hot scientific debate (3), but it is certain that clusters participate in the catalytic cycle (4). Our interest in these clusters is two-fold: First, if clusters are indeed the catalysts, then ligand-free applications may be envisaged (5). Second, combining different metals may result in synergistic effects, as we demonstrated for Suzuki cross-coupling (6, 7).

The activity of metal clusters led us to think that selective heating of the catalytic sites may be possible. If catalysis occurs on the clusters' surface, this could give higher substrate conversion and/or better product selectivity. The clusters are small pieces of metal, so they could be selectively heated in solution using microwave (m/w) irradiation. There have been several reports of so-called 'non-thermal microwave effects' in organic reactions (8, 9). However, Strauss and co-workers showed that microwaves merely create 'hot spots' in the solution (10, 11). In any case, using microwaves is a versatile and efficient heating technique (12), and we reasoned that the metal clusters could serve as focal points for these hot spots, resulting in a 'hot catalyst' and a favourable reaction environment (13). Another incentive is the possiblity of parallel screening, which is a must if one wants to investigate numerous cluster combinations. Here, we present our first results on microwave activation of nanocluster catalysts, using the Heck reaction as a case study.

Results and Discussion

Our first objective was to construct a simple, inexpensive experimental system for the synthesis and rough-quality parallel screening of nanocluster catalysts. For this, we adapted the 96 well plate format to a conventional m/w oven. The setup was tested on the Heck cross-coupling of various aryl halides with *n*-butylacrylate. Reactions were perfomed in 2 mL glass vials set in a PTFE matrix and monitored by GC. We screened twelve different substrates, **1–12**, and eight different catalysts, **A–H** (see Figure 1). The nanocluster suspensions **A–D** were prepared by reducing stock solutions of their metal chloride precursors using tetra-*n*-octylammonium formate (TOAF) in DMF at 65 °C. Row **E** was used as a control (no metal catalyst). Row **G** was prepared using NaEt₃BH as reducing agent and poly(N-vinylpyrrolodinone) (PVP) as stabiliser. A homogeneous palladium complex, Pd(PPh₃)₄, was placed in row **H** for comparison.

Substrate reactivity was as expected (ArI > ArBr >> ArCl). In contrast to the Suzuki cross-coupling, however, Cu and Ru clusters were not active in the Heck reactions, and the activity of Cu/Pd clusters was lower than that of pure Pd clusters. Note the higher activity of Pd clusters prepared *in situ* (row **F**) compared to pre-prepared clusters (rows **B** and **G**). This increased activity tallies with our findings for Suzuki cross-coupling (7). After reaction, palladium black was observed in all the vials in rows **B** and **G**, but not in row **F**.

Control experiments confirmed that DMF solutions containing small amounts of PVP-stabilised metal clusters are heated much faster under m/w irradiation than pure DMF or DMP/PVP solutions. Thus, pure DMF and DMF/PVP took 7 min to reach boiling (153 °C), while a 10 mM soln of PVP-stabilised Pd clusters in DMF boiled after 2 min (all other parameters were identical, including reaction vessel position in the oven, *vide infra*).

Further, we examined the Heck reaction between *n*-butyl acrylate and 4-bromobenzotrifluoride **5** in the presence of 2 mol% Pd clusters in a single-vessel monomode m/w oven fitted with an infrared thermometer. 100% conversion with quantitative yield to the cinnamate was obtained after 5 min irradiation at 75 W/240 °C. We then repeated the reaction under conventional heating at 240 °C. After 3.5 min a black tarry gel formed. Extraction followed by GC analysis showed only cinnamate, but the tarry material (probably acrylate polymers/oligomers) could not be analysed. These experiments show that when clusters are present different results are obtained depending whether m/w heating or conventional heating is used. In principle, this could be the result of 'hot spots' created on the metal clusters.

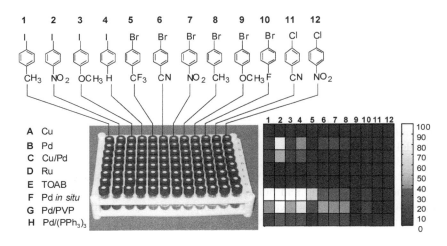

Figure 1 Rectangular 8×12 PTFE matrix containing 96 2 mL glass reaction vials (left) and the resulting conversions of aryl halides (**1–12**) using the catalysts **A–H** (right). Conditions: 0.062 mmol aryl halide, 0.12 mmol *n*-butylacrylate, 0.18 mmol KOAc, catalyst (2 mol% total metal relative to aryl halide), 0.5 mL DMF, 240 W m/w irradiation for 15 min.

We also examined the spatial reproducibility of the parallel 96-vessel matrix. Two groups of twenty-one reactions were performed simultaneously: the coupling of *n*-butylacrylate with iodobenzene using pre-prepared Pd clusters, and with 4-bromobenzotriflouride using Pd clusters prepared *in situ*. For easy distinction, the first group was covered with red septa and the second with white septa. The vials were arranged to study the influence of the positioning on the reactor plate, the presence/absence of neighbours, and the type of these neighbours. The reproducibility in the 'white' group was better than that of the

Table 1 Spatial reproducibility tests.

Entry	Position	Pre-prepared Pd clusters		*In situ* prepared Pd clusters	
		% Mean conversion	Stdev	% Mean conversion	Stdev
1	'edge'	22.1	10.8	7.1	3.2
2	'middle'	78.5	20.4	23.8	9.7
3	'lonely'	6.0	1.7	0	0

'red' group. For both reaction sets, rather to our surprise, the vials placed without any neighbours gave lower conversion, regardless of where they were placed on the matrix, while vials with many neighbours gave better results. For each reaction, we can divide the vials into fourteen 'edge' vials, four 'middle' vials, and three 'lonely' vials. Table 1 shows the mean conversion values and the standard deviations for the three types of positions.

The high standard deviation values reflect the inhomogeneous heating in the multimode m/w oven. Nevertheless, the advantages of fast and inexpensive parallel screening should not be neglected, for it can be used as a rough 'go/no go' indicator. Monomode ovens offer uniform heating, but they are harder to adapt to parallel experiments. We are now working on a system that will combine parallel testing and homogeneous energy distribution. There are two possible approaches: one way is to incorporate in the multimode cavity a rotating metal propeller that breaks up the standing waves. Alternatively, a monomode cavity can be constructed that has uniform energy distribution and sufficient space for multiple reaction vessels. Research on both applications is ongoing in our laboratory.

In conclusion, metal nanoclusters in DMF interact strongly with microwaves. In reactions catalysed by these clusters, the microwave heating may be tantamount to preferentially heating the catalytic site, which can lead to more effective catalysis. Such cluster-catalysed reactions can be in principle screened in parallel in multimode m/w ovens reducing both time and operational costs. However, the ovens must be adapted so that the parallel reactors are uniformly heated.

Experimental Section

Materials and instrumentation. A detailed description of the analytics and procedures for the synthesis of TOAF and catalysts **A–D** were published previously (6). Parallel m/w experiments were performed using a Samsung domestic microwave oven (M1712N, 2.45 GHz source, 800W o/p). Reactions were carried out in 2 mL glass vials sealed with PTFE septa. The reaction volume was filled not exceeding 1/4th of the total volume of the vial, thereby allowing headspace for pressure buildup. Holes were punched in the septa using a 15-gauge needle to prevent the caps of the vials from blowing. Care was taken to place identical quantity of DMF in each vial to avoid uneven heating. All reactions were performed without stirring. *In situ* preparation of Pd clusters (Figure 1, row **F**) was performed according to literature procedures (2).

Pd nanoclusters, sodium triethylborohydride (Na(Et)$_3$BH) method. This is a modification of Bönnemann's method (14): A Schlenk-type vessel was evacuated and refilled with N_2. The vessel was charged with PVP (0.22 g, 2 mmol) and a PdCl$_2$ solution in DMF (10 mL, 10 mM). 0.3 mL of a 1 M Na(Et)$_3$BH solution in THF was added in one portion at 25 °C, and the mixture was stirred for 2 h under a slight N_2 overpressure. The color changed from yellow-orange to black. The suspension was taken as is for the catalysis studies.

Heck coupling of aryl halides with *n*-butylacrylate under m/w irradiation. Stock solutions of the reactants were prepared in DMF. Aryl halide (125 μL, 495 mM), *n*-butylacrylate (125 μL, 1.0 M), KOAc (base, 3 equiv) and the nanocluster suspension (125 μL, 10 mM) were placed in the reaction vials. The 96 reactor PTFE matrix was then placed in the m/w oven on the turntable glass plate and irradiated in five-minute bursts at 240 W for 15 min, separated by 5 min pauses for cooling. The vials were cooled to room temperature. Pentadecane solution (external standard, 125 μL, 234 mM) was added and the suspension was filtered through a plug of Al_2O_3 and analysed by GC.

Heck coupling using conventional heating. 4-bromobenzotrifluoride (0.5 mL, 495 mM), *n*-butylacrylate (0.5 mL, 1.0 M), KOAc (approx. 3 equiv), and nanocluster suspension (0.5 mL, 10 mM) were sealed in a 5 mL glass tube and heated (240 °C, 3.5 min) in a water-cooled oven. Workup and analysis was performed as above.

References

1. F. Diederich, and P. J. Stang, Metal-Catalyzed Cross-coupling Reactions. Wiley-VCH, Weinheim, (1997).
2. M. T. Reetz, and E. Westermann, *Angew. Chem. Int. Ed.*, **39**, 165 (2000).
3. A. Biffis, M. Zecca, and M. Basato, *J. Mol. Catal. A: Chem.*, **173**, 249 (2001).
4. N. E. Leadbeater, and M. Marco, *J. Org. Chem.*, **68**, 5660 (2003).
5. V. M. Wall, A. Eisenstadt, D. J. Ager, and S. A. Laneman, *Platinum Met. Rev.*, **43**, 138 (1999).
6. M. B. Thathagar, J. Beckers, and G. Rothenberg, *J. Am. Chem. Soc.*, **124**, 11858 (2002).
7. M. B. Thathagar, J. Beckers, and G. Rothenberg, *Adv. Synth. Catal.*, **345**, 979 (2003).
8. P. Lidstrom, J. Tierney, B. Wathey, and J. Westman, *Tetrahedron*, **57**, 9225 (2001).
9. N. Kuhnert, *Angew. Chem. Int. Ed.*, **41**, 1863 (2002).
10. C. R. Strauss, *Angew. Chem. Int. Ed.*, **41**, 3589 (2002).
11. C. R. Strauss, and R. W. Trainor, *Aust. J. Chem.*, **48**, 1665 (1995).
12. M. Larhed, and A. Hallberg, *J. Org. Chem.*, **61**, 9582 (1996).
13. J. R. Thomas, *Catal. Lett.*, **49**, 137 (1997).
14. H. Bönnemann, and W. Brijoux, *Nanostruct. Mater.*, **5**, 135 (1995).

b. General Papers

25. Do You Need A Transition Metal for Biaryl Synthesis?

A. S. Kotnis,* K. Zhu, B. L. Lotz, K. J. Natalie, J. H. Simpson, D. Kacusr, D. Thurston, J. S. Prasad, S. Mathew, and A. K. Singh

Process Research and Development, Bristol-Myers Squibb Pharmaceuticals,
New Brunswick, NJ 08903-0191

atul.kotnis@bms.com

Abstract

The focus of this paper is on the preparation of biaryl moieties which are commonly present in a wide variety of pharmaceutically active compounds. Two different and practical approaches to synthesize the biaryl substructure will be discussed.

Introduction

Nature has an abundance of biologically interesting and pharmaceutically active natural products possessing the biaryl moiety. Synthetic chemists have thus devised a number of different methods to assemble the biaryl functionality.

Among the many approaches, the Pd(0)-catalyzed cross-coupling reaction has become the most widely used method for the preparation of biaryls. In particular, the Suzuki-Miyaura[1] coupling reaction involving an aryl boronic acid, and an aryl halide or triflate has emerged as one of the most popular. In the process research and development area it has virtually replaced the Migita-Kosugi-Stille coupling reaction for biaryls which involves the use of organotin compounds. The major difference between the two is that the Stille coupling reaction is neutral in nature, while the Suzuki coupling reaction is a base-mediated reaction. This may pose as a significant problem as base sensitive functionalities (like esters) may not survive the coupling reaction. The utility of this coupling reaction has been greatly enhanced by the successful use of inexpensive aryl chlorides as partners in Suzuki couplings. Aryl boronic acids required for these couplings are very easily synthesized by transmetallation of aryl halides, or via directed metallation chemistry or, also, organopalladium chemistry. Aryl boronic acids can be isolated as their pinacol derivatives or used directly in coupling reactions to provide biaryls.

Results and Discussion

In connection with one of our drug candidates, an endothelin receptor antagonist (VI), which was being developed for congestive heart failure, the need arose to prepare large quantities of aryl boronic acid (II). Suzuki coupling between iodoxazole (III) and aryl boronic acid (II) provided coupled biaryl (IV), as its sodium salt after hydrolysis (Scheme 1). The biaryl sodium salt (IV) was activated with Vilsmeier reagent to form the biaryl sulfonyl chloride, and reacted with the appropriate amine to provide the final drug.

Scheme 1

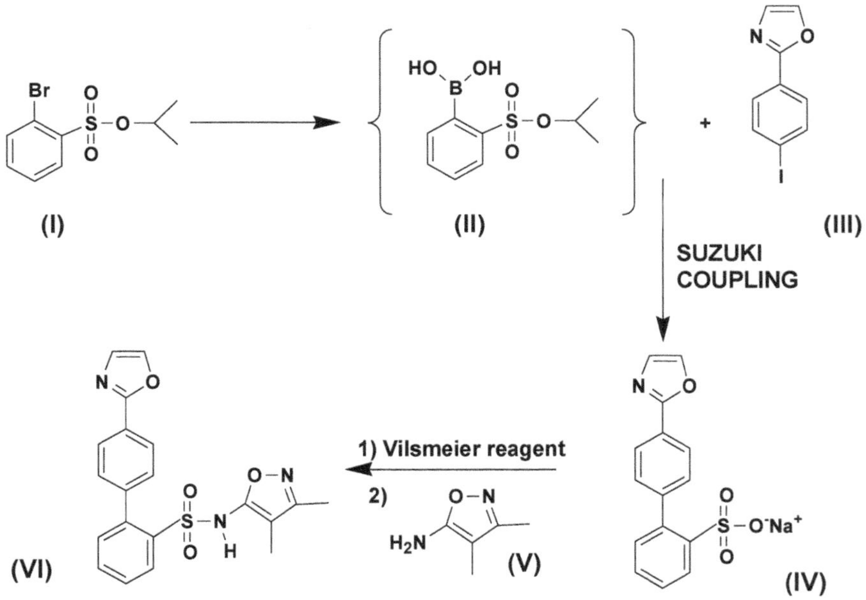

In recent years, a variety of aryl boronic acids are commercially available, albeit in some cases they may be expensive for large scale purposes. During our work in the mid-1990's boronic acid (II) was not commercially available and so two different protocols were used to prepare this acid. The first approach involved the transmetallation with n-butyl lithium of aryl bromide (I) and trapping the lithio species generated with trialkyl borate followed by an acid quench. Aryl bromide (I) is easily prepared by reaction of o-bromobenzenesulfonyl chloride with 2-propanol in the presence of pyridine as a base. The second approach was a directed metallation of isopropyl ester of benzene sulfonic acid (VII), to generate the same lithio species and reaction with trialkyl borate. The sulfonyl ester is prepared by reaction of 2-propanol with benzenesulfonyl chloride. From a long-term strategy the latter approach is

obviously preferred due to the low cost of the respective starting materials as
shown in Scheme 2.

Scheme 2

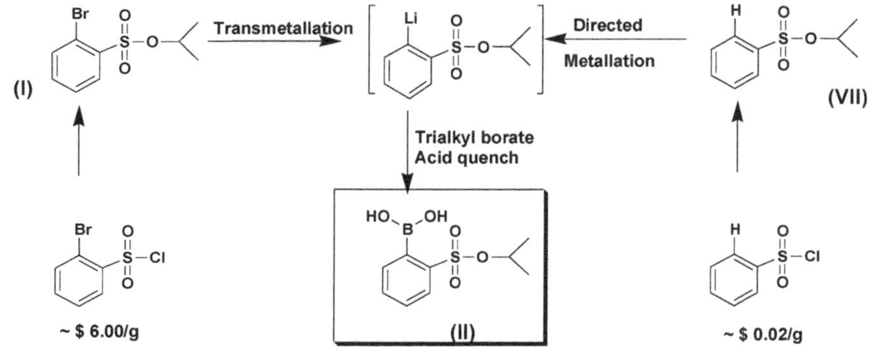

Preparations of Aryl boronic acid (II).   The isopropyl ester (I) is
transmetallated with n-butyl lithium (instead of using LDA [2b]or the more
expensive LiTMP[2c]), followed by boronation with a trialkyl borate.   Initially,
boronic acid (II) was prepared by the *sequential addition* outlined above to give
the aryl boronate, which on quenching with a mineral acid gives boronic acid
(II).   This protocol to prepare boronic acid (II) exhibits a distinct color change
when the lithio species is formed, and the color is discharged when the lithio
species reacts with electrophilic boron to form the aryl boronate.   This protocol
gives generally high quality of boronic acid (II) containing about 2-3% of de-
brominated compound (VII).   Though, the formation of this de-brominated
compound (VII) does not cause any problems in the subsequent Suzuki
coupling, its formation is detrimental to the purity of the product.   The isolated
yield for this approach typically ranged from 89-92 mole % with a mass balance
of > 97 mole %.

An alternative approach to reduce the levels of impurity (VII) would be to
have a *"transient"* existence of the lithio species, so that it reacts
instantaneously with trialkyl borate to form the aryl boronate, prior to being
quenched by any extraneous proton source to form (VII).   Thus, the preparation
of boronic acid (II) was improved by changing the order of the reagents.   The
slow addition of n-butyl lithium also controls the exotherm of the reaction.
There was no reaction observed between n-butyl lithium and triisopropyl borate
(to form any butyl boronic acid), nor was there any formation of 2-butyl
derivative of (VII) formed by reaction between butyl bromide and the lithio
species.   The reaction is very fast and as soon as the addition of n-butyl lithium
is completed the reaction is finished.   This indicates a rapid transmetallation and
instantaneous boronation of the lithio species.   The reaction is very much a

titration of (I) with n-butyl lithium to form aryl boronate. Experiments were conducted to evaluate portion-wise addition of butyl lithium. The results show an equivalent amount of aryl boronate formed to the amount of butyl lithium added. The level of de-brominated compound (VII) was found to be < 0.5% by this protocol, and the overall yield of the boronic acid (II) was 95-96 mole %. The largest scale for demonstration of this protocol was on was 19.0 kg of (I) to provide 15.0 kg of boronic acid (II) with a HPLC quality of 99.8 AP.

We have broadened the scope of this *reverse addition* protocol to prepare a variety of boronic acids bearing different functional groups for use in Suzuki coupling reactions. The yield and quality of the boronic acid prepared by this *reverse addition* protocol is usually better than the sequential approach. The boronic acids can be used without further purification (formation of pinacols) in Suzuki coupling reactions.

For the second approach involving a directed metallation of (VII), the reaction involves generation of the same lithio species and treatment with trialkyl borate to provide the same aryl boronate and boronic acid (II) on acid quench. This preparation is similar to the sequential addition as it involves the complete generation of lithio species prior to addition of the trialkyl borate. In the formation of aryl boronic acid and a slight excess of (VII) is always used to avoid bismetalation occurring at both the "*ortho*" positions of the sulfonyl group generating "bisaryl boronic acids" as impurities. Aryl boronic acid (II) prepared by either approach was used without further purification directly in the Suzuki coupling with iodoxazole (III) to generate coupled product (IV).

Suzuki Coupling Reaction (IV). The coupling of boronic acid (II) with iodooxazole (III) was the first Suzuki reaction done on-scale at Bristol Myers Squibb. Our hope from the beginning was to isolate isopropyl ester of the coupled biaryl, so that issues related to removal of Pd can be separated from other quality issues of the coupling reaction as outlined in Scheme 3.

The coupling reaction being heterogeneous in nature could pose sampling problems on scale-up. Additionally we also observed that boronic acid (II) readily de-boronates on heating it at temperatures around 40 °C and exposure to high pH of 10 to 10.5. For a successful Suzuki coupling reaction of (II) and (III), maintaining a basic pH in the range of 9.5 to 10.5 and a temperature ~ 45 °C was very essential. It was also important to have the reactants (II) and (III) to be compatible to the Pd catalyst to be used at ambient or temperature of the reaction in the absence of the base used for the reaction.

The process variables for the coupling reaction in addition to the temperature and the pH of the reaction, also involved the equivalents of iodoxazole, % of Pd catalyst required for the reaction, choice of base, and also the choice of reaction solvent. After many experiments the choice of Pd catalyst was $Pd_2(dba)_3$ instead of $Pd(PPh_3)_4$. A wide variety of bases were examined

Scheme 3

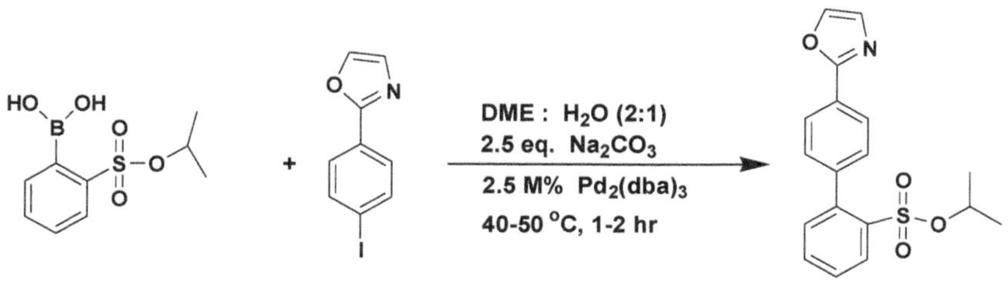

(KHCO$_3$, NaOAc, K$_2$HPO$_4$, Na$_3$PO$_4$, triethyl amine) and finally sodium carbonate was selected as the base of choice. Fluoride initiated Suzuki coupling with KF was unsuccessful. Dimethoxy ether was selected as the solvent after screening a variety of solvents (acetone, tetrahydrofuran, methanol, isopropyl alcohol, and methyl-*t*-butyl ether).

Due to the thermal and basic lability of boronic acid (II), after experimenting with various addition sequences for efficient mixing of the heterogeneous reaction a *slow or portion-wise addition of boronic acid (II)* to a mixture of iodoxazole, Pd$_2$(dba)$_3$ and sodium carbonate in dimethoxy ether-water at around 40 °C to 50 °C was the preferred mode of addition. The optimum amount of base was 2.8 to 3.2 equivalents and normally the reaction was performed using ~ 3.0 equivalents of a 1.5 M solution of sodium carbonate and 2-3 M% of Pd$_2$(dba)$_3$ was used. The amount of boronic acid (II) required for the coupling reaction is 1.2 –1.25 equivalents relative to the iodoxazole (III). One important aspect of the Suzuki reaction is all the solvents of the reaction have to be completely degassed to avoid deactivating the catalyst and stopping the reaction.

All efforts in the laboratory to isolate the coupled product as its isopropyl ester were unsuccessful. The high pH of the reaction required for the coupling reaction renders partial hydrolysis of the coupled isopropyl ester. Thus the reaction mixture on complete consumption of the iodoxazole (III) has a mixture of nearly 1:1 of the coupled isopropyl ester to its hydrolyzed product. Additional water was added to the reaction and the reaction was further heated at 40-50 °C for 4-16 hours when complete hydrolysis to the sodium salt is observed (Scheme 4).

Our fundamental goal was to separate the two quality issues: a) lowering the levels of Pd (<10 ppm) and boron (<150ppm) and b) quality of isolated sodium salt.

Scheme 4

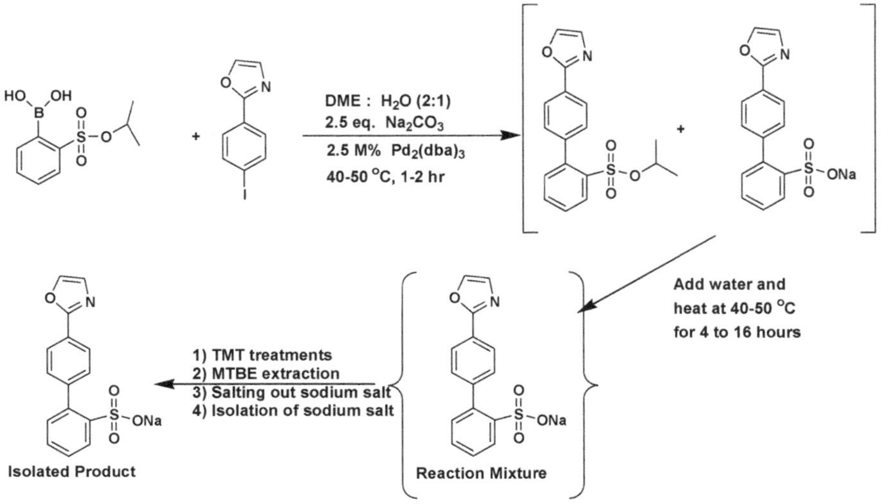

The reaction mixture was filtered to separate it from solid wastes containing Pd. The reaction mixture was then treated with 1,3,5-trimethyl thiocyanuric acid (TMT) for removal of the Pd used in the reaction. The treatment of TMT was performed at pH of ~ 10, by adding 5% TMT relative to iodoxazole used for the reaction and stirring the reaction mixture for 30 minutes at room temperature. This reduces the level of Pd in the range of 1000-3000 ppm in the reaction mixture. The reaction mixture is filtered and the pH is adjusted to around 2 by adding concentrate hydrochloric acid. A second treatment with 5% TMT followed by stirring for 30 minutes at ambient temperature. The level of Pd drops into a range of 200-500 ppm in the reaction mixture after filtration. The reaction stream is fairly dark and passing it through a Darco pad decolorizes minimally but reduces the level of Pd to below our desired level of 10 ppm. The levels of boron were always found to be below 125 ppm.

The pH of the reaction stream was adjusted to neutral and the dark stream was washed with methyl-t-butyl ether to remove neutral organic impurities. The major impurity observed in the MTBE wash is phenyl oxazole formed by de-iodination of the starting iodoxazole. Phase separation of this wash is a difficult one due to the dark color of the reaction stream.

All efforts of trying to extract the sodium salt into an organic solvent were unsuccessful, and, thus, a salting out procedure using the common ion effect was used. The product at this stage is in the aqueous phase and addition of 15 g of sodium chloride /100 mL of the reaction mixture, completely precipitates all the coupled sodium salt which is filtered. The filtered crude product is dewatered by washing with isopropyl alcohol or drying under vacuum. The mother liquors of this isolation of crude product contain benzene sulfonic acid as the major product. The isolated crude dried product contains some sodium chloride and needed to be separated from the coupled product. The crude product is dissolved by warming in SDA3A at 50-60 °C. Then a hot polish filtration is performed to remove the sodium chloride which does not dissolve in hot SDA3A (95% ethanol and 5% isopropyl alcohol). The filtered solution is then concentrated and addition of MTBE crystallizes the product which is filtered, washed and dried to give sodium salt, which can be transformed successfully to the final drug substance. The largest scale the coupling reaction has been performed is on 12 kg of iodoxazole providing very high quality of coupled sodium salt (99.9 HPLC AP). The range for the yield of the coupling reaction is 73-79 M%.

The iodoxazole required for the coupling is prepared in two steps from readily available 4-iodobenzoyl chloride and amino dimethyl acetaldehyde. Formation of the amide followed by reaction with Eaton's reagent generates the iodoxazole (Scheme 5). In an effort to further reduce the cost of the coupling reaction, the two step was applied to provide 4-bromoxazole starting from 4-bromo benzoyl chloride which costs 70% less than the corresponding iodo compound. The bromoxazole works to provide the sodium salt, but requires either higher amount of catalyst, and boronic acid (II) and longer reaction times. The reaction was not scaled up using bromoxazole.

In connection with another project being developed we needed large quantities of a biaryl phenol or biaryl aryl methyl ether which was being prepared by a Suzuki coupling reaction (Scheme 6).

Scheme 5

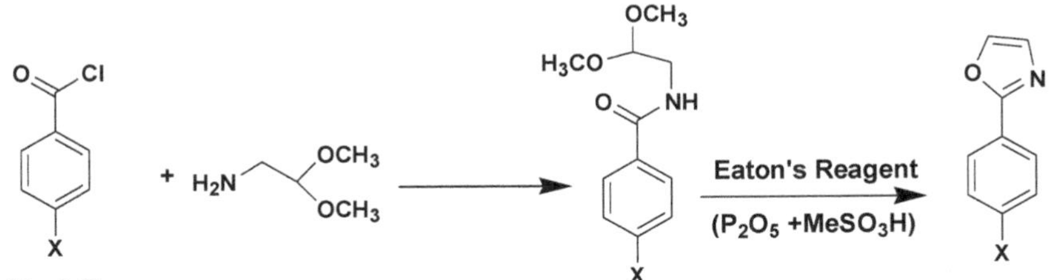

X = I, Br

Scheme 6

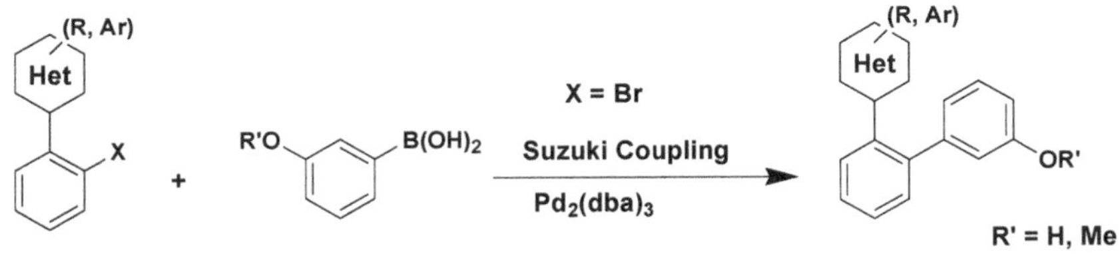

Unlike in the earlier example, boronic acid (IX) is readily available (R' = H, or Me). The coupling partner is easily prepared in two-steps from readily available starting materials, providing a highly convergent synthesis of biaryl (X). The desired compound was the biaryl phenol (X) which was initially prepared from 3-methoxy phenyl boronic acid and after the Suzuki coupling a demethylation was performed using boron tribromide or trimethylsilyl iodide. Though both these reagents have been employed at Bristol-Myers Squibb on other projects on-scale, but they are not environmentally friendly. The second phase involved a Suzuki coupling with 3-hydroxyphenylboronic acid which directly gives the desired biaryl phenol (X) thereby eliminating one step which employs unfriendly reagents. The coupling reaction also works on the chloro analog of coupling partner (VIII).

The biaryl phenol (X) was the penultimate intermediate in the synthesis of this final drug substance. Thus after the Suzuki coupling reaction is performed to give the phenol, the levels of Pd have to be lowered to < 10 ppm. In the pharma industry this can be a significant problem. Additionally there is always batch to batch variability observed when catalysts like $Pd_2(dba)_3$ have been used in Suzuki coupling reactions.

The biaryl phenol (X) is converted in one step to the final drug substance which was being developed for diabetes. The dose requirement for the phase studies for this program was very high. The patients were to be treated with two 500 mg tablets per day. In addition, the cost of the two coupling partners (VIII) and (IX) were very high. Even on bulk the 3-hydroxyphenyl boronic acid (IX) was $3500/kg and the bromo (or chloro) analog was in the range of $4500/kg. Thus, an alternative, inexpensive approach was developed which we call the *"out-of-the-box"* approach. This protocol would significantly reduce the overall cost of synthesis of biaryl phenol (X).

Several years ago in connection with the synthesis of a milbemycin β3 we had the occasion to develop a new route to prepare *p*-methoxybenzoates moiety present in milbemycins. The approach involved preparing the appropriate Hagemann's ester and aromatizing with iodine and methanol to the corresponding *p*-methoxybenzoates in very high yields.[3] Hagemann's ester bearing a phenyl group provided biaryl methyl ether on aromatization (Scheme 7).

Scheme 7

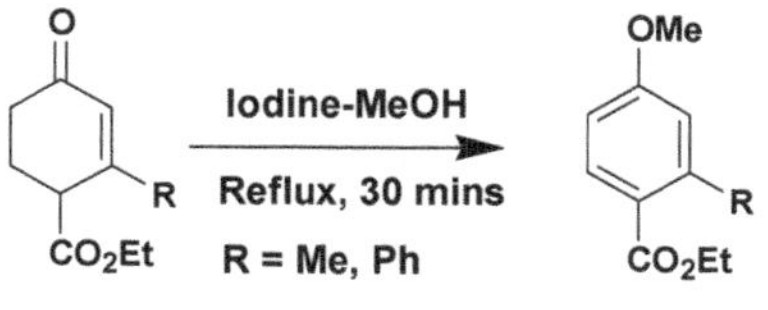

Scheme 8

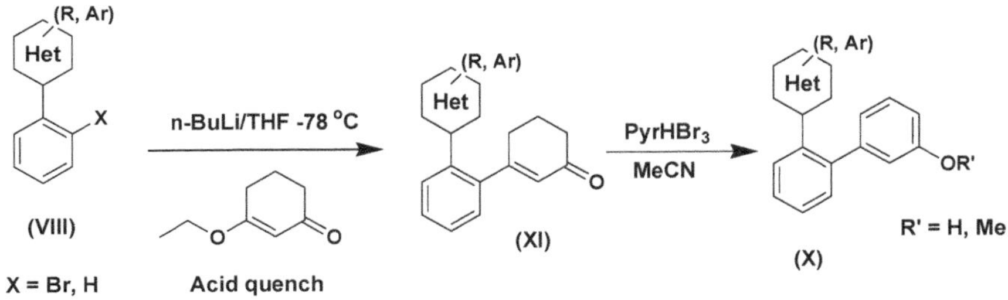

Transmetallation of aryl halide (VIII) with n-butyl lithium and treating it with readily available and inexpensive 3-ethoxycyclohexen-1-one followed by acid work-up would provide enone (XI) (Scheme 8). Enone (XI) can be then aromatized to phenol (X) or any other biaryl alkyl ether. By using this protocol, 3-ethoxycyclohexen-1-one is a synthon for 3-hydroxyphenyl boronic acid. The bulk price of 3-ethoxycyclohexen-1-one is $85/kg compared to $3500 for the boronic acid, thereby significantly reducing the cost to prepare biaryl (X).

Interestingly, when the chloro analog was transmetallated and treated with 3-ethoxy cyclohexen-1-one, the expected enone (XI) was not observed, but an enone with a mass of 34 units greater than (XI) was noticed. It also indicated the enone carried the chloro analog. It was presumed that the hetero atoms in the heterocycle present in the starting material (VIII) had performed a directed metallated lithiation providing a different enone bearing the chloro moiety.

Based on this information the preparation of enone was examined from the unhalogenated (VIII)(X=H). Deprotonation can be performed with n-butyl lithium in THF at 0-5 °C followed by treatment with 3-ethoxy cyclohexen-1-one, followed by an acid quench provides the same enone (XI). This deprotonation also avoids the cryogenic conditions required to prepare enone (XI) when the bromo analog is used. Pyridinium tribromide used for aromatization of enone (XI) to biaryl phenol (X) is an inexpensive reagents ($80/kg).

The unhalogenated compound is also prepared in two steps using the same chemistry employed for the bromo analog (VIII). The cost of the bromo starting material is 100 times the cost of unhalogenated starting material prior to the two-steps to introduce the heterocyclic ring. This provides an additional cost savings to the overall process which avoids the use of Pd metal used as a catalyst in preparing the biaryl phenol (X) using the Suzuki coupling.

226

References

1. (a) N. Miyaura and A. Suzuki, *Chem. Rev.*, **95**, 2457-2483 (1995). (b) A. Suzuki, *J. Organometal. Chem.*, **576**, 147 (1999). (c) A. Suzuki, In Metal-Catalyzed Cross-Coupling Reactions; F. Diedrich, P. J. Stang, Eds.; VCH: Weinheim, 1998; pp 49-97, and references cited therein.
2. (a) Portions of this work were first presented at the Bristol-Myers Squibb Process Review meeting held on 14th August 1996 at New Brunswick, NJ; and also recently at the 85th Canadian Society for Chemistry, June 2002 at Vancouver (b) A. Singh, C. K. Chen, J. A. Grosso, E. Delaney, X. Wang, R. P. Polniaszek, and J. K. Thottathill, PCT Int. Appl. WO 0056685, 2000. (c) S. Caron and J. M. Hawkins, *J. Org. Chem.*, **63**, 2054 (1998), (d) P. Veds, M. Lysen, J. Kristensen, M. Begtrup, *Org. Lett.* **3**, 1435 (2001).
3. A. S. Kotnis, *Tetrahedron Lett.*, **31**, 481(1990).

26. Enzyme-like Acceleration in Catalytic Anti-Markovnikov Hydration of Alkynes to Aldehydes

Douglas B. Grotjahn and Daniel A. Lev

Department of Chemistry, San Diego State University, 5500 Campanile Drive, San Diego, CA 92182-1030

grotjahn@chemistry.sdsu.edu

Abstract

We are applying the principles of enzyme mechanism to organometallic catalysis of the reactions of nonpolar and polar molecules; for our early work using heterocyclic phosphines, please see ref. 1.(1) Here we report that whereas uncatalyzed alkyne hydration by water has a half-life measured in *thousands of years*, we have created improved catalysts which reduce the half-life to *minutes*, even at neutral pH. These data correspond to enzyme-like rate accelerations of >3.4 x 10^9, which is 12.8 times faster than our previously reported catalyst and 1170 times faster than the best catalyst known in the literature without a heterocyclic phosphine. In some cases, practical hydration can now be conducted at room temperature. Moreover, our improved catalysts favor *anti-Markovnikov* hydration over traditional Markovnikov hydration in ratios of over 1000 to 1, with aldehyde yields above 99% in many cases. In addition, we find that very active hydration catalysts can be created *in situ* by adding heterocyclic phosphines to otherwise inactive catalysts. The scope, limitations, and development of these reactions will be described in detail.

Eq. 1

$$R\text{---}\equiv\text{---}H \;+\; H_2O \xrightarrow[\substack{25\text{-}70\,^\circ C, \\ \text{neutral pH}}]{\substack{\text{2 to 5 mol \%} \\ \text{Ru catalyst}}} R\diagup\!\!\diagdown\overset{O}{\diagdown}H \quad \text{not} \quad R\diagup\!\!\overset{O}{\diagdown}\!CH_3$$

>99 % yields
ratios >1000 to 1

Introduction

Nature has developed exquisite catalysts for a variety of reactions, for example, the hydrolysis of an amide bond. In the absence of a catalyst, amide bonds are extremely stable, making them ideal bonds to link amino acids in order to create stable, large molecules of life. In recent years, sensitive detection methods have allowed one to determine that the half-life for hydrolysis of an amide bond as found in a protein is 168 years at pH 9, (2) or 300 to 600 years at pH 7.(3) Remarkably, enzymes such as carboxypeptidase accelerate these reactions so that they are done in fractions of a second. This corresponds to rate enhancements of 4×10^{11}. Here we report using the design features of enzymes to create organometallic catalysts capable of similar rate enhancement of a reaction for which enzymes are not suited.

Scheme 1. Cooperativity in carboxypeptidase catalysis of amide hydrolysis

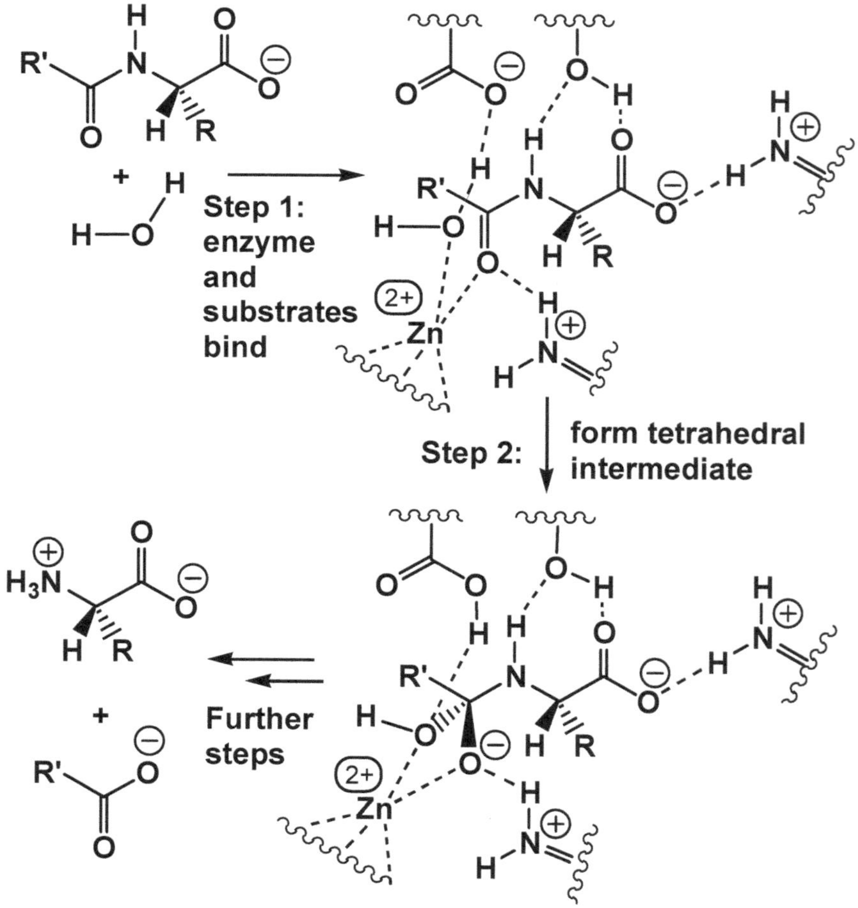

Such large rate enhancements naturally excite our curiosity: how does Nature achieve such incredible feats? Knowing something about the design features of enzymes, how can we improve other catalysts, to achieve reactions which are unknown for enzymes? A conceptually different approach along these lines is the generation of catalytic antibodies. Here, however, we focus on a generally-accepted mechanism for carboxypeptidase-catalyzed hydrolysis of amides (Scheme 1).(4) Hydrogen bonding, electrostatic interactions, metal-substrate binding and activation are common themes, as shown by several basic steps, the first two of which are shown in Scheme 1. Step 1 shows the binding of a water molecule and the carboxy terminus of the protein to be hydrolyzed, through a number of hydrogen bonds and complexation of the water molecule to a Zn(II) ion. Step 2 shows intramolecular proton transfer from a water molecule to a carboxypeptidase carboxylate group, with attack of the hydroxide ion thus formed on the amide bond; the developing negative charge on the tetrahedral intermediate's alkoxide group is dispersed by bonding to the Zn(II) center. For simplicity here, further steps leading to the amide hydrolysis products are not shown in detail, yet for these steps proton transfers aided by the enzyme active site are thought to be crucial. Similar concepts apply to other systems, such as phosphatases, which hydrolytically cleave phosphate linkages on modified proteins or small carbohydrates. Comparison(5) of the catalytic domains of metal-containing and metal-free phosphatases points to the importance of positively charged species on the enzyme, either a metal(6) or a protonated arginine side chain, which stabilize negative charges during hydrolysis.

It is no wonder that enzymes have inspired many chemists in their search for new artificial receptors or catalysts. Multipoint recognition of nucleobases by Zn(II) complexes featuring several NH groups for hydrogen bond donation has been documented.(7) A related hydrogen bond donation from a Ru-coordinated NH group was shown as a key feature in catalytic enantioselective hydrogenation of ketones.(8) Electrostatic association of ions can lead to molecular recognition and selective catalysis.(9) Organic compounds with arrays of proton-donating and accepting groups have been shown to catalyze hydrolysis of phosphate diesters.(10, 11) In summary, because Lewis acidity, hydrogen bonding and proton transfer have been shown to be so important in both enzymatic and artificial catalysis, our work focuses on using these concepts, but in the non-enzymatic context of organometallic chemistry. We call this *biologically inspired organometallic chemistry.*

Whereas a variety of enzymes are known to catalyze hydrolysis of an amide, ester, or phosphate ester by water, there are only a handful of enzymes known to catalyze the addition of water to an alkyne, and the few known enzymes either act slowly on acetylene itself,(12) or on special alkynes, for example those featuring one or more strongly electron-withdrawing groups on the triple bond.(13) Hydration of an alkyne could be a valuable way to add oxygenated functionality to an organic molecule. In principle, from a terminal alkyne, two isomeric products could be produced, an aldehyde or a ketone. Until 1998, known catalysts for this reaction [e.g. Hg(II)] produced only the ketone, the product of Markovnikov addition of water, unless there were special features of the alkyne, for example a strongly electron-withdrawing R group. In order to make the isomeric aldehyde, non-catalytic methods had to be applied; a premier example would be the addition of a stoichiometric amount of a borane, followed by oxidation of the addition product.

We became interested in applying enzyme design principles to organometallic catalysis by the 1998 report (Scheme 2) of the first catalytic anti-Markovnikov addition of water to terminal alkynes.(14) The arene-Ru catalysts described gave either the aldehyde or the ketone as major but not exclusive product, depending on the identity of the phosphine used. It was not clear to us what role the phosphine played. Moreover, the catalyst loading of the optimal aldehyde-producing catalyst was high: in order to hydrate one mole of alkyne, one would have to use 165 grams of catalyst A!

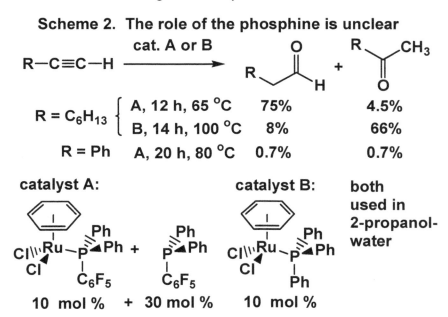

Scheme 2. The role of the phosphine is unclear

Space limitations here preclude a lengthy discussion of mechanism of this interesting reaction. For further details and background, we refer the reader to our other paper in this volume.(15) However, clearly at some stage of aldehyde production, the oxygen atom of water must form a bond to C-1 of the alkyne. Until 2001, this key step was proposed to be nucleophilic attack of a water molecule on C-1 of a vinylidene ligand ($M=C^1=C^2HR$) derived from the alkyne H-CC-R.(16, 17) Indeed, others had shown that isolable vinylidene complexes of the metal fragment $[CpRu(PPh_3)_2]^+$ react with methanol or less cleanly, with water,(16) albeit in stoichiometric reactions. We showed (1) that $[CpRu(PPh_3)_2]^+$-related complexes **1** and **2** (Chart 1) were completely ineffective as catalysts for addition of water to 1-hexyne, not even giving one turnover (Table 1, entries 1 and 2). Our working hypothesis was that we could accelerate addition of water to the putative vinylidene ligand by using phosphines with heterocyclic substituents as internal bases.

We note that while our work was in progress, Tokunaga and Wakatsuki independently showed the ineffectiveness of **1** and suggested that phosphine loss was the problem, on the basis of observing significant conversion of the catalyst to a catalytically inactive monophosphine complex and one mole of free PPh$_3$.(18, 19) The Japanese groups reported high yields of aldehyde using complexes with either chelating phosphines or small, electron-rich phosphines – both expected to lose one phosphine less readily than **1**. One of their optimal catalysts was **3** at 2 mol %, though it required heavier catalyst loading (10%) for arylalkynes or a nitrile-substituted alkyne. Below, we report that our catalysts are 90 to 1150 times faster than **3**.

Chart 1

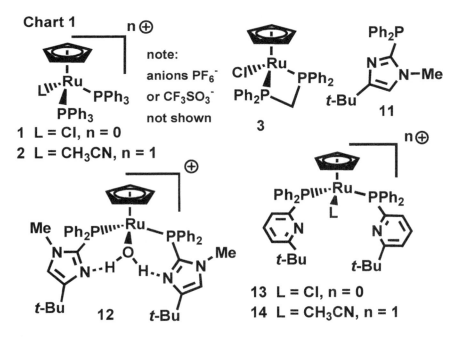

1 L = Cl, n = 0
2 L = CH$_3$CN, n = 1

3

11

12

13 L = Cl, n = 0
14 L = CH$_3$CN, n = 1

Results and Discussion

We examined two known analogs of PPh$_3$, one with an imidazole ring (**4**, Scheme 3), the other with a pyridine ring (**5**). In our hands, when two moles of each phosphine were allowed to react with the tris(acetonitrile) complex **6**,(20) either **7** or **8** were formed in high yields. Each complex showed diagnostic ^{31}P NMR spectra, with two doublets ($^2J_{PP} = 37$ Hz) consistent with the presence of two cis- coordinated phosphines. Complexes **7** and **8** are interesting because of the presence of the same ligand in two different coordination modes, η^2-*P,N* and

η^1-*P*; the chelating phosphine forms a four-membered ring, leading to unusually upfield chemical shifts (20 to 50 ppm upfield).

When either **7** or **8** were applied as catalysts for the hydration of 1-hexyne, neither hexanal nor 2-hexanone were detected. Intriguingly, however, because we monitored the reactions carefully by ^{1}H and ^{31}P NMR spectroscopy, we realized that both **7** and **8** were converted cleanly to another species. Repeating the reactions of **7** or **8** with 1-hexyne on larger scale in the absence of added water led to isolation of metallacycles **9** and **10**, which were fully identified by NMR spectroscopy, and ultimately, by X-ray diffraction.

Scheme 3.

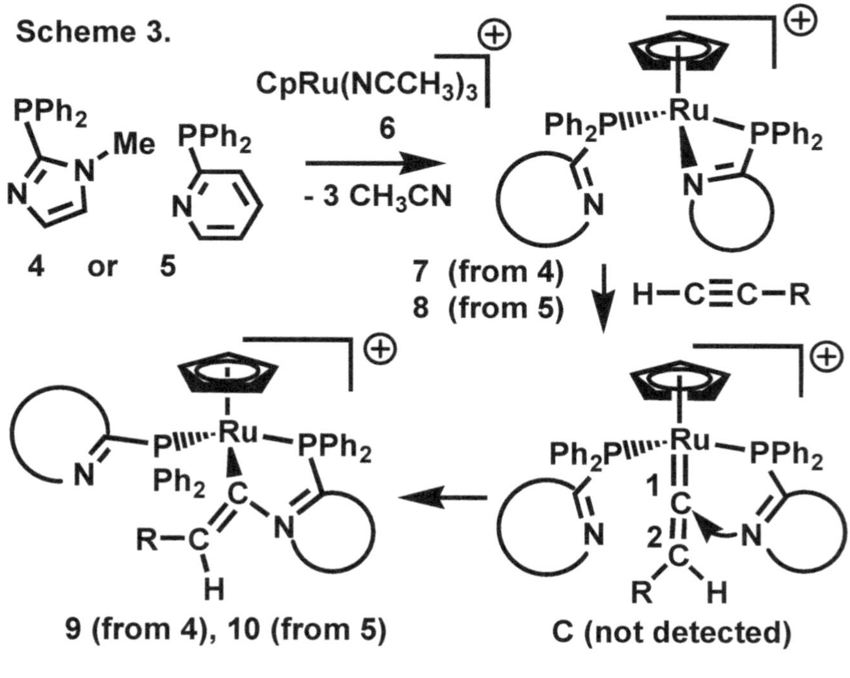

Although the stoichiometric production of **9** and **10** was initially disappointing, it is consistent with the formation of a vinylidene ligand from 1-hexyne (see **C**, Scheme 3), and attack of a pendant heterocyclic nitrogen on C-1 of the vinylidene. In this process, the heterocyclic nitrogen approaches C-1. Thus, we decided to slow the rate of heterocycle approach by adding a bulky substituent next to the nitrogen.

Complex **12** (Chart 1) is prepared by displacing two acetonitrile ligands from **6** using 2 equiv of the corresponding phosphine **11**;(1) in the presence of as little as 5 equiv of water the third acetonitrile ligand is also displaced.

Spectroscopic data and subsequent isolation by crystallization point to the formation of **12** in 98% yield. The ^{1}H NMR spectrum of **12** shows a slightly broad singlet near 9.8 ppm for the two protons of the aquo ligand. Moreover, results of an X-ray diffraction study showed binding of the water molecule in the catalytic site of the complex, through a Ru-O bond as well as two hydrogen bonds between the O-H bonds of the aquo ligand and the imidazole nitrogens.(1)

Table 1 Selected results on catalytic hydration of terminal alkynes RCCH.[a]

Entry	Catalyst (Chart 1)	Alkyne R	Aldehyde yield (%) (time, h)		
			earlier	after 20-24 h	later
1	**1**	n-C_4H_9		1.0 (21)	
2	**2**	n-C_4H_9		0.3 (21)	
3	**1 + 11**[b]	n-C_4H_9		57 (21)	
4	**3**	n-C_7H_{15}		1.3 (20)	5.2 (96)
5	**12**	n-C_4H_9		92 (21)	96 (36)
6		$PhCH_2CH_2$		88 (21)	92 (46)
7		t-Bu		21 (21)	91 (108)[c]
8		Ph		64 (21)[d]	75 (42)[d]
9		t-$BuMe_2SiOCH_2$		91 (21)	96 (36)
10		$THPOCH_2$		83 (21)	86 (50)
11		$NC(CH_2)_3$		96 (21)	98 (40)
12	**13**	n-C_7H_{15}	19.6 (3)	65.0 (20)	98.3 (48)
13	**14**	n-C_7H_{15}	99.9 (3)		
14		Ph	95.2 (8)	99.8 (20)	
15		4-$MeOC_6H_4$	90.3 (9)	99.7 (24)	

[a] Standard conditions: acetone (or -d_6), water (5 equiv) and 2 mol % catalyst, initial alkyne concentration 0.50 M, reaction temperature 70±2 °C, reaction monitored by ^{1}H NMR or GC using appropriate internal standards.
[b] *In situ* generation of catalyst; see text for further details.
[c] Yield after heating at 90 °C.
[d] Yield using 10 mol % catalyst.

To our delight, **12** is an excellent catalyst at the 2 mol % level: within 1 to 2 days at 70 °C, 1-hexyne and similar terminal alkynes give >90% yields of the corresponding aldehydes (Table 1, entries 5 and 6). That this occurs under neutral conditions is underscored by the ability of **12** to hydrate two alkynes bearing acid-sensitive protecting groups (entries 9 and 10). Interestingly, a nitrile-bearing alkyne has no significant effect on the rate of hydration (entry 11). Hindrance of the alkyne does slow conversion, though the alkyl-substituted case (t-Bu, entry 7) succeeds at higher temperature.

It would be more convenient if **12** or a similar catalyst could be made *in situ* from precursors which are more air-tolerant than **6** and require fewer steps to make. (Phosphine **11** is air-sensitive but no more so than PPh_3, which can be weighed and transferred in open air.) Complex **1** seemed a suitable candidate, for it is air-stable and can be made in one step from ruthenium trichloride, CpH, and PPh_3,(21) and ligand substitutions are facile, particularly in protic solvents.

Thus, in an NMR tube a mixture of **1** and **11** in a molar ratio of 1 to 2 were heated in acetone-d_6 containing water. After 1 h at 70 °C, a mixture of roughly equal amounts of **1**, **11**, PPh$_3$, and a new CpRu complex containing two different phosphines were seen. In addition, a very small amount of **12** was detected. Gratifyingly, this mixture (made from 2 mol % **1** and 4 mol % **11**) functioned as an effective catalyst for hydration of 1-hexyne, albeit at about half the rate shown by isolated **12** (see table 1, entries 3 and 5). This is an intriguing result, because clearly there were only traces of **12** in the solution, yet the rate of catalysis was half that using pure **12**. Thus, creating mixed complexes containing heterocyclic and nonheterocyclic phosphines may be a promising strategy for catalyst optimization, an avenue which we are pursuing further.

One important class of alkynes for which **12** does not work so well are aryl-substituted alkynes, also problematic using the Tokunaga-Wakatsuki groups' optimal catalysts (e.g. **3**), which also required higher catalyst loading. We now report a solution, in the form of catalysts derived from a hindered pyridine ligand reported in 2000.(22) As seen in Tables 1 and 2, complexes **13** and **14** are the fastest catalysts now known for alkyne hydration, the latter being more than 1000 times faster than **3**, in both isopropanol-water and acetone-water mixtures. Significantly, with 5 mol % of **14**, 1-nonyne is hydrated *at room temperature* within 2 d in 98% yield. Moreover, even arylalkynes work well. Taken together, **14** is not only the fastest catalyst but the most general.

These results prompted us to ask: how do these catalysts compare with enzymes? What is the rate of uncatalyzed alkyne hydration? Our search of the literature thus far has not revealed a study of the uncatalyzed reaction.

Table 2 More detailed comparison of four catalysts and uncatalyzed reactions.

	Cat.	Initial rate of aldehyde formation (mol/mol cat./h)			Rel. rates[e]
		1-nonyne[a]	1-nonyne[b]	PhCCH [a]	
1	**3**	0.0201	0.0344		1
2	**12**	1.88			94
3	**13**	2.45		1.89	122
4	**14**	23.6	36.1	5.85	1170
5	**14**[c]	6.04			300
6	None[d]	$< 6.9 \times 10^{-9}$		$< 6.9 \times 10^{-9}$	$< 3 \times 10^{-6}$

[a] $[alkyne]_0 = 0.50$ M, water (5 equiv), catalyst (2 mol%), acetone, 70 °C.
[b] $[alkyne]_0 = 0.50$ M, i-PrOH – water (3:1), catalyst (2 mol%), 70 °C.
[c] Room temperature with 5 mol % catalyst.
[d] See text. Same as [a] except no catalyst. Rate = mol ketone/mol of alkyne/h.
[e] For 1-nonyne in acetone-water. Rate for **3** defined as 1.

We decided to subject 1-nonyne and phenylacetylene in separate experiments to reaction with water (5 equiv) in acetone containing hexadecane as internal standard at 70 °C. In order to avoid any chance of spurious results from traces of acid on the glass, the vessel used was previously passivated using hexamethylsilazane, and then dried in an oven at 100 °C before use. The mixtures were analyzed by GC; using varying known amounts of the expected aldehydes and ketones and hexadecane, it was determined that our lower limit of detection for oxygenated products was about 5 ppm. In each experiment, the vessel's contents as described above were deoxygenated and *heated for one month*. The result? Neither ketone (expected product of protic catalysis) nor aldehyde were seen at the 5 ppm level. Thus, we calculate that the half-life for hydration to the ketones is at least 8200 years! Compare this with half-life using **14**, which would be 1.3 minutes with one mole of catalyst. This corresponds to an acceleration of $> 3.4 \times 10^9$, within the realm of enzymatic performance.

Whatever the mechanistic explanation for this remarkable result, we hope that we have given the reader a taste of the fruits of considering both the fields of enzyme mechanism and organometallic chemistry. We are exploring the acceleration of other reactions using bifunctional catalysts, and these results will be described in due course.

Experimental Section

Reactions were conducted under N_2 using glovebox and Schlenk techniques; conditions and analytical methods are described in Tables and text. See also ref. 1 for details of how hydration yields were determined by NMR.

Acknowledgments

NSF and Université Pierre et Marie Curie (Dr. Hani Amouri) supported D.G.'s stay in Paris, during which this manuscript was written. SDSU also contributed.

References

1. D. B. Grotjahn, C. D. Incarvito, and A. L. Rheingold, *Angew. Chem., Int. Ed. Engl.,* **40**, 3884 (2001).
2. R. A. R. Bryant and D. E. Hansen, *J. Am. Chem. Soc.,* **118**, 5498 (1996).
3. A. Radzicka and R. Wolfenden, *J. Am. Chem. Soc.,* **118**, 6105 (1996).
4. T. E. Creighton, "Proteins: Structures and Molecular Properties." W. H. Freeman, New York, 1993.
5. W. P. Taylor and T. S. Widlanski, *Chemistry and Biology* **2**, 713 (1995).
6. J. A. Cowan, *Chem. Rev.,* **98**, 1067 (1998).
7. E. Kimura and M. Shionoya, *in* "Metal Ions in Biological Systems" (A. Sigel and H. Sigel, eds.), Vol. 33, p. 29. Dekker, New York, 1996.
8. Y. Jiang, Q. Jiang, and X. Zhang, *J. Am. Chem. Soc.,* **120**, 3817 (1998).

9. P. J. Smith, E. Kim, and C. S. Wilcox, *Angew. Chem., Int. Ed. Engl.,* **32**, 1648 (1993).
10. M. Olivanen, S. Kuusela, and H. Lönnberg, *Chem. Rev.,* **98**, 961 (1998).
11. T. Kato, T. Takeuchi, and I. Karube, *J. Chem. Soc., Chem. Commun.,* 953 (1996).
12. C. Kisker, H. Schindelin, D. Baas, J. Retey, R. U. Meckenstock, and P. M. H. Kroneck, *FEMS Microbiology Reviews,* **22**, 503 (1998).
13. O. D. Alipui, D. Zhang, and H. Schulz, *Biochem. Biophys. Res. Comm.,* **292**, 1171 (2002).
14. M. Tokunaga and Y. Wakatsuki, *Angew. Chem., Int. Ed. Engl.,* **37**, 2867 (1998).
15. D. A. Lev and D. B. Grotjahn, Ch 27, *this volume.*
16. M. I. Bruce and A. G. Swincer, *Aust. J. Chem.,* **33**, 1471 (1980).
17. M. Tokunaga, T. Suzuki, N. Koga, T. Fukushima, A. Horiuchi, and Y. Wakatsuki, *J. Am. Chem. Soc.,* **123**, 11917 (2001).
18. T. Suzuki, M. Tokunaga, and Y. Wakatsuki, *Org. Lett.,* **3**, 735 (2001).
19. P. Alvarez, M. Bassetti, J. Gimeno, and G. Mancini, *Tetrahedron Lett.,* **42**, 8467 (2001).
20. D. B. Grotjahn and H. C. Lo, *Organometallics,* **15**, 2860 (1996).
21. M. I. Bruce, C. Hameister, A. G. Swincer, and R. C. Wallis, *Inorg. Synth.,* **21**, 78 (1982).
22. J. Baur, H. Jacobsen, P. Burger, G. Artus, H. Berke, and L. Dahlenburg, *Eur. J. Inorg. Chem.,* 1411 (2000).

27.   An Enzyme-like Mechanism Involving Bifunctional Ru Catalysts

Daniel A. Lev and Douglas B. Grotjahn

Department of Chemistry, San Diego State University, 5500 Campanile Drive, San Diego, CA 92182-1030

Abstract

In our investigations of ligands capable of donating/accepting hydrogen bonds in important catalytic applications, we have found that heterocyclic phosphine ligands, namely 2-pyridyl and 2-imidazolyl phosphines, greatly influence the reactivity of $CpRu^+$ catalysts. This has led to highly efficient *anti-Markovnikov* hydration of a variety of terminal alkynes, including aryl alkynes and trimethylsilyl alkynes. Enzyme-like rate enhancements of $>10^9$ have facilitated experimental and computational mechanistic studies, supporting multipoint cooperativity involving the metal and the heterocyclic phosphine ligands.

Introduction

Since Bruce's pioneering work in the area of ruthenium vinylidene chemistry (1), it has been well known that isomerization of a terminal alkyne to a vinylidene on a metal center is not only favorable but also effects a reversal in the reactivity of the carbon atoms. However, hydration catalysis was not possible, because alkyl migration from a proposed acyl intermediate led to an

inactive carbonyl complex. Recently, Wakatsuki et al. developed two catalytic systems based on traditional organometallic approaches, i.e. electronic manipulation of the metal center (2) and use of chelate ligands (3).

In an effort to apply the cooperative principles of metalloenzyme reactivity, involving a combination of metal-ligand and hydrogen bonding, we have reported a ruthenium catalyst incorporating imidazolyl phosphine ligands that efficiently and selectively hydrates terminal alkynes (5). We subsequently found that application of pyridyl phosphines to the reaction resulted in a >10-fold rate enhancement and complete anti-Markovnikov selectivity, even in the case of aryl alkynes (6).   In this paper we present an experimental and computational study of alkyne hydration by our bifunctional catalysts.

Results and Discussion

Our initial experiments examined the isotope labeling pattern observed when D_2O and hexyne-d_1 replace water and hexyne in the reaction.   Our results demonstrate complete incorporation of protons from the nucleophile at C-2 of the product.   This is analogous to the observation of Wakatsuki et al. (4), which contradicts the expectation of a 1,2-hydrogen shift to form the presumed vinylidene intermediate.   An alternative mechanism is further supported by the retention of the alkynyl proton at C1 in the product.   The partial deuteration of this position in the presence of D_2O and vice versa is evidence for the breaking of the alkynyl C-H(D) bond in the catalytic mechanism.   If the mechanism involved a 1,2-hydrogen shift, one would expect incorporation of these protons or deuterons at C-2 of the product.   While we can rule out a direct 1,2 shift, other mechanisms involving the breaking of this bond are not differentiated, including oxidative addition, alkyne deprotonation, and alkyne protonation.

Table 1 Isotope labeling and kinetic effects.

Entry		Alkyne	C-2 of hexanal		C-1 of hexanal		k_{obs} (h^{-1})
			% H	% D	% H	% D	
1	H_2O	hexyne	100	0	100	0	8.70
2	H_2O	hexyne-d_1	100	0	86.5	13.5	7.84
3	D_2O	hexyne	<1	>99	26.3	73.7	4.62
4	D_2O	hexyne-d_1	0	100	0	100	4.26

0.500 M hexyne, 2.50 M water, 0.0250 M [**1**][OTf] in acetone at 50 °C.

In addition to labeling studies, we measured the kinetic isotope effect for both nucleophile and alkyne protons.   The primary isotope effect observed for the H-OH bond, $k_H/k_D = 1.9$, is consistent with the breaking of this bond in the transition state.   In proton catalyzed alkyne hydrations kinetic isotope effects between 1.5 and 5 have been reported (7).   The determination of a kinetic isotope effect for the alkynyl hydrogen is complicated by the H/D exchange observed in the product.   However, in contrast to the H/D exchange reported for terminal alkynes with Ag (8) or Pd (9) in acidic medium, during the course of the hydration no exchange was observed in the terminal position of the unreacted substrate.   Therefore, we can conclude that the small kinetic isotope effect, $k_H/k_D = 1.1$, indicates H-CCR bond breaking is not involved in the rate determining step of the catalysis, but a rate determining step occurring after H-CCR bond breaking is possible.   This value is consistent with secondary isotope effects observed for terminal alkynes in acid-catalyzed hydration (10) and inconsistent with a marked inverse isotope effect in silver catalyzed H/D exchange (8). The combination of isotopic labeling and kinetic effects is inconsistent with mechanistic pathways involving rate determining C-H bond breaking and supportive of pathways involving alkyne protonation. In the future we will examine the effects of exogenous protons on the reaction and key intermediates in the catalytic cycle.

Figure 1 Activation involving H-OH bond breaking and a vinyl intermediate.

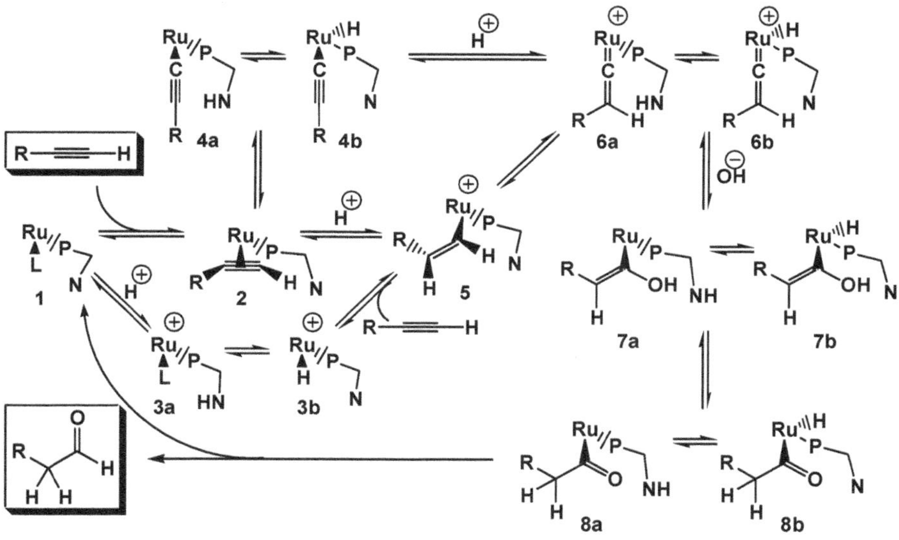

Figure 1 presents a view of the mechanism consistent with the experimental evidence. The three possible rate determining steps in the catalytic cycle are formation of **5** from **2** by alkyne protonation, formation of **3a,b** from **1** and subsequent alkyne insertion to **5**, and protonation of **4a,b** to yield **6a,b**. While it is unlikely that **1** and/or the pendant base is basic enough to deprotonate H_2O to form **3a,b**, this mechanism is consistent with all of the experimental evidence. Of the remaining two mechanisms, protonation of **2** by H_2O is more consistent with the isotopic labeling pattern, because intermediate **4a,b** would exchange with aquo protons and lead to scrambling of the C-2 protons. If the pre-equilibrium in the catalysis is between **2** and **4a,b**, scrambling of the alkynyl position in the substrate would be observed. Thus, while protonation of **4a,b** cannot be ruled out, protonation of **2** is the favored rate determining step. Both pathways were examined by density functional calculations on possible intermediates and transition states.

Wakatsuki et al. (4) proposed vinyl complex, **5**, and presented DFT results supporting isomerization to a vinylidene hydride as the rate determining step. Our results indicate that the rate determining step involves H-OH bond breaking and that protonation of a bound alkyne is the rate determining step in this

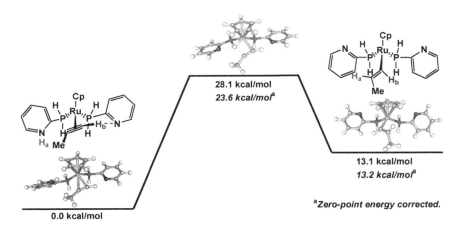

Figure 2 B3LYP/LANL2DZ** structures in preferred reaction pathway.

process, with an activation energy of 23.6 kcal/mol. This leads to a vinyl complex that is 13.2 kcal/mol higher in energy than the alkyne complex. The key transition state in this process is shown in Figure 2. In a manner similar to metalloenzymes, this structure involves a combination of proton transfer by a pendant heterocycle and organometallic bonding to activate the substrate. Interestingly, two types of hydrogen bonding are seen: (1) the first, between the terminal alkyne H_b and a single pyridine with an N-H distance of 2.06 Å; (2) the second, involving *three* interactions of H_a, with the other pyridyl nitrogen (2.16 Å), the internal sp carbon of the alkyne (1.72 Å), and with the Ru center (1.68 Å). The Ru-H_a interaction for **2** → **5** may shed light on the transition state for the conversion **3a,b** → **5**. An examination of the HOMO of the transition state reveals overlap between the d_{z2} orbital on the Ru center and the proton and an interaction between the $\pi_\perp$ orbital of the alkyne ligand and the proton.

This study supports rate-determining H-OH bond breaking, which constrasts with previous reports that identified vinylidene isomerization as the key step in catalytic alkyne activation. The results indicate an enzyme-like mechanism is operative involving cooperative substrate activation by a metal center and proximal hydrogen bond donor/acceptors. In the future we will apply these principles to the activation of additional species.

Experimental Section

Reaction rate experiments were conducted in NMR tubes sealed with Teflon valves. In an inert atmospere glovebox, catalysts and internal standard, TMS_4C, were weighed into the tube, followed by addition of solvents and reactants. The tube was immediately inserted in the preheated (50 °C) probe of a 500 MHz Varian UnityInova spectrometer. To acquire spectra the sample was irradiated twice with a 30° pulse, 5 sec acquisition time, and 120 sec delay.

Computed structures were optimized with the B3LYP hybrid functional method as it is implemented in the Gaussian 98 and 03 programs (11). The LANL2DZ basis set was used with a single polarization function added to all H, C, N, and P atoms, designated LANL2DZ** in the text. Transition states were located using the STQN method and confirmed by IRC and frequency analysis (NIMAG = 1). Minima were confirmed by frequency analysis (NIMAG = 0).

Acknowledgements

We thank Dr. Andrew Cooksy and the SDSU Computational Science Center for assistance and resources and the SDSU Foundation for financial support.

References

1. M. I. Bruce, and A. G. Swincer, *Aust. J. Chem.*, **33**, 1471 (1980).
2. M. Tokunaga, and Y. Wakatsuki, *Angew. Chem., Int. Ed. Engl.*, **37**, 2867 (1998).
3. T. Suzuki, M. Tokunaga, and Y. Wakatsuki, *Org. Lett.*, **3**, 735 (2001).
4. M. Tokunaga, T. Suzuki, N. Koga, T. Fukushima, A. Horiuchi, and Y. Wakatsuki, *J. Am. Chem. Soc.*, **123**, 11917 (2001).
5. D. B. Grotjahn, C. D. Incarvito, and A. L. Rheingold, *Angew. Chem., Int. Ed. Eng.*, **40**, 3884 (2001), and references therein.
6. D. B. Grojahn and D. A. Lev, Ch. 26, *this volume*.
7. A. D. Allen, Y. Chiang, A. J. Kresge, and T. T. Tidwell, *J. Org. Chem.*, **47**, 775 (1982), and references therein.
8. G. S. Lewandos, J. W. Maki, and J. P. Ginnebaugh, *Organometallics*, **1**, 1700 (1982).
9. A. Scrivanti, V. Beghetto, E. Campagna, M. Zanato, and U. Matteoli, *Organometallics*, **17**, 630 (1998).
10. D. S. Noyce and M. D. Schiavelli, *J. Am. Chem. Soc.*, **90**, 1023 (1968).
11. M. J. Frisch, et al., Gaussian, Inc., Pittsburgh PA, 2003.

28.   Activation of Dendrimer Encapsulated Pt Nanoparticles for Heterogeneous Catalysts

HuiFang Lang, R. Alan May, Brianna L. Iversen, and Bert D. Chandler

Department of Chemistry, Trinity University, One Trinity Place, San Antonio, TX, 78213

Abstract

We report on the preparation, deposition, and activation of dendrimer stabilized nanoparticles as a new and versatile method for the preparation of supported metal catalysts. Using polyamidoamine (PAMAM) dendrimers as metal nanoparticle templates and stabilizers, Pt nanoparticles with very narrow size distributions can be prepared in aqueous solution. After deposition onto an appropriate support, the organic dendrimer must be removed to prepare active heterogeneous catalysts. Activation (removal of the dendrimer) under O_2, H_2, and He was examined using *in-situ* infrared spectroscopy and infrared spectroscopy of adsorbed CO. Activation of the supported precursors in solution was also investigated by refluxing the supported DENs in solvent/acid mixtures. The solvent, acid, and reflux time were all varied. The activated catalysts were characterized with Atomic Absorption Spectroscopy, Infrared Spectroscopy of adsorbed CO, CO chemisorption, CO oxidation catalysis, and toluene hydrogenation catalysis. Attempts to remove the dendrimer in solution generally resulted in loss of Pt, although choosing appropriate reflux conditions minimized the leaching. Oxidation under flowing O_2/He followed by reduction under flowing H_2/He was found to be the best means of activating the DENs.

Introduction

Supported Pt nanoparticles are arguably the most versatile materials for the catalysis of organic reactions. Reports from the 19th Conference on the Catalysis of Organic Reactions include the application of supported Pt catalysts to selective oxidations (1), selective hydrogenations (2-4), and environmental catalysis (5). Using traditional preparation routes, Pt nanoparticles are generally prepared *in-situ* on the support via wetness impregnation or adsorption methods, followed by thermal activation. These catalyst preparation methods are relatively fast and simple, but they offer only limited control over particle size and particle size distributions.

Polyamidoamine (PAMAM) dendrimers have drawn considerable interest in recent years due to their potential applications in medicine, nanotechnology, and catalysis (6,7). The ability to control dendrimer interior/exterior functionalities and the macromolecular architecture of PAMAM dendrimers (open spaces within the interior) create an ideal environment for trapping guest species (8,9). Further, PAMAM dendrimers can bind a defined number of transition metal cations and thus template and stabilize metal oxide or metal nanoparticles (6, 10).

The potential to ultimately control nanoparticle size and composition makes dendrimer encapsulated metal nanoparticles (DENs) extremely attractive as potential precursors for studying supported metal catalysts. Because the nanoparticles are prepared *ex situ* and can be deposited onto a substrate or support, DENs offer the opportunity to bridge the gap between surface science studies of model systems and real world catalysts on high surface area supports. The possibility of varying particle size and composition through nanoparticle preparation schemes makes DENs uniquely suited to exploring and understanding the relative importance of these effects on catalytic reactions using materials comparable to those employed as industrial catalysts.

Before DENs can be utilized as catalyst precursors, appropriate methods must be developed to remove the organic portion of the DENs. If activation conditions are too harsh, particle agglomeration may negate the potential advantages of the dendrimer method. If activation conditions are too mild, incomplete removal of the dendrimer may leave organic residues on the particles and poison the catalyst. This study examines a variety of activation procedures for removing the organic portion of the catalyst precursors – a critical step for the application of these materials to the preparation and study of heterogeneous catalysts.

Experimental Section

Nanoparticle Preparation and Catalyst Characterization: All solutions were prepared using deionized water. Hydroxy-terminated generation 5 polyamidoamine (PAMAM) dendrimers (G5-OH, 5% w/w in methanol), K_2PtCl_4, and $NaBH_4$ were purchased from Aldrich and used as received. Dialysis was performed with 12 KD MW cutoff cellulose dialysis sack (250-7U, Sigma). Silica (DAVICAT SI-1403, 245 m^2/g) was supplied by W. R. Grace. O_2, H_2, He gasses are all UHP grade from Airgas. Carbon monoxide (99.8% in aluminum cylinder) was purchased from Matheson Tri-Gas.

Dendrimer encapsulated Pt nanoparticles (DENs) were prepared via literature methods (1, 11). $PtCl_4^{2-}$ and dendrimer solutions (20:1 Pt^{2+}:dendrimer molar ratio) were mixed and stirred under N_2 at room temperature for 3 days. After reduction with 30 equivalents of BH_4^- overnight, dialysis of the resulted light brown solution (2 days) yielded Pt_{20} nanoparticle stock solution. The stock solution was filtered through a fine frit and Pt concentration was determined with Atomic Absorption Spectroscopy (11). Details on catalyst characterization and activity measurements have been published previously (11).

DEN Deposition: The Pt_{20} DEN stock solution was deposited on silica via wet impregnation. SiO_2 masses were adjusted to prepare ~0.3 wt % Pt catalysts. The purified dilute nanocomposite stock solution was first condensed to roughly 1/5 of its original volume at 38°C using a Rotary Evaporator. This condensed solution (roughly the pore volume of the carrier) was immediately added to a sample vial containing SiO_2. Excess water was slowly evaporated in a vacuum oven at 50 °C overnight.

Catalyst Activation: Gas phase activation of supported DENs was examined using *in-situ* FTIR spectroscopy and FTIR spectroscopy of adsorbed CO. For *in-situ* dendrimer decomposition studies, the spectra were collected under a gas flow composed of 20% O_2/He or 20% H_2/He. The supported DEN sample was pressed into a self-supporting wafer, loaded into a controlled atmosphere IR cell, and collected as the sample background. The temperature was raised stepwise and spectra were collected at each temperature until little or no change was observed. After oxidation, the sample was reduced in 20% H_2/He flow with various time/temperature combinations. The sample was then flushed with He for 1hr at the reduction temperature. After cooling under He flow, a background spectrum was collected at room temperature. A 5% CO/He mixture was flowed over the sample for 15 minutes, followed by pure He. IR spectra of CO adsorbed on the catalyst surface were collected after the gas phase CO had been purged from the cell.

Solution Activation: 50 mg of supported Pt_{20} DEN was mixed with a solvent/acid mixture (see Table 1) in a 50 ml round bottom flask and refluxed for 2 – 6 hrs. Solid samples were separated from solution by vacuum filtration and dried in vacuum oven at 50°C overnight. To prepare sample 6 (see Table 1), supported Pt_{20} DENs were mixed with HNO_3/H_2O (volume ≈ pore volume of SiO_2) in a sample vial, heated to 70°C for 2hrs, and dried in a vacuum oven at 50°C overnight.

Results and Discussion

Supported, intact DENs do not bind CO and are not active catalysts. Presumably, in the absence of solvent, the dendrimer collapses onto the nanoparticles preventing even small substrates from accessing the metal surface (11,12). This means that the organic dendrimer must be removed in order to prepare active catalysts.

Activation conditions for the supported Pt_{20} DENs were chosen based on the thermal decomposition experiments in a controlled atmosphere transmission IR cell. *In-situ* destruction of the dendrimer amide linkages was followed under O_2/He flow by monitoring IR spectral changes in the $1750 - 1450$ cm^{-1} region (Figure 1). Under O_2, PAMAM dendrimer amide bonds began decomposing near 50 °C and continued changing as the temperature increased to 300 °C. The spectrum collected at 300 °C is similar to that obtained after 300 °C for 2 or 3hrs, indicating that the amide bonds are mostly destroyed at this temperature.

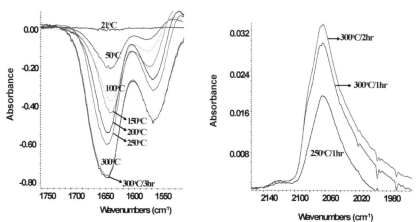

Figure 1 FT-IR spectra of G5-OH(Pt$_{20}$)/SiO$_2$ in flowing O$_2$/He at different temperatures.

Figure 2 FT-IR spectra of CO adsorbed on oxidized (300 °C for 4hrs) & reduced Pt$_{20}$/SiO$_2$ catalysts.

However, after oxidation at 300 °C in flowing O$_2$/He for several hours, the supported catalysts do not bind CO and are relatively inactive for CO oxidation. Further treatment with flowing H$_2$ was examined as an additional treatment to activate catalysts. Figure 2 contains the infrared spectra of CO adsorbed to the activated catalysts after oxidation at 300 °C (4hrs) and reduction with different temperatures and times. The increasing peak area with increasing reduction temperature indicates that more Pt sites are available after more extreme reduction conditions. Figure 2 suggests that the reduction of the catalyst is largely complete after treatment under H$_2$ for 2hrs at 300 °C. Based on these results, an activation protocol of oxidation at 300 °C/4hrs in flowing O$_2$/He and reduction at 300 °C/2hrs in flowing H$_2$/He was chosen as the baseline activation protocol to which other activation procedures are compared.

A variety of gas phase activation procedures were examined for the Pt$_{20}$/SiO$_2$ catalysts. IR spectroscopy of adsorbed CO was used to evaluate these activation procedures by using the Pt-CO peak area as a measure of Pt availability. Figure 3 shows the effects of higher temperature oxidation (400 °C/2hrs under O$_2$/He followed by 300 °C/2hrs under H$_2$/He) as well as reduction only activation (300 °C/5hrs under H$_2$/He). Compared to the baseline protocol, the smaller Pt-CO peak after reduction only activation indicates that the oxidation step is necessary to remove PAMAM dendrimers. The smaller CO peak after oxidation at 400 °C likely results from particle sintering at higher temperatures. Particle agglomeration processes have been directly observed with similar materials activated at higher temperatures (12,13) and sintering process are expected to be much faster at higher temperatures (14). This result highlights the importance of keeping activation conditions as mild as possible in order to maintain the possible advantages of using the dendrimer method to prepare heterogeneous catalysts.

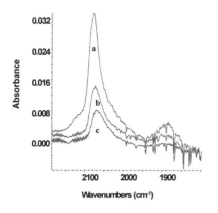

Figure 3 FT-IR spectra of CO adsorbed on Pt_{20}/SiO_2 catalysts: (a) after activation at 300 °C/4hr in flowing O_2/He followed by 300 °C/2hr in flowing H_2/He, (b) after activation at 400 °C/2hr in flowing O_2/He followed by 300 °C/2hr in flowing H_2/He, (c) after activation at 300 °C/6hr in flowing H_2/He.

Since mild activation conditions appear to be important, a number of solution activation conditions were tested. PAMAM dendrimers are comprised of amide bonds, so the favorable conditions for *retro*-Michael addition reactions, (low pH, high temperature and the presence of water) may be able to cleave these bonds. Table 1 shows a series of reaction tests using various acid/solvent combinations to activate the dendrimer amide bonds. Characterization of the solution-activated catalysts with Atomic Absorption spectroscopy, FTIR spectroscopy and FTIR spectroscopy of adsorbed CO indicated that the solution activation generally resulted in Pt loss. Appropriate choice of solvent and acid, particularly EtOH/HOAc, minimized the leaching. FTIR spectra of these samples indicate that a substantial portion of the dendrimer amide bonds was removed by solution activation (note the small y-axis value in Figure 4 relative to Figure 1). However, IR spectra of adsorbed CO showed less Pt available than that with the baseline activation protocol

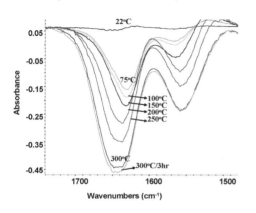

Figure 4 FT-IR spectra of G5-OH(Pt_{20}) /SiO_2 catalyst activation in flowing O_2/He at different temperatures after refluxing in H_2O/acid for 6hrs.

Table 1 Characterization of supported Pt_{20} catalysts after solution treatments.

ID	Solvent	Temp.	Time	%Pt	IR (cm^{-1})	CO Peak Area
1	-----	------	-----	0.30	2072	3.4
2	H$_2$O/HOAc (10:1)	100 °C	6hr	0.17	2077	1.9
3	EtOH/HOAc (10:1)	78 °C	2hr	0.28	2077	1.7
4	EtOH/H$_2$O/HOAc (20:1:2)	78 °C	2hr	0.24	2082	1.8
5	1-propanol/HOAc (10:1)	97 °C	2hr	0.15	2082	1.6
6	HNO$_3$/H$_2$O (pH=1)	impregnation	2hr	0.30	2068	1.7

Note: all catalysts were activated with the baseline activation protocol after any solution treatments.

Catalysis and characterization studies were conducted with catalysts activated using the baseline activation protocol. CO chemisorption data (34% dispersion) indicate that the surface of the Pt_{20}/SiO$_2$ catalyst is not completely clean after the baseline activation protocol. Consistent with this assessment, CO oxidation catalysis experiments on the activated Pt_{20}/SiO$_2$ catalyst showed that it is somewhat less active than a traditionally prepared platinic acid catalyst (PtCl/SiO$_2$) activated using the same protocol (Figure 5). Pt_{20}/SiO$_2$ catalysts were also tested using toluene hydrogenation (Figure 6). TOFs (based on CO chemisorption) for toluene hydrogenation were essentially the same for the Pt_{20}/SiO$_2$ and PtCl/SiO$_2$ catalysts, consistent with the structure insensitivity of this reaction (15). Apparent activation energies for this reaction (Figure 6) were also the same for both catalysts.

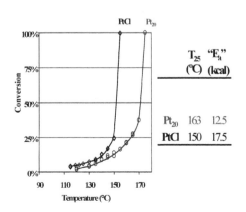

Figure 5 CO oxidation catalysis for after oxidation at 300 °C (20% O$_2$/He, 4hrs) and reduction at 300 °C (20% H$_2$/He, 2hrs).

TEM data for the Pt_{20}/SiO_2 catalysts indicated particles with $d_{ave} \approx 2.4$ nm. This is somewhat larger than expected and previous results (11) indicate that particle agglomeration occurs during deposition of the DENs onto the support rather than during the thermal activation. Nevertheless, the TEM particle size is substantially smaller than predicted by chemisorption data ($d_{ave} \approx 3.5$ nm based on 34% dispersion), which suggests that the surface of the catalyst is still contaminated with organic residues.

The characterization and catalysis results with Pt_{20}/SiO_2 are generally consistent with previous studies using Pt_{50}/SiO_2 and Pt_{100}/SiO_2 catalysts prepared from DENs. Toluene hydrogenation TOFs and apparent activation energies are internally consistent with CO chemisorption data for the DEN and platinic acid catalysts. CO oxidation with the Pt_{50}/SiO_2 and Pt_{100}/SiO_2 catalysts, however, showed the dendrimer-derived catalysts to be more active than the traditionally prepared catalyst. The results with Pt_{20}/SiO_2 show that the baseline activation protocol does not completely remove the organic residues from all catalysts. Although FTIR spectroscopy detected no changes after several hours of oxidation, this does not necessarily prove that no organic species remain on the Pt surface. The Pt nanoparticles themselves appear to play an important role in catalyzing the dendrimer decomposition, so changes in metal and dendrimer loading as well as metal:dendrimer ratio appear to influence the conditions required to completely remove the PAMAM dendrimers from the catalyst. Consequently, activation conditions likely need to be optimized for individual catalysts in order to completely remove the dendrimer without sintering the metal. Continuing studies in our labs are directed towards this end.

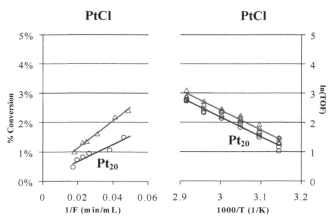

Figure 6 Toluene hydrogenation catalysis for after oxidation at 300 °C (20% O_2/He, 4hrs) and reduction at 300 °C (20% H_2/He, 2hrs)

Conclusions

This study demonstrates that dendrimer encapsulated nanoparticles (DENs) can be used as metal nanoparticle precursors to prepare heterogeneous catalysts. Attempts at activating the DENs with solution treatments generally resulted in Pt leaching. Gas-phase dendrimer removal requires both oxidation and reduction treatment. The *in-situ* activation of supported DENs with flowing O_2/He for several hours decomposes the dendrimer's amide bonds, but may leave significant organic residues on the metal. Therefore, the activation conditions need to be individually optimized based on dendrimer loading and metal:dendrimer ratio in order to prepare active catalysts.

References

1. K. G. Griffin, P. Johnston, S. Bennet, and S. Kaliq, *Chem. Ind.* (Dekker) **85**, (*Catal. Org. React.*) 169 (2003).
2. E. Toukonitty, M.-A. Pälvi, V. Nieminen, M. Hotokka, J. Pälvärinta, T. Salmi, D. Y. Murzin, *Chem. Ind.* (Dekker) **85**, (*Catal. Org. React.*) 341 (2003).
3. N. K. Nag, *Chem. Ind.* (Dekker) **85**, (*Catal. Org. React.*) 415 (2003).
4. K. G. Griffin, S. Hawker, P. Johnston, and M-L. Palacios-Alcolado, *Chem. Ind.* (Dekker) **85**, (*Catal. Org. React.*) 529 (2003).
5. B. Chen, J. Carson, J. Gibson, R. Renneke, T. Tacke, M. Reisinger, R. Hausmann, A. Geisselmann, G. Stochniol, and P. Panster, *Chem. Ind.* (Dekker) **85**, (*Catal. Org. React.*) 179 (2003).
6. R. M. Crooks, B. I. Lemon, L. Sun, L. K. Yeung, M. Zhao, *Top. Curr. Chem.*, **212**, 82-135 (2001).
7. J. M. J. Fréchet and D. A. Tomalia, (Eds.) Dendrimers and Other Dendritic Polymers, Wiley, West Sussex, England, 2001.
8. M. Fisher, F. Vogtle, *Angew. Chem. Int. Ed.*, **38**, 884 (1999).
9. A. I. Cooper, J. D. Londono, G. Wignall, J. B. McClain, E. T. Samulski, J. S. Lin, A. Dobrynin, M. Rubinstein, A. L. C. Burke, J. M. J. Frechet, J. M. DeSimone, *Nature*, **389**, 368-371 (1997).
10. R. M. Crooks, M. Zhao, L. Sun, V. Chechik, L. K. Yeung, *Accts. Chem. Res.*, **34**, 181-190 (2001).
11. H. F. Lang, A. R. May, L. B. Iversen, D. B. Chandler, *J. Am. Chem. Soc.* (2003), in press.
12. S. D. Deutsch, G. Lafaye, H. Lang, B. D. Chandler, D. Liu, C. T. Williams, J. Gao, C. J. Murphy, M. D. Amiridis, *Langmuir*, (2003), submitted.
13. L. Sun, R. M. Crooks, *Langmuir*, **18**, 8231-8236 (2002).
14. V. Ponec, and G. C. Bond in Catalysis by Metals and Alloys. Studies in Surface Science and Catalysis, **95** (1995).
15. S. D. Lin and M. A. Vannice, *J. Catal.*, **143**, 554-162 (1993).

29. Avada — a New Green Process for the Production of Ethyl Acetate

Flora T.T. Ng,* Te Mure, Ming Jiang, Manzoor Sultan,
Jin-hai Xie, and Pierre Gayraud
Department of Chemical Engineering, University of Waterloo, Waterloo,
Ontario, Canada, N2L 3G1

Warren J. Smith, Matthew Hague, Robert Watt, and Steve Hodge
BP Chemicals, Hull Research and Technology Center, Saltend,
Hull, East Yorkshire, HU12 8DS, United Kingdom

fttng@cape.uwaterloo.ca

Abstract

BP Chemicals is the world's largest producer of ethyl acetate. A common route for the production of ethyl acetate is *via* the esterification of ethanol with acetic acid. In 1997 BP Chemicals disclosed a new technology to produce ethyl acetate directly from ethylene and acetic acid using supported heteropoly acid catalysts. In 2001 the world's largest ethyl acetate plant (220,000 tonnes per annum) was successfully commissioned on BP Chemicals site, in Hull, UK. This innovative green technology is trademarked as Avada process (**Ad**Vanced **A**cetates by **D**irect **A**ddition). The Avada process is more energy efficient and environmentally friendly than other routes to ethyl acetate. It won the prestigious 2002 AstraZeneca award for the best Green Chemistry and Engineering Process from the UK Institute of Chemical Engineers. This paper describes some of the studies carried out at the University of Waterloo, which were designed to provide a fundamental understanding of this novel Avada process.

Introduction

Ethyl acetate is an oxygenated solvent widely used in the inks, pharmaceuticals and fragrance sectors. The current global capacity for ethyl acetate is 1.2 million tonnes per annum. BP Chemicals is the world's largest producer of ethyl acetate. Conventional methods for the production of ethyl acetate are either *via* the liquid phase esterification of acetic acid and ethanol or by the coupling of acetaldehyde also known as the Tischenko reaction. Both of these processes require environmentally unfriendly catalysts (e.g. p-toluenesulphonic acid for the esterification and metal chlorides and strong bases for the Tischenko route). In 1997 BP Chemicals disclosed a new route to produce ethyl acetate directly from the reaction of ethylene with acetic acid using supported heteropoly acids

(HPAs) (1). This new direct addition route to ethyl acetate has considerable benefits over conventional technologies since it eliminates the intermediate step for the production of ethanol and also avoids environmental hazards associated with the other processes by using a supported heteropoly acid on silica. Although heteropoly acids are known to be strong acids, there are hardly any known commercial applications for supported heteropoly acids due to rapid catalyst deactivation. Based on a combination of catalyst and process development using microreactors and a pilot plant, BP Chemicals successfully brought on stream a 220,000 tonnes per annum plant in Hull, UK using the novel direct addition technology trademarked Avada (for **Ad**Vanced **A**cetates by **D**irect **A**ddition) which produces ethyl acetate with 99.98 % purity. This Avada process won the highly prestigious AstraZeneca award in 2002 for the best Green Chemistry and Engineering Process from the UK Institute of Chemical Engineers.

Catalysis by HPAs is a field of increasing industrial interest. HPAs have several advantages that make them more economically and environmentally attractive as commercial catalysts. HPAs have very strong Brönsted acidity, approaching that of superacids, and can undergo multi-electron redox transformation under rather mild conditions. Solid HPAs have a discrete ionic structure, comprising of fairly mobile structure units, *i.e.*, heteropoly anions and counter ions such as H^+, Cs^+, NH_4^+ and other cations. This unique structure exhibits extremely high proton mobility and "pseudo-liquid" phase behaviour while the heteropoly anions can stabilize cationic intermediates (2). The fairly high thermal stability of the HPAs in the solid state renders HPAs suitable for heterogeneous catalysis. The Keggin type HPAs with PW, SiW, PMo and SiMo heteropoly anions decompose at 465, 445, 375 and 350 °C, respectively (3). Decomposition of HPAs result in a loss of acidity due to the fact that these solids have discrete and mobile ionic units. Solid HPAs can absorb a large amount of polar molecules, *e.g.* alcohols, ethers, amines, water, in the catalyst bulk, forming HPA solvates. However, non-polar molecules such as alkanes and alkenes adsorb primarily on the surface. Despite their acidic properties, the activities of HPAs are quite low due to their low surface areas. To improve the activities of HPAs as catalysts, it is necessary to increase the available surface area of HPAs. There are two possible approaches, the use of high surface area salts of HPAs (*e.g.*, the Cs salts that have surface areas of the order of 100 m^2/g) or to disperse HPAs on high surface area supports such as silica, carbon, or MCM- 41. This paper is focused on the study of stability and reactivity of HPAs dispersed on silica. Comparison has been made between different types of HPA catalysts for the direct addition reaction and the effect of different HPAs on the reaction has been discussed.

Experimental Section

Keggin phosphotungstic acid $H_3PW_{12}O_{40} \cdot nH_2O$ and Keggin silicotungstic acid $H_4SiW_{12}O_{40} \cdot nH_2O$ were obtained from Sigma-Aldrich and denoted as HPW and HSiW respectively. Keggin germanotungstic acid $H_4GeW_{12}O_{40} \cdot nH_2O$ and Dawson phosphotungstic acid $H_6P_2W_{18}O_{62} \cdot nH_2O$ were synthesized in the laboratory according to the literature (4,5) and denoted as HGeW and HP_2W respectively. Silica-supported HPAs were prepared by a wet impregnation method.

Activities of the catalysts were measured on a microreactor. About 3 g of catalyst was charged into a reactor and heat-treated in nitrogen at reaction temperature. Acetic acid was added to the process and the reaction was initiated by switching nitrogen to ethylene. Reaction product analyses were performed by an online gas chromatograph equipped with a flame ionization detector (Perkin Elmer Auto System II).

XRD patterns were recorded on a Siemens D500 diffractometer using Cu K_α (λ=0.15406 nm) monochromated radiation source.

Diffuse reflectance FTIR (DRIFT) spectra were recorded on a Bio-Rad FTIR spectrometer (EXCALIBUR FTS3000). A high-temperature cell was attached to a flow system that allows in-situ sample treatment, adsorption and desorption of probe molecules at different temperatures.

Temperature programmed desorption (TPD) of NH_3 adsorbed on the samples was carried out on an Altamira TPD apparatus. NH_3 adsorption was performed at 50°C on the sample that had been heat-treated at 120°C in a helium flow. After flushing with helium, the sample was subjected to TPD from 50 to 600°C ($\Delta T = 10$°C min^{-1}). The evolved NH_3, H_2O and N_2 were monitored by mass spectroscopy by recording the mass signals of m/e = 16, 18 and 28, respectively using a VG Trio-1 mass spectrometer.

Thermal gravimetric (TG) analyses were carried out on a TG / DTA instrument (TA Instruments). N_2 was used as carrier gas and alumina was used as reference. The sample was heated from 40 to 700°C ($\Delta T = 10$°C min^{-1}).

Results and Discussion

Process Description

There are a number of routes to produce ethyl acetate (Figure 1). Ethylene is generally the common starting feedstock for the Tischenko (*via* acetaldehyde), esterification and Avada processes. Ethyl acetate can also be produced through

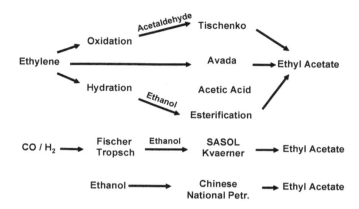

Figure 1 Routes for ethyl acetate production.

Fischer-Tropsch reaction of CO/H_2 to produce ethanol followed by the Sasol Kvaerner process.

Reactivity of a number of solid acid catalysts that include zeolites, resin, nafion and HPAs was determined for the direct reaction of ethylene with acetic acid to produce ethyl acetate (Table 1). It was established that the Keggin HSiW supported on silica is very active for the vapour phase reaction of acetic acid with ethylene at about 180°C, 145 psig with a high molar ratio of ethylene to

Table 1 Solid acid catalysts for the reaction of ethylene with acetic acid to produce ethyl acetate (6).

Catalyst	$C_2H_4/AcOH$ mol/mol	T °C	P psig	time sec.	H_2O in feed %mol	EA STY* g/l/h
H-montmorillonite	5/1	200	725	4	0	144
XE386 resin	5/1	155	725	4	0	120
Nafion	5/1	170	725	4	0	102
H-zeolite Y	5/1	200	725	4	0	2
H-Theta-1	5/1	200	725	4	0	0
$H_4SiW_{12}O_{40}/SiO_2$	12/1	180	145	2	6	380

* EA: ethyl acetate; STY: space time yield.

acetic acid (6). The high ratio of ethylene to acetic acid is required since acetic acid readily absorbs into the bulk of HPAs while a non-polar molecule such as ethylene is adsorbed primarily on the surface of the HPAs. The direct reaction is mildly exothermic and is carried out in a series of adiabatic fixed bed reactors. Inter-bed cooling limits the temperature increase through the reaction system and brings a number of advantages. For example, lower average temperatures reduce the rate of catalyst deactivation.

In 2001 a 220,000 tonnes per annum plant for the production of ethyl acetate using a silica-supported HPA was successfully commissioned on the BP Chemicals site in Hull, UK. A schematic of the Avada plant is shown in Figure 2 (7). The Avada process is superior to other traditional processes in terms of

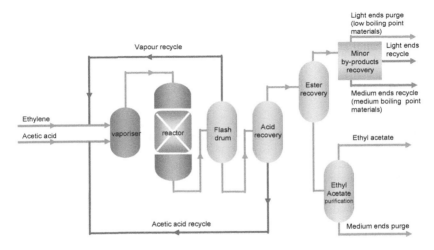

Figure 2 Schematic of BP Chemicals' Avada process (reproduced from *BP Frontiers*, August 2002).

environmental protection since it uses a solid acid and therefore there is less requirement for the treatment and disposal of aqueous effluent compared to the traditional esterification reaction that produces as much water as ethyl acetate. Since the Avada process eliminates the intermediate step to the production of ethanol, energy costs are about 20% lower and feedstock losses some 35% less. The Avada process represents a new trend towards development of green processes in chemical industry by utilizing the principles of process intensification, that is, reduction of processing steps, being more energy efficient, using environmentally benign reagents and producing high purity products (high selectivity to the desired product) which reduces the production of by-products or waste treatment streams.

Reactivity

Activity of the Keggin HPW/SiO$_2$ catalyst in terms of the conversions of ethylene and acetic acid and production of ethyl acetate *vs.* reaction time is displayed in Figure 3. Besides ethyl acetate, ethanol and diethyl ether are also produced. It can be seen that the catalyst is quite stable over the 17 hour period on stream. Activities of the other silica-supported Keggin

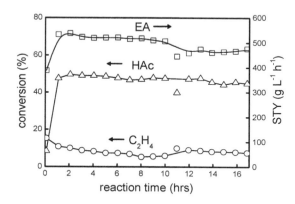

Figure 3 Performance of Keggin HPW/SiO$_2$ (T=180°C, P=150 psig).

HPAs containing heteroatoms such as Si and Ge were also measured and the results are outlined in Table 2. From the table, the activity of silica-supported Keggin HPAs is in an order of HPW ~ HSiW > HGeW, which corresponds to the order of the acidity of the bulk Keggin HPAs (8). This is not surprising since the direct addition of the ethylene to acetic acid is an acid catalysed reaction.

Table 2 Comparison of performance of different standard HPAs.

Catalyst	EA selectivity (wt%)	EA STY* (g/gHPA/h)
Keggin-HPW/SiO$_2$	92	2.7
Keggin-HSiW/SiO$_2$	94	2.4
Keggin-HGeW/SiO$_2$	97	2.0
Dawson-HP$_2$W/SiO$_2$	85	4.85

* see footnote on Table 1.

The silica-supported Dawson HP_2W was also used for the direct addition reaction. Interestingly, the activity of the Dawson catalyst is much higher than that of the Keggin catalysts (Table 2). In general, the catalytic activity for an acid-catalyzed reaction is related to the acidity of the catalysts. However, it is known that the acidity of Dawson HP_2W is lower than that of Keggin HPW (9). In an effort to understand the higher activity of the silica-supported Dawson HP_2W catalyst for the direct addition reaction, characterizations of these HPA catalysts were carried out.

Characterizations

Although there is a lot of information reported in the literature on bulk HPAs, information on silica-supported HPAs, especially on supported Dawson HP_2W, is quite limited. XRD studies on silica-supported Keggin HPW and Dawson HP_2W are shown in Figure 4. In the case of the Keggin sample, the HPA crystalline phase can be detected. The particles of HPA tend to aggregate during the reaction as revealed by the increase in intensity of the diffraction peaks arising from HPA. For the fresh and spent Dawson HP_2W/SiO_2 catalysts, crystalline phases were hardly

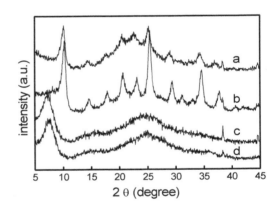

Figure 4 XRD patterns of fresh HPW/SiO_2 (a), spent HPW/SiO_2 (b), fresh HP_2W/SiO_2 (c) and spent HP_2W/SiO_2 (d).

detectable. This suggests that the Dawson HP_2W species is more dispersed and the dispersed species is not prone to aggregation during reaction.

The hydration state of unsupported Keggin HPW and Dawson HP_2W catalysts were investigated by means of diffuse reflectance FTIR and TG analysis. It has been found that affinity of Keggin polyanions to water is much higher than that of the Dawson HP_2W. However, after the HPAs are loaded on the support, the difference is significantly reduced. Mass spectroscopy-monitored thermal dehydration of the catalysts in helium also confirms the above conclusion. For the unsupported HPAs, water in the Keggin structure is removed at much higher temperatures than in the Dawson. After being loaded on SiO_2, water in both structures is removed at comparable temperatures.

Investigations of the acidity and thermal stability of the catalysts were carried out by using NH_3-TPD monitored by mass spectroscopy. For both the unsupported and supported HPAs, the weakly held or physisorbed ammonia and water desorb at low temperatures ($<200°C$). The thermal stability and acidity of the HPAs are characterized by the species desorbing at higher temperatures ($>300°C$). The representative results are summarized in Table 3. Ammonia chemisorbed on the unsupported Keggin HPW and HSiW desorbs at temperatures higher than $550°C$, while on the Dawson HP_2W, chemisorbed ammonia desorbs at about $465°C$. During ammonia desorption, apparently a reaction occurs between the polyanions of HPA and ammonia producing N_2 and water which desorb at the same temperature. It can be seen that water and nitrogen also evolve at much lower temperatures for the Dawson HP_2W than for the Keggin HPAs. The results clearly demonstrate that the Dawson HP_2W has much lower acidity and thermal stability than the Keggin HPAs. After the Keggin HPAs are supported on silica, two types of HPA species are apparently formed on the support. One species is characterized by the desorption of NH_3, H_2O and N_2 at high temperatures ($> 500°C$), similar to that of the bulk or unsupported HPAs. The other species is a more dispersed species that has lower acidity as seen in the lower temperature for NH_3 desorption. The broad peak of water desorption between $400 \sim 500°C$ is due to the degradation and elimination of the acidic protons from the dispersed HPA species. The desorption temperatures for both of these two species are lower than for the unsupported HPAs which indicate that the thermal stability and acidity of the supported HPAs are decreased because of the support effect. In the case of the silica-supported Dawson HP_2W catalyst, only one species represented by desorption peaks at about $410°C$ is observed. The desorption temperature is close to that for

Table 3 Summary of the TPD results.

Species	Unsupported HPAs			Silica-supported HPAs		
	HPW	HSiW	HP_2W	HPW	HSiW	HP_2W
NH_3 at T>400°C	~580	~575	465	425~520 555	400~500 530	410
H_2O at T > 400°C	~580	~565	475 540	435~500 555	400~500 530	410
N_2	~580	~585	545	555	530	---

the unsupported Dawson catalyst, indicating that the thermal stability and acidity of the supported Dawson HP_2W is essentially the same as the unsupported Dawson HP_2W. From Table 3, it can also be seen that the supported Keggin and Dawson HPA catalysts have comparable thermal stability and acidity.

Effect of Process Parameters on the Avada Process

Over the range of HPA loading varying from 28 to 40 wt%, the yield of ethyl acetate increases with HPA loading (Figure 5), which suggests that the reaction occurs in the HPA bulk phase or the pseudo-liquid phase. Misono et al. investigated the synthesis of methyl-tert-butyl ether on unsupported Keggin HPW and Dawson HP_2W. They proposed that a substantial amount of methanol can be absorbed into the HPA bulk behaving like a concentrated solution described as the "pseudo-liquid phase". Under high pressure reaction conditions, the absorption of iso-butylene into the pseudo-liquid phase will facilitate the reaction. They also found that Dawson HP_2W is more active than Keggin HPAs and attributed the higher activity to the fact that Dawson HP_2W has a less rigid structure and weaker acidity than Keggin HPW, which gives rise to a faster absorption - desorption rate of polar solvents in the pseudo-liquid phase of the HPA.

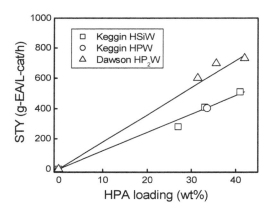

Figure 5   Effect of HPA loading on ethyl acetate yield.

The first step of the reaction is likely to be the protonation of ethylene to produce a carbocation that undergoes the direct addition of acetic acid to produce ethyl acetate. The successive addition of ethylene to the carbocation leading to the production of alkene oligomers is a likely side reaction. Formation and accumulation of these oligomers could eventually deactivate the catalyst. Detailed studies for a better understanding of the complex reaction mechanism are in progress.

Conclusion

In summary, the Avada process is an excellent example of process intensification to achieve higher energy efficiency and reduction of waste streams due to the use of a solid acid catalyst. The successful application of supported HPAs for the production of ethyl acetate paves the way for future applications of supported HPAs in new green processes for the production of other chemicals, fuels and lubricants. Our results also show that application of characterization techniques enables a better understanding of the effects of process parameters on reactivity and the eventual rational design of more active catalysts.

Acknowledgements

Funding from BP Chemicals (UK) and Natural Sciences and Engineering Research Council of Canada, Cooperative Research and Development Program are gratefully acknowledged.

References

1. M. P. Atkins and B. Sharma, EP 757027 A1, 1997.
2. T. Okuhara, N. Mizuno, and M. Misono, *Adv. Catal.*, **41**, 113 (1996).
3. I.V. Kozhevnikov, *Chem. Rev.*, **98**, 71 (1998) (reference therein).
4. R. Thouvenot, *Inorg. Chem.*, **22**, 207 (1983).
5. H. Wu, *Biol. Chem.*, **43**, 189 (1920).
6. M. J. Howard, G. J. Sunley, A. D. Poole, R. J. Watt and B. K. Sharma, *Stud. Surf. Sci. Catal.*, **121**, 61 (1998).
7. *BP Frontiers*, August, 2002, p 12-15. Available on line at www.bp.com.
8. Y. Izumi, K. Matsuo and K. Urabe, *J. Mol. Catal.*, **18**, 299 (1983).
9. S. Shikata, T. Okuhara and M. Misono, *J. Mol Catal. A*, **100**, 49 (1995).

30. A Catalytic Distillation Process for the One Step Synthesis of Methyl Isobutyl Ketone from Acetone: Liquid Phase Kinetics of the Hydrogenation of Mesityl Oxide

W. K. O'Keefe, M. Jiang, F. T. T. Ng*, and G. L. Rempel

Department of Chemical Engineering, The University of Waterloo, 200 University Avenue West, Waterloo, Ontario, Canada. N2L 3G1

fttng@cape.uwaterloo.ca

Abstract

The present economic and environmental incentives for the development of a viable one-step process for MIBK production provide an excellent opportunity for the application of catalytic distillation (CD) technology. Here, the use of CD technology for the synthesis of MIBK from acetone is described and recent progress on this process development is reported. Specifically, the results of a study on the liquid phase kinetics of the liquid phase hydrogenation of mesityl oxide (MO) in acetone are presented. Our preliminary spectroscopic results suggest that MO exists as a diadsorbed species with both the carbonyl and olefin groups coordinated to the catalyst. An empirical kinetic model was developed which will be incorporated into our three-phase non-equilibrium rate-based model for the simulation of yield and selectivity for the one step synthesis of MIBK via CD.

Introduction

MIBK is a valuable industrial solvent used primarily in the paint and coating industry and in metallurgical extraction processes. It is also used as a precursor in the production of specialty chemicals such as pesticides, rubber anti-oxidants as well as antibiotics and pharmaceuticals (1). Historically, MIBK has been produced commercially from acetone and hydrogen feedstock in three stages. First, acetone is dimerized to produce diacetone alcohol (DAA). Second, DAA is dehydrated to produce MO and water. Finally, the carbon-carbon double bond of MO is selectively hydrogenated to produce MIBK. These consecutive reactions are outlined in equations (1-3).

Current state-of-the-art technology for the production of MIBK involves one-step liquid phase processes in trickle bed reactors at 100-160°C and 1 to 10 MPa utilizing various multifunctional catalysts including Pd, Pt, Ni or Cu supported on, metal oxides, cation exchange resins, modified ZSM5 and other zeolites with full energy integration (2,3,4). However, the MIBK

$$\underset{\text{(acetone)}}{CH_3\overset{O}{\overset{\|}{C}}CH_3} \; \overset{k_1}{\rightleftharpoons} \; \underset{\text{(DAA)}}{(CH_3)_2\overset{OH}{\overset{|}{C}}CH_2\overset{O}{\overset{\|}{C}}CH_3} \qquad (1)$$

$$\underset{\text{(DAA)}}{(CH_3)_2\overset{OH}{\overset{|}{C}}CH_2\overset{O}{\overset{\|}{C}}CH_3} \; \overset{k_2}{\rightleftharpoons} \; \underset{\text{(MO)}}{(CH_3)_2C=CH\overset{O}{\overset{\|}{C}}CH_3} \; + \; H_2O \qquad (2)$$

$$\underset{\text{(MO)}}{(CH_3)_2C=CH\overset{O}{\overset{\|}{C}}CH_3} \; + \; H_2 \; \overset{k_3}{\longrightarrow} \; \underset{\text{(MIBK)}}{(CH_3)_2CH_2CH_2\overset{O}{\overset{\|}{C}}CH_3} \qquad (3)$$

concentration in the effluent is typically less than 30 wt% necessitating additional purification steps and the high pressure operation and its complicated operation are a disadvantage (2,3). Consequently, there remains incentive to further develop improved one-step processes for MIBK production.

MIBK has been successfully produced from hydrogen and acetone feedstock in a single stage in a CD reactor (5,6). The CD reactor consists of a packed distillation column with solid catalyst located in discreet reaction zones. The reaction occurs in the liquid phase at the solid catalyst surface. MIBK has a high boiling point relative to acetone and is transported from the reaction zone by distillation to the bottoms product. Unreacted acetone is concentrated in the distillate and is refluxed. The continuous removal of product from the reaction zone breaks the equilibrium limitations and promotes product formation in accordance with Le Châtelier's Principle while averting possible consecutive reactions. The catalyst is either supported on or enclosed within engineered structures in order to prevent an excessive pressure drop. A hydrogenation catalyst such as Pd, Ni, Pt or Cu is typically dispersed onto a support that is catalytically active for the conversion of acetone to MO. Numerous catalysts have been investigated for this purpose including zeolites (7,8,9), hydrotalcites (10,11), metal oxides (2,12) and cationic exchange resins (1,4). Alternatively, MIBK synthesis by CD may be achieved using a mixture of catalysts in the same reaction zone or by separate catalysts in separate reaction zones (4). We are currently investigating the effects of these reactor configurations on MIBK productivity and selectivity. Two possible CD reactor configurations are illustrated in Figure 1.

A CD process for the production of DAA from acetone was developed previously in our laboratory and the kinetics of the aldol condensation of acetone were characterized (13,14). CD is a green reactor technology that provides enhanced yield and selectivity in addition to significant energy savings (15). The one-step synthesis of MIBK by CD appears to be a simple extension of this process. However, the introduction of hydrogen to this system opens numerous possible reaction pathways. A fundamental understanding of the kinetics of these reactions is required. To this end, the liquid phase kinetics of the selective hydrogenation of MO in acetone were investigated (16). The resulting kinetic model will be integrated into our rate-based three-phase non-equilibrium model for the simulation of CD processes, which will aid in the CD process design and development (17).

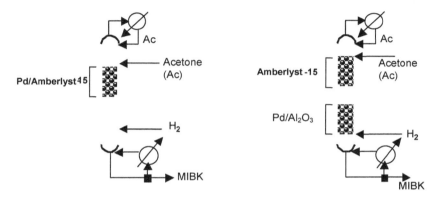

Figure 1 CD processes for the one-step synthesis of MIBK from acetone.

Experimental Section

The liquid phase kinetics of the selective hydrogenation of MO in acetone over a 0.52 wt% Pd/Al_2O_3 commercial catalyst (Aldrich 20,574-5) were investigated for temperatures ranging from 100 to 120°C, hydrogen pressures ranging from 0.7 to 6.2 MPa and initial MO concentrations ranging from 0.1 to 1.0 M in a 300 mL Parr 4560 series microreactor with a PID temperature controller (14). Typically 1 g of catalyst was used. Liquid samples were obtained periodically and their compositions were ascertained by GC/FID using an Agilent Technologies' 6890N gas chromatograph equipped with a 7683 Series autosampler injector and a J&W Scientific DB-WAX capillary column (30m X 0.53mm I.D. X 1μm film thickness). 1-Propanol (Fisher A414-500), was used as an internal standard. In situ measurements of the Diffuse Reflectance Infrared Fourier Transform (DRIFT) spectra MO, MIBK and acetone pulse adsorbed on the Pd/Al_2O_3 catalyst were obtained using a Bio-Rad FTS 3000 FTIR.

Results and Discussion

The liquid phase kinetics of the selective hydrogenation of mesityl oxide in acetone were studied for the purpose of developing a robust kinetic model to be integrated into an existing non-equilibrium rate-based model for the simulation of the CD process for MIBK production. A typical concentration versus time profile is illustrated in Figure 2. MIBK was produced with very high selectivity with essentially all of the MO converted to MIBK. Products from the consecutive reaction of MIBK such as methyl isobutyl carbinol were rarely observed and were only found in trace amounts after essentially all of the MO had been consumed. DAA was present as an impurity in the MO stock solution. However, its concentration remained invariant throughout each experiment since the catalyst was inert for the aldol condensation of acetone. Isomesityl oxide, the isomer of MO, showed identical kinetic behaviour as MO having the same reaction rate constant. The concentration of isomesityl oxide was very low and was neglected from Figure 2 for clarity. The hydrogenation of acetone to produce isopropyl alcohol (IPA) was found to be the only significant competing reaction. Since acetone was in excess, the rate of IPA production was found to be constant.

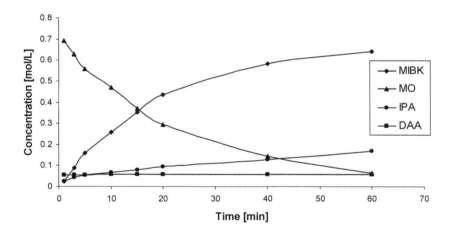

Figure 2 A typical concentration versus time profile for the liquid phase hydrogenation of MO [T=100°C, P_H=6.2 MPa; ω=300 RPM, W=1g 0.52 wt% Pd/Al$_2$O$_3$].

Table 1 illustrates the effect of hydrogen pressure on the selectivity to MIBK based on the initial rate of MIBK production and the rate of IPA production. Palladium gives very high selectivity to MIBK, typically in excess of 93% with the selectivity improving significantly with decreasing pressure. This result is of particular importance since the CD process for MIBK production is carried out at relatively low pressure (< 1MPa). In contrast, alternative one-step processes for MIBK production are carried out in the liquid phase in trickle-bed reactors at pressures as high as 10 MPa.

Table 1 The effect of hydrogen pressure on
the selectivity to MIBK for the liquid phase hydrogenation of MO.

Hydrogen Pressure	Average Selectivity	Maximum Selectivity	Minimum Selectivity	Std. Dev.
$P_H > 3.79$ MPa	93.54	97.80	81.68	3.744
$P_H < 3.79$ MPa	97.54	98.09	96.97	0.4875
all data	94.51	98.09	81.68	3.687

The MO concentrations versus time profiles were fitted to second order polynomial equations and the parameters estimated by nonlinear regression analysis. The initial rates of reactions were obtained by taking the derivative at t=0. The reaction is first order with respect to hydrogen pressure changing to zero order dependence above about 3.45 MPa hydrogen pressure. This was attributed to saturation of the catalyst sites. Experiments were conducted in which HPLC grade MIBK was added to the initial reactant mixture, there was no evidence of product inhibition.

In contrast to kinetic models reported previously in the literature (18,19) where MO was assumed to adsorb at a single site, our preliminary data based on DRIFT results suggest that MO exists as a diadsorbed species with both the carbonyl and olefin groups being coordinated to the catalyst. This diadsorption mode for α-β unsaturated ketones and aldehydes on palladium have been previously suggested based on quantum chemical predictions (20). A two parameter empirical model (equation 4) where - r_A refers to the rate of hydrogenation of MO, C_A and P_H refer to the concentration of MO and the hydrogen partial pressure respectively was developed. This rate expression will be incorporated in our rate-based three-phase non-equilibrium model to predict the yield and selectivity for the production of MIBK from acetone via CD.

$$-r_A = \frac{\kappa_1 C_A P_H}{\left[1 + \kappa_2 C_A^2\right]^4} \tag{4}$$

$\kappa_1 = 8.408 \times 10^{-5}$ [min/psi.gcat]
$\kappa_2 = 0.1634$ [L^2/mol^2]

Conclusions

An empirical kinetic model for the hydrogenation of MO was developed, and its parameters were estimated. The reaction was found to be first order with respect to hydrogen up to about 3.45 MPa changing to zero order kinetics. Preliminary DRIFT data suggest that MO coordinates to the catalyst via both the carbonyl and the olefin groups. MIBK does not appear to inhibit the reaction. The Pd/Al$_2$O$_3$ catalyst gave very high selectivity to MIBK with the selectivity improving with decreasing hydrogen pressure. The hydrogenation of acetone to produce isopropyl alcohol was found to be the only significant competing reaction under the conditions investigated.

Acknowledgements

Funding for this project provided by the Natural Science and Engineering Research Council (NSERC) of Canada, Strategic Grants Program is gratefully acknowledged.

References

1. J. Braithwaite, Kirk-Othmer Encyclopedia of Chemical Technology, 4th Ed. J. I. Kroschwitz and M. Howe-Grant, Ed., vol. **4**, 1991, p. 978-1021.
2. K. H. Lin and A. N. Ko, *Appl. Catal. A.,* **147**, 1259 (1996).
3. J. I. DiCosimo, G. Torres and C. R. Apesteguia, *J. Mol. Catal.,* **208**, 114 (2002).
4. P.Y. Chen, S. J. Chu, K. C. Wu and W. C. Lin, U.S. Patent 5,684,207, (1997) Assigned to Industrial Technology Research Institute (Hsinchu, TW).
5. K. H. Lawson and B. Nkosi, U.S. Patent 6,008,416. (1999) assigned to Catalytic Distillation Technologies (Pasadena, TX).
6. N. Saayman, G. J. Grant and S. Kindermans, U.S. Patent 6,518,462 (2003), Assigned to Catalytic Distillation Technologies (Pasadena, TX).
7. Melo, G. Magnoux, Giannetto, F. Alvarez and M. Guisnet, *J. Catal.,* **124**, 155 (1997).
8. L. V. Mattos, F. B. Noronha, and J. L. F. Monteiro, *J. Catal.,* **209**, 166 (2002).
9. S. M. Yang and Y. M. Wu, *Appl. Catal. A,* **192**, 211 (2000).
10. J. J. Spivey, A. A. Nikolopoulos, and G. B. Howe, Chem. Ind. (Dekker) XX (*Catal. Org. React.*) (2000).
11. Y. Z. Chen, C. M. Hwang, and C. W. Liaw, *Appl. Catal. A.,* **169**, 207 (1998).
12. V. Chikan, A. Molnar, and K. Balazsik, *J. Catal.,* **184**, 134 (1999).
13. G. G. Podrebarac, F. T. T. Ng, and G. L. Rempel, *Chem. Eng. Sci.,* **53**, 1067 (1998).
14. G. G. Podrebarac, F. T. T. Ng, and G. L. Rempel, *Chem. Eng. Sci.,* **52**, 2991 (1997).
15. F. T. T. Ng and G. L. Rempel, Encyclopedia of Catalysis (John Wiley), 477-509 (2003).
16. W. K. O'Keefe, M. Jiang, F. T. T. Ng, and G. L. Rempel, *Unpublished data* (2003).
17. Y. Zheng, F. T. T. Ng and G. L. Rempel, *Ind. Eng. Chem. Res.,* **40**, 5342 (2001); *ibid.* **42**, 3962 (2003).
18. S. Kishida and S. Teranishi, *Bull. Chem. Soc. Jpn.,* **41**, 2528 (1968).
19. Y. Watannabe, et al., *Bull. Chem. Soc. Jpn.,* **50**, 1539 (1977).
20. V. Ponec, *Appl. Catal. A.,* **149**, 27 (1997).

31.   A Strategy for the Preparation of Isolated Organometallic Catalysts on Silica Supports – Towards Single-Site Solid Catalysts

Michael W. McKittrick, Shara D. McClendon, and Christopher W. Jones*

School of Chemical and Biomolecular Engineering, Georgia Institute of Technology, Atlanta, GA 30332

Abstract

A well-defined amine functionalized silica surface displaying evidence of isolated amine sites is prepared using a molecular patterning technique. These aminosilicas serve as a support for immobilized organometallic catalysts, specifically constrained geometry catalysts for olefin polymerizations. The patterned catalysts show an activity of 27-32 kg polymer/mol Ti hr for the polymerization of ethylene with methylaluminoxane as cocatalyst. This activity is significantly higher than catalysts made via traditional supporting techniques on a densely loaded amine-functionalized silica, which show an activity of 4-10 kg polymer/mol Ti hr. This activity increase, when combined with reactivity and probe molecule studies, provides further evidence of the site-isolated behavior of the patterned amine sites and the unique nature of active sites that can be assembled on these aminosilica materials.

Introduction

Organometallic complexes are extremely versatile catalysts because of the chemical properties which can be incorporated into them – the symmetry, sterics and electronics of the active center can all be controlled using molecular design principles. However, since homogeneous catalyst recovery and recyclability is a costly industrial problem, the implementation of these catalysts is not straightforward. For this reason, transition metal complex catalysts have been immobilized on solid supports for decades (1). Supporting a catalyst has the potential to result in a material with all the attributes of both homogeneous (high rates and selectivities) and heterogeneous catalysts (ease of separation). However, in practice this is rarely achieved as the performance of the supported catalysts is usually significantly inferior compared to their homogeneous analogues. In reality, the supported catalysts typically display both the drawbacks of heterogeneous (low rates and selectivities) and homogeneous catalysis (difficult separation due to metal leaching) instead of the attributes of both systems. While this phenomenon is due to many factors, a primary cause is the formation of multiple types of ill-defined potential active sites on the solid support (1). As an example, consider single-site transition metal complexes such as Group 4 metallocenes and related complexes, which are a new class of olefin polymerization catalysts that have been recently developed (2). Although the homogeneous systems are well-understood, the preparation of well-defined supported single-site polymerization catalysts has not progressed as quickly, as protocols which would lead to the synthesis of uniform, isolated organometallic species on surfaces are poorly developed (3). For example, Ti and Zr constrained geometry-inspired catalysts (CGCs) (4) have been supported on a variety of different solids using two different immobilization strategies. In the first approach, the organometallic precatalyst is contacted with a support (5). However multiple types of surface species are formed due to solid–metal atom interactions. In the alternate approach, the complex is assembled step-wise on the support (6). Steric crowding on the support and the heterogeneous nature of the support surface commonly also results in multiple types of sites.

The above example outlines a general problem in immobilized molecular catalysts – multiple types of sites are often produced. To this end, we are developing techniques to prepare well-defined immobilized organometallic catalysts on silica supports with isolated catalytic sites (7). Our new strategy is demonstrated by creation of isolated titanium complexes on a mesoporous silica support. These new materials are characterized in detail and their catalytic properties in test reactions (polymerization of ethylene) indicate improved catalytic performance over supported catalysts prepared via conventional means (8). The generality of this catalyst design approach is discussed and additional immobilized metal complex catalysts are considered.

Experimental Section

General Considerations. The following chemicals were commercially available and used as received: 3,3,3-Triphenylpropionic acid (Acros), 1.0 M LiAlH$_4$ in tetrahydrofuran (THF) (Aldrich), pyridinium dichromate (Acros), 2,6 di-tert-butylpyridine (Acros), dichlorodimethylsilane (Acros), tetraethyl orthosilicate (Aldrich), 3-aminopropyltrimethoxysilane (Aldrich), hexamethyldisilazane (Aldrich), tetrakis (diethylamino) titanium (Aldrich), trimethyl silyl chloride (Aldrich), terephthaloyl chloride (Acros), anhydrous toluene (Acros), and n-butyllithium in hexanes (Aldrich). Anhydrous ether, anhydrous THF, anhydrous dichloromethane, and anhydrous hexanes were obtained from a packed bed solvent purification system utilizing columns of copper oxide catalyst and alumina (ether, hexanes) or dual alumina columns (tetrahydrofuran, dichloromethane) (9). Tetramethylcyclopentadiene (Aldrich) was distilled over sodium metal prior to use. p-Aminophenyltrimethoxysilane (Gelest) was purified by recrystallization from methanol. Anhydrous methanol (Acros) was further dried over 4-Å molecular sieves prior to use. All air- and moisture-sensitive compounds were manipulated using standard vacuum line, Schlenk, or cannula techniques under dry, deoxygenated argon or in a drybox under a deoxygenated nitrogen atmosphere.

Synthesis of Flexible Tritylimine Patterning Agent ([3-(Trimethoxysilanyl)-propyl]-(3,3,3-triphenylpropylidene)amine) (1a). 3,3,3-Triphenylpropanal was synthesized from 3,3,3-triphenyl propionic acid under argon using standard Schlenk techniques (10) Triphenylpropionic acid (4.373 g, 14.46 mmol) was added to a flask with THF. The flask was then cooled in an ice-water bath and 8.97 g of pre-cooled 1.0 M LiAlH$_4$ in THF was added dropwise. After the evolution of hydrogen completed, the mixture was stirred at room temperature for 30 min. The solution was then added to a suspension of pyridinium dichromate (6.00 g, 15.95 mmol) in methylene chloride and allowed to stir for 6 h. The reaction mixture was subsequently diluted with diethyl ether and filtered through a silica column, which was washed with several portions of diethyl ether. The filtrate was collected, and through rotary evaporation, the aldehyde was isolated in 60% yield. 1H NMR (300 MHz, CDCl3): 3.6 (d, 2H), 7.1-7.3 (multi, 15 H), 9.5 (t, 1H).

The aldehyde (1.0 g, 3.5 mmol) was then refluxed with 3-aminopropyltrimethoxysilane (0.501, 2.8 mmol) in dry methanol for 24 h. The methanol was removed in vacuo. The imine product was then recrystallized from methanol (38% yield). NMR data: ^{1}H NMR (300 MHz, CD$_3$OD): 0.4 (2 H), 1.51 (2 H), 3.14 (2 H), 3.49 (9 H), 3.61 (2 H), 7.22 (15 H), 7.42 (1 H).

Synthesis of Rigid Tritylimine Patterning Agent ([3-(Trimethoxysilanyl)propyl]-(3,3,3-triphenylpropylidene)amine) (1b). The same aldehyde used in the synthesis of **1a** was used in this synthesis. The aldehyde (1.0 g, 3.5 mmol) was refluxed with p-aminophenyltrimethoxysilane (1.35 g, 2.8 mmol) in dry methanol for 24 h. The methanol was removed in vacuo. The imine product was then recrystallized from methanol (yield). NMR data: ^{1}H NMR (300 MHz, CD$_3$OD): 2.2 (2 H), 3.6 (9 H), 7.1-7.4 (19 H), 7.6 (1 H).

Synthesis of SBA-15 Silica Support. SBA-15 was synthesized by literature methods (11). The as-prepared material was calcined using the following temperature program: (a) increasing the temperature (1.2 C/min) to 200 C, (b) heating at 200 C for 1 h, (c) increasing at 1.2 C/min to 550 C, and (d) holding at 550 C for 6 h. Prior to functionalization, the SBA-15 was dried under vacuum at 150 C for 3 h and stored in a drybox.

Synthesis of SBA-15 Functionalized with Patterning Agent (2). Patterning agent 1a (0.5 g/1.12 mmol) was added to 2 g of SBA-15 with anhydrous toluene and stirred at room temperature under argon for 24 h. The resulting solid was filtered and washed with toluene in a drybox, dried under vacuum at room temperature overnight, and then stored in a drybox. TGA showed 0.39 mmol/g SiO_2 of patterning agent was immobilized on the SBA-15.

Silanol Capping Reaction (3). The synthesis of **3** was carried out by contacting a large excess of hexamethyldisilazane with **2** in anhydrous toluene at room temperature under argon for 24 h. The resulting solid was filtered and washed with toluene and hexanes in a drybox, dried under vacuum at room temperature overnight, and then stored in a drybox.

Hydrolysis of 1a (4). **3** (0.5 g) was added to 50 g of a 2:2:1 solution of H2O/MeOH/HCl (38 wt %). The mixture was stirred in air at room temperature for 5 h. The solid was collected via filtration, washed with water, methanol, and THF, and then dried under vacuum at room temperature overnight.

Second Silanol Capping (5). **4** was reacted with HMDS using the same procedure as in preparing **3**.

Synthesis of Densely Loaded APTMS-Functionalized SBA-15 (6)-Control Sample. Excess 3-aminopropyltrimethoxysilane (0.5 g, 2.79 mmol; APTMS) was added to 1 g of SBA-15 in anhydrous toluene. The mixture was allowed to stir for 24 h at room temperature under argon. The resulting solid was filtered, washed with toluene, dried under vacuum at room temperature overnight, and then stored in a drybox. TGA showed 1.15 mmol/g SiO_2 of APTMS was immobilized on the SBA-15.

Reaction of Chlorodimethyl(2,3,4,5-tetramethyl-2,4-cyclopentadien-1-yl)silane with Silica Materials. Chlorodimethyl-(2,3,4,5-tetramethyl-2,4-cyclopentadien-1-yl)silane (excess) was added to a mixture of amine-functionalized silica with hexanes in a drybox. 2,6-Di-tert-butylpyridine (excess) was added as a proton sponge. The mixture was allowed to react while stirring for 24 h. The solid was filtered and washed with hexanes and THF in the drybox. The solid was then contacted with another aliquot of chlorodimethyl-(2,3,4,5-tetramethyl-2,4-cyclopentadien-1-yl)silane and the procedure was repeated.

Metallation of cyclopentadienyl-functionalized silica materials with tetrakis (diethylamino) titanium.
The cyclopentadienyl-functionalized solid was added to a flask with excess tetrakis (diethylamino) titanium in toluene. The mixture was stirred under reflux for 24 hours. The resulting solid was filtered, washed with toluene, and dried under vacuum at room temperature overnight, and then stored in a drybox.

Test Reaction – Ethylene Polymerization

In a typical polymerization, the immobilized precatalyst, toluene, and methylaluminoxane (MAO) at a ratio of 800:1 Al:Ti were added to the reactor in a drybox. The mixture was allowed to stir for 20 minutes to allow for sufficient activation of the catalyst. The reactor was then removed from the glovebox, placed in a 25 C water bath, then ethylene at 60 psi was introduced. The polymerization was allowed to continue for 10 minutes, then terminated by adding acidic ethanol. The precipitated polymers were washed with ethanol, then dried at 70 C.

Characterization. Cross-polarization magic angle spinning (CP-MAS) NMR spectra were collected on a Bruker DSX 300-MHz instrument. Samples were spun in 7-mm zirconia rotors at 5 kHz. Typical ^{13}C CP-MAS parameters were 10000 scans, a 90 pulse length of 4 s, and a delay of 4 s between scans. Typical ^{29}Si CP-MAS parameters were 2000 scans, a 90 pulse length of 5 s, and a delay of 10 s between scans. FT-Raman spectra were obtained on a Bruker FRA-106. At least 256 scans were collected for each spectrum, with a resolution of 2-4 cm^{-1}. Thermogravimetic analysis (TGA) was performed on a Netzsch STA409. Samples were heated under air from 30 to 1000 C at a rate of 5 C/min. The organic loading was measured by determining the weight loss from 200 to 650 C. Melting point determinations used performed on the same Netzsch STA409. Samples were heated under nitrogen from 30 to 150 C at a rate of 2 C/min, then cooled to 75 C at 2 C/min, followed by cycle of heating to 150 C then cooling to 75 C. The melting point was determined by using the 2nd derivative of the differential scanning calorimetry curve. Nitrogen physisorption measurements were conducted on a Micromeritics ASAP 2010 at 77 K. Samples were pretreated by heating under vacuum at 150 C for 8 h. Gel Permeation Chromatography was performed with Polymer Laboratories PL-220 high temperature GPC equipped with a Wyatt MiniDawn (620 nm diode laser) high temperature light scattering detector and refractive index detector at 135 °C using 1,2,4-trichlorobenzene as solvent and calibrated using polystyrene standards. The dn/dc value for polyethylene used was -0.11. Polyethylene was extracted from silica at 130 °C using TCB as the solvent prior to the GPC analysis.

Results and Discussion

Aminosilicas have been widely studied for use in catalysis, either as a base catalyst or as a support for metal complexes (12). For example, amine functionalized silica can be used to catalyze the Knoevenagel condensation, an important C-C bond forming reaction. Also, the amine sites on the silica can be further functionalized to form supported imines, guanidine, and other species which are effective catalysts for a variety of organic transformations. In addition to using the aminofunctionalized silicate as catalyst itself, aminosilicas have been used as the building blocks for organometallic catalysts. Olefin polymerization catalysts based on Group IV transition metal complexes can be supported on these materials (6a-d). Catalysts for small molecule transformations can also be immobilized on aminosilicas, for example rhodium hydrogenation catalysts have been supported previously (13).

A major impediment to using the aminosilicas made via traditional techniques as a support for organometallic catalysts is the difficulty in effectively functionalizing the amine in a uniform manner to create the desired site. There are several reasons for this difficulty. For example, the densely functionalized amine groups can show inconsistent reactivity due to their varying accessibility and chemical environment. Amine groups on a densely functionalized aminosilica can adopt a number of different orientations. For instance, some amines will hydrogen bond with a neighboring amines due to the proximity of other surface immobilized groups, others will interact with silanol groups on the surface via weak acid/base interactions, while others may be isolated and relatively non-interacting with the surface. This array of possible structures is especially troublesome when a flexible linker such as the typical propyl linker is used. These difficulties can be potentially overcome by modifying the amine functionality on the silica surface. Of particular interest is synthesizing a material with isolated amine functionalities that do not strongly interact with the silanols of the silica surface. Also of interest is decreasing the flexibility of the linker. While capping agents can be used to discourage silanol-amine interactions, having a more rigid surface structure would further limit this undesired phenomenon as well, making the functionalities even more accessible as a result of their forced orientation from the surface.

We have previously reported the synthesis and use of a bulky tritylimine (**1a**) for use as a patterning molecule which allows for the effective spacing of aminosilane groups on a silica surface (7). The patterning agent, ([3-(trimethoxy-silanyl)-propyl]-(3,3,3-triphenyl-propylidene)-amine), has a flexible aminopropyl silane capped with a trityl group. An additional patterning agent, [4-(trimethoxy-silanyl)-phenyl]-(3,3,3-triphenyl-propylidene)-amine, with a more rigid aminophenyl silane capped with the same trityl group has been designed and synthesized (**1b**). The patterning agents were synthesized via condensation of 3,3,3 triphenylpropanal and the corresponding amine, either 3-aminopropyltrimethoxysilane (**1a**) or p-aminophenyltrimethoxysilane (**1b**).

Scheme 1 shows the general patterning methodology developed. The patterning molecule is contacted with a hexagonal mesoporous silica material such as SBA-15 (**7**) with an average pore diameter of ~50 A to produce a functionalized material (**2**). The large trityl groups on the patterning agent prevent incorporation of the silane on the surface at sites immediately adjacent to each other. After this functionalization, any additional, unreacted silanol groups on the surface are covered via treatment with hexamethyldisilazane yielding material (**3**). The imine bond is then selectively hydrolyzed to remove the trityl groups from the functionalized surface, leaving aminopropyl species on material (**4**). A final capping step is used to cover any additional silanol groups that might be produced in the hydrolysis step, giving material (**5**).

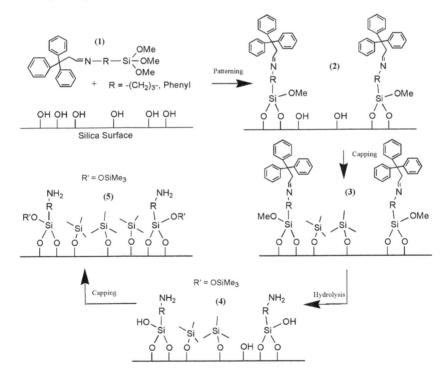

Scheme 1 Molecular patterning protocol.

The organic loadings of the functionalized surfaces were quantified by thermogravimetric analysis (TGA). The loading of the propyl patterning agent (**1a**) was determined to be 0.39 mmoles/g solid, compared to 1 mmole/g for densely functionalized materials made via traditional grafting techniques. After the initial hydrolysis, TGA showed 0.345 mmoles amine/g solid. 89% of the imine patterning agent is hydrolyzed to the amine. This hydrolysis treatment routinely hydrolyzes 89-93% of the imine. A second hydrolysis treatment can be used to remove additional imine, resulting in 93-95% total hydrolysis.

The loading of the phenyl patterning agent (**1b**) was determined to be 0.22 mmoles/g solid, compared to 0.71 mmoles/g solid for dense materials made via traditional grafting techniques. Hydrolysis attempts using similar conditions used in the conversion of **1a** resulted in poor conversion of the imine to the amine. The initial treatment for 5 hours showed via TGA 16% of the imine was hydrolyzed to the amine. A second hydrolysis treatment for 3 hours resulted in 56% total hydrolysis. As these results show the phenyl patterning agent is more difficult to hydrolyze than the propyl analog, various conditions were tested to achieve a more complete conversion from imine to amine. Instead of the 2:2:1 water/methanol/HCl mixture used previously, a 1:1:1 water/methanol/HCl mixture was found to work satiafactorily. Analysis via TGA showed 80-95% hydrolysis of the phenyl patterning agent. Further characterization studies are being undertaken to ensure the acid treatment is not removing the entire patterning agent molecule from the surface.

The patterned propyl amines have been characterized by ^{13}C and ^{29}Si CP-MAS NMR, FT-Raman spectroscopy, and nitrogen adsorption. In particular, the combination of ^{13}C NMR and FT-Raman allows for the development of the organic functionalities to be monitored. The imine carbon-nitrogen bond and the trityl aromatic carbons are the characteristic signals used to show the presence of the patterning agent on a silica surface. Upon hydrolysis, both of these functionalities are greatly reduced (which confirms partial hydrolysis). In addition, ^{13}C CP-MAS NMR shows the presence of an amine carbon bond in the hydrolyzed material. ^{29}Si NMR and nitrogen adsorption are used to ensure the surface functionalizations do not destroy or alter the silica support or disrupt the attachment of the ligands to the surface. (7).

The patterned aminopropyl functionalities have been tested for uniform reactivity in an effort to probe degree of site isolation (7). For example, the patterened aminopropyl functionalities quantitatively reacted with chlorodimethyl-(2,3,4,5-tetramethyl-2,4-cyclopentadien-1-yl) silane (7). In our hands, a densely-loaded control material was found to have only 66% of the amine sites react with the silane functionality. This value is in line with literature values of ~75% (6). These results and others (7) suggest the patterning technique allows for the synthesis of supported amine sites that exhibit behavior that is consistent with isolated, single-site materials. Similar studies are currently underway with the phenyl amine patterned material and its densely functionalized analog.

The patterned amino-silica is used as a support for immobilizing a titanium constrained geometry catalyst, shown in Scheme 2. The amine sites are first reacted with chlorodimethyl-(2,3,4,5-tetramethyl-2,4-cyclopentadien-1-yl) silane. The bifunctional solid is then metallated via an amine elimination method (14). The solid is stirred with excess tetrakis (diethylamino) titanium under reflux to generate surface titanium species. The functionalities on the patterned material were almost quantitatively metallated via the amine elimination method, as determined by elemental analysis. From elemental analysis, it was determined the metal loading was 0.38 mmols/g solid. This corresponds to 108% of the amine sites being metallated (15). Under identical conditions, the densely loaded control material contained a titanium loading that was consistent with an 80% metallation yield, which is consistent with literature results (6). The cyclopentadienyl functionalized material and the metallated solid were characterized by ^{13}C CP-MAS NMR. The spectra showed the signals corresponding to the carbons in the tetramethyl cyclopentadienyl ligand, as well as the carbons in the aminopropyl linkage (8).

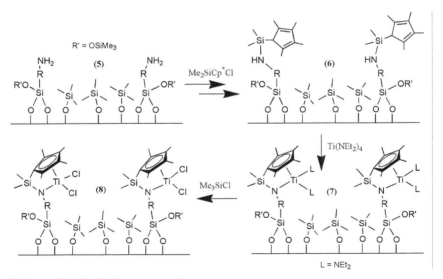

Scheme 2 Metallation protocol.

The immobilized Ti-CGC precatalyst derived from the patterned silica support was evaluated in the catalytic polymerization of ethylene using methylaluminoxane as a co-catalyst. Preliminary polymerization results are listed in Table 1. The patterned catalysts (entries 1 and 2) were compared to three control materials. One assembled using the identical protocol as the patterned material, with the difference being the complexes were assembled on a densely functionalized aminosilica surface (entry 3). The second control material (entry 4) was assembled on a dense aminosilica using a literature method that includes nBuLi treatments (6), which are known to destroy the surface of silica, creating multiple types of surface species. The third control (entry 5) was the homogeneous analog of the Ti-CGC precatalyst, made via literature methods (5). Molecular weights, polydispersity indices (PDI), and melting points (Tm) were determined for each of the polymers produced. As the catalyst particle is incorporated into the polyethylene as it is produced, post-polymerization analysis of the structure of the metal center is impossible.

Table 1 Polymerization results.

Entry	Catalyst	Ti loading (mmol/gcat)	Activity (kg PE/mol Ti-hr)	Tm (°C)	Molecular Weight (g/mol) x 10^5	PDI
1	Patterned	0.38	32.3	133.7	5.7	2.4
2	Patterned	0.38	27.7	133.1	8.9	2.3
3	Control 1	0.53	4.5	132.9	2.5	5.1
4	Control 2	0.65	5.1	131.5	ND	ND
5	Control 3	NA	6.0	133.4	7.8	2.0

The patterned catalyst shows a substantial increase in activity compared to the control materials. In addition, the patterned catalyst also exhibits increased activity even when compared to the homogeneous analogue. There are several possible explanations for the improved performance of the patterned materials. First, the patterned active sites will likely be more accessible than the metal centers on the control materials. This increase in accessibility may result in more effective activation by the co-catalyst, meaning a greater percentage of the immobilized metal centers are active catalytic sites. Second, the patterned sites may be less likely to undergo bimolecular termination, a common method of deactivation of homogeneous metal complex catalysts and potentially for supported catalysts with a dense loading of metal centers. In addition, the patterning technique promotes the formation of a single type of active site, eliminating or reducing the number of inactive metal centers present in the material (16).

When combined with the isolation and reactivity studies of the patterned aminosilica (7), the increased activity of the patterned catalysts provide further evidence that the patterning technique developed allows for the synthesis of aminosilicas which behave like isolated, single-site materials (although a true single site nature has not been proven). As the olefin polymerization catalysts supported by the patterned materials show a marked improvement over those materials supported on traditional aminosilicas, these patterned materials should be able to improve supported small molecular catalysis as well. Future improvements in catalysis with immobilized molecular active sites could be realized if this methodology is adopted to prepare new catalysts with isolated, well-defined, single-site active centers.

Summary

A new, general patterning methodology that can be utilized to create site-isolated organometallic catalysts on a silica surface is reported here. Using the new approach, primary amine sites with propyl or phenyl tethers have been immobilized on a silica surface. The patterned amine sites that have been created from a propyl imine patterning agent have been shown to exhibit behavior consist with materials which have isolated, single-site functionalities (7).

The patterned amine materials have been used to construct CGC-inspired sites that were evaluated in the catalytic polymerization of ethylene after activation with MAO. The complexes assembled on a porous silica surface using this methodology are more active than previously reported materials prepared on densely-loaded amine surfaces. This increased activity further suggests the isolated, unique nature of the metal centers. Work is continuing in our laboratory to further characterize the nature of the active sites, as well as to obtain more detailed kinetic data on the catalysts. The patterning methodology is also being applied to the creation of immobilized catalysts for small molecule reactions, such as Heck and Suzuki catalysis.

Acknowledgments

CWJ thanks the National Science Foundation for support of this work through the CAREER program (CTS-0133209). MWM thanks the Georgia Institute of Technology Molecular Design Institute (under prime contract N00014-95-1-1116 from the Office of Naval Research) for partial support of this research through a graduate fellowship. We thank the Coughlin group at UMass for GPC analysis.

References

1. (a) A. G. Pomogailo, Catalysis by Polymer-Immobilized Metal Complexes, Gordon and Breach Science Publishers, Australia, 1998. (b) F. R. Hartley, Supported Metal Complexes: A New Generation of Catalysts, Dordrecht, Boston, 1995.
2. (a) G. J. P. Britovsek, V. Gibson, and D. F. Wass, *Angew. Chem. Int. Ed.*, **56**, 428 (1999).
3. G. G. Hlatkey, *Chem. Rev.*, **100**, 1347 (2000) and references cited therein.
4. The catalysts are referred to here as "CGC-inspired" because there is no evidence in the literature that illustrates that the immobilized catalysts have the same structure as homogeneous CGC catalysts.
5. M. Galan-Fereres, T. Koch, E. Hey-Hawkins, and M.S. Eisen, *J. Organometal. Chem.*, **580**, 145 (1999).
6. (a) H. Juvaste, E. I. Iiskola, and T. T. Pakkanen, *J. Organometal. Chem.*, **587**, 38 (1999). (b) H. Juvaste, T. T. Pakkanen, and E. I. Iiskola, *Organometallics,* **19**, 4834 (2000). (c) H. Juvaste, T. T. Pakkanen, and E. I. Iiskola, *Organometallics,* **19**, 1729 (2000). (d) H. Juvaste, E. I. Iiskola, and T. T. Pakkanen, *J. Mol. Catal. A.*, **150**, 1 (1999). (e) R. M. Kasi and E. B. Coughlin, *Organometallics*, **22**, 1534 (2003).
7. M. W. McKittrick and C. W. Jones, *Chem. Mater.*, **15**, 1132 (2003).
8. M. W. McKittrick and C. W. Jones, manuscript submitted.
9. A. B. Pangborn, M. A. Giardello, R. H. Grubbs, R. K. Rosen, and F. J. Timmers, *Organometallics*, **15**, 1518 (1996).
10. J. S. Cha, J. H. C. Chun, J. M. Kim, O. O. Kwon, S. Y. Kwon, and J. C. Lee, *Bull. Korean Chem. Soc.,* **20**, 400 (1999).
11. D. Zhao, Q. Huo, J. Feng, B. F. Chmelka, and G. D. Stucky, *J. Am. Chem. Soc.,* **120**, 6024 (1998).
12. A. P. Wight and M. E. Davis, *Chem. Rev.*, **102**, 3589 (2002).
13. M. G. L. Petrucci and A. K. Kakkar, *Chem. Mater.,* **11**, 269 (1999).
14. D. W. Carpenetti, L. Kloppenburg, J. T. Kupec, and J. L. Petersen, *Organometallics,* **15**, 1572 (1996).
15. The slight excess of titanium determined by elemental analysis may be within the experimental error of the TGA and elemental analysis.
16. MAO induces leaching of some of the metal complexes from the support and this issue is currently under investigation in our laboratory.

Ch 32   Ring Opening of Naphthenes for Clean Diesel Fuel Production

David Kubička,[a] Narendra Kumar,[a] Päivi Mäki-Arvela,[a] Marja Tiitta,[b] Tapio Salmi,[a] and Dmitry Yu. Murzin[a]

[a]*Laboratory of Industrial Chemistry, Process Chemistry Centre, Åbo Akademi University, FIN-20500 Åbo/Turku, Finland*
[b]*Fortum Oil and Gas Oy, POB 310, FIN-06101 Porvoo, Finland*

Abstract

The activity of H-Beta-25, H-Beta-75, H-Y-12, and of their Pt-impregnated counterparts has been investigated in the ring opening of decalin at 523 K in the presence of hydrogen. Ring opening was found to take place with an increased rate and selectivity on Pt-impregnated zeolites as compared to proton-form zeolites. Prior to ring opening, decalin underwent skeletal isomerization; cracking of ring-opening products occurred consecutively.

Introduction

The new tighter diesel fuel specifications aiming at decreasing the amount of harmful emissions from diesel engines compel the development of advanced refinery technologies. The use of mainly secondary refinery streams, such as light cycle oil (LCO) from fluid catalytic cracking (FCC), as a blending stock in the diesel pool, is constrained by their high sulfur and aromatics content and low cetane number. Due to the legislation, either less LCO can be used for diesel blending or it has to be upgraded so that the final diesel fuel satisfies the specifications. Cetane number (CN) is a very important parameter being directly related to the reduction of emissions. Diesel fuels with higher CN emit less CO, particulate matter and unburned hydrocarbons. Emissions of NOx can be reduced as well. The cetane number is increased by hydrogenation of aromatics in the fuel into naphthenes. However, the enhancement is not always sufficient and further upgrading to alkylnaphthenes and/or paraffins is required. This can be achieved by ring opening of naphthenes (1, 2).

Ring opening of naphthenes can be accomplished by employing supported metal catalysts (3-7), such as Pt or Ir, and/or acidic catalysts, namely zeolites. The most thoroughly studied ring-opening reaction is the one of methylcyclopentane and cyclohexane (3-7). The metal-catalyzed ring opening was reviewed in detail by Gault (3). Two different mechanisms were recognized: the dicarbene mechanism with molecules adsorbed perpendicularly to the metal surface, ensuring selective rupture of secondary-secondary carbon bonds, and the multiplet mechanism, according to which the naphthenes adsorb flat on the metal surface resulting in statistical product distribution. The former selective mechanism is observed over Ir- or Rh-catalysts, whereas the latter non-selective one is found over Pt-catalysts (3). Recent studies with metals supported on zeolites have shown the importance of the catalyst acid function for cyclohexane ring contraction to cyclopentane ring, which makes the ring opening more facile (4-6). The significance of non-branching ring contraction of alkyl-substituted cyclohexanes for ring opening over Ir-supported catalysts has been reported by McVicker et al. (6, 7).

Besides the metal-catalyzed ring opening, naphthenic rings can be opened on the acid sites of zeolites. As the ring opening of naphthenes relies on cracking of endocyclic C-C bonds, the principles applicable for cracking and isomerization reactions, which have been studied extensively mostly for short alkanes both experimentally (8-17) and theoretically (8-10, 18-19), can be used. In accordance with these principles, a naphthene is protonated by a Brønsted acid site (BAS) and undergoes either protolytic cracking, forming an aliphatic carbenium ion, or protolytic dehydrogenation followed by scission of the resulting cyclic carbenium ion (8-10, 13, 17).

Ring opening of more complex molecules containing two fused rings, typical for hydrocarbons in the diesel fuel range, is investigated considerably less. Simultaneous hydrogenation and ring opening of 1-methylnaphthalene and tetrahydronaphthalene over Pt-modified Beta and Y zeolite catalysts in liquid phase yielded alkyl-substituted benzenes, and decalin and its isomers as the dominant products (20, 21). However, detailed information about the different ring-opening and decalin-isomer products was not provided. Our group has reported ring opening of decalin over proton-form zeolites results in a complex mixture of skeletal isomers of decalin and alkyl-substituted C_{10}- cyclohexanes and cyclopentanes (22). Differences in the product distribution over H-Y and H-Beta zeolites were observed and attributed to the shape selective properties of H-Y zeolite (22).

The main focus of this work is on the influence of metal introduction into zeolites on the ring-opening activity and selectivity. The effects of zeolite acidity and structure on the product distribution will be discussed as well. Decalin is used as a representative molecule for ring opening of dinaphthenes formed during hydrogenation of diaromatics in middle distillates.

Experimental Section

The NH_4-Y (CBV712, $a_0 = 24.35$ Å), H-Beta (CP811E-75), NH_4-Beta (CP814E) zeolites were obtained from Zeolyst International. The NH_4-Y and Beta zeolites were transformed to proton forms through step calcination procedure in a muffle oven. Zeolites containing 1 wt-% platinum were prepared by wet-impregnation method using hexachloroplatinic acid as the Pt-source.

The specific surface area of the fresh and used catalysts was measured by nitrogen adsorption method (Sorptometer 1900, Carlo Erba Instruments). The catalysts were outgassed at 473 K prior to the measurements and the Dubinin equation was used to calculate the specific surface area. The acidity of investigated samples was measured by infrared spectroscopy (ATI Mattson FTIR) by using pyridine ($\geq99.5\%$, a.r.) as a probe molecule for qualitative and quantitative determination of both Brønsted and Lewis acid sites (further denoted as BAS and LAS). The amounts of BAS and LAS were calculated from the intensities of corresponding spectral bands by using the molar extinction coefficients reported by Emeis (23). Full details of the acidity measurements are provided elsewhere (22).

A mixture of decalin (bicyclo[4.4.0]decane) isomers (Fluka, $\geq98\%$) with a cis-to-trans ratio of 2-to-3 was used as a starting material. The experiments were performed in an electrically heated 300-mL stainless steel autoclave (Parr Industries) at 523 K and 2 MPa. The stirring rate and the starting material-to-catalyst ratio were kept at constant values equal to 1500 rpm and 22 (w/w), respectively. The screened catalysts were crushed and the fraction bellow 63 μm was used in the experiments to suppress internal diffusion.

The reaction products were analyzed with a gas chromatograph (Agilent 6890N) equipped with a capillary column (DB-Petro 50 m x 0.2 mm x 0.5 μm) and a FI detector. Helium was used as the carrier gas. The following temperature program was applied: dwelling for 2 min at 323 K, heating 1.2 K/min to 393 K followed by heating 2 K/min to 473 K and dwelling at 473 K for 20 min. GC/MS technique was applied to distinguish the wide variety of isomerization and ring-opening products. The GC/MS analysis was carried out in an HP 6890-5973 instrument. The same column and separation conditions as for the neat GC analysis were used.

Results and Discussion

Product analysis

The reaction of decalin led to a complex mixture of products containing more than 200 components. In order to facilitate the basic evaluation of kinetic results the products have to be reasonably grouped. Any C_{10}-bicyclic structures other than decalin (cis- and trans-isomers) were referred as decalin isomers or simply isomers (Iso). C_{10}-monocyclic products, i.e. alkylsubstituted cyclopentanes and cyclohexanes, were denoted as ring-opening products (ROP). All products with a lower molecular weight than decalin were called cracking products (CP) and all products having more than 10 carbon atoms in the molecule and C_{10}-aromatics were named heavy products (HP).

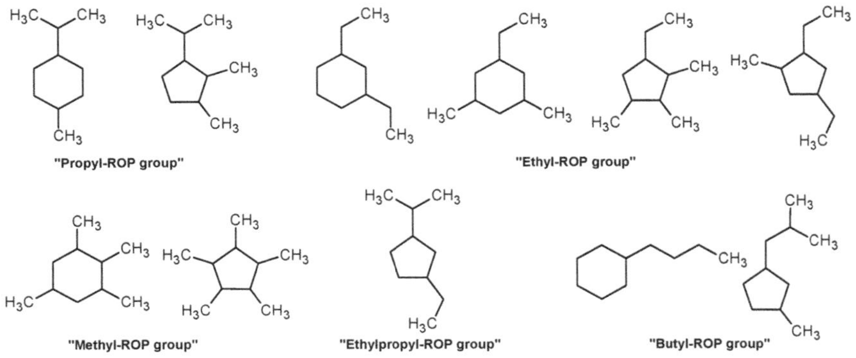

"Propyl-ROP group"   "Ethyl-ROP group"

"Methyl-ROP group"   "Ethylpropyl-ROP group"   "Butyl-ROP group"

Figure 1 Classification of ring opening products into ROP-groups and typical structures belonging to these groups.

The ROP were divided, based on GC/MS data, into 5 ROP groups as follows: Methyl-ROP, Ethyl-ROP, Propyl-ROP, Ethylpropyl-ROP, and Butyl-ROP, where the first part of the name characterizes the main alkylsubstituent. Typical structures of all ROP groups are given in Figure 1. The following 6 groups were established for isomers: methylbicyclo[4.3.0]nonanes (MBCN1), methylbicyclo[3.3.1]nonanes (MBCN2), dimethylbicyclo[3.3.0]octanes (DMBCO1), dimethylbicyclo[3.2.1]octanes (DMBCO2), ethylsubstituted bicyclooctanes (EBCO) and trimethylbicycloheptanes (TMBCH). The last group consists of trimethylbicyclo[2.2.1]heptanes, trimethylbicyclo[3.1.1]heptanes and dimethylbicyclo[2.2.2]octanes. Some representative structures of the Iso groups are shown in Figure 2. Major cracking products (CP) were identified as cyclic hydrocarbons C_5–C_9 and isobutane. The main heavy products (HP) were methyl- and dimethyldecalins, or their isomers. No olefins and only traces of aromatics were found in the products. Detailed information about the product analysis is reported elsewhere (22).

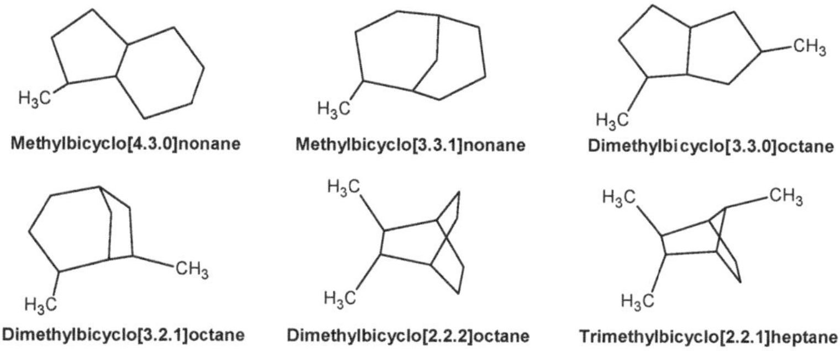

Methylbicyclo[4.3.0]nonane   Methylbicyclo[3.3.1]nonane   Dimethylbicyclo[3.3.0]octane

Dimethylbicyclo[3.2.1]octane   Dimethylbicyclo[2.2.2]octane   Trimethylbicyclo[2.2.1]heptane

Figure 2 Typical isomer structures.

Activity of catalysts

Pt-impregnated zeolites were found to be more active than the corresponding proton-form zeolites. This is especially clear from the conversion increase over Pt/H-Y and Pt/H-Beta-25 as compared to H-Y and H-Beta-25, respectively (Figure 3). It can be observed as well that the initial activity, expressed as the conversion of decalin after 1 h, increases with the increasing acidity (Figure 3, Table 1) over H- and Pt-zeolites. Nevertheless, a clear difference is found for the behavior of proton and platinum forms. While the activity of Pt/H-Y and Pt/H-Beta-25 is basically the same, which can be explained by the same loading of Pt and similar concentration and strength distribution of Brønsted acid sites (BAS) (Table 1), the activity of H-Y is lower than the one of H-Beta-25. The trend is more profound at longer reaction times and can be explained by faster deactivation of H-Y associated with formation of large organic molecules inside the cavities of H-Y causing thus the blockage of the zeolite channels. However, as follows from comparison of the specific surface areas of fresh and used catalysts given in Table 2, spent Pt- and H-form zeolites show similar specific surface areas indicating the same extent of deactivation in course of the 9 h experiment. Nevertheless, the trends of decalin conversion (Figure 3) point to the deactivation of H-zeolites being faster than the deactivation of Pt-zeolites. Furthermore, Pt increases the reaction rates of isomerization and ring opening, as discussed below.

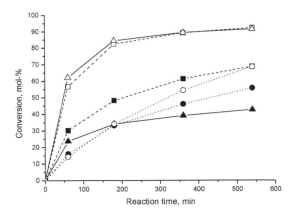

Figure 3 Comparison of decalin conversion over proton-form zeolites (filled) and Pt-zeolites (open). H-Beta-25 (■, □), H-Beta-75 (●, ○), H-Y-12 (▲, △).

Table 1 Brønsted and Lewis acidity of fresh proton-form catalysts.

Catalyst	Brønsted acid sites, µmol/g			Lewis acid sites, µmol/g		
	523 K	**623 K**	**723 K**	**523 K**	**623 K**	**723 K**
H-Beta-25	269	207	120	162	128	113
H-Beta-75	147	135	114	39	29	16
H-Y-12	255	205	129	123	75	58
Pt/H-Beta-25	290	88	0	164	13	0
Pt/H-Beta-75	159	8	0	36	4	0
Pt/H-Y-12	311	194	0	112	11	0

General product distribution

The characteristic trends of product concentrations for proton-form (H-Beta-25) and Pt-impregnated (Pt/H-Beta-25) zeolite are displayed in Figures 4A and 4B, respectively. The transformation of decalin into reaction products has been described for proton-form zeolites (22) by a series of consecutive reactions where decalin is first isomerized yielding a complex mixture of its skeletal isomers (Iso). These react further producing the desired ring-opening products (ROP), which can undergo subsequent cracking reactions forming thus cracking products (CP). The simplified scheme is shown in Figure 5. It can be unequivocally seen from Figure 4B that the same scheme applies also for the Pt-impregnated zeolites.

Table 2 Specific surface area of fresh and used catalysts, conditions: 523 K, 2 MPa, and 540 min.

| Catalyst | SiO$_2$/Al$_2$O$_3$ mol/mol | Specific surface area, m^2/g | | | |
| | | Proton form | | Platinum form | |
		Fresh	Used	Fresh	Used
H-Beta-25	25	807	461	730	412
H-Beta-75	75	664	566	744	557
H-Y-12	12	1218	496	982	388

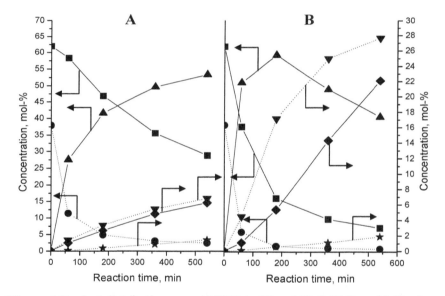

Figure 4 Product distribution over **A)** H-Beta-25 and **B)** Pt/H-Beta-25. Product groups: Trans-decalin (■), cis-decalin (●), Iso (▲), ROP (▼), CP (◆), and HP (★).

The changes in the product concentrations are more pronounced in case of Pt-zeolites. Particularly the rate of decalin isomerization is considerably enhanced by the addition of platinum. The ratio of Iso/ROP decreases from the value of 12 after 1 h to the value of 1.4 after 9 h. The ratio of ROP/CP decreases from the values close to 3 within the first 3 h to the value of 1.3 after 9 h. This is in contrast to H-Beta-25, where the ratio ROP/CP is almost constant during the entire experiment and its values do not exceed 1.5. The same trends are observed for the other investigated catalyst, too, which indicates that Pt-impregnated zeolites are more selective for ring opening than the proton-form counterparts.

Figure 5 Simplified reaction scheme.

It is helpful to look at the comparison of ROP and CP at the same conversion levels to explain this phenomenon (Figure 6). While the concentrations of Iso and ROP are, at comparable conversion, the same for proton- and Pt-form of zeolites, the concentrations of CP are substantially lower over Pt-zeolites. Heavy products (HP) concentrations are lower with Pt-zeolites as well, even though the difference is not as marked as in the case of CP. The different cracking behavior of proton-form and Pt-impregnated zeolites is clearly seen in Figure 7A. The observed lower cracking activity of Pt-zeolites can be ascribed to the lower amount of the strongest acid sites, i.e. sites that retain pyridine at 723 K, in Pt-impregnated zeolites (Table 1). The decreased strength of acid sites in Pt-zeolites may be explained by interactions of small Pt particles with the acid sites. As a result of this interaction, the strength of acid sites is reduced and Pt particles become slightly positively charged. Furthermore, interactions between the hydrogen activated on Pt and the reaction intermediates may result in less cracking due to hydrogenation of the reaction intermediates prone to cracking.

It could be expected that ring-opening reaction will be, similarly to cracking, suppressed due to the absence of the strongest acid sites. This happens, however, only to a limited extent and the concentration of ROP shows almost the same dependency on the concentration of Iso for H- and Pt-zeolites (Figure 7B). The difference can be explained by Pt taking place in the ring opening of naphthenes, as described in the literature (3–7), and consequently compensating the decreased ring-opening activity caused by the lower acidic strength of Pt-impregnated catalysts. Although Pt is less active in ring opening than other metals, e.g. Ir or Rh, it is claimed to be better in ring opening of substituted C–C bonds than Ir (6). Therefore it seems plausible to ascribe the higher selectivity to ROP over Pt-zeolites, as compared to H-zeolites, to ring opening occurring on the metal. The similar trends for ROP concentration over H-and Pt-zeolites suggest that the rates of ring opening are of similar order of magnitude for the ring opening on the acid sites and on Pt, respectively. Nevertheless, as the concentration of ROP rises, the successive cracking becomes more important and the concentration of CP sharply increases resulting in the reduction of ROP/CP ratio, as described above.

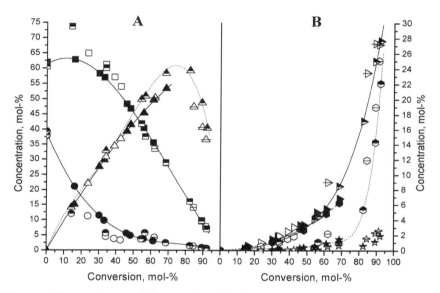

Figure 6 Dependence of the product distribution on conversion for H-Beta zeolites (filled), H-Y-12 (open), Pt/H-Beta zeolites (half-filled), and Pt/H-Y-12 (open with dash). **A)** Trans-decalin (■, □, ◪, ⊟), cis-decalin (●, ○, ◐, ◒), Isomers (▲, △, ◮, ◭); **B)** ROP (▶, ▷, ◭, ◮), CP (◆, ○, ◑, ◒), HP (★, ☆, ⯪, ⯪).

Furthermore, while no significant difference in the product distribution, with the exception of the trans-decalin/cis-decalin ratio, is observed for the tested proton-form zeolites, dissimilarity between Pt/H-Y on one hand and both Pt/H-Beta zeolites on the other hand is found (Figure 6). More ROP and CP, accompanied by less Iso, are formed on Pt/H-Y than on Pt/H-Beta zeolites. This implies that the consecutive ring opening and cracking are faster over Y-zeolite than over Beta-zeolites resulting in lower concentration of isomers and higher concentrations of ROP and CP.

Isomers and ring-opening products distribution
The complexity of skeletal isomerization of decalin is depicted in Figures 8A and 8B. It shows that methylbicyclo[4.3.0]nonanes (MBCN1) are the

primary isomerization products followed by methylbicyclo[3.3.1]nonanes (MBCN2), dimethylbicyclo[3.3.0]octanes (DMBCO1) and dimethylbicyclo-[3.2.1]octanes (DMBCO2). Trimethylbicycloheptanes (TMBCH) and ethyl-substituted bicyclooctanes (EBCO) are formed last. Variations between the structures both in H- and Pt-form can be discovered and are discussed in detail below. On the contrary, no differences are found when comparing the Beta zeolites hinting that rather structural effects than acidity cause the different isomers distribution. Furthermore, Pt influences the distribution as well (Figures 8A-B).

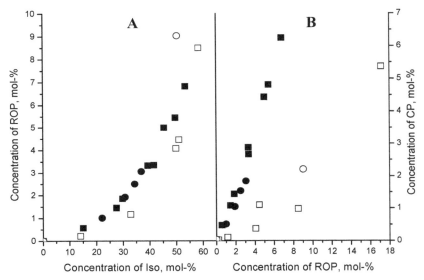

Figure 7 A) Dependence of ROP concentration on the concentration of Iso, **B**) Dependence of CP concentration on the concentration of ROP. H-zeolites (filled), Pt-zeolites (open); Beta-zeolites (■, □), Y-zeolite (●, ○).

A general trend of lower concentration of methylbicyclononanes (MBCN1 and MBCN2) on Y-zeolite than on Beta-zeolites can be seen for both H- and Pt-forms. It is more pronounced with the more abundant MBCN1. In contrast to H-Y, where the lower concentration of MBCN1 was accompanied by higher concentration of DMBCO2 resulting thus in the same total concentration of isomers over Y and Beta zeolites, on Pt-Y concentrations of all major isomers are lower than over Pt-Beta zeolites leading therefore to the lower total concentration of isomers, as discussed above. The lower concentration of methylbicyclononanes (MBCN) implies either higher rate of direct ring opening of MBCN or higher isomerization rate of MBCN into secondary isomerization products (dimethylbicyclooctanes (DMBCO)) as compared to Beta-zeolites. The behavior of Beta-zeolites, however, points toward the latter explanation. Despite the overall higher isomerization activity of Pt-zeolites, slightly less MBCN are formed over the Pt-Beta zeolites than over H-Beta zeolites. This decrease is compensated by higher concentration of DMBCO2, the successive skeletal isomers, rather than by increased formation of ROP (Figures 6 and 8). Nevertheless, at high conversions (>80%) the ring opening is more favored on Pt/H-Y than on Pt/H-Beta-25. Interestingly, more trimethylbicycloheptanes, the most complex isomers, are formed over Pt-zeolites than over H-zeolites (Figure 8).

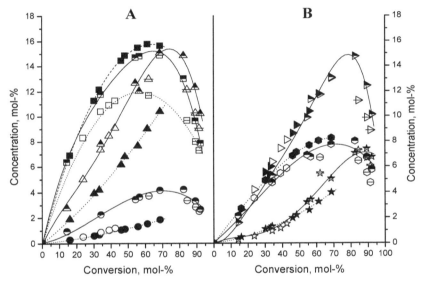

Figure 8 Dependence of the isomers groups concentration on conversion for H-Beta-25 (filled), H-Y-12 (open), Pt/H-Beta-25 (half-filled), and Pt/H-Y-12 (open with dash). **A)** Methylbicyclo[4.3.0]nonanes (■, □, ◪, ⊟), Dimethylbicyclo[3.2.1]octanes (▲, △, ◮, ◭), trimethylbicycloheptanes (●, ○, ◗, ⊖); **B)** Dimethylbicyclo[3.3.0]octanes (▶, ▷, ◮, ▷), Methylbicyclo[3.3.1]nonanes (⬣, ○, ⬡, ⊖), Ethylsubstituted bicyclooctanes (★, ☆, ★, ☆).

The discussion on differences in ROP distribution is hindered not only by the fact that each isomers group can yield several ROP belonging to different groups, but also because of the isomerization of the resulting ROP, which is speeded up due to the presence of Pt. This is well illustrated by the high concentration of Methyl-ROP (Figure 9), which principally cannot be primary

ROP. Their concentration is lower than the one of Ethyl- and Propyl-ROP at low conversions, but with increasing conversion they become the most abundant ROP group. Nevertheless, a noteworthy trend is observed for concentrations of all ROP groups, except Butyl-ROP, which increase rapidly after reaching conversion higher than 60%. This links the ROP formation to the secondary formed skeletal isomers of decalin (DMBCO) rather than to MBCN, which are present already at lower conversions in sufficient amounts.

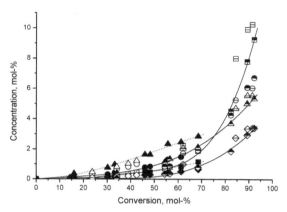

Figure 9 Dependence of the ROP groups concentration on conversion for H-Beta-25 (filled), H-Y-12 (open), Pt/H-Beta-25 (half-filled), and Pt/H-Y-12 (open with dash). Methyl-ROP (■, □, ◧, ▤), Ethyl-ROP (●, ○, ◑, ◔), Propyl-ROP (▲, △, ◮, ◭), Ethylpropyl-ROP (◆, ◇, ◈, ◒).

Conclusions

The activity of H-Beta-25, H-Beta-75, H-Y-12, and of their Pt-impregnated counterparts has been investigated in ring opening of decalin at 523 K in the presence of hydrogen. The main results can be summarized as follows:

The ring opening of decalin consists of consecutive isomerization steps leading to several types of decalin skeletal isomers. In succession of isomerization, ring opening followed by cracking take place. The overall activity of the proton-form zeolites is increased by impregnating them with platinum. The higher overall activity is mainly caused by the enhancement of the isomerization rate that has increased 3–4 times. At the same time, the yield of ring-opening products increases from 8% for proton-form zeolites to 30% for Pt-impregnated zeolites. However, this is accompanied by the increased yield of cracking products. The initial activity of all zeolites increases with their increasing acidity.

The successive cracking of ring-opening products (ROP) is diminished by the presence of Pt resulting thus in higher ring-opening selectivity. This is explained by the lower strength of acid sites in Pt-impregnated zeolites consequently decreasing the rate of the consecutive cracking reaction. The ring-opening reaction takes place on both Pt and acidic sites. As the concentration of ring-opening products was the same, at comparable conversion, regardless of using the H- or Pt-zeolite, it was concluded that the rates of ring opening on Pt and on acid sites are of similar order of magnitude.

No differences in the product distribution of ROP and isomers were observed for the investigated Beta zeolites with different acidity implying that they are independent of the catalyst acidity. The zeolite structure seems, however, to influence these product distributions by promoting the successive ring opening over Y-zeolite as compared to Beta-zeolites. This could be explained by faster isomerization of the primarily formed methylbicyclononanes into dimethylbicyclooctanes, which can undergo more easily the ring opening owing to the presence of cyclopentane rings containing secondary C-C bonds in the molecule. Furthermore, they have more possibilities to form tertiary carbenium ions as well. As a result of the high isomerization activity of the Pt-zeolites, Methyl-ROP were found to be the most abundant ROP followed by Ethyl- and Propyl-ROP.

Acknowledgements

This work is part of the activities at Åbo Akademi Process Chemistry Centre within the Finnish Centre of Excellence Programme (2000–2005) by the Academy of Finland. The authors express their gratitude to Markku Reunanen (Åbo Akademi) and Kim Wickström (Fortum) for their contribution to the GC/MS analysis. Economic support from Fortum and TEKES is gratefully acknowledged.

References

1. T. C. Kaufmann, A. Kaldor, G. F. Stuntz, M. C. Kerby, and L. L. Ansell, *Catal. Today,* **62**, 77 (2000).
2. B. H. Cooper and B. B. L. Donnis, *Appl. Catal. A: General,* **137**, 203 (1996).
3. F. G. Gault, *Adv. Catal.,* **30**, 1 (1981).
4. T. V. Vasina, O. V. Masloboishchikova, E. G. Khelkovskaya-Sergeeva, L. M. Kustov, and P. Zeuthen, *Stud. Surf. Sci. Catal.,* **135**, CD-ROM 26-O-03 (2001).
5. G. Onyestyák, G. Pál-Borbély, and H. K. Beyer, *Appl. Catal. A: General,* **229**, 65 (2002).

6.   G. B. McVicker, M. Daage, M. S. Touvelle, C. W. Hudson, D. P. Klein, W. C. Baird Jr., B. R. Cook, J. G. Chen, S. Hantzer, D. E. W. Vaughan, E. S. Ellis, and O. C. Feeley, *J.Catal.*, **210**, 137 (2002).
7.   M. Daage, G. B. McVicker, M. S. Touvelle, C. W. Hudson, D. P. Klein, B. R. Cook, J. G. Chen, S. Hantzer, D. E. W. Vaughan, and E. S. Ellis, *Stud. Surf. Sci. Catal.*, **135**, CD-ROM 26-O-04 (2001).
8.   F. C. Jentoft and B. C. Gates, *Topics Catal.*, **4**, 1 (1997).
9.   J. A. Martens and P. A. Jacobs, *Stud. Surf. Sci. Catal.*, **137**, 633 (2001).
10. Y. Ono, *Catal. Today*, **81**, 3 (2003).
11. S. T. Sie, *Ind. Eng. Chem. Res.*, **31**, 1881 (1992); **32**, 397 (1993); and **32**, 403 (1993).
12. A. Corma, J. Planelles, J. Sánchez-Marín, and F. Tomás, *J. Catal.*, **93**, 30 (1985).
13. A. Corma, F. Mocholi, A. V. Orchillés, G. S. Koermer, and R. J. Madon, *Appl. Catal.*, **67**, 307 (1991).
14. A. Corma, P. J. Miguel, and A. V. Orchillés, *J. Catal.*, **145**, 171 (1994).
15. E. M. Lombardo, R. Pierantozzi, and W. K. Hall, *J. Catal.*, **110**, 171 (1988).
16. H. S. Cerqueira, P. C. Mihindou-Koumba, P. Magnoux, and M. Guisnet, *Ind. Eng. Chem. Res.*, **40**, 1032 (2001).
17. J. Abbot, *J. Catal.*, **123**, 383 (1990).
18. F. V. Frash and R. A. van Santen, *Topics Catal.*, **9**, 191 (1999).
19. A. M. Rigby, G. J. Kramer, and R. A. van Santen, *J. Catal.*, **170**, 1 (1997).
20. M. A. Arribas, J. J. Mahiques, and A. Martínez, *Stud. Surf. Sci. Catal.*, **135**, CD-ROM 26-P-13 (2001).
21. M. A. Arribas and A. Martínez, *Appl. Catal. A: General*, **230**, 203 (2002).
22. D. Kubička, N. Kumar, P. Mäki-Arvela, M. Tiitta, V. Niemi, T. Salmi, and D.Yu. Murzin, *J. Catal.*, **222**, 65 (2004).
23. C. A Emeis, *J. Catal.*, **141**, 347 (1993).

33.

The Use of Copper Catalysts in the HydrogenTransfer Reduction of 4-*Tert*-butylcyclohexanone

Federica Zaccheria,[a] **Achille Fusi,**[b] **Rinaldo Psaro,**[c] **and Nicoletta Ravasio**[c]

[a]*Consorzio INSTM, via B. Varchi 59, I-50132 Firenze, Italy*
[b]*Dipartimento Chimica IMA, Università degli Studi di Milano, via G. Venezian 21, I-20133 Milano, Italy*
[c]*CNR-ISTM, via C. Golgi 19, I-20133 Milano, Italy*

Federica.Zaccheria@unimi.it

Abstract

Several low loading supported copper catalysts were tested in the hydrogen transfer reduction of 4-*tert*-butyl-cyclohexanone showing good activity under very mild conditions (2-propanol as solvent, 83°C, N_2). Results obtained by using different alcohols as donors suggest that a two step dehydrogenation-hydrogenation mechanism is operating over Cu/SiO_2.

Introduction

Catalyzed hydrogen transfer from a hydrogen donor other than H_2 is attractive industrially because of safety, engineering and economic concerns (1). This reaction has been extensively studied in the homogeneous phase (2) and under Meerwein-Ponndorf-Verley conditions (3).

In the fine chemical industry, reduction of carbonyl groups mainly relies on the use of complex metal hydrides; sodium dihydrobis-(2-methoxyethoxy)-aluminate, commercialized as RedAl® or Vitride® is one of the most used (4).

We recently reported that Cu/SiO_2 is an effective catalyst for the hydrogenation of cyclohexanones under very mild experimental conditions. Thus, a series of cyclohexanones with different substituents, including 3-oxo-steroids, could be reduced under 1 atm of H_2 at 40-90°C, with excellent selectivity (5). The catalyst is non-toxic and reusable. This prompted us to investigate the reduction of cyclohexanones over a series of supported copper catalysts under hydrogen transfer (h.t.) conditions (2-propanol, N_2, 83°C) and to compare the results with those obtained under catalytic hydrogenation (n-heptane, 1 atm H_2, 40-90°C) conditions. Here we report the results obtained in the hydrogenation of 4-*tert*-butyl-cyclohexanone, a molecule whose reduction, also under Meerwein-Ponndorf-Verley conditions, has been investigated in detail because of the interest in the corresponding axial alcohol (6).

Moreover, we here report some results obtained by using different secondary alcohols as donors in order to verify the possibility of tuning the stereochemistry of the product.

Results and Discussion

Catalyst support effect
Results obtained in the hydrogenation of 4-*tert*-butyl-cyclohexanone over seven different pre-reduced copper catalysts are reported in Table 1. The productivities, expressed as mmole converted/g_{cat}·h, are compared in Figure 1.

Table 1 Hydrogen transfer from 2-propanol to 4-*tert*-butyl-cyclohexanone in the presence of different copper catalysts.

Entry	Catalyst	t (h)	Conv.(%)	Sel. (%)	Eq/ax
1	Cu/MgO	4	100	100	64/35
2	Cu/SiO$_2$	4	100	90	47/53
3	Cu/Aerosil	7	100	98	46/54
4	Cu/SiAl 0.6	7	100	96	51/49
5	Cu/SiZr	12	100	92	57/43
6	Cu/ SiAl 13	24	39	98	56/44
7	Cu/Sepiolite	24	25	73	30/70
8	CuO/MgO	24	96	100	98/2
9	MgO	8	97	100	97/3

Cu/MgO and Cu/SiO$_2$ gel were found to be the most active, followed by the catalysts obtained supporting copper on fumed silica and silica modified with small amount of a second oxide containing Lewis acid sites. It is also worth noting that the reaction over Cu/MgO is selective while small amounts of secondary products are formed over the other catalysts.

As far as the activity of this catalyst is concerned, the possible contribution of magnesia to the transfer reaction has to be considered. Thus, it is well known that MgO activated at 350°C can effectively catalyze the hydrogen transfer from 2-propanol to unsaturated compounds in the gas phase (7), although no reports are available on MgO activated when used under milder conditions.

To gain insight into the contribution of MgO, we carried out the hydrogenation of 4-*tert*-butyl-cyclohexanone under the same experimental conditions over pure MgO pre-treated as the supported catalyst at 270°C and over the catalyst pre-treated at 270°C but unreduced (CuO/MgO).

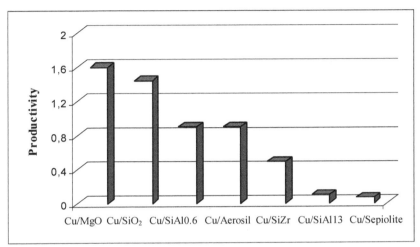

Figure 1 Productivities of different copper catalysts (mmol prod/g_{cat} h) in hydrogen transfer reactions.

Results are reported in Table 1 and Figure 2. The pure oxide shows in fact activity, but it is much lower than that observed over Cu/MgO. On the other hand the presence of CuO inhibits greatly the transfer reaction, as shown by very low activity of the unreduced CuO/MgO catalyst. Moreover the equatorial alcohol is almost exclusively formed, suggesting that the basicity of MgO is mainly responsible for the stereochemistry of the product.

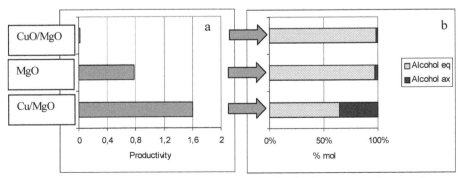

Figure 2 (a) Productivity (mmol prod/g_{cat} h) in the hydrogen transfer reaction obtained with Cu/MgO, MgO and CuO/MgO; (b) Stereoselectivity obtained with Cu/MgO, MgO and CuO/MgO, respectively.

The higher activity observed when working with reduced Cu/MgO suggests that there is a synergic effect between the support and the metallic phase, while the MgO may play a major role in determining the product's stereochemistry.

Among the other catalysts no significant changes were found except when working over Cu/Sepiolite which gives the highest amount of axial alcohol although at very low conversion. This may be due to the ordered structure of sepiolite that contains a continuous, two-dimensional, tetrahedral silicate sheet and would suggest that the transfer reaction occurs inside the channels.

If we compare these productivities with those obtained under catalytic hydrogenation conditions (Table 2, Figure 3), we can see that the trend is different; Cu/MgO shows very low activity, particularly if compared with that of the catalyst supported on silica and modified silicas.

Table 2 Results obtained in hydrogenation of 4-*tert*-butyl-cyclohexanone with different copper catalysts (n-heptane, 1 atm H_2, 90°C).

Entry	Catalyst	t (h)	Sel (%)	Eq/ax
1	Cu/MgO	16	92	64/36
2	Cu/SiO$_2$	1	90	58/42
3	Cu/Aerosil	1	95	50/50
4	Cu/SiZr	1	92	55/45
5	Cu/SiAl 0.6	1	92	50/50
6	Cu/SiAl 13	9	86	63/27
7	Cu/Sepiolite	6	95	48/52

Detailed characterization of Cu/SiO$_2$ samples has shown that the preparative method used, that is chemisorption-hydrolysis (CH) from a Cu(NH$_3$)$_4$$^{2+}$ solution, not only allows formation of catalysts with very high surface area, but also has a strong influence on the metallic phase morphology. Thus, the main feature of CH prepared catalysts is the presence of well formed crystallites exposing a significant fraction of step and edge sites, particularly (111) microfacets very efficient in H$_2$ dissociation on reduced samples (8). The high activity of this catalyst in a wide range of hydrogenation reactions (9) may well be due to these features.

Donor alcohol effect
In order to explore the possibility of tuning product stereochemistry by changing the experimental conditions, we carried out the reaction using different secondary alcohols as donor, namely di-isopropyl-carbinol (2,4-dimethyl-2-pentanol), 2-octanol, 3-octanol and cyclohexanol.

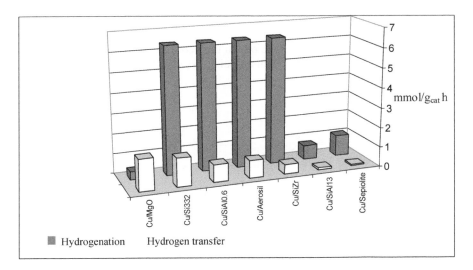

Figure 3 Productivity (mmol prod/g_{cat} h) of hydrogen transfer and hydrogenation reactions with different catalysts.

We have already reported that the use of cyclohexanol at 140°C in the hydrogenation of ergosterol over Cu/Al$_2$O$_3$ gives the 5β derivative with an 81% selectivity owing to an intramolecular hydrogen transfer reaction, whereas direct H$_2$ addition over the same catalyst gives the 5α isomer with 89% selectivity (10) (Scheme 1).

The synthetic value of this reaction should be mentioned. Thus, current production of 5β-steroids from Δ^5-3β-ols, readily available and cheap starting materials, requires a preliminary Oppenauer oxidation or fermentation to the Δ^4-3-keto derivative followed by catalytic hydrogenation under alkaline conditions, as direct catalytic hydrogenation with both heterogeneous and homogeneous systems gives only the 5α isomer.

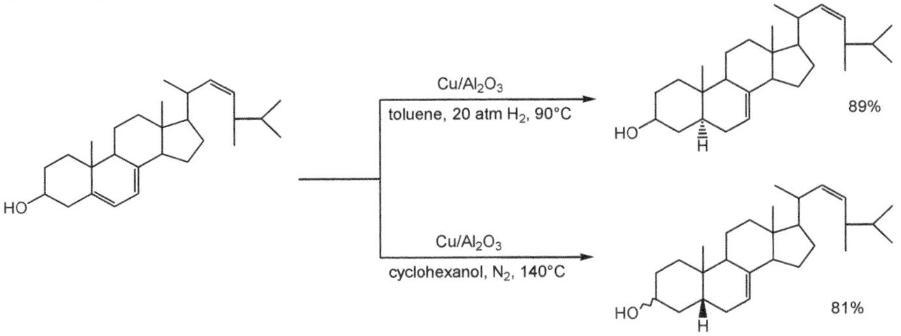

Scheme 1

Transfer hydrogenation in the alcohol-ketone system on metal catalysts was investigated by Patterson et al. In particular, by studying the reaction between 2-propanol and butanone on Cu they concluded that it must be a direct surface reaction (11), the mechanism being essentially a proton transfer in the adsorbed phase (Scheme 2).

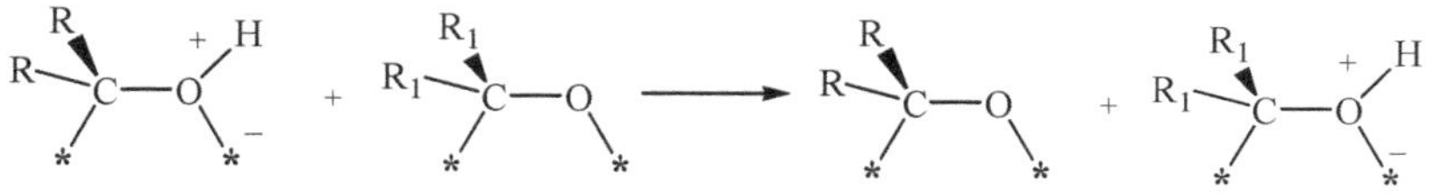

Scheme 2

On the other hand a direct hydrogen transfer through a Meerwein-Ponndorf mechanism, involving coordination of both the donor alcohol and the ketone to the copper site may also be considered. In this case, by using alcohols other than 2-propanol, we could expect some difference in stereochemistry. This would also imply the possibility of carrying out the enantioselective reduction of a prochiral ketone with a chiral alcohol as donor.

A third possibility is represented by a two-step mechanism where the donor alcohol is dehydrogenated and the ketone reduced by the H_2 produced. In this case, the easier the donor alcohol is dehydrogenated, the higher is the hydrogen availability on the catalyst surface and the faster is the reaction. If the donor is slowly dehydrogenated, the hydrogen availability is lower.

When the reaction was carried out in the presence of Cu/silica, all the alcohols tested were capable of transferring hydrogen to the substrate even at 90°C (Table 3). At higher temperature, the reaction was also faster. Analysis of the reaction mixtures suggests that at least over Cu/SiO_2 the dehydrogenation-hydrogenation mechanism plays a major role. Thus, in the case of $(iPr)_2CHOH$ and 3-octanol, not only formation of the corresponding ketones was observed, but this continued also after complete conversion of 4-*tert*-butylcyclohexanone. Moreover the stereochemistry observed in di-isopropyl-carbinol is coincident with that obtained under catalytic hydrogenation conditions, strongly suggesting that hydrogen availability on the catalyst surface is comparable both in the presence of molecular H_2 and when using this particular alcohol as a donor.

In the case of 2-octanol and cyclohexanol, differences in stereochemistry might be due to a lower hydrogen availability. In the case of cyclohexanol, its high viscosity, modifying the substrate adsorption step on the surface, may also play a role.

It is worth noting that we have previously observed a sharp increase in the amount of the axial epimer when reducing 5α-androstan-3,17-dione with 2-

octanol and cyclohexanol over Cu/Al_2O_3 (12). Results obtained with Cu/MgO are reported in Table 4. With this catalyst transfer at 90°C is very difficult except for 2-propanol.

Table 3 Results obtained for hydrogen transfer and hydrogenation reactions in different conditions using Cu/SiO_2 as catalyst.

Donor	90°C			140°C		
	t(h)	Conv.	% ax	t(h)	Conv.	% ax
H_2	1	100	42			
2-PrOH	4	100	53			
$(iPr)_2CHOH$	10	100	43	1	100	31
3-octanol	20	100	55	1	100	41
2-octanol	24	100	70	3	100	57
Cyclohexanol				5	100	72

Table 4 Results obtained for hydrogen transfer and hydrogenation reactions in different conditions using Cu/MgO as catalyst.

Donor	90°C			140°C		
	t (h)	Conv	%ax	t (h)	Conv	%ax
H_2	16	100	36			
2-PrOH	4	100	35			
$(iPr)_2CHOH$	24	0	-	24	91	33
2-octanol	48	60	56	5	100	47
Cyclohexanol	48	84	82	4	100	76

Over Cu/MgO the transfer from 2-propanol is much faster than molecular H_2 addition (entry 1, Table 2), thus excluding the dehydrogenation-hydrogenation mechanism. For the other donors, the reaction needs to carried out at 140°C to reach reasonable rates.

Results are not easily rationalized. On one hand, the trend of the product's stereochemistry is comparable with that observed over Cu/SiO_2, but the reaction rate seems to depend on the steric hindrance of the donor. Moreover, as already mentioned, MgO play a significant role in this reaction. Such observations make it impossible to draw any conclusion about the mechanism.

Cu-Mg and Cu-Zn-Al catalysts are well known for the industrial dehydrogenation of cyclohexanol at 220-260°C (13), whereas Cu/SiO_2 catalysts are much less studied.

Recently Cu/MgO and Cu/SiO$_2$ samples have been compared in the dehydrogenation of 2-butanol to methyl-ethyl-ketone MEK in the range 240-390°C. The MEK selectivities for the optimum MgO supported catalyst ($\approx$100%) were better than those for the optimum SiO$_2$ supported catalyst (>93%) that produced small amounts of the dehydration product butene. However the MEK yields were much higher for the silica supported catalysts, showing their higher dehydrogenation activity (14).

Therefore, the high activity of Cu/SiO$_2$ in transferring hydrogen from a donor alcohol may be due not only, as already mentioned, to its ability to activate molecular H$_2$, but also to its dehydrogenation activity.

Experimental Section

Supports: SiO$_2$ gel (BET=320 m^2/g, PV=1.75 ml/g), SiO$_2$-ZrO$_2$ (herein referred to as SiZr, 4.7% ZrO$_2$, BET=320 m^2/g PV=1.60 ml/g) SiO$_2$-Al$_2$O$_3$ 0.6 (SiAl 0.6, 0.6% Al$_2$O$_3$, BET=500 m^2/g PV=1.45 ml/g), SiO$_2$-Al$_2$O$_3$ 13 (SiAl 13, 13% Al$_2$O$_3$, BET=475 m^2/g PV=0.77 ml/g) were obtained from Grace Davison (Worms, Germany); pyrogenic silica (Aerosil, BET=380 m^2/g) from Degussa (Hanau, Germany); sepiolite (BET=240 m^2/g , PV=0.4 ml/g) from Tolsa (Madrid, Spain); and MgO (BET=30 m^2/g) from Merck.

The copper catalysts used were prepared as reported (15), via the CH method and all have an 8% copper content except for Cu/MgO (5.5%).

4-*tert*-butyl-cyclohexanone (99%), 2,4-dimethyl-3-pentanol (99%), 2- and 3-octanol (99%) were purchased from Aldrich; 2-propanol (99.5%) and cyclohexanol (99%) from Fluka.

The catalysts were treated at 270°C for 20 min in air, for 20 min under vacuo, and then reduced at the same temperature with H$_2$ at atmospheric pressure. MgO and CuO/MgO underwent only the dehydration step.

For the hydrogen transfer reactions, the substrate (0.100 g, 0.64 mmol) was dissolved in anhydrous n-heptane (8 mL) and the solution transferred under N$_2$ into a glass reaction vessel where the catalyst (0.100 g) had been previously treated. Catalytic tests were carried out with magnetic stirring under N$_2$ at boiling point temperature with 2-propanol and 90°C or 140°C with other donor alcohols.

Reaction mixtures were analysed by GC (with mesitylene as internal standard) using a SP-2560 (100m) capillary column and the equatorial/axial ratio determined by ^{1}H NMR (Bruker, 300 MHz).

Conclusions

In the particular case studied in this paper, it is not worth carrying out the reaction under hydrogen transfer conditions to increase the amount of axial epimer, as up to 65% of the thermodynamically unfavoured alcohol can be obtained over Cu/SiO$_2$ at 60°C and 1 atm of H$_2$ (5). However, this work shows that the use of secondary alcohols as donors is possible under very mild conditions over the same catalyst. This can be useful both for safety reasons and for operating under mild experimental conditions in order to convert sensitive molecules (such as the ones used in the synthesis of speciality chemicals that can not withstand gas phase conditions).

Work is in progress to better elucidate the potential of these systems as dehydrogenation catalysts.

References

1. M. D. Le Page, D. Poon, and B. R. James, Chemical Industries (Dekker), **89**, (*Catal. Org. React.*) 61-72, (2003).
2. G. Zassinovich, G. Mestroni, and S. Gladiali, *Chem. Rev.*, **97**, 1051 (1992).
3. C.F. de Graauw, J. A. Peters, H. van Bekkum, and J. Huskens, *Synthesis,* **10**, 1007 (1994).
4. M. Capka, V. Chvalovsky, K. Kochloefl, and M. Kraus, *Coll. Czech. Chem. Commun.*, **34**, 118 (1969).
5. N. Ravasio, R. Psaro, and F. Zaccheria, *Tetrahedron Lett.,* **43**, 3943 (2002).
6. A. Corma, M. E. Domine, and S. Valencia, *J. Catal.,* **215**, 294 (2003) and references therein.
7. J. Kaspar, A. Trovarelli, F. Zamoner, E. Farnetti, and M. Graziani, *Stud. Surf. Sci. Catal.*, **59**, 253 (1991).
8. F. Boccuzzi, A. Chiorino, M. Gargano, and N. Ravasio, *J. Catal.*, **165**, 140 (1997).
9. F. Zaccheria, R. Psaro, and N. Ravasio, Recent Research Developments in Catalysis Vol. 2, S. G. Pandalai, Ed., Research Signpost, Trivandrum, 2003, p. 23-34.
10. N. Ravasio, M. Gargano, and M. Rossi, *J. Org. Chem.*, **58**, 1259 (1993).
11. W. R. Patterson, J. A. Roth, and R. L. Burwell, *J. Am. Chem. Soc.*, **93**, 839 (1971).
12. N. Ravasio, M. Gargano, V.P. Quatraro, and M. Rossi, *Stud. Surf. Sci. Catal.,* **59**, 161 (1991).
13. V. Z. Fridman and A. A. Davydov, Chemical Industries (Dekker) **75**, (*Catal. Org. React.*) 495-506 (1998).
14. J. N. Keuler, L. Lorenzen, and S. Miachon, *Appl. Catal. A: General,* **218**, 171 (2001).
15. N. Ravasio, M. Antenori, M. Gargano, and P. Mastrorilli, *Tetrahedron Lett.,* **37**, 3529 (1996).

34.

The Effect of Ultrasound on the Isomerization versus Reduction Reaction Pathways in the Hydrogenation of 3-Buten-2-ol and 1,4-Pentadien-3-ol on Pd-Black

R. S. Disselkamp and C. H. F. Peden

Environmental Molecular Sciences Laboratory, Pacific Northwest National Laboratory, Richland, WA 99352

robert.disselkamp@pnl.gov

Abstract

Ultrasound at 20 kHz was applied to the isothermal (298±2 K) heterogeneous catalytic hydrogenation of the aqueous phase β-unsaturated alcohols 3-buten-2-ol and 1,4-pentadien-3-ol and compared to a conventional experiment in which magnetic stirring replaced the sonic probe in facilitating the room temperature reaction. Hydrogenation employed hydrogen gas at 6.8 atm and a catalyst (Pd-black) pre-reduced by ultrasound. This study was undertaken to examine the effect of ultrasound on the competing reaction processes of isomerization to the enol that then undergoes tautomerization to the corresponding ketone, versus hydrogenation to the saturated alcohols (2-butanol and 3-pentanol). Sampling of the reacting solutions at pre-determined time intervals followed by *ex-post-facto* GC/MS analyses yielded time-dependent concentration information. The concentrations of all products and reagent were graphed with respect to time of reaction, enabling mechanistic information to be obtained from the experimental data. For 3-buten-2-ol, the final ratios of 2-butanone to 2-butanol were 0.67 (ultrasound) and 1.67 (stirred). Also for this system, a kinetic modeling of the stirred system revealed that the 3-buten-2-ol reagent was depleted by two primary reaction pathways: 1. isomerization to 2-butanone, and 2. reaction with 2-butanone to yield 3-buten-2-one plus 2-butanol. For the 1,4-pentadien-3-ol system, the 3-pentanone to 3-pentanol ratios were 0.33 (ultrasound) and 0.43 (stirred). Although the olefins studied here are relatively simple, an analysis of the kinetic data indicates that the actual reaction processes can be complex.

Introduction

The hydrogenation of β-unsaturated alcohols is a classical example of competing reaction processes in heterogeneous catalysis. In principle, either isomerization to an enol followed by tautomerization to the ketone can occur, or direct hydrogenation to the saturated alcohol is possible (1). It is fair to state that interest in this type of chemistry is derived from both basic scientific curiosity, but more importantly as a practical matter of ketone and alcohol production applicable to industrial processes.

The primary production pathway of β-unsaturated alcohols is through the sequential processes of hydroformylation followed by selective carbonyl reduction, yielding in excess of 5 million tons globally of aldehydes and aldehyde-derived chemicals (2). With this quantity of production, it is important to examine the hydrogenation/isomerization properties of this class of compounds as possible synthetic routes to other high commodity chemicals. For 3-buten-2-ol and 1,4-pentadien-3-ol in particular, a review of the literature reveals no prior studies of hydrogenation/isomerization on Pd-black, either for conventional or ultrasound-assisted heterogeneous catalysis approaches. This lack of research effort examining the effect of ultrasound on competing reaction processes serves as motivation for our investigation here. Ultrasound as a synthetic tool is well-known (3-6), thus our work here is an extension of prior investigations.

Experimental Section

The reagents selected as model β-unsaturated alcohol compounds for hydrogenation were 3-buten-2-ol and 1,4-pentadien-3-ol (Aldrich, 97% and 99% purity, respectively). Palladium-black catalyst was used (Alfa Aesar, 99.9% purity metals basis). The specific surface area of the catalyst was determined using the nitrogen adsorption BET and O_2-H_2 titration method. The effect of sonication was to increase specific surface area from 7.3 to 7.6 m^2/g (BET) and 12.2 to 16.7 m^2/g (titration). Deionized water (18 MΩ-cm) was used as the solvent. Hydrogenations were performed with hydrogen gas (A&L specialty gas, 99.99% purity) at a pressure of 6.8 atm (100 psia). All components used in the reaction apparatus are commercially available. The apparatus consisted of a Branson Ultrasonics model 450 Sonifier II unit (20 kHz, up to 400W deliverable) with a probe extending into a jacketed reaction cell (Branson model 101-021-006) (3). The cell was connected to a NesLab RTE-140 bath circulation unit to allow isothermal heterogeneous catalysis at 298±2 K. Analysis of samples collected during an experiment were performed on a Hewlett-Packard GC/MS (5890 GC and 5972 MS) with injections done by an Hewlett-Packard GC/SFC automated injector. Authentic standards were employed in the calibration of mass area counts.

The general experimental procedure employed in the study here has been described previously (7), thus only a brief overview is presented here. For all experiments, 45 mL deionized water and catalyst (50 mg Pd-black for 3-buten-2-ol and 25 mg for 1,4-pentadien-3-ol) were added to the reaction cell. For ultrasound-assisted, as well as stirred (blank) experiments, the catalyst was reduced with hydrogen (6.8 atm) in water for 5 minutes at an average power of 360 W (electrical; 90% amplitude). The reagents (320 mg 3-buten-2-ol or 360 mg 1,4-pentadien-3-ol) were added to the reduced catalyst solution to achieve

2.0 M/g-catalyst initial concentrations. The first sample for each experiment was taken for time equal to zero minutes and filtered through a 0.45 μm hydrophilic Millipore filter to remove catalyst powder into a capped vial for subsequent analysis.

At this point the experiments followed one of two protocols. For stirred (blank) experiments, the cell was connected to the probe again and pressurized with hydrogen (14 pressure-vent cycles). Stirring was commenced and hydrogen pressure was occasionally applied so as to maintain constant reactor pressure. After the allotted time interval had passed, stirring was stopped, the cell was vented and purged with nitrogen gas, and the cell was disassembled. A sample was collected and filtered. Further samples were taken for subsequent time intervals using the same method of pressurizing and stirring.

For ultrasound-treated experiments, the cell was connected, vented, and pressurized with hydrogen as just described. The solution in the cell was irradiated with sound for the set amount of time, hydrogen gas pressure was occasionally applied so as to maintain constant reactor pressure, then sonification was stopped, the cell was vented and purged with nitrogen, and a sample was collected and filtered. During sonication, an amplitude of 90% was employed, resulting in 300±15 W delivered to solution as calibrated from calorimetric measurements. The horn diameter was 1.27 cm. The cell was assembled again and the next sonification interval began. Samples were analyzed qualitatively and quantitatively using an Hewlett Packard GC/MS. The column selected for separation was a 30 meter, 0.5 micron DB-1 column. An injection volume of 5.0 μL (200 mL split flow) was used by an automated injector in triplicate to enable an average and estimation of standard deviation in mass counts. The samples were run through the column at a ramp rate of 6.0 K/min.. Analyses of accuracy among injections for the autosampler was +5.0% (2σ), whereas the partitioning of species within a chromatogram was ±2.0%, as discussed in previous studies (7).

Results

3-Buten-2-ol
Figure 1 below presents the hydrogenation results of 3-buten-2-ol for the (a) ultrasound-assisted and (b) stirred systems. The figure caption defines the labeling scheme we have adopted. The lines in Figure 1(a) simply connect the experimental data for the ultrasound-assisted experiment, however in Figure 1(b) the lines correspond to a kinetic model for the stirred experiment (*vide infra*). Several conclusions can be drawn by examining the figure, in particular in comparing the two means of assisting the chemistry. For the ultrasound experiment it is seen that the reagent 3-buten-2-ol (3BEN2OL) undergoes a nearly constant depletion rate until all reagent is depleted. Also, of even more curiosity is the increase in the product 2-butanone (2BONE) with time, where the increasing slope suggests an auto-catalytic reaction in product(s) species 2-

butanone or 2-butanol. The data accuracy (deduced by repeating experiments) is ±3.0% (2σ), thus the observed phenomena is real and not an artifact. We have attempted to fit the ultrasound experiment to various kinetic schemes employing pseudo first-order and second-order reactions but were unsuccessful in converging to an acceptable result. What can be said about the reaction mechanism for the ultrasound experiment of Figure 1(a) is that no less than three heterogeneous reactions are needed to describe the kinetic data.

The stirred experiment of Figure 1(b) is more typical in that all concentrations behave as expected (reagent decreasing with pseudo-first order kinetics; products monotonically increasing; intermediate undergoing well-behaved rise and fall concentration). We have performed kinetic analyses of the data in Figure 1(b) according to the equations:

$$3BEN2OL \rightarrow 2BONE \qquad\qquad [1]; k1=0.074 \text{ min.}^{-1}$$
$$3BEN2OL + 2BONE \rightarrow 3BEN2ONE + 2BOL \quad [2]; k2=0.088 \text{ min.}^{-1} (M/g)^{-1}$$
$$3BEN2ONE \rightarrow 2BONE \qquad\qquad\qquad [3]; k3=0.560 \text{ min.}^{-1}$$

The root-mean-square error in the kinetic fit was an acceptable 2.83% and was minimized by the SIMPLEX method discussed elsewhere (8). An important result for the stirred system is the following: The presence of the bimolecular reaction [2] with a comparable reaction rate to that of reaction [1] reveals that the system exhibits a strong propensity to form the conjugated ketone-unsaturated 3-buten-2-one species. For both the ultrasound-assisted and stirred systems it is important to note that when the reaction is complete (extent of reaction at 100%) the 2-butanone to 2-butanol ratios are 0.67 for the ultrasound compared to 1.67 for the stirred systems. Thus ultrasound is seen to favor the hydrogenation reaction pathway.

1,4-Pentadien-3-ol
A second system we have chosen for study is 1,4-pentadien-3-ol. The rationale for choosing this species is based on the hypothesis that its conversion to the enol (1,3-pentadien-3-ol) and subsequent tautomerization to the ketone may be enhanced relative to the hydrogenation pathway because the enol here is conjugated, an energetically favorable process. Therefore the ketone to saturated alcohol product ratio for this species may be expected to be larger than that for the 3-buten-2-ol system just described. Figure 2 below presents results of 1,4-pentadien-3-ol hydrogenation, again for the (a) ultrasound-assisted and (b) stirred (blank) systems. The caption of Figure 2 identifies the chemical species observed during the course of the experiment.

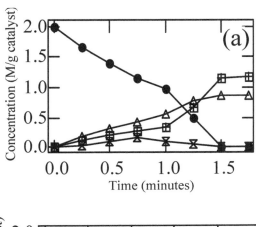

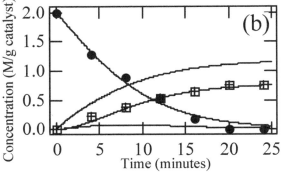

Figure 1 Time-dependent composition data is shown for the hydrogenation of aqueous 3-buten-2-ol for both (a) ultrasound irradiated and (b) magnetically stirred systems. The symbols correspond to experimental measurements (3-buten-2-ol: 3BEN2OL–solid circles; 3-buten-2-one: 3BEN2ONE–open hourglass; 2-butanone: 2BONE-open triangles; 2-butanol: 2BOL-crossed squares). The lines in the ultrasound experiment simply connect the data points, whereas for the stirred experiment the lines correspond to a modeled fit (see text).

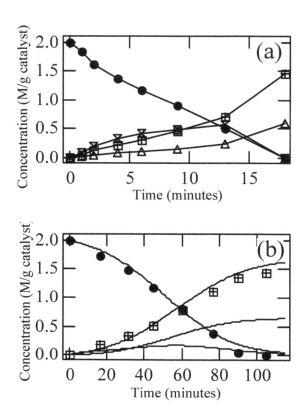

Figure 2 Time-dependent composition data is shown for the hydrogenation of aqueous 1,4-pentadien-3-ol for both (a) ultrasound irradiated and (b) magnetically stirred systems. The symbols correspond to experimental measurements (1,4-pentadien-3-ol: 14PD3OL–solid circles; 1-penten-3-ol: 1P3OL–open hourglass; 3-pentanone: 3PONE-open triangles; 3-pentanol: 3POL-crossed squares). The lines in the ultrasound experiment simply connect the data points, whereas for the stirred experiment the lines correspond to a modeled fit (see text).

The ultrasound-assisted experiment of Figure 2(a) is again not typical in that the reagent concentrations have an inflection point mid-way through the reaction. We have performed a kinetic analyses of the stirred (blank) data in Figure 2(b) and found the following equations reproduce the data well with a root-mean-square error of 2.8%:

14PD3OL → 3PONE	[4]; $k4=5.58E-5$ min.$^{-1}$
14PD3OL + 3PONE → 1P3OL + 3POL	[5]; $k5=5.49E-2$ min.$^{-1}$ (M/g)$^{-1}$
1P3OL → 3POL	[6]; $k6=0.171$ min.$^{-1}$
1P3OL + 1P3OL → 3PONE + 3POL	[7]; $k7=8.11E-2$ min.$^{-1}$ (M/g)$^{-1}$

14PD3OL + 3POL → 1P3OL + 3PONE   [8]; k8=7.42E-2 min.$^{-1}$ (M/g)$^{-1}$
14PD3OL → 1P3OL   [9]; k9=4.23E-3 min.$^{-1}$

The similarity of the magnitudes in reaction rates of reactions [4]-[9] supports the assertion that the chemical mechanism is complex. We believe that a fewer number of different reactions will not model the data equally well, but this remains to be proven. The ketone-to-alcohol ratios at the completion of the reactions are 0.33 for ultrasound-assisted and 0.43 for the stirred system. Again the ultrasound is lower, as was the case for the 3-buten-2-ol system, which perhaps is due to ultrasound enhancing the hydrogenation pathway by increasing the surface diffusion of chemisorbed hydrogen. Contrary to our hypothesis, however, is that 1,4-pentadien-3-ol has overall smaller ratios as compared to the 3-buten-2-ol system, suggesting that the propensity of conjugation enhancing ketone formation is not operative.

Discussion

An important aspect of this work that needs to be addressed is the specific mechanism by which ultrasound enhances reaction rate and changes product state distribution. Unfortunately, since we do not have a microscopic probe of the catalyst surface, we can only present indirect measurements for such effects. Firstly, we contend that our chemistry is occurring under non-cavitating, but high power ultrasound, conditions. How do we know this? We have performed beaker experiments examining the onset of cavitation in aqueous alcohol (5 wt.% 1-propanol) solutions in ambient air. We've observed that upon cavitation the volume doubles, the applied ultrasound power drops from ~350 W to ~200 W, the audible sound changes from being annoyingly loud to being relatively silent, and the solution color changes from being clear to milky white. (The power drop can be rationalized by the increased acoustic impedance mismatch between titanium horn and solution by incorporating gas (air or hydrogen) into the liquid.) Using our ultrasound reactor, in experiments not presented here, we've observed the same phenomena (becoming silent, power decrease) as well as during venting after sonication a gas/liquid spray effluent is emitted from reactor (e.g., evidence of gas incorporated into liquid). We can only surmise that whatever the mechanism for observed ultrasound and stirred chemistry differences, it is not the effect of pure cavitation or its effects in the classical sense.

Something that will become useful are the hydrogenation reaction schemes for 3-buten-2-ol and 1,4-pentadien-3-ol are illustrated below.

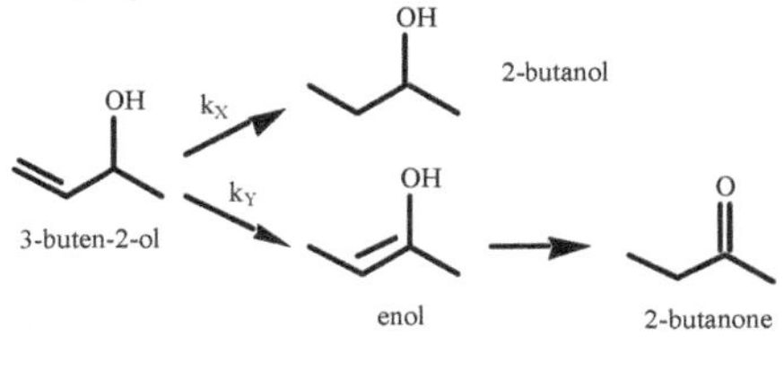

We assume (without proof) that our experimental data is a probe of the k_Y/k_X reaction rate coefficient ratios of 0.67 (ultrasound) and 1.67 (stirred). Similarly, the k_B/k_A ratios would be 0.33 (ultrasound) and 0.43 (stirred). For the 3-buten-2-ol system, one plausible explanation is that ultrasound, in a mechanism that remains unclear, enhances the diffusion of chemisorbed atomic hydrogen on the catalyst surface, thus increasing the relative yield of 2-butanol compared to 2-butanone in the ultrasound compared to stirred systems. This effect is substantial here (~250% enhancement in alcohol-to-ketone yield for ultrasound compared to stirred). The situation for 1,4-pentadien-3-ol is slightly more complex. Examining the data of Figure 2(a) and (b), since the 1-penten-3-ol (intermediate) concentration is a factor of ~3 larger in the ultrasound compared to stirred systems, this suggests an enhancement in the rate coefficient ratio $k_C/(k_A+k_B)$ for the ultrasound compared to stirred systems. Further, the ultrasound is seen to enhance the 3-pentanol-to-3-pentanone ratio, but only by ~30%. As a final note, comparing the 3-buten-2-ol to 1,4-pentadien-3-ol systems, the ultrasound experiment proceeded roughly 10 times faster, whereas the stirred system reacted ~4 times faster. The much slower reacting 1,4-pentadien-3-ol system is perplexing. A possible cause is that the C4-alkyl radical 1-penten-3-ol surface bound species may be the entity that is difficult to hydrogenate (to 1-penten-3-ol) because of surface steric effects. We simply do not know, however.

A final item to mention is that by comparing our time equal to zero mass concentrations to those during the course of the reactions for all the above experiments has allowed us to track mass conservation in the experiments. For both the ultrasound and stirred experiments the mass conservation is ~100%. This is important in demonstrating that ultrasound can be a synthetic tool, since in principle ultrasound induced cavitation can degrade analytes non-selectively. Evidence for this is based on experiments in our laboratory using GC/MS detection of the disappearance of aqueous 1-propanol solutions, as well as prior work on tert-butanol that observed carbon dioxide, water, and hydrogen gas evolution in solutions that were sonicated (9) and that of ether aquasonolysis (10). It is plausible that the reducing environment under which these experiments were performed curtail oxidation processes (peroxides, hydroxyl radical moieties) from reacting with and degrading the analyte. It is plausible that the presence of the solid catalyst affects the cavitation process so as to either limit or deter bubble growth processes.

Acknowledgements

This work was funded under a Laboratory Directed Research and Development (LDRD) grant administered by Pacific Northwest National Laboratory (PNNL). PNNL is operated by Battelle for the US Department of Energy. We wish to offer appreciation to the peer reviewers and Dr. John Holladay of PNNL for a thorough review of this paper.

References

1. M. Jones Jr., Organic Chemistry, W. W. Norton & Co., New York, 1997.
2. G. Ertl, H. Knozinger, J. Weitkamp, Handbook of Heterogeneous Catalysis, vol. 5, Wiley-VCH, Weinheim, Germany, 1997.
3. K. S. Suslick, in Handbook of Heterogeneous Catalysis, vol. 3, G. Ertl, H. Knözinger, and J. Weitkamp, Eds., Wiley-VCH, 1997, p. 1350.
4. T. J. Mason, Sonochemistry, Oxford University Press, Oxford, 2000.
5. J. L. Luche, Ed., Synthetic Organic Sonochemistry, Plenum Press, New York, 1998.
6. J. W. Chen, J. A. Chang, and G. V. Smith, *Chem. Eng. Symp. Ser.*, 67 (1971).
7. R. S. Disselkamp, K. M. Judd, T. R. Hart, C. H. F. Peden G. J. Posakony, L. J. Bond, *J. Catal.*, in press, (2003).
8. W. H. Press, B. P. Flannery, S. A. Teukolsky, W. T. Vetterling, Numerical Recipes – The Art of Scientific Computing, Cambridge University Press, New York, 1989.
9. A. Tauber, G. Mark, H. P. Schuchmann, C. von Sonntag, *J. Chem. Soc., Perkins Trans.*, **2**(6), 1129 (1999).
10. J. Lifka, J. Hoffmann, B. Ondruschka, *Water Sci. and Tech.*, **44**(5), 139 (2001).

35. Lessons Learned: Batch Processing, Scaleup from Laboratory to Plant

John D. Super

Dixie Chemical Company, Inc., 10601 Bay Area Blvd., Pasadena, TX 77507

Jsuper@dixiechemical.com

Abstract

Reaction kinetics, catalyst handling, mass and heat transfer, corrosion and many other practical industrial chemistry and engineering considerations impact the success of scaleup from lab to commercial for batch processing. Since the starting point for scaleup studies is the ultimate intended commercial unit, the professional should "scaledown" from the design parameters and constraints of the proposed commercial unit.

Introduction

Around the world chemical professionals continually commercialize new products and processes. Much of this activity results in batch processing. Fine and custom chemicals can involve as many as ten to twenty batch reactions in series, sometimes with multi-step parallel paths, with various separation technologies between reaction steps. This paper is an attempt to reflect the experience of many individuals as seen through the author's eyes over almost four decades, with several very typical situations.

What is Scaleup?

One of the best definitions is by Attilio Bisio (1): "The successful startup and operation of a commercial size unit whose design and operating procedures are in part based on experimentation and demonstration at a smaller scale of operation." He also points out that Smith (2) argued in 1968 that the starting point for scaleup studies is the ultimate intended commercial unit. The professional should "scaledown" from the design parameters and constraints of that commercial unit so that the smaller scale experiments were most useful in reducing the uncertainties of the commercial run. Smith wrote that scaleup from small-scale studies is a misleading concept.

To be successful, every project needs an advocate or champion. Mukesh (3) presented 15 tips for producing a successful project, the most important one being that one needs a champion.

Reaction Kinetics

There are two "scaledown" concepts that even experienced professionals relearn: reactions take longer to run in the plant than lab, and batch distillation times are also noticeably longer in the plant compared with lab.

A typical reason for longer reaction times in the plant is slower overall heat removal rates. And a typical case is batch hydrogenation, with a neat reactant or in solvent. Here all reactants are charged before the batch reaction is initiated. Typical set of reactions is:

$$A + H_2 \xrightarrow{\text{catalyst}} P$$

$$P + A \xrightarrow{\text{catalyst}} BAD$$

The desired reaction is A reacting with H_2 to form P, the product. But the product reacts with the reactant A to form an undesirable molecule, BAD. With slower overall heat removal rates in the plant, primarily because the cooling surface area/liquid volume ratio is much larger in the lab, the batch yield in the plant is noticeably less than in the lab. Figures One and Two show in relative terms the quantity of A, P and BAD with time for the two situations. At 97-98% conversion of A, the longer plant reaction time case has four times as much unwanted by-product BAD as the lab case.

One should go to the lab knowing that the large-scale reaction time is three hours vs. one hour in the lab, for example. If the reaction of A with P occurs at any significant rate, by running the lab experiment as long as the plant process, the yields will be similar. If one obtains a poor yield with long reaction times, the dilemma is resolved by feeding A into the product or into a solvent over time, and at a rate which does not allow significant concentrations of A to build. See Tom Johnson's ORCS paper (4) where the catalyst activity was even higher with low concentrations of reactant.

Difficulties often occur with scaleup of batch distillations, whether one runs straight single stage flash distillation or fractionation with trays. An historical perspective on the proper definitions of these two operations, was given over 50 years ago in "The Chemical and Engineering Dictionary" (5). It properly defined the terms: distillation: the process of vaporizing a liquid and condensing the vapors by cooling. And fractional distillation was defined as the separation of the components of a liquid mixture by collection of fractions condensing in certain temperature ranges. This is best accomplished by means of fractionating columns. Often when an engineer is told that one has distilled a material, the operation is misinterpreted as involving a column. It does not.

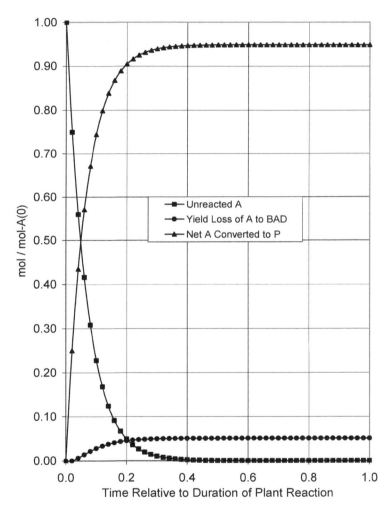

Figure 1. Hydrogenation and Degradation of A
in Lab (Fast) Reaction

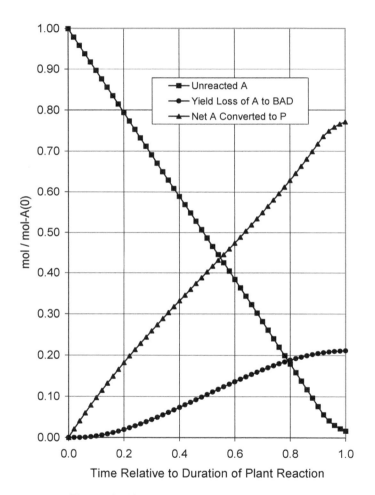

Figure 2. Hydrogenation and Degradation of A
in Plant (Slow) Reaction

Lab batch distillation is always much quicker than the plant, again due to overall heat-input rate being much higher in the lab. And in a plant setting, heating the last 10-20% of the volume is often very inefficient or impractical due to heating jacket or coil design. So again one really must scaledown from knowledge of the plant equipment, then run the lab in a similar manner. The penalty for long plant distillation is often high boiling tar formation, from either the product reacting with itself or decomposing and then reacting with itself:

$$P + P \longrightarrow TAR$$

To illustrate the relative quantities vs time curve, see Figures 3 and 4. Here we assume that the product is distilled from a high boiling solvent or high boiling by-products. With 5% P remaining as bottoms, the plant case took four times as long and had three times as much tar when compared with the lab case.

Another cause of long plant distillation time is poorer vacuum. Poor vacuum can be caused by a number of factors, all of which will raise the boiling temperature that can cause additional undesired reactions. The lab distillation should be run as long as the plant and also at vacuum achievable in the plant.

If long distillation time is a problem, one can move to continuous distillation with conventional shell and tube heaters accompanied by a typical column bottom (often called a sump) which is a high temperature holdup, or better yet a short path evaporator (falling film, thin film, or wiped film) with usually a smaller receiver (called an accumulator in this case). The most chemical damage is in the thin liquid film at the heat transfer surface, so the short path evaporators do the least thermal damage.

Catalyst Handling

There are several practical scaleup lessons with heterogeneous catalysts in batch slurry reactions. One often uses four times the catalyst concentration in the lab to achieve the same results in the plant reactor. Plant charge is 1 wt. part for 1000 wt. part, where as in the lab one uses 5 grams or more per 1000 grams. A common error is the confusion between wet (gross) weight and dry weight (net)

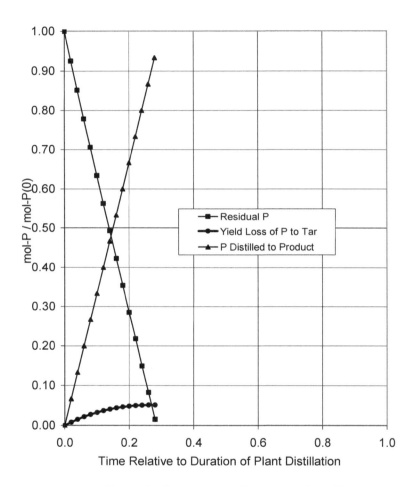

Figure 3. Recovery and Decomposition of P
in Lab (Fast) Distillation

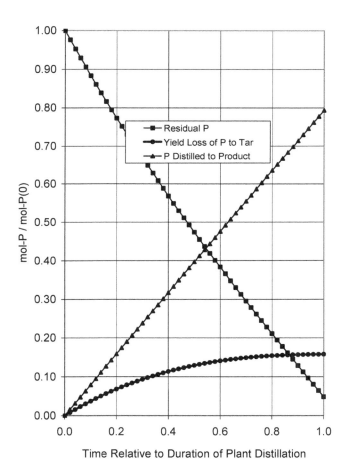

Figure 4. Recovery and Decomposition of P
in Plant (Slow) Distillation

with sponge metal or carbon supported catalysts. Most sponge metal is 50% water; supported carbon can be 50% to 67% water. Since catalyst purchases are based on net weight I recommend using net weight in the lab and specifying net weight on batch sheet charge quantities with a section in the batch sheet where the plant technicians calculate the gross charge.

The most significant catalyst handling problems are filtration. If filtration is slow in the lab, a general rule of thumb is that it will be worse in the plant. Best plant processes are ones where catalyst is single use and removed from the batch completely when transferred. Excellent process operation also is associated with attrition resistant sponge metal catalyst which settles well and then the product is decanted from the reactor and fines are filtered from the product. If economics demand recovery and recycle, often the only meaningful scaleup plan is to run plant trials. Catalyst attrition varies between individual large reactors, and is difficult to scaledown to the lab. When using large amounts of catalyst, the recovery and recycle filter must have significant filter area and volume.

Precious metal catalysts on carbon are often used. The entire cycle of catalyst manufacture, use, removal, and metal refining (recovery) typically loses 20-25% of the metal when the process operates well.

The crude from a heterogeneous reaction step must be filtered very well, and then one should always filter well again before feeding a distillation column. The active catalyst can cause undesirable reactions at the higher temperatures experienced in a column, either in the packing or the bottoms heater. And when one suspects catalyst in the column, oxidizing and/or passivating is often not effective, rather the catalyst can be reduced by the organics in the system and is negatively active again.

Processes with homogeneous catalysts have been scaled-up directly by a factor of 100,000.

Heat Transfer

Often in the lab with exothermic reactions, one can combine the two reactants without solvent, heat until reaction begins and cool during the exotherm and not experience a runaway reaction. Most of the time this cannot be transferred into the plant. With a strongly exothermic reaction, say 20-40 kcal/gmole of product, one must incrementally feed one of the reactants in order to manage the heat release, and also add a non-reactive solvent, because in the commercial reactor the heat transfer is poor at low level (10-20% volume) and the level must be higher initially to have good cooling area/liquid volume ratio to achieve reasonable heat removal. Often there is a small bottom jacket to heat to reach the "light-off" temperature, and the side jacket is used for cooling, more reason to add solvent and incrementally feed one of the reactants. Given these realizations, often plant reaction times are 4-8 hours instead of 1 hour in the lab.

Using pencil and paper, or a batch simulator, a capable engineer can estimate the likely plant timeline, which should then be run in the lab, in order to conduct a successful scaleup.

As a side bar, in private discussions with T A Koch on September 24, 2003, continuous fixed bed reactions are challenging with a 45-50 deg C adiabatic temperature rise. Furthermore if one is attempting such a reaction with a 100 deg C rise, a fluid bed should be selected as the commercial unit operation, therefore also run this type of reactor system in the experimental program. When Ted and I worked together at DuPont, we used this as a guideline to define new programs and recently we attempted to find this guidance in writing, but we were unsuccessful.

Mass Transfer, Mixing

Mixing is a very extensively published topic. The commercial situation is often a glass lined steel reactor with a retreat curve agitator and simple baffle, or an engineered stainless steel reactor with turbine agitator and serious wall mounted baffles. The first poorly mixes two liquid phases with large specific gravity differences, which can be corrected in glass to a certain extent with the newer glass turbines and improved baffles. The later can mix so well that emulsions result with the two immiscible phases and separation times can be 24 hours long instead of 15 minutes in the lab. Heat and gentle continuous or intermittent mixing may break a stubborn emulsion. Attempt a simple lab centrifuge experiment with an emulsion to determine if more g force would effect a quick separation. These engineered reactors should be equipped with a variable speed drive agitator and one can design lab kettles to simulate the plant conditions for lab experimentation.

No matter what the situation, the specific gravity difference is a very, very important variable. Aerstin & Street (6) have published the simple decanting equation with viscosity of the continuous phase in cp and time in hours:

Time to separate = 0.1 (viscosity/sp gr difference)

Plugging in a few values at a viscosity of one cp:

Gravity diff	Hours
.001	100
.01	10
.1	1
.2	0.5

When taking a phasing step into the plant, having a specific gravity difference of 0.1 or greater is a very good plan.

When reacting two phases that are not very soluble in each other, for example when carrying out nucleophilic substitution reactions, phase transfer catalysts should be considered when scaling down from equipment with poor mixing characteristic, rather than buying new equipment.

Mass Transfer, Fractional Distillation

Packed fractional distillation columns run in the batch mode are often used for low-pressure drop vacuum separation. With a trayed column, the liquid holdup on the trays contributes directly to the hydraulic head required to pass through the column, and with twenty theoretical stages that static pressure drop is very high, e.g., as much as 100-200 mm Hg.

A typical packed column with adequate reflux has about 1 mm Hg pressure drop per theoretical stage, so plant delta P in a 20 stage column is about 20 mm Hg. But a lab column with "ProPac" and 20 stages has only a few mm Hg pressure drop. The increase in boiling temperature over 20 mm Hg can be significant under vacuum. For adiponitrile, the boiling temperature at 1 mm Hg is 102 deg C, but at 20 mm Hg, 172 deg C. So, when running the lab experiment, one should impose a back pressure on the system to establish the same pot temperature as expected in the plant, and discover if bad reactions occur in the pot. And before approaching any temperature condition that could result in decomposition with gas generation, on should run a differential scanning calorimetry to establish the incipient decomposition temperature with pressure generation.

Mass Transfer, Crystallization and Cake Washing

Wayne Genck (7 & 8) has recently published several useful articles about batch crystallization. Often lab filtration after crystallization is done with a thin cake and no problem is observed. But when taken to the plant, this operation takes days to build and wash a cake. To avoid this problem it is best to operate a crystallizer that is properly seeded and cooled according to a profile that follows the equation in reference (7), slow at first and fastest at the end. The other reference (8) discusses the challenges without seeding. Experience by the author confirms that a large amount of seed crystal is required, about 1-2 % wt of the final crystal yield.

Another area of concern is cake washing. In the lab, one can easily reslurry a cake, and wash again to obtain the desired color and purity. Many plant-dewatering devices can not reslurry, and there will be a significant fraction of the original liquor remaining in the cake using a simple filter with washing. In the 1970's, DuPont developed the guidelines shown on Figure 5. The wash ratio,

N(w), is the amount of wash liquor used divided by the amount of liquor in the cake at the start of washing. Note that in practice results do not improve beyond about N(w)=3, fine solids wash better than coarse solids, and that less than 10% fraction of original liquor remaining in the cake is seldom achieved. Understanding this general pattern will result in a scaleup with either a reslurry step and a second filtration, or the use of a filter like a Rosenmund that can reslurry the cake internally, and filter and wash again.

Other Brief Topics, Including Economics

Glass lined vessels are very common in the batch business. Both hot strong acid and hot strong base can damage this equipment, at best removing the fire polish and generating porosity, and at worst dissolving the glass and eating through the vessel. See references 9 and 10 for data.

Often the faith in analytical methods gives a false sense of high yields, when the tars or heavy by-products are not part of the quantification. One should always get an "in-hand" yield, actually isolating a known weight of a pure product from a known amount of starting materials.

Front running the commercial starting materials, including the catalysts, is a very import checkpoint. And as a minimum, technical grades should be ordered from the lab supply houses.

Finally, we are in the business to produce products and profits. Broadly, if a product is made in 3-4 reaction steps in the batch-manufacturing environment, the market value should be 10 times the materials. If market value is only twice the raw material cost, the project should be redefined or stopped. In a 10-step process, it is not uncommon that materials are $1/20^{th}$ of the selling price. A comparison of a two reaction step product using discounted cash flow methods (11) showed that a process with a market value 4 times the materials was greatly preferred over one at 3 times materials.

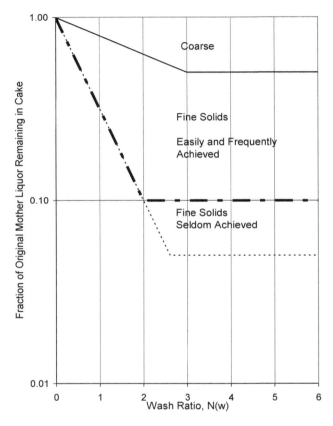

Figure 5. Filter Cake Wash Efficiency

Acknowledgments

Far too many to list. I thank all the many scientists and engineers that shared their knowledge and experience with me, and for the successes and time together that we enjoyed.

References

1. Attilio Bisio and Robert L Kabel, Scaleup of Chemical Processes, John Wiley & Sons, New York, 1985, p. 2, 10.
2. J. M. Smith, Scaledown to Research, *Chem. Eng. Prog.*, **64** (8), 78-82 (1968).
3. D. Mukesh, Improve Your Chances for Successful Process Development, *Chem. Eng. Prog.*, 110-112 (February 1996)
4. T. A. Johnson and D. P. Freyberger, Lithium Hydroxide Modified Sponge Catalysts for Control of Primary Amine Selectivity in Nitrile Hydrogenations, Chemical Industries (Dekker), **82** (*Catal. Org. React.*), 201-227 (2000).
5. The Chemical and Engineering Dictionary, The Chemical Publishing Company of NY, New York, NY, date unknown, pp. 55, 66.
6. F. Aerstin and G. Street, Applied Chemical Process Design, Plenum Press, New York, 1978, pp. 35-38.
7. W. J. Genck, Better Growth in Batch Crystallizers, *Chem. Eng.,* 90-95 (August 2000).
8. W. J. Genck, Crystal Clear, *Chem. Processing,* 63-67 (October 2003).
9. Pfaudler, Inc., Worldwide "Glasteel" 9100 (a product bulletin), Rochester NY, January 2000.
10. De Dietrich Equipement Chimique, 3009 Glass (a product bulletin), France, February 1999.
11. E. L Gaden Jr., AIChE Symposium Series (American Institute of Chemical Engineers), **87** (285), (*Investment Appraisal for Chemical Engineers.*), 73-79 (1991).

III. Symposium on Acid-Base Catalysis

36.   Improved Solid-Acid Catalysis for Sustainable, High 2-Phenyl, LAB Production

John F. Knifton
Shell Chemical Company, Houston, TX 77251
john.knifton@shell.com

Prakasa Rao Anantaneni, and Melvin E. Stockton
Huntsman Corporation, Austin, TX 78761

Abstract

Linear alkylbenzenes (LAB's) comprising >80% 2-phenyl isomer content have been prepared in sustainable high yields from detergent-range linear olefins via regioselective benzene alkylation. HF-treated mordenites, acidic β-zeolites, and fluorided montmorillonite clays each provide enhanced shape-selective alkylation performance when employed as heterogeneous catalysts in the subject syntheses. Both individual C_8-C_{12} α-olefins, as well as mixed, plant-derived, C_{10} through C_{14} paraffin dehydrogenate feedstocks containing ca. 8.5% internal olefins, have been successfully alkylated.

Catalyst lifetimes exceeding 3000 hours on stream have been demonstrated with the fluorided mordenite catalysts using reactive distillation technology. When feeding typical C_{10}-C_{14} paraffin dehydrogenate, average C_{10}-C_{14} olefin conversions are near quantitative, while 2-phenyl LAB selectivities are in the range 74-84%. The bromine numbers of these LAB products are typically <0.5, with only trace formation of undesirable dialkylated benzene homologues or tetralin derivatives. The acidic, dealuminized, mordenite zeolite catalysts have high silica/alumina ratios (10:1 to 50:1) and low sodium contents, and after HF treatment, their fluoride contents are generally in the range 0.1 to 4%. Performances of these acidic zeolites and clays have been examined as a function of the benzene/olefin feed composition, total liquid feed rate, operating temperature and pressure, reactor configuration, as well as the level of fluoride pretreatment.

Introduction

In the commercial production of linear alkylbenzenes (LAB), typically through the alkylation of benzene with C_{10} through C_{14} olefins, the 2-phenylalkanes are the most desirable. This is because their 2-phenylalkane sulfonate (LAS) derivatives are more readily biodegradable and the 2-phenyl LAS have the highest aqueous solubilities (lowest cloud points).[1]   The practice of this chemistry on a commercial scale can, however, still present peculiar challenges, particularly regarding controlling the regioselectivity of the alkylation step, strictly limiting the formation of unwanted polyalkylaromatic and tetralin by-products, and preventing the accumulation of an aqueous component in the reactor system.

Traditionally, the production of LABs has been practiced commercially using either Lewis acid catalysts, or liquid hydrofluoric acid (HF).[2]   The HF catalysis typically gives 2-phenylalkane selectivities of only 17-18%.   More recently, UOP/CEPSA have announced the Detal[R] process for LAB production that is reported to employ a solid acid catalyst.[3]   Within the same time frame, a number of papers and patents have been published describing LAB synthesis using a range of solid acid (sterically constrained) catalysts, including acidic clays,[4] sulfated oxides,[5] plus a variety of acidic zeolite structures.[6-9]   Many of these solid acids provide improved 2-phenylalkane selectivities.

We have demonstrated that linear alkylbenzenes comprising >80% 2-phenyl isomer content can be prepared in **sustainable** high yields from detergent-range linear α-olefins via regioselective benzene alkylation using reactive distillation technology in combination with HF-treated mordenite catalysis. The acid-modified mordenite catalysts provide enhanced shape-selective alkylation performance when employed as heterogeneous catalysts in the subject syntheses. Both individual C_{10}-C_{12} α-olefin cuts, as well as mixed, commercial plant-derived, C_{10} through C_{14} paraffin dehydrogenate feedstocks containing ca. 8.5% internal olefins, have been successfully alkylated.   Selective alkylation of 1-decene to 2-phenyldecane is illustrated by eq 1.

$$CH_3(CH_2)_7CH{=\!=}CH_2 \quad + \quad \text{(benzene ring)} \quad \longrightarrow \quad \text{(phenyl)}{-}CH \begin{array}{c} {}^{CH_3}\\ {}_{(CH_2)_7CH_3} \end{array} \qquad (1)$$

Catalyst lifetimes exceeding 3000 hours on stream have been demonstrated with the fluorided mordenite catalysts using reactive distillation technology.   When feeding typical C_{10}-C_{14} paraffin dehydrogenate, average C_{10}-C_{14} olefin conversions are near quantitative, while 2-phenyl LAB selectivities are in the range 74-84%. The bromine numbers of these LAB products are typically <0.5, with only trace formation of undesirable dialkylated benzene homologues and tetralin derivatives. The reactive distillation mode ensures that the catalyst bed is continuously washed of any heavy organics – thereby extending the catalyst life. Additional improvements in catalyst performance are realized by continuous removal of any aqueous fraction from the catalyst area.

The fluoride-treated acidic mordenite catalysts have, in fact, the triple advantages of:

1. High alkylation activity that allows quantitative alkene conversions at adequate feed rates.
2. Exceptionally high regioselectivity for the more desirable 2-phenyl LAB.
3. Sustainable alkylation activity that ensures long catalyst life, ease of handling, and ready regeneration.

Experimental Section

The mordenite zeolites used in this study were purchased from both PQ Corporation (CBV-20A, silica/alumina molar ratio 20, Na_2O content 0.02 wt%, surface area 550 m^2/g, in ca. 1.5 mm extruded form) and from Union Carbide Corporation (LZM-8, silica/alumina molar ratio 17, Na_2O content 0.02%, surface area 517 m^2/g in powder form). All samples were calcined at 540 °C prior to use.

The following example illustrates the preparation of a hydrogen fluoride-modified mordenite:

To a 500 g sample of acidic, dealuminized mordenite (CBV-20A from PQ Corporation, 1.5 mm diameter extrudates that had been calcined at 540 °C, overnight) was added a solution of 33 ml of 48% HF in 1633 ml of distilled water, and the mix cooled in ice, stirred on a rotary evaporator overnight, then filtered to recover the extruded solids. The extrudates were further washed with distilled water, dried in vacuo at 100 °C, and then calcined at 540 °C, overnight. Analyses of the treated mordenite showed: 1.2% fluoride, 0.49 meq/g acidity. Samples were charged to the reactive distillation unit either as 20/40 mesh granules, or as ca. 1.5 mm extrudates.

Individual α-olefin feedstocks were purchased from Aldrich Chemical Company. The C_{10}-C_{14} paraffin dehydrogenate was obtained from a commercial LAB plant and contained 8.5% C_{10}-C_{14} olefins, primarily a mixture of internal olefins.

Continuous benzene alkylation was conducted in a reactive distillation column of the type illustrated in Figure 1. The process unit comprises the following principal elements: a double column of solid catalyst **32**, packing columns above and below the catalyst bed, a liquid reboiler **42** fitted with a liquid bottoms product takeoff **44**, a condenser **21** fitted with a water collection and takeoff, and a feed inlet

14 above the catalyst column, plus the necessary temperature and pressure controls. The feed mixture of benzene and olefin flows from a feed pump **10** to a feed inlet **14** via line **12** and falls to the packed acidic catalyst bed **32** where alkylation takes place. Typically the catalyst bed comprises two parts, one 24 cm in length, the other ca. 55 cm, and has a total volume of about 250 cc. In the catalyst bed **32**, the falling feed also contacts rising vapors of unreacted benzene which have been heated to reflux in reboiler **42** by heater **40**. Such rising vapors pass over the thermocouples **28** and **38**, which monitor temperature and provide feedback to heater **40**. The rising vapors also pass through standard, woven-mesh, stainless-steel, Goodloe packing **36** that holds the catalyst beds in place. A small plug of glass wool/Goodloe packing above the catalyst bed **32** prevents the smaller particles from reaching the feed inlet **14**.

Prior to start-up, the reboiler is charged with 100-500 ml of benzene/olefin feed mix, and the complete system is flushed with nitrogen, which enters via line **54**. Also, prior to start-up, it may be desirable to heat the catalyst bed so as to remove any residual moisture. Trace moisture in the feed mix is collected in water trap **24** after liquification at condenser **21**. The reboiler mix is then heated to reflux and the benzene/olefin mix continuously introduced into the unit above the catalyst bed as described above. When the LAB content in the reboiler **42** rises to the desired level, the bottoms LAB product is removed from the system via line **47** and valve **44**. The dip tube **46** is employed to slightly increase the pressure in the reboiler **42**. Alternatively, a pressure generator **56** may be employed. Control mechanisms for heat shut-off **50** and pump shut-off **52** serve to close down heat/feed if the liquid level in the system rises too high. Line **60** connects the pump shut-off **52** to the system above the condenser **21**. With typical benzene/olefin feed rates of 50-500 ml/hr, and benzene-to-olefin feed molar ratios in the range 5:1 to 40:1, the temperature profile in the two-part catalyst beds **32** is generally within the range of 80-120 °C, according to the applied unit pressure. The reboiler temperature is maintained, for the most part, below 150 °C.[10]

Results and Discussion

Aromatic alkylation is illustrated in this paper for both individual α-olefin cuts (e.g. 1-decene and 1-dodecene) as well as typical samples of C_{10}-C_{14} paraffin dehydrogenate that contain ca. 8.5% C_{10}-C_{14} olefins, primarily internal olefins. Syntheses have, for the most part, been conducted in a continuous reactive distillation unit of the type illustrated in Figure 1, operated slightly above atmospheric pressure. Operational details for this unit may be found in the previous section.

As part of our studies into the use of solid acid catalysis for LAB production, we have screened the performances of at least four potential candidates, including fluorided montmorillonite clays, β-zeolite, dealuminized mordenite, and fluorided, dealuminized mordenites.[10-12] The fluorided mordenite catalysts generally provided the highest regioselectivity for a variety of olefin feedstocks.[13] Their preparation is detailed in the Experimental Section. The dealuminized, mordenite precursors typically have silica/alumina ratios in the range 10:1 to 50:1, plus low sodium contents, and after HF treatment, their fluoride contents are generally in the range 0.1 to 4%. The HF treatment is likely to decrease the total Bronsted acid site density, but increase the strength of the remaining acid sites in the mordenite. This, in turn, would be expected to raise the reactivity of the carbocation intermediates formed during LAB synthesis. Surface fluorination of H-mordenite has been reported previously to increase the catalyst alkylation activity and enhance the stability.[14]

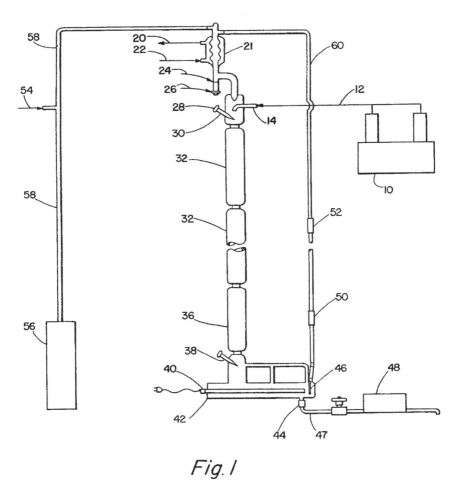

Fig. 1

Figure 1 Schematic of reactive distillation unit used for LAB synthesis.

Under batch, refluxing benzene conditions (ca. 80 °C), where the starting benzene /1-dodecene molar ratio is ca. 25,[10] we observe:

- Dodecene conversions are close to quantitative after 3 hr.
- 2-Phenyldodecane isomer selectivity is 79.9%.
- Heavies constitute only ca. 0.2%.

A detailed analysis of the LAB isomer distribution in this experiment is provided in Table 1. The results are very similar for LAB production from benzene/1-decene mixtures using the same batch procedures and equipment (see also Table 1).

Table 1 Benzene alkylation with 1-dodecene and 1-decene , using fluorided-mordenite catalyst.

Dodecene Conv. (%)	LAB isomer distribution (%)					Heavies (%)	Linear LAB (LLAB)
	2-Ph	3-Ph	4-Ph	5-Ph	6-Ph		
99.7	79.9	16.6	0.8	1.3	1.3	0.2	95.9

Decene Conv. (%)	LAB isomer distribution (%)			
	2-Ph	3-Ph	4-Ph	5-Ph
99	84	13	2	1

The fluorided mordenite catalysts are particularly effective for benzene alkylation with typical, plant-derived, C_{10}-C_{14} paraffin dehydrogenate. Figure 2 illustrates continuous alkylation over 250 hr using a 1.2% fluorided mordenite catalyst (250 cc) in the reactive distillation unit of Figure 1. Feeding a benzene/paraffin dehydrogenate mix (10:1 molar C_6H_6/C_{10}-C_{14} olefin) the 2-phenyl isomer selectivity remains in the range 75-83%, while total olefin conversions are above 95%. The recovered HF-treated mordenite shows <10% loss of fluoride and a maintained acidity (1.1% F, 0.29 meq/g titratable acidity). One of the intrinsic advantages of the reactive distillation configuration (Figure 1), is that the rising benzene vapors from the reboiler continually wash the HF/mordenite catalyst of heavy organics, thereby increasing the lifetime of the solid acid. Improved catalyst life and performance is further enhanced through continuously removing any water fraction from the catalyst area. By better mixing of C_{10}-C_{14} olefin and benzene reactants and increasing the instantaneous C_6H_6 concentration in the alkylation zone, we are able to achieve selective benzene monoalkylation, with only trace quantities of undesirable dialkylated benzene homologues and tetralin derivatives (see Tables 1 and 2).

Figure 2 also includes a comparative experiment, where the solid acid catalyst is a sample of non-fluorided (but calcined), acidic mordenite. Here we see: a) a significant loss of alkylation activity with time on stream and b) a measurably lower

2-phenyl isomer content. The 2-phenyl isomer selectivity (70-75%) observed in this second experimental series is, however, within the range reported earlier by Travers et al., who also used non-fluorinated, but dealuminized, mordenite catalysts for benzene/1-dodecene batch alkylation.[6]

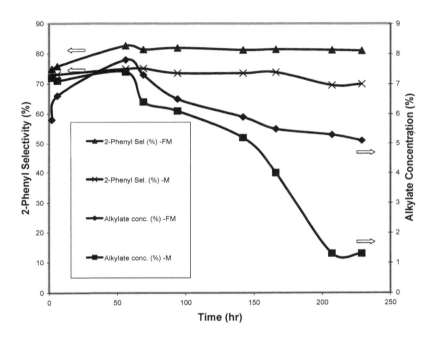

Figure 2 Benzene alkylation with C_{10}-C_{14} paraffin dehydrogenate; MF, using fluorinated mordenite catalyst; M, using the non-fluorinated mordenite precursor.

Table 2 Benzene alkylation with 1-dodecene, fluorided mordenite catalysts.

Catalyst	Fluoride level (%)	Dodecene Conv. (%)	LAB isomer distribution (%)					Heavies (%)
			2-Ph	3-Ph	4-Ph	5-Ph	6-Ph	
A	0.25	>90	71.6	22.9	2.9	1.1	1.3	0
B	0.5	>90	73	22.2	2.9	1.2	0.8	0
C	1	>90	74.3	19.5	2.9	3.4		0
D	4.8	<1						

Performances of these acidic mordenite zeolites have been examined as a function of the level of fluoride treatment. For mordenites treated with HF to fluoride levels of 0.25, 0.5, and 1.0%, batch alkylation at ca. 80 °C of typical benzene/1-dodecene mixtures show (after 1 hr) very similar levels of activity and 2-

phenyl selectivity – see Table 2. A more heavily loaded, 4.8% fluorided mordenite (catalyst D), however, was inactive. Analyses of the freshly-treated mordenites by MAS NMR indicated that the mildly-treated catalysts A – C still retained their tetrahedral Al, whereas there was no framework tetrahedral Al remaining in the case of catalyst D and the sample was completely dealuminized.[15] Soled et al. report that severe fluoridation also leads to the conversion of non-framework aluminum species to AlF_3.[16]

Catalyst lifetimes well in excess of 3000 hours on stream have been demonstrated with the fluorided mordenite catalysts using the reactive distillation technology.[17,18] When feeding typical C_{10}-C_{14} paraffin dehydrogenate at LHSV of 0.2, average C_{10}-C_{14} olefin conversions are near quantitative, while 2-phenyl LAB selectivities are in the range 74-84%. The bromine numbers of these stripped LAB products are typically $0.1 \rightarrow 0.5$. Illustrative data may be found in Figure 3. Elemental analyses of the recovered catalyst in this case shows 1.0% fluoride and a titratable acidity of 0.34 meq/g. MAS NMR studies indicate some loss of tetrahedral alumina.[15] Generally, for typical used HF/mordenite samples, carbon is fairly evenly distributed throughout the sample. For partially deactivated catalysts, the deposits are mainly polyalkylated aromatics. Polymer coke is prevalent in fully deactivated samples. Nevertheless, a simple calcination will fully restore alkylation activity and sustainable lifetimes in C_{12}-C_{14} paraffin dehydrogenate/benzene service have thereby been demonstrated.

It is envisioned that this new LAB technology would be most readily applicable industrially as a retrofit to existing LAB commercial units, possibly as a slip-stream reactor.[19,20] The manufacturer would then have the flexibility to tailor the 2-phenyl content of their LAB and LAS products according to the customer's specific needs. Alternative solid acid reactor configurations have been presented at an earlier Organic Reactions Catalysis Meeting.[21]

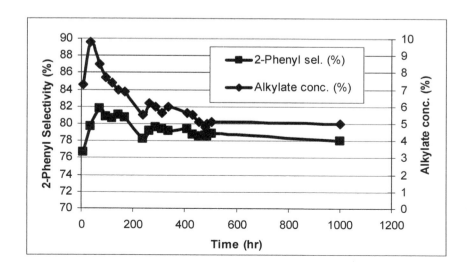

Figure 3 Benzene alkylation with C_{10}-C_{14} paraffin dehydrogenate catalyzed by fluorinated mordenite.

References

1. P. R. Pujado, Handbook of Petroleum Refining Processes, 2nd Ed., McGraw-Hill, New York, 1996, Chapter 1.5.
2. Encyclopedia of Chemical Technology, Wiley, New York, 1999, Vol. 2, p. 58.
3. T. Imai, J. A. Kocal, and B. V. Vora, *Sci. & Tech. Catal.*, 339 (1994).
4. H. Ming-Yuan, L. Zhonghui, and M. Enze, *Catal. Today*, **2**, 321(1988).
5. J. H. Clark, G. L. Monks, D. J. Nightinggale, P. M. Price, and J. F. White, *J. Catal.*, **193**, 348 (2000).
6. C. Travers, F. Raatz, B. Juguin, and G. Martino, European Pat. 466,558 to Institute Francais du Petrole (1992).
7. Y. Cao, R. Kessas, C. Naccache, and Y. B. Taarit, *Appl. Catal.*, **184**, 231 (1999).
8. Z. Da, P. Magnoux, and M. Guisnet, *Catal. Lett.*, **61**, 203 (1999).
9. L. B. Young, US Patent 4,301,317 to Mobil Oil Corporation (1981).
10. J. F. Knifton, P. R. Anantaneni, and M. Stockton, US Patent 5,770,782 to Huntsman Petrochemical Corporation (1998).
11. J. F. Knifton and P. R. Anantaneni, US Patent 5,777,187 to Huntsman Petrochemical Corporation (1998).
12. J. F. Knifton, P. R. Anantaneni, and P. E. Dai, US Patent 5,847,254 to Huntsman Petrochemical Corporation (1998).
13. J. F. Knifton, P. R. Anantaneni, P. E. Dai, and M. E. Stockton, *Catal. Lett.*, **75**, 113 (2001).
14. A. Kurosaki and S. Okazali, *Bull. Chem. Soc. Jpn.*, **63**, 2367 (1990).
15. J. F. Knifton and J. C. Edwards, unpublished results.
16. S. Soled, G. B. McVicker, J. J. Ziemiak, G. DeMartin, L. Yacullo, K. Strohmaier, and J. Miller, Division of Petroleum Chemistry, Preprints, 206th Am. Chem. Soc. Meeting, Chicago, 1993, p. 546.
17. J. F. Knifton, P. R. Anantaneni, and M. E. Stockton, US Patent 6,315,964 to Huntsman Petrochemical Corporation (2001).
18. J. F. Knifton, P. R. Anantaneni, and M. E. Stockton, US Patent pending.
19. P. R. Anantaneni, US Patent 6,133,492 to Huntsman Petrochemical Corporation (2000).
20. P. R. Anantaneni, US Patent 6,166,281 to Huntsman Petrochemical Corporation (2000).
21. J. F. Knifton, R .J. Taylor, and P.E. Dai, Diisopropyl ether one-step generation from acetone-rich feedstocks, Chemical Industries (Dekker) **85** (*Catal. Org. React.*) 131-141 (2002).

37.   Recent Advances in Thermally Stable Resins for Olefin Hydration

Ajit Pujari,[1] Setrak Tanielyan,[1] Robert L. Augustine,[1] Dorai Ramprasad,[2*] and Pat Crawford[2]

[1]Center for Applied Catalysis, Seton Hall University, South Orange, NJ 07079
[2]Rohm And Haas Company, 727 Norristown Road, Spring House, PA 19477

DRamprasad@rohmhaas.com

Abstract

The hydration of olefins is an industrially important reaction. It is used to convert alkenes such as 1-butene into 2-butanol which can then be dehydrogenated to give methyl ethyl ketone. Currently, a number of plants still use an indirect hydration process wherein the butenes are esterified with liquid sulfuric acid and the resulting butyl esters are hydrolyzed to produce the alcohol. The disadvantage to such a process is the neutralization waste, which has to be treated and disposed. Between 1972-1984, leaders in the industry have pioneered the application of a heterogeneous catalyst for this hydration. This catalyst is a polymer containing sulfuric acid groups. The operating conditions for the *sec*-butanol process involve supercritical conditions with temperatures between 150-170 °C and pressures between 50-70 bar. Under these conditions, the catalyst can desulfonate resulting in the formation of corrosive components and in a loss of activity.

This paper describes our efforts to synthesize a series of thermally stable sulfonated resins and to evaluate their activity in a model reaction of 1-butene hydration to *sec*-butanol.

Introduction

The use of sulfonated polystyrene resins (SPS), crosslinked with divinylbenzene (DVB) as catalysts for the hydration of olefins such as propylene or butene-1 has been generally described in the literature. [1-4] Olefins such as n-butene have a limited solubility in water and are less reactive, necessitating elevated temperatures in the range of 150-170 °C. However, the use of elevated temperatures has the significant drawback in the resulting loss of SO_3H groups leading to the gradual reduction in the catalyst activity. [5] In that respect, there is an ever-growing need for thermally stable acidic ion exchange resins, capable of promoting the olefin hydration at high pressures and high temperatures. One of the methods to improve the thermal stability of the acidic resin catalysts is to incorporate halogen atoms, such as chlorine, into the aromatic ring of the polymeric backbone. [6] However, indiscriminate chlorination of the sulfonated polymeric precursors can produce weakly bound halogen atoms, particularly those, in the benzylic position to the aromatic ring. Under the operating conditions commonly used in the olefin hydration, this type of halogen will easily be hydrolyzed to produce HCl, and consequently, this would lead to accelerated formation of acid during use and corrosion to the equipment used. [7] This work describes the preparation of a series of chlorinated polymers with high thermal stability with respect to desulfonation and reduced chloride leaching. Additionally, the catalytic activity of the newly synthesized thermally stable resins is evaluated in the model reaction hydration of butene-1 to *sec*-butanol.

Experimental Section

1. Polymer synthesis. Styrene divinylbenzene (DVB) co-polymers used as catalyst precursors are commercially available in their sulfonated or polysulfonated form. For example, a sulfonated polystyrene divinylbenzene co-polymer resin (SST-7DVB), containing 7%DVB was converted to catalyst A according to the reaction sequence outlined in Scheme 1, path A/B. Similarly, using the same reaction sequence, a second sulfonated polymer SST-12DVB, containing 12%DVB was used to prepare catalyst B. The third precursor resin used was the base polystyrene divinylbenzene polymer ST-12DVB, which was first polysulfonated with oleum and then converted to catalyst D, following the second modification path D/E. For the preparation of resin E, the precursor was poly 4-chlorostyrene divinylbenzene, containing 12%DVB (CST-12DVB), and the modification was done also along path D/E. [6] For reference purposes, the intermediate resin (4), obtained after chlorination of the precursors SST-12DVB was isolated and used without further treatment (in the text denoted as catalyst F). A second reference resin was also prepared by hydrothermal treatment of the same intermediate resin (4) by a known procedure[7] to produce the reference catalyst C.

2. Chlorination of the polymeric precursors. A two-liter jacketed glass reactor, equipped with magnetic stirrer and thermowell was charged with 1180 g of water and 490 ml of wet resin PSST-12DVB (precursor to the synthesis of catalyst D). The autoclave was purged for ten minutes with nitrogen, the stirring was initiated and the temperature of the suspension raised to 35 °C. The nitrogen flow was replaced with chlorine gas and the pressure set to 15 psig. The progress of the reaction was monitored by the chlorine consumption, measured by the weight loss of the lecture bottle. Since the incorporation of chlorine atom into the polymeric backbone is accompanied by stoichiometric release of HCl into the aqueous phase, by titrating aliquots of the solution one can calculate the degree of chlorination with time. The reaction was continued for 15 hours, the resin isolated by filtration and washed with DI water until free of Cl⁻. The elemental analysis of the resulting Cl-polysulfonated resin gave 11.9% Cl and 16.7% S.

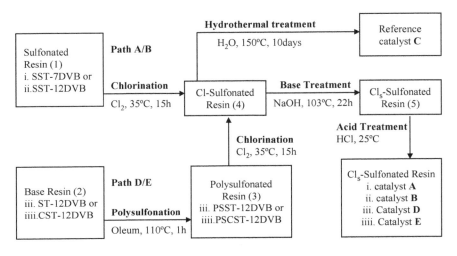

Scheme 1

3. Base treatment step. The following general base treatment procedure was used to prepare resins A, B, D and E. Approximately 275 g of chloro-sulfonated or chloro-polysulfonated resin (4) was first washed with 150 ml of deionized water and then placed into a three necked round-bottomed flask, equipped with a mechanical stirrer, water condenser and a Thermowatch. The flask was additionally charged with 1192 ml solution of 2 N NaOH, the mixture was heated to the reflux temperature of 103 °C under stirring and held at that setting for a total of 22 hrs. The aqueous solution was extracted, the resin was washed with 100 ml of DI water and the combined solution was titrated with silver nitrate for chloride determination. The resulting resin was additionally washed with 3x500 ml portions of deionized water, transferred into a column and alternately flushed with 3 L of water followed by 3 L of 4% HCl solution and finally again with 3 L water or until free of Cl⁻.

4. Standard test for hydrolytic stability. The hydrolytic stability of the chlorinated resins was determined by the following test procedure. An acid digestion autoclave having a volume of 125 ml is charged with 40 ml of resin and 28 ml of deionized water. The bomb is sealed and transferred into an oven, pre-heated to 200 °C. The test is continued for 24 hours. The bomb is removed and cooled to ambient temperature. The liquid is separated from the resin and the chlorine content analyzed while the resin is washed thoroughly and its acid capacity is determined as described in section 5. The test results are shown in Table 2.

5. Determination of the Exchange Capacity. The acid capacity of the resins is determined by the following procedure. Approximately 10-12 g of a wet resin was placed in a beaker and dried overnight at 110 °C. The beaker was then placed in a desiccator, allowed to cool and weighed. The resin was transferred afterwards into a column, flushed first with approximately 300 ml of deionized water for over half hour and later, with 1 L of 4.5% aqueous solution of $NaNO_3$ at a rate of 10 ml/min. An aliquot of the flush solution was titrated with 0.1 N NaOH and the exchange capacity calculated as mmol H+ /g dry resin.

6. Hydration of 1-Butene Over Thermally Stabilized Resins as a Catalytic Test Reaction. The hydration experiments were conducted in a stainless still EAZY SEEL reactor (Autoclave Engineers), which was equipped with units for an accurate speed and temperature control. In a typical hydration experiment, a known amount of catalyst and water was charged in the reactor cup. The 1-butene substrate was metered in a calibrated glass flask and injected into the reactor under 60 psig positive pressure. The reaction temperature, the reactor pressure and the stirring rate were continuously monitored and recorded. A typical hydration charge consisted of 1 g catalyst, 10 ml 1-butene and 10 ml water. After pressurizing the rector with argon to 550 psig the reactor was heated to the target temperature of 155 °C, the stirring rate was set to 1200 RPM while the pressure was adjusted to 900 psig. The reaction continued for the pre-determined time, the reactor was cooled to ambient temperature and the liquid phase released slowly into a chilled 25 ml volumetric flask.

7. Analysis. After diluting to the mark with dioxane, the reaction mixture was analyzed for the product *sec*-butyl alcohol. The GC analysis was done on Stabiliwax column (50 m, 0.35 mm) with FID using N,N-dimethylformamide as an internal standard.

Results and Discussion

1. Modification of the resins. The synthetic strategy used in preparing thermally stable and highly acidic resins, which are of commercial interest for high temperature acid catalyzed reactions, is shown in Scheme 1. It is based on chlorination and subsequent base treatment of commercially available sulfonated styrene-divenylbenzene co-polymers (ST-7DVB or ST-12DVB), producing either resin A (7% DVB) or resin B (12%DVB) catalysts. The chlorination step incorporates chlorine in the aromatic ring and in the side chain, presumably in benzylic position.[6] The introduction of chlorine atom on the aromatic rings dramatically improves the thermal stability of the resin with respect to loss of sulfonic acid groups.

The subsequent base treatment of the chlorinated resins in the second step is required to remove the weakly bound chlorine in the side chain thus leading to thermally stable 'low bleed' resin catalysts A/B. As a reference, the intermediate resin 4 (12% DVB also denoted as catalyst F) was isolated after the chlorination step with the aim of evaluating the effect of the base treatment. The same intermediate resin was hydrothermally treated according to a known procedure[7] to produce the second reference resin catalyst C having also 12% DVB as catalyst B.

It is well known that resins with more than one sulfonic acid group per ring (polysulfonated) show higher thermal stabilities[8] than conventional resins. It was hypothesized that chlorination of a polysulfonated resin should lead to enhanced thermal stability. Catalyst D is a polysulfonated chlorinated resin containing 12% DVB prepared according to path D/E in Scheme 1.

Table 1 Reduction in chlorine content after base treatment of resins A-E.

Resin	Precursor	DVB	Base treatment				Loss of Cl
			Before		After		
		%	% Cl	% S	% Cl	% S	%
A	SST-DVB[c]	7	29.7	8.2	26.8	8.6	9.8
B	SST-DVB	12	22.0	10.1	19.6	10.2	10.9
F[a]	SST-DVB	12	22.0		22.0	na	-
C[b]	SST-DVB	12	22.0		19.3	na	12.3
D	PS ST-DVB[d]	12	11.9	16.7	10.0	16.7	16.0
E	PS CST-DVB[e]	12	18.9	14.0	16.0	14.0	15.3

[a] Reference catalyst without base treatment (intermediate 4 in Scheme 1).
[b] Reference catalyst hydrothermal treatment (see Scheme 1).
[c] Sulfonated styrene divinylbenzene.
[d] Polysulfonated styrene divinylbenzene.
[e] Polysulfonated 4-clorostyrene divinylbenzene.

Contrary to this, a different chemistry is expected for the last example of resin catalyst E, where the precursor used was p-chlorostyrene-12%DVB (CST-12DVB) co-polymer. It is known in the literature that styrenic groups containing SO_3H groups in the meta position are more thermally stable[9,10] than those in the para position. By using a p-Cl styrene DVB co-polymer, the para position in the styrene moiety was blocked. Since chlorine is an ortho para director it was hypothesized that oleum sulfonation could lead to styrenic groups with enhanced stability as a result of some SO_3H groups going to the meta position. Further chlorination of the sulfonated polymer (PS CST-12DVB) as in Scheme 1 is expected to incorporate chlorine in the DVB units and thus stabilize the SO_3H groups on the DVB part of the resin. .

The results for the chlorine and sulfur content before and after the base treatment are listed in Table 1. As one can see, the least amount of chlorine removed after the base treatment of the resins was recorded for the precursor of catalyst A. Precursors of polysulfonated resins D and E released a larger fraction of the incorporated chlorine during the base treatment step, thus producing the final catalyst, which were next subjected to the test for hydrothermal stability and catalytic activity in the model reaction of hydration of 1-butene.

2. *Hydrothermal stability (HTS)*. The stability of the chlorinated resins was determined by a test procedure described in the experimental. The resin according to the standard test is treated with water in a sealed flask at 200 °C for 24 hours to determine the loss in acid functionality and additionally, the level of chlorine released into the aqueous phase.

Table 2 Hydrolytic stability of resins A-E.

Resin	Cl	S	Exchange capacity[a]		SO_3H/g[b]	Cl[c]
	%	%	Before	After	% lost	% in liquid
A	26.8	8.6	2.75	2.49	9.4	0.20
B	19.6	10.2	3.31	2.50	24.6	0.19
F	22.0	na	3.14	2.10	33.4	1.34
C	19.3	na	3.06	2.12	30.4	0.08
D	10.0	16.7	4.78	3.79	20.7	0.11
E	16.0	14.0	4.16	3.52	15.3	0.12
**	<0.05	16.2	5.08	0.92	81.8	0.002

[a] Measured as $[SO_3H]$, Mmol/g.
[b] Calculated as amount lost from the resin after the test.
[c] %Cl found in the aqueous phase after the test.
**SST 7DVB

The data in Table 2 show that Resin A (7% DVB) with the largest chlorine content has the highest stability with respect to desulfonation as compared to all resins. When A is compared to resin B (12% DVB) shows three times higher stability (based on the % sulfonic acid groups lost in the test) while the level of chlorine bleed was the same for both resins. Among the resins with 12%DVB, resin F is a conventional chlorinated resin, which is used as a precursor to reference resin C (base treated by a known procedure) and to resin B, which is base treated by the new modification protocol. The data in the same table show that chlorinated only resin F has lost 33.4% of the acidic functionality under the test conditions and releases large amount of chlorine into the aqueous phase. In contrast, base treated Resin B shows much higher exchange capacity, significantly higher hydrothermal stability and low level of chlorine bleed in the liquid phase. Reference resin C, while showing lower level of chlorine bleed, has lower exchange capacity and shows lower HTS.

Oleum sulfonation of a styrene DVB co-polymer results in a polysulfonated polymer where more than one SO_3H group might be present in the aromatic ring. Resin D, which is created *via* Path E/D is both polysulfonated and chlorinated resin. The results from the test study show that the same resin, with only 10% chlorine content produces exceptional performance in terms of high acid capacity, high stability and low chlorine leaching.

In the last example, where p-chlorostyrene-12% DVB resin was sequentially polysulfonated and chlorinated and base treated to give resin E, characterized by high exchange capacity, stability and low bleed. We hypothesize that blocking the para position to sulfonation in the polymer leads to enhanced stability.

3. Catalytic Activity in 1-Butene Hydration. The resins A-E were tested as polymeric thermostable catalysts in a model reaction for the hydration of 1-butene to *sec*-butyl alcohol (SBA). In general, both a flow system and a batch reactor can be used to evaluate the catalyst activity and selectivity of hydration. Most of the information regarding the kinetic parameters of the reaction and the data on the thermodynamic equilibrium can be found in reference 3 and in some patent sources such as references 1,6-7. It has been shown, that the hydration over strong ion-exchange resins, as well as in the presence of an acid catalyst, takes place through a common carbenium ion, which is shown in Scheme 2. The rate-limiting step is believed to be the protonation of the three isomeric butenes to form the carbenium ion. Along with the formation of *sec*-butanol, a large fraction of the initial 1-butene isomerizes to the corresponding cis- and trans-2-butene. The carbenium ion can also react with the *sec*-butanol to form di-*sec*-butyl ether, a by-product which forms in measurable levels at higher operating temperatures. Industrially, the hydration of 1-butene is carried out in an up-flow fixed bed mode at temperature 155 °C and 1000 psig. At this temperature, the supercritical butene phase co-exist with a water rich liquid phase. The *sec*-butanol formed is partitioned between the aqueous and the supercritical phase in close to a 1:3 ratio, which allows one to carry the hydration in multiple extraction mode.[2]

One of the most critical parameters in the SBA production is the process temperature. Since the catalyst performance is slowly degrading because of SO_3H loss, to maintain constant productivity, the temperature of the catalyst bed needs to be gradually raised to approximately 170 °C. This leads to further loss of acid functionality and an increase in the level of di-*sec*-butyl ether.

The initial screening of the resin catalysts was done in a batch reactor at supercritical for butene-1 conditions of temperature 155 °C, pressure of 1000 psig and at molar ratio of 1-butene:water of 5.5. The reaction was stopped after predetermined period of time and the products analyzed. It was found that under the standard reaction conditions, for all of the catalysts studied, a constant concentration in the sec-butanol concentration was achieved within a 1-2 hour reaction time. Using only the linear section of the concentration–time plot, the one hour result was used to evaluate the catalyst activity, which was normalized as mmol of SBA/ per proton/ per hour (a), as mmol of product/ per gram of dry catalyst/ per hour (b) and mmol of product/ per ml of wet catalyst/ per hour (c).

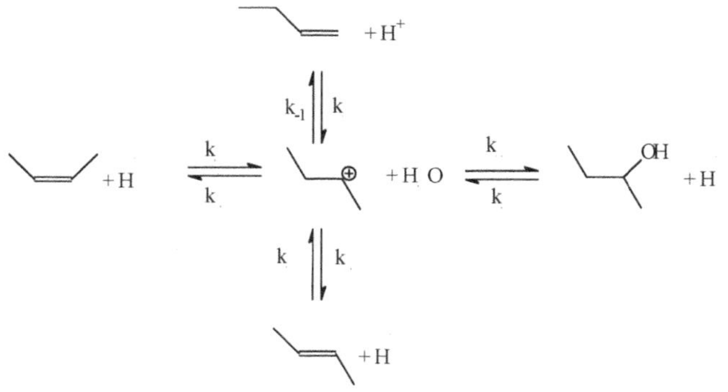

Scheme 2

Table 3 Catalyst activity based on the conversion at 1 hour reaction time.

Resin	Mmol/H+/hr	Mmol/g/hr	Mmol/cc/hr
A	4.0	10.9	4.1
B	2.4	7.8	2.9
C	2.8	8.8	3.0
D	2.4	11.7	3.4
E	2.1	8.8	2.2

The results of the catalyst testing are shown in Table 3. The data listed in the table show, that on a per proton basis, catalyst A (based on 7% DVB) has higher activity as compared to resin materials, crosslinked with 12% DVB. This result is in accord with the finding by Petrus et al.,[3] that at temperatures higher than 120 °C the hydration is under intra particle diffusion limitation and as such, a more flexible polymeric matrix will provide better access to the acidic sites. On a dry weight basis, catalyst D showed the highest activity, which correlates well with the high acid site density found for this resin (Table 2). On a catalyst volume basis, catalyst A has the best performance characteristics followed by catalyst D.

Acknowledgements

The authors thank Bruce Rosenbaum, Rudolf Weinand, Mark Vandersall, David Lam, Ann Beaulieu, Robert Olsen, and Jim Barrett for useful discussions

References

1. F. Henn, W. Neier, G. Strehlke, and W. Webers, US Pat. 4,831,197, to Deutsche Texaco Aktiengesellschaft (1989).
2. S. M. Mahajani, M. M. Sharma, and T. Sridhar, *Chem. Eng. Sci.*, **56**, 5625 (2001).
3. L. Petrus, R. W. Dee Roo, E. J. Stamhius, and G. E. H. Joosten, *Chem. Eng. Sci.*, **41**, 217 (1986).
4. D. Kallo, R. M. Mihalyi, *Appl. Catal. A*, **121**, 45-56 (1995).
5. M. Prezelj, W. Koog, and M. Dettmer, *Hydrocarbon Processing*, **67/11**, 75 (1988).
6. British Pat. 1,393,594, to Rohm and Haas Company (1975).
7. G. Brandes, W. Neier, and W. Webers, US Patent 4,705,808, to Deutsche Texaco Aktiengesellschaft (1987).
8. M. Hart, G. Fuller, D. R. Brown, C. Park, M. A. Keane, J. A. Dale, C. M. Fougret, and R. W. Cockman, *Catal.Lett.*, **72**, 135 (2001).
9. A. Chakrabarti, M. M. Sharma, *React. Polym.*, **20**, 1 (1993).
10. J. Klein, S. Cao, K. Teng, and H. Widdecke, *Chem. Ing. Tech.*, **61**, 826 (1989).

38. Enhancement of Regioselectivity in the Gas-phase Methylation of m-Cresol with Multifunctional Heterogeneous Catalysts

F. Cavani,[1*] L. Dal Pozzo,[1] C. Felloni,[1] L. Maselli,[1] D. Scagliarini,[1] C. Flego,[2] and C. Perego[2]

[1]Dipartimento di Chimica Industriale e dei Materiali, Viale Risorgimento 4, 40136 Bologna, Italy. *INSTM, research unit of Bologna.
[2]EniTecnologie SpA, via Maritano 26, 20097 S. Donato MI, Italy

cavani@ms.fci.unibo.it

Abstract

Mg/Me (Me=Al, Fe) mixed oxides prepared from hydrotalcite precursors were compared in the gas-phase m-cresol methylation in order to find out a relationship between catalytic activity and physico-chemical properties. It was found that the regio-selectivity in the methylation is considerably affected by the surface acid-basic properties of the catalysts. The co-existence of Lewis acid sites and basic sites leads to an enhancement of the selectivity to the product of ortho-C-alkylation with respect to the sole presence of basic sites. This derives from the combination of two effects. (i) The H^+-abstraction properties of the basic site lead to the generation of the phenolate anion. (ii) The coordinative properties of Lewis acid sites, through their interaction with the aromatic ring, make the mesomeric effect less efficient, with predominance of the inductive effect of the $-O^-$ species in directing the regio-selectivity of the C-methylation into the ortho position.

Introduction

The literature on basic- and acid-catalyzed alkylation of phenol and of its derivatives is wide [1,2], since this class of reactions finds industrial application for the synthesis of several intermediates: 2-methylphenol as a monomer for the synthesis of epoxy cresol novolac resin; 2,5-dimethylphenol as an intermediate for the synthesis of antiseptics, dyes and antioxidants; 2,6-dimethylphenol used for the manufacture of polyphenylenoxide resins, and 2,3,6-trimethylphenol as a starting material for the synthesis of vitamin E. The nature of the products obtained in phenol methylation is affected by the surface characteristics of the catalyst, since catalysts having acid features address the electrophilic substitution in the ortho and para positions with respect to the hydroxy group (steric effects in confined environments may however affect the ortho/para-C-alkylation ratio), while with basic catalysts the ortho positions become the preferred ones, due to the repulsive effect between the aromatic ring and oxygen atoms of the metal oxide. This repulsive effect forces the phenolate to adopt a vertical position with respect to the catalyst surface, making the para position further from the active sites and hence less available for nucleophilic attack onto activated methanol [3]. In the present work we compare the performance in the gas-phase m-cresol methylation of Mg/Al and Mg/Fe mixed oxides, both prepared starting from hydrotalcite precursors. The peculiarities of these two systems are (i) the presence of only basic sites in the case of Mg/Al/O, and (ii) the co-existence of Lewis acid and basic sites in the case of Mg/Fe/O [4].

Results and Discussion

Mg/Al/O and Mg/Fe/O catalysts have been prepared by thermal treatment of the corresponding hydrotalcite precursors. Main characteristics of the catalysts are reported in Table 1. Specifically, the surface area of the calcined materials, and the amount of acid and of basic sites, as determined by pyridine and by carbon dioxide adsorption, respectively, are reported. For reference, the main characteristics of MgO, Al_2O_3 and Fe_2O_3 are also given. In both Mg/Al/O and Mg/Fe/O a solid solution of either Al^{3+} or Fe^{3+} in MgO (periclase) develops, in which the replacement of trivalent cations for Mg^{2+} in the lattice co-generates cationic vacancies. Oxygen anions having unsaturated ligands act as strongly nucleophilic sites, while O anions coordinated to both Mg and to the guest cation are less basic, due to the higher electronegative property of Al and Fe as compared to that of Mg. Furthermore, the presence of acid sites in the Mg/Fe/O catalyst can be associated to the coordinative properties of Fe [4].

Table 1 Main characteristics of catalysts used in the present work.

Catalyst	Mg/Me	Structure	Acid sites, $\mu mol/g^a$	Basic sites, $\mu mol/g^b$	Surface area, m^2/g
Mg/Al/O	1.8	defective MgO: $Mg_{0.64}Al_{0.36}O_{1.18}$	-	263	185
Mg/Fe/O	2.0	defective MgO: $Mg_{0.67}Fe_{0.33}O_{1.17}$	142	159	149
Mg/O	∞	Periclase	-	356	206
Al/O	0	γ-Al_2O_3	-	264	203
Fe/O	0	Hematite	nd^c	151	32

[a] by pyridine adsorption; [b] by CO_2 adsorption; [c] not detectable, due to the poor transparency of this material in the IR region.

The distribution of basic strength of reference and mixed oxides was different, as experimentally determined in [4]. While in the reference oxides the majority of basic sites was strong, in the case of mixed oxides (Mg/Al/O and Mg/Fe/O) the main fraction of basic sites had medium strength. Acid sites of Mg/Fe/O sample were exclusively medium-strength Lewis-type [4].

The catalytic activity of Mg/Al/O sample in m-cresol gas-phase methylation is summarized in Figure 1, where the conversion of m-cresol, and the selectivity to the products are reported as a function of the reaction temperature. Products were 3-methylanisole (3-MA, the product of *O*-methylation), 2,3-dimethylphenol and 2,5-dimethylphenol (2,3-DMP and 2,5-DMP, the products of ortho-*C*-methylation), 3,4-dimethylphenol (3,4-DMP, the product of para-*C*-methylation), and poly-*C*-methylated compounds. Other by-products which formed in minor amounts were dimethylanisoles, toluene, benzene and anisole (not reported in the Figure).

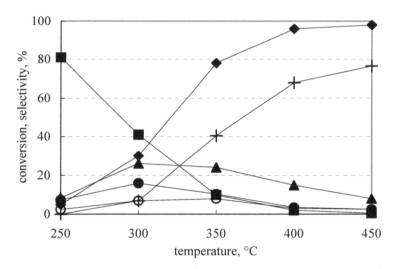

Figure 1 Effect of temperature on catalytic performance of Mg/Al/O catalyst in m-cresol methylation; m-cresol conversion (υ), selectivity to 3-MA (ν), 2,3-DMP (λ), 2,5-DMP (σ), 3,4-DMP (μ), polyalkylates (:).

Data of Figure 1 show that at low temperature the prevailing product was 3-MA, while the formation of *C*-methylated compounds was considerably lower. On increasing the reaction temperature, and hence the conversion of m-cresol, the selectivity to 3-MA rapidly decreased, while that to *C*-methylated compounds correspondingly increased. Amongst the latter, products of ortho *C*-methylation (2,3- and 2,5-DMP) were clearly favored with respect to that of para methylation (3,4-DMP). However, the formation of mono-*C*-methylated compounds was accompanied by consecutive polyalkylations, and the corresponding products became soon the prevailing ones, when the reaction temperature was raised above 350°C.

Results suggest that the activation energies associated to the two reactions of *O*- and *C*-methylation are different. The *O*-methylation is kinetically favored at low temperature, while the *C*-methylation becomes faster when the reaction temperature is increased. However, a contribution of the consecutive reaction occurring on 3-MA to yield the *C*-methylated compounds can not be excluded. This latter reaction may occur either via intramolecular rearrangement, or via intermolecular reaction, in which the methyl group associated to the methoxy moiety in 3-MA constitutes the alkylating agent. In order to distinguish between these different hypotheses, 3-MA was fed into the reactor at the same reaction conditions employed for tests of m-cresol methylation, but in the absence of methanol. In Figure 2 the conversion of 3-MA and the selectivity to the products are plotted as a function of the reaction temperature.

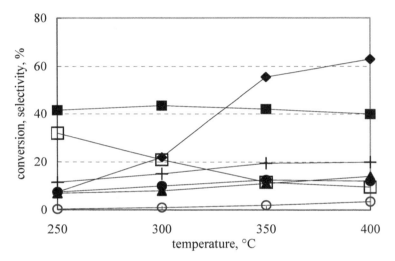

Figure 2 Effect of the reaction temperature on catalytic performance of Mg/Al/O catalyst in 3-MA transformation. 3-MA conversion (υ), selectivity to m-cresol (v), 2,3-DMP (λ), 2,5-DMP (σ), 3,4-DMP (μ), DMAs (o), polyalkylates (:).

At low temperature, an almost equimolar amount of m-cresol and of DMAs were produced (selectivity of 41% and 32%, respectively), with minor amounts of DMPs. Therefore, it is likely that the primary mechanism for the transformation of 3-MA consists of the intermolecular methylation with co-generation of m-cresol and DMA. Also, the contribution of hydrolysis of 3-MA to yield m-cresol can not be excluded. On increasing the reaction temperature, the selectivity to DMAs decreased, with a corresponding increase in the formation of DMPs and polyalkylates. This is likely due to the reactions between DMAs and m-cresol to yield DMPs, and between two molecules of DMAs (a sort of intermolecular disproportionation) to yield polyalkylates and DMPs. The selectivity to m-cresol remained approximately constant on increasing the reaction temperature.

The results of Figure 1 may in part be attributed to a consecutive reaction on 3-MA, the contribution of which becomes relevant at high m-cresol conversion. However, the comparison between the distribution of products obtained when starting either from m-cresol (Figure 1) or from 3-MA (Figure 2) evidences some important differences, in particular the selectivity to polyalkylates. This considerably increases in m-cresol methylation at above 300°C, and can be attributed to the consecutive *C*-methylation occurring on DMPs and 3-MA. This reaction becomes the dominant one at high temperature.

The effect of temperature on the catalytic performance of Mg/Fe/O is reported in Figure 3. The behavior was quite different from that of the Mg/Al/O catalyst. The conversion of m-cresol with Mg/Fe/O was always lower than that with Mg/Al/O. The selectivity to 3-MA was almost negligible in the whole range of temperature. The selectivity to polyalkylates and to 3,4-DMP was also much lower than that observed with Mg/Al/O. Therefore, the catalyst was very selective to the products of ortho-*C*-methylation, 2,3-DMP and in particular 2,5-DMP. This behavior has to be attributed to specific surface features of Mg/Fe/O catalyst, that favor the ortho-*C*-methylation with respect to *O*-methylation. A different behavior of Mg/Al/O and Mg/Fe/O catalysts, having Mg/Me atomic ratio equal to 4, has also been recently reported by other authors for the reaction of phenol and o-cresol methylation [5]. The effect was attributed to the different basic strength of catalysts. This explanation does not hold in our case, since a similar distribution of basic strength was obtained for Mg/Al/O and Mg/Fe/O catalysts [4].

The reaction pathway for the gas-phase methylation of m-cresol, as inferred from catalytic data here reported, can be summarized as shown in Scheme 1. Methanol and m-cresol react through two parallel reactions, yielding either 3-MA or DMPs. The relative contribution of the two reactions is a function of the physico-chemical features of the catalysts, and of the reaction temperature as well, *C*-methylation being kinetically favored at high temperature. Consecutive reactions occur on 3-MA, which acts as a methylating agent yielding DMPs, DMAs and polyalkylates (with co-production of m-cresol in all cases) by reaction with m-cresol, 3-MA and DMPs, respectively. Consecutive reactions may also occur on DMPs to yield polyalkylates.

Selectivity parameters can be used to compare the catalytic performance of the different catalysts, and to find relationships between catalysts performance and physico-chemical features. Specifically, the following parameters were chosen: (a) the *O/C*-methylation ratio, that is the ratio between the selectivity to 3-MA and that to 2,3-DMP+2,5-DMP+3,4-DMP; (b) the ortho/para-*C*-alkylation ratio, that is the ratio between the selectivity to 2,3-DMP+2,5-DMP and the selectivity to 3,4-DMP; (c) the 2,5-DMP/2,3-DMP selectivity ratio. Table 2 compares these parameters for MgO, Mg/Al/O and Mg/Fe/O catalysts. Data were reported at 30% m-cresol conversion, thus under conditions of negligible consecutive reactions. In this way it is possible to compare the ratio of the sole parallel reactions, and to search for relationships between catalyst features and reactivity. The comparison evidences that the regioselectivity of methylation is dramatically affected by the catalyst features. MgO was poorly active, despite the high surface area and the high number of basic sites (Table 1). This is due to the poisoning of the majority of basic sites, caused by the almost irreversible interaction of the strong basic sites with acid molecules present in the reaction environment [4]. Since the *O/C*-ratio is also affected by the sites strength [1,2], strong basic sites make the *C*-methylation preferred over the *O*-methylation [6], and consequently the *O/C*-methylation ratio was largely lower than 1.

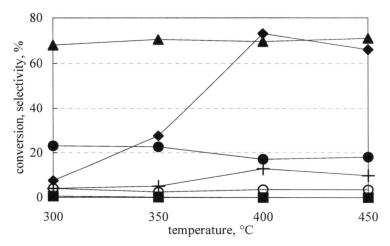

Figure 3 Effect of temperature on catalytic performance of Mg/Fe/O catalyst in m-cresol methylation. Symbols as in Figure 1.

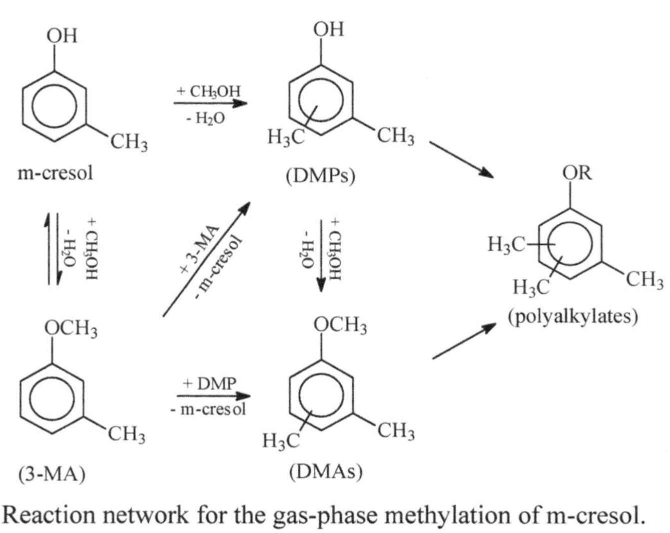

Scheme 1 Reaction network for the gas-phase methylation of m-cresol.

Table 2 Summary of catalytic performance at 30% m-cresol conversion.

Catalyst	O/C-methylation ratio	ortho/para- C-methylation ratio	2,5-DMP / 2,3-DMP selectivity ratio
Mg/Al/O	0.85	6.0	1.6
Mg/Fe/O	0.002	37.2	3.1
Mg/O	0.02	7.3	3.0

On the contrary, Mg/Al/O and Mg/Fe/O had a comparable number of basic sites, with prevalence of medium-strength sites. In this case the large difference in selectivity parameters is attributed to the presence of the Lewis acid sites. It is proposed here that the contemporaneous presence of (i) medium-strength basic sites, and (ii) electrophilic Lewis acid sites, may address the reaction pathway towards the C-methylation. The medium-strength basic sites are responsible for the generation of the phenolate ion by proton abstraction from the hydroxyl moiety of m-cresol. The interaction of the aromatic ring with Lewis sites associated to the Fe^{3+} cations make the high electron density of the – O^- ion shift towards the aromatic ring by inductive effect, lowering its nucleophilic character. The same phenomenon can also explain the very high ortho/para-C-methylation ratio of the Mg/Fe/O catalyst. In fact, the interaction of the aromatic ring with the Lewis acid site decreases the mesomeric effect, which would otherwise favor the delocalization of the negative charge in the para position. The inductive effect becomes dominant, and causes the electron density at the C atoms in the ortho positions to be quite relevant. These two positions are finally those which perform the nucleophilic attack onto methanol adsorbed on adjacent sites, with preferential formation of 2,3- and 2,5-DMPs. Less clear is the reason for the considerable effect on the 2,5-DMP/2,3-DMP selectivity ratio; in fact the Mg/Fe/O catalyst is much more selective to 2,5-DMP than Mg/Al/O.

In order to explain the experimental differences in the regio-selectivity for C-methylated products, we have calculated with a semi-empirical approach the enthalpy of formation for the different transition states which develop by (i) reaction between methanol and the anion generated by proton abstraction in m-cresol to yield the different C-methylated compounds (basic mechanism), and (ii) reaction between methanol and the anion generated by proton abstraction in m-cresol, when a coordination of the π-electrons in the aromatic ring establishes with a generic Lewis acid site (in this case, the latter has been approximated by $AlCl_3$) (bi-functional mechanism). In both cases the reaction conditions were approximated by assuming that reactants were in the gas-phase, in a vacuum environment, and that therefore not any specific interaction with the catalyst (i.e., adsorption geometry on catalyst surface) developed. Table 3 summarizes the results of the simulation, and also reports the experimental selectivity to the different products of C-methylation, as inferred from results obtained at low m-cresol conversion on the basic (Mg/Al/O) and on the bi-functional catalyst (Mg/Fe/O).

Table 3 Results of the calculation of energies for the formation of transition states in basic and bi-functional mechanism, and experimental selectivity to the primary products of C-methylation, in gas-phase methylation of m-cresol.

Mechanism (catalyst)	Product of C-methylation	Experimental selectivity, %	Energy of formation of the transition state, kcal/mole
Basic (Mg/Al/O)	2,5-DMP	8.5[a], 26[b]	-33.7
	2,3-DMP	7.5[a], 16[b]	-33.5
	3,4-DMP	2.5[a], 7[b]	-34.5
Bi-functional (Mg/Fe/O)	2,5-DMP	68[a], 70.5[b]	-55.8
	2,3-DMP	23[a], 22.5[b]	-49.2
	3,4-DMP	4[a], 2.5[b]	-49.2

[a]conversion $\approx$ 10%. [b]conversion $\approx$ 30%

In the absence of any localized adsorptive effect on the catalyst surface, the ortho and para positions in the aromatic ring are calculated to be substantially equivalent in regard to the nucleophilic attack onto methanol. When the interaction between the aromatic ring and the Lewis site is also taken into account, the transition state which develops for the nucleophilic attack of position 6 in m-cresol becomes more stable. This result agrees with the most significant effect on the selectivity experimentally found, that is the considerable improvement in selectivity to 2,5-DMP with respect to both 2,3-DMP and 3,4-DMP, and with the overall increase of the ortho/para-C-methylation ratio. Moreover, the vertical adsorption [3] makes the experimental ortho/para-C-methylation ratio higher than 2 (with a 2,5-DMP/3,4-DMP selectivity ratio higher than 1), and the experimental 2,5-DMP/2,3-DMP selectivity ratio higher than 1. This ratio is influenced by the steric effect of the methyl group which makes position 2 in m-cresol less available for the reaction.

This interpretation is further confirmed by the results obtained in the liquid-phase methylation of m-cresol with Mg/Fe mixed oxides having different Fe contents: $Mg_{1-x}Fe_xO_{1+1/2x}$. In these catalysts the number of Lewis acid sites was proportional to the Fe content [4]. Figure 4 plots the selectivity parameters and the number of Lewis acid sites as a function of the Mg/Fe ratio in catalysts (selectivity was calculated at very low m-cresol conversion, thus in the absence of any consecutive reaction).

Increasing the number of acid sites, the C-methylation becomes largely preferred over the O-methylation (the O/C-methylation ratio increased), while between the two ortho-C-methylated compounds, 2,5-DMP becomes more preferred than 2,3-DMP, in agreement with the calculated energies for the corresponding transition states. Disagreement between results obtained in gas- and liquid-phase methylation with the Mg/Fe/O catalyst (Table 2) concerns the ortho/para-C-methylation ratio (which in liquid phase was not varied when the Fe content was increased), and the 2,5-DMP/2,3-DMP ratio (which in gas phase was not different for MgO and Mg/Fe/O). These discrepancies may be attributed to the role of reaction temperature, and to the different contributions of strong basic sites in the gas- and liquid-phase conditions.

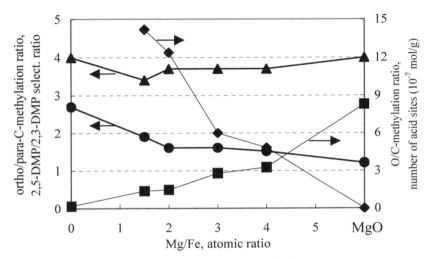

Figure 4 Effect of the Mg/Fe atomic ratio in Mg/Fe/O catalysts on the number of acid sites, as determined by pyridine adsorption (υ), on the *O/C*-methylation ratio (v), on the ortho/para-*C*-methylation ratio (σ), and on the 2,5-DMP/2,3-DMP selectivity ratio (λ), in the liquid-phase methylation of m-cresol [4].

Experimental Section

Mg/Al and Mg/Fe mixed oxides were prepared following the conventional procedure of co-precipitation, as described elsewhere [7]. The dried samples were calcined at 450°C for 8 hours. Catalysts were characterized by means of (i) X-ray diffraction (Philips PW 1050/81), (ii) surface area measurement (single point BET, Sorpty 1700 Carlo Erba), (iii) adsorption and thermal-programmed-desorption (TPD) of CO_2 (PulseChemisorb 2705, Micromeritics), (iv) atomic absorption spectroscopy (Philips PU 9100), and (v) adsorption at 21°C and stepwise desorption (1h, dynamic vacuum) of pyridine, followed by FT-IR spectroscopy (Perkin Elmer, 2000). The gas-phase catalytic tests were carried out in a continuous-flow reactor operating at atmospheric pressure in a temperature range between 250 and 450°C. A mixture of m-cresol and methanol (1/5 molar ratio) was injected using a syringe pump in a N_2 flow of 60 mL/min (N_2/reactants = 5/1 molar ratio). The catalytic bed contained 1.5 g catalyst (30-60 mesh) which resulted in a W/F ratio of 50 g·h·mol^{-1}. The reaction products were collected during 60 min time-on-stream. The reaction mixture (liquid and gas phases) was analyzed by GC, with a HP-5 capillary column.

Catalysts were activated before reaction at 450°C in N_2 flow for 3 hours. Software Spartan Pro has been used for the quantum chemical semi-empirical (type AM 1) calculation of transition states energies.

References

1. R. F. Parton, J. M. Jacobs, D. R. Huybrechts, and P. A. Jacobs, *Stud. Surf. Sci. Catal.*, **46**, 163 (1989).
2. K. Tanabe and W. F. Hölderich, *Appl. Catal. A*, **181**, 399 (1999).
3. K. Tanabe, *Stud. Surf. Sci. Catal.*, **20**, 1 (1985).
4. M. Bolognini, F. Cavani, C. Felloni, D. Scagliarini, C. Flego and C. Perego, *Chem. Ind.* (Dekker), **89**, (*Catal. Org. React.*), 115-127 (2002).
5. S. Velu and C.S. Swamy, *Appl. Catal.*, A, **162**, 81 (1997).
6. M. Bolognini, F. Cavani, D. Scagliarini, C. Flego, C. Perego, and M. Saba, *Catal. Today*, **75**, 103 (2002).
7. F. Cavani, F. Trifirò, and A. Vaccari, *Catal. Today*, **11**, 173 (1991).

39.   Hydroxymethylation of 2-Methoxyphenol with Formaldehyde to Yield Vanillic Alcohols: A Comparison Between Homogeneous and Heterogeneous Acid Catalysis

M. Ardizzi, F. Cavani, L. Dal Pozzo, and L. Maselli

Dipartimento di Chimica Industriale e dei Materiali, Viale Risorgimento 4,
40136 Bologna, Italy

cavani@ms.fci.unibo.it

Abstract

The hydroxymethylation of 2-methoxyphenol (guaiacol) with aqueous solutions of formaldehyde (formalin) to yield 3-methoxy-4-hydroxybenzyl alcohol (p-vanillol) has been carried out under both homogeneous conditions, at acid pH, and with a H-mordenite zeolite as the heterogeneous catalyst. In the former case, good selectivity to vanillic alcohols could be achieved only when a pH was used at which the conversion of guaiacol was low. The use of the zeolite led to high selectivity to vanillol, due to the low formation to undesired diaryl by-products even at high guaiacol conversion. This was attributed to the zeolite hydrophilic/hydrophobic properties, which made possible the selective interaction with specific reactants, depending on (i) molecules polarity, and (ii) zeolite Si/Al ratio.

Introduction

The hydroxyalkylation of activated arenes (containing functional groups such as methoxy or hydroxy groups) [1,2] with aldehydes and ketones is a reaction of interest for the production of drugs, polymers, and food additives. Typical for these reactions are the condensation of a carbonyl compounds with arenes, aimed at the production of diarylmethanes. The hydroxymethylation of 2-methoxyphenol (guaiacol), represents one step in the multi-step synthesis of the 3-methoxy-4-hydroxybenzaldehyde (vanillin; Scheme 1), an environmentally friendly process for the production of this important food additive [3,4]. Hydroxyalkylations are usually catalyzed by Lewis type acids, like $AlCl_3$, and by mineral Brönsted acids. Some papers and patents have proposed the use of zeolitic materials as catalysts for this reaction [4-7]. Solid acid materials are highly desirable catalysts, since the environmental impact of the process takes advantage from an easier separation of the catalyst, the absence of liquid wastes containing inorganic salts, and fewer corrosion problems. In the present work we compare the performance of homogeneous and heterogeneous catalysis; in the latter case, H-mordenites have been used, which are claimed to be the optimal catalyst for the reaction of liquid-phase hydroxymethylation of guaiacol [4-6].

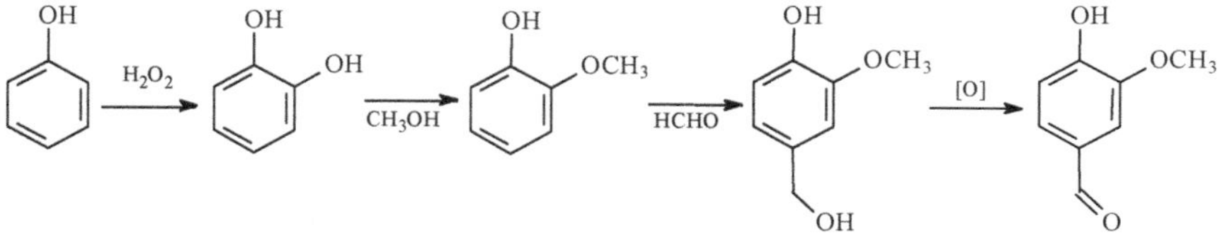

Scheme 1 Multi-step synthesis of vanillin.

Results and Discussion

Figure 1 plots the effect of the pH of reaction on the yields to products under homogeneous conditions. The top figure reports results obtained with commercial formalin solution (containing 36 wt.% formaldehyde and 15% wt. methanol), while the bottom figure shows results with a non-standard formalin solution having approximately 1 wt.% of methanol. Products identified were: (*a*) isomers of vanillic alcohols, o-vanillol (2-hydroxy-3-methoxybenzyl alcohol), p-vanillol (3-methoxy-4-hydroxybenzyl alcohol), and m-vanillol (only 3-hydroxy-4-methoxy benzyl alcohol); (*b*) monoaryl by-products, having molecular weight MW 168 and 198, obtained by reaction of vanillols (mainly p-vanillol) with methanol or by acetalization of vanillol with hemiformal, respectively; (*c*) diaryl by-products, mainly diarylmethanes obtained by reaction of one vanillol with one guaiacol, having MW 260, and minor amounts of diarylmethanes having MW 290 obtained by reaction between two vanillol molecules. Traces of other by-products having MW 138 and 152 were assigned to methylmethoxyphenols and vanillaldehydes.

At pH between 2 and 6, no conversion of guaiacol was detected; this is due to the fact that at pH higher than 2-3 formaldehyde is not protonated, and thus is not activated for the electrophilic substitution on the aromatic ring. At pH < 2, the guaiacol conversion was not affected by methanol content in solution; the effect of methanol was instead remarkable on the distribution of products, when the pH was equal to 1. At pH 2, vanillols were the main products, and only traces of monoaryl and diaryl by-products were found; the overall conversion was very low. At pH 1 the conversion was close to 35%, both in the presence and in the absence of methanol; however, when the commercial formalin solution was used, an equimolar amount of vanillols and of monoaryl by-products were found. The latter compounds were instead substantially absent when the methanol-lean formalin solution was used, with a corresponding higher yield to vanillols. Under these pH conditions, however, the prevailing by-products were diarylmethanes; the selectivity to these compounds was substantially unaffected by the presence of methanol. At pH 2, p-vanillol was favored over the ortho isomer (para/ortho selectivity ratio was equal to 3-4), while at pH 1 the selectivity ratio between the two isomers was close to 1, due to the transformation of p-vanillol to diaryl by-products [7].

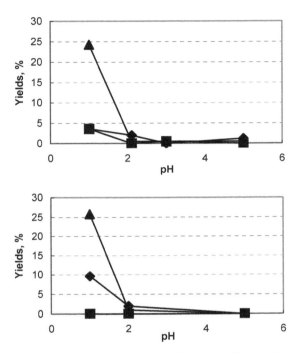

Figure 1 Effect of the solution pH on yields to vanillols (υ), monoaryl by-products (v) and diaryl by-products (σ). Top: formalin with 15 wt.% methanol. Bottom: formalin with 1 wt.% methanol. The pH has been varied by addition of controlled amounts of H_2SO_4. Reaction temperature 80°C, reaction time 1 h, molar ratio formaldehyde/guaiacol 18.

The results indicate that in homogenous medium it is not possible to reach high guaiacol conversion. Under the conditions which are necessary to activate formaldehyde, the condensation reaction between vanillols or between p-vanillol and guaiacol is very quick. This is likely due to the high stability of the benzyl carbocation which forms by protonation of the hydroxymethyl group on p-vanillol [8].

One way to accelerate the rate of formaldehyde activation, while limiting the rate of the undesired consecutive reaction on vanillols, is through an effective control of the relative reactants concentration. This might be achieved by realizing (i) an higher concentration of formaldehyde in the reaction medium, and (ii) a lower concentration of guaiacol, so to decrease the probability of reaction between vanillol and either guaiacol or vanillol, to yield diaryl by-products. It is worth mentioning that the concentration of formaldehyde (or of its hydrated form, methylene glycol) in formalin is usually not higher than 40%, to limit the formation of oligomers. Methanol is added to formalin to hinder the oligomerization process; it is also a co-reactant, which favours the consecutive transformation of vanillol to ethers, and is therefore detrimental for selectivity. On the other hand, the presence of water leads to large volumes of reactants, in which the aromatic substrate is very diluted (due to the low solubility of guaiacol in water). Therefore, the optimal ratio between guaiacol and formaldehyde is different from that one which indeed can be achieved by using the aqueous solution of formaldehyde.

Results obtained with H-mordenites having different Si/Al ratio are summarized in Figure 2, for tests run under the same reaction conditions as for Figure 1, with both methanol-rich and methanol-lean formalin solutions. On increasing the zeolite Si/Al ratio, that is on decreasing the Al content and hence the number of acid sites, the conversion of guaiacol decreased. This indicates that the main parameter affecting the catalytic activity was not the acidity, but rather the hydrophilic properties of the material. Higher Al contents mean higher affinity for polar molecules, and this leads to a competition of methanol/water with formaldehyde for the protonation on active sites. Hence more hydrophobic materials give higher rate of guaiacol conversion, due to the lower competition effect. On the other hand, a higher hydrophobicity means a higher concentration of aromatics, and the consequence of this is a higher probability for the formation of diaryl by-products. This explains the higher selectivity to diaryl compounds for the catalyst having the highest Si/Al ratio. The H-mordenite having the intermediate Si/Al ratio represents the best compromize between the two opposite effects; it gave high conversion while keeping high selectivity to vanillols.

Methanol played a role not only on selectivity but also on activity. With the methanol-lean solution the conversion was higher than with the commercial, methanol-rich solution. This confirms that methanol competes with formaldehyde for adsorption on active sites in the zeolite cavities. The ratio between isomers was not very different from that achieved at pH 2 under homogeneous conditions.

The results indicate that the zeolite can selectively extract specific compounds from the reaction medium, due to the different affinity towards each of them. This makes possible to develop reactant concentrations inside pores which are different from the bulk ones. This property is a function of the zeolite hydrophobic characteristics, which are affected by the Si/Al ratio. The best zeolite is that one which does not interact too strongly neither with more polar molecules, so to allow activation of formaldehyde to proceed faster, nor with the least polar ones. The intermediate Si/Al ratio in H-mordenites is able to develop the optimal concentration ratio between reactants inside the pores, and to reach the highest yield to vanillols.

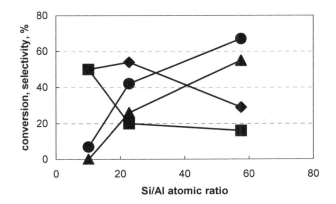

Figure 2 Conversion of guaiacol (λ), selectivity to vanillols (υ), to monoaryl by-products (v) and to diaryl by-products (σ) as functions of the Si/Al ratio of H-mordenites. Reaction time, 2 h; other reaction conditions as in Figure 1. H-mordenites were supplied by Engelhard.

References

1. A. Corma and H. Garcia, *J. Chem. Soc. Dalton Trans.*, 1381 (2000).
2. R.A. Sheldon and H. van Bekkum, Fine Chemicals through Heterogeneous Catalysis, R.A. Sheldon and H. van Bekkum, Eds., Wiley-VCH, 2001, p. 1.
3. P. Metivier, Fine Chemicals through Heterogeneous Catalysis, R.A. Sheldon and H. van Bekkum, Eds., Wiley-VCH, 2001, p. 173.
4. C. Moreau, S. Razigade-Trousselier, A. Finiels, F. Fazula, and L. Gilbert, WO Pat. 96/37452 to Rhone-Poulenc Chimie (1996).
5. C. Moreau, F. Fajula, A. Finiels, S. Razigade, L. Gilbert, R. Jacquot, and M. Spagnol, *Chem. Ind.* (Dekker), **75**, (*Catal. Org. React.*), 51 (1998).
6. F. Cavani and R. Mezzogori, *Chem. Ind.* (Dekker), **89**, (*Catal. Org. React.*), 483 (2002).
7. F. Cavani, L. Dal Pozzo, L. Maselli, and R. Mezzogori, *Stud. Surf. Sci. Catal.*, **142**, 565 (2002).

40. The Aldol Condensation of Acetone Over a CsOH/SiO₂ Solid Base Catalyst

Arran S. Canning, S. David Jackson, Eilidh McLeod, and Elaine M. Vass

Department of Chemistry, The University, Glasgow G12 8QQ, Scotland

Abstract

The solid base catalysed aldol condensation of acetone was performed over a CsOH/SiO₂ catalyst using a H₂ carrier gas. The products observed were diacetone alcohol, mesityl oxide, phorone, iso-phorone and the hydrogenated product, methyl isobutyl ketone. Deuterium tracer experiments were performed to gain an insight into the reaction mechanism. A mechanism is proposed.

Introduction

The aldol condensation reactions of aldehydes and ketones are commonly used in organic synthesis reactions for preparing chemicals containing a double bond conjugated with a carbonyl group. Industrially this process is carried out in the liquid phase with homogeneous alkaline bases. However, solid base catalysts are becoming more interesting due to several advantages that they offer. Solid base catalysts allow the condensation reactions to occur in the vapour phase, without the production of large basic waste streams. As well as this heterogeneous base catalysis allows an easier separation of catalyst from products compared to homogenous base catalytic system (1,2).

The aldol condensation of acetone to diacetone alcohol is the first step in a three-step process in the traditional method for the production of methyl isobutyl ketone (MIBK). This reaction is catalysed by aqueous NaOH in the liquid phase. (3) The second step involves the acid catalysed dehydration of diacetone alcohol (DAA) to mesityl oxide (MO) by H₂SO₄ at 373 K. Finally the MO is hydrogenated to MIBK using Cu or Ni catalysts at 288 – 473 K and 3- 10 bar (3).

In liquid phase reactions the base catalysed reaction may only proceed as far as its self-condensation to DAA with minimal dehydration to MO, however when operating in the gas phase at elevated temperatures, secondary reactions can take place (4).

This paper investigates the acetone condensation reaction in the vapour phase over a CsOH/SiO₂ solid based catalyst over a range of reaction temperatures using hydrogen and deuterium as carrier gases.

Experimental Section

The solid base catalysts were prepared by dissolving $Cs(NO_3)_2$ (Aldrich, 99%) in the minimum amount of distilled water before addition to the silica support by spray impregnation; a method used to give a high dispersion of the metal salt on the support. The amount of each precursor added was calculated in order to give a 10% loading of metal on each catalyst. The catalyst was then dried in an oven overnight at 373 K. Prior to the reaction the catalyst was calcined *in situ* in a flow of N_2 (BOC, O_2 free N_2) at 10 cm^3 min^{-1} for 2 hours at 723 K.

Carbon dioxide chemisorptions were carried out on a pulse-flow microreactor system with on-line gas chromatography using a thermal conductivity detector. The catalyst (0.4 g) was heated in flowing helium (40 cm^3min^{-1}) to 723 K at 10 $Kmin^{-1}$. The samples were held at this temperature for 2 hours before being cooled to room temperature and maintained in a helium flow. Pulses of gas ($\sim1.53 \times 10^{-5}$ moles) were introduced to the carrier gas from the sample loop. After passage through the catalyst bed the total contents of the pulse were analysed by GC and mass spectroscopy (ESS MS).

After the catalyst was saturated with carbon dioxide, a temperature programmed desorption (TPD) was carried out by heating the sample in helium (40 cm^3min^{-1}) from room temperature to 873 K (10 $Kmin^{-1}$). The mass spectrometer was used to follow water (mass 18), carbon monoxide (mass 28), carbon dioxide (mass 44) and oxygen (mass 32).

The condensation reaction was carried out in a continuous flow microreactor. The liquid reactant, acetone (Fisher Scientific HPLC grade >99.99%), was pumped via a Gilson HPLC 307 pump at 5 ml hr^{-1} into the carrier gas stream of H_2 (50 cm^3 min^{-1}) (BOC high purity) where it entered a heated chamber and was volatilised. To ensure good mixing between the hydrogen and the acetone vapour the gases were taken through a static mixer. The carrier gas and reactant then entered the reactor and passed through the catalyst (0.50g). The reactor was run at 5 bar pressure and at reaction temperatures between 373 and 673 K. Samples were collected in a cooled drop out tank and analysed by a Thermoquest GC-MS fitted with a CP-Sil 5CB column. The deuterium tracer reactions were carried out using the same microreator system, however the hydrogen gas feed was substituted with deuterium (Lynde, 99.98%) at a flow rate of 50 cm^3 min^{-1}.

Results

CO_2 chemisorption experiments on $CsOH/SiO_2$ revealed that 1.6 % of the Cs present on the catalyst existed as basic sites. CO_2-TPD experiments (Figure 1) revealed a major desorption peak at 373 K along with a small peak at 437 K.

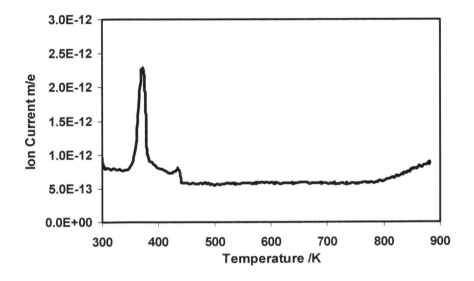

Figure 1 CO_2-Temperature programmed desorption from $CsOH/SiO_2$.

The aldol condensation reaction of acetone was performed over $CsOH/SiO_2$ at a range of reaction temperatures between 373 and 673 K (a typical product distribution is shown in Figure 2). Table 1 displays the conversion of acetone along with the selectivities for the products produced once steady state conditions were achieved. Figure 3 presents the effect of temperature on the yield of the products. The activation energy for acetone conversion was calculated to be 24 kJ. mol^{-1}.

Deuterium tracer reactions were performed at 573 K with acetone in a stream of deuterium. Samples were taken throughout the experiment and analysed by GC-MS. The resultant chromatographic peaks were analysed by mass spectroscopy to determine the mass of the products formed during the reaction. Table 2 presents the different isotopic products formed during the reaction at 573 K, along with the mass of the molecular ion observed during a typical reaction under H_2. The majority of the acetone in the samples was in the form C_3H_5DO, at a relative amount of ca. 60 %. $C_3H_4D_2O$ was present at 24 %, $C_3H_3D_3O$ at 8 %, while C_3H_6O itself only existed at 8 %.

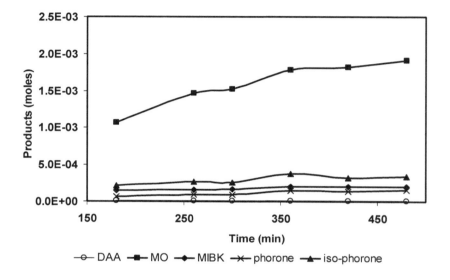

Figure 2 Reaction profile for acetone condensation over Cs/SiO$_2$ at 573 K.

Table 1 Conversion and selectivities for acetone condensation over Cs/SiO$_2$ over a range of temperatures under a hydrogen stream.

Temperature / K	Conversion (%)	Selectivities (%)				
		DAA	MO	MIBK	phorone	iso-phorone
373	0.14	20.1	76.1	3.8	0.0	0.0
473	1.25	0.4	85.0	8.3	4.5	1.8
573	2.9	0.3	70.9	7.9	5.9	15.0
673	2.7	0.1	65.3	11.8	4.4	18.5

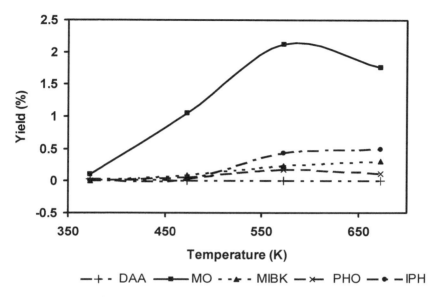

Figure 3 Yields of products formed for acetone condensation reaction as a function of temperature.

Table 2 The molecular ions observed during acetone condensation reactions under H_2 and D_2.

Products	Mass of molecular ion		Relative amount (%) in D_2 reaction
	H_2	D_2	
Acetone	58	58 (d_0)	8
		59 (d_1)	60
		60 (d_2)	24
DAA	116	-	-
MO	98	99 (d_1)	65
		100 (d_2)	25
		101 (d_3)	10
MIBK	100	101 (d_1)	27
		102 (d_2)	31
		103 (d_3)	23
Phorone	138	139/140*/141	
Iso-phororne	138	138/139*/140	

The overall percentage may not total 100% as other isomers were also present in small amounts
* represents the major isomer

Discussion

The self-condensation of acetone over $CsOH/SiO_2$, showed a conversion of acetone to products of less than 3% under the conditions studied. At lower temperatures the products formed were diacetone alcohol (DAA), mesityl oxide (MO) and methyl isobutyl ketone (MIBK). The production of MIBK, the hydrogenation product from MO, was observed as the reaction was performed using H_2 as the carrier gas. The production of DAA and MO is consistent with a number of studies [4-8] for acetone condensation reactions over solid base catalysts. Once the base catalysed self condensation occurs, then at a sufficiently high enough temperature, dehydration takes place to produce MO. The low selectivity towards DAA at higher temperatures (above 473 K), suggests rapid formation of MO and that DAA itself is an unstable intermediate at these temperatures. At elevated temperatures of above 473 K for $CsOH/SiO_2$, the production of phorone and iso-phorone is observed via the addition of an additional acetone molecule. This is consistent with the work by Di Cosimo et al. [8], who characterised the formation of phorone and isophorone as precursors to non-volatile organic compounds. These non-volatile organic compounds resulted in the formation of coke species and hence the deactivation of MgO catalysts.

The reaction mechanism is shown in Figure 4 and is adapted from work by Fiego et al. [9] on the acid catalysed condensation of acetone by basic molecular sieves. The scheme has been modified to include the hydrogenation of mesityl oxide to MIBK. The scheme begins with the self-condensation of acetone to form diacetone alcohol as the primary product. The dehydration of DAA forms mesityl oxide, which undergoes addition of an addition acetone to form phorone that then can cyclise, via a 1,6-Michael addition to produce isophorone. Alternatively, the mesityl oxide can hydrogenate to form MIBK.

The condensation of acetone can also occur over acidic sites as shown by a number of authors [1,9]. Generally, when this occurs other products are formed such as isobutene and acetic acid, by the cracking of DAA. Additionally mesitylene can be formed by the internal 2,7-aldol condensation of 4,6-dimethylhepta-3,5-dien-2-one which is in turn obtained by the aldol condensation of MO with a deprotonated acetone molecule [7, 8]. As these species are not observed we can concluded that any acidic sites on the silica support are playing no significant role in the condensation of acetone.

The CO_2 chemisorption results reveal that the equivalent of only 1.6% of the Cs present on the catalyst is an active basic site, accounting for the low reaction conversions observed. CO_2-TPD shows that the basic sites that do exist are weak, with desorption peaks observed at 373 and 437 K. An investigation into Cs-promoted α-$NiMO_4$ catalysts by Madeira et al. [10] demonstrated with CO_2-

TPD that Cs loadings above 3% showed a significant decrease in the basicity of the catalyst, suggesting the growth of caesium oxide particles.

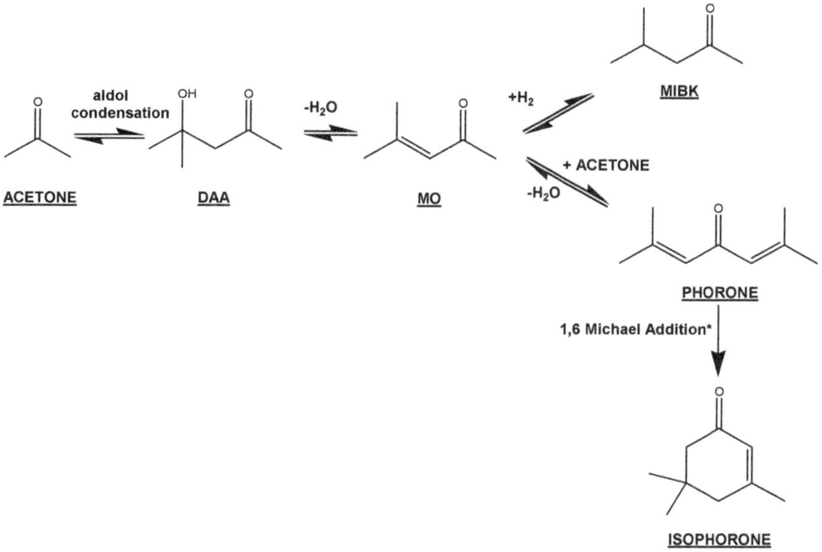

Figure 4 Mechanism for the formation of various species during acetone condensation under H_2 conditions over $CsOH/SiO_2$.

The selectivity of DAA (Table 1) decreases with increasing temperature, which is expected, as there would be a higher level of dehydration to form MO as the temperature increases. The formation of MO reaches a maximum at 473 K, however the selectivity decreases as the temperature increases to 673 K due to the formation of phorone and iso-phorone. The selectivity for MIBK increases with increasing temperature. The iso-phorone/phorone ratio increases with increasing reaction temperature. Therefore, the 1,6-Michael addition of the phorone occurs more readily at increased temperatures. Di Cosimo et al. [8] have implied that the formation of iso-phorone from phorone is irreversible and that it is the stable terminal product for acetone condensation in the vapour phase using solid base catalysts. Our results are in agreement we detected no higher molecular weight species such as tetramers.

The condensation reaction in deuterium revealed that up to 92% of the acetone undergoes deuterium exchange to form various states of deuterated acetone, with C_3H_5DO present as the most populated species at ca. 60%. These results indicate that the exchange process is extremely facile even on a catalyst only containing weakly basic sites. Various deuterated products were observed in all of the reaction products, implying that there is significant H-transfer taking place. It is usual to perform a control experiment using only the support to be able to assess the contribution of the support to the exchange process. A control experiment was performed over the silica support and the acetone underwent no exchange.

Once acetone adsorbs on the catalyst surface, it rapidly undergoes H-D exchange and can either desorb or undergo the subsequent aldol condensation to produce DAA and MO and so on. The major MO product produced is the singly deuterated species, C_6H_9DO (approximately 65 %) with a molecular ion mass of 99, along with this $C_6H_8D_2O$ (~ 25 %) with a mass of 100 is produced. If these two main MO species were to undergo hydrogenation with D_2 the hydrogenated MIBK species would have relative masses of 103 and 104. However, the main MIBK species observed are at masses 101 and 102, indicating that the hydrogen added is protium rather than deuterium. We have attributed this to hydrogenation of MO with protium previously exchanged on the surface from the H-D exchange of acetone, however other causes such as a cage effect could also be attributed to this observation The triply deuterated MIBK isomer present at 23% may arise from the hydrogenation from the triply deuterated MO and possibly by some minor deuteration of C_6H_9DO. Therefore the following reactions represent the likely hydrogenation processes occurring under deuterium conditions. The number in parenthesis represents the molecular weight of the deuterated MIBK species.

$$C_6H_9DO \ + \ 2H \ \rightarrow \ C_6H_{11}DO \qquad (101)$$
$$C_6H_8D_2O \ + \ 2H \ \rightarrow \ C_6H_{10}D_2O \qquad (102)$$
$$C_6H_7D_3O \ + \ 2H \ \rightarrow \ C_6H_9D_3O \qquad (103)$$
$$C_6H_9DO \ + \ 2D \ \rightarrow \ C_6H_{11}DO \qquad (103)$$

There was no evidence of a kinetic isotope effect for the hydrogenation of MO to MIBK, which further supports that the hydrogenation step favours the addition of protium on the catalyst surface as apposed to deuterium. A relatively high H-D exchange (51 %) was observed by Chikán et al. [11] in a similar isotope exchange experiment for deuterium and acetone over Cu-MgO. They also noticed a lower than expected deuterium content in the produced MIBK attributing this to a high dilution in the surface deuterium pool as they were using a high acetone:deuterium ratio. In our system this surface dilution effect is seen even though we have massive excess in D_2 in the system.

Phorone was also detected in a range of isotopic species that formed from the addition of various combinations of isotopic MO and acetone to give D_1, D_2 and D_3 – exchanged phorone, with phorone-D_2 existing as the major product. For a 1,6-Michael addition to occur a strong basic site first abstracts a proton/deuterium from a terminal methyl group, a proton is then donated back to form iso-phorone through ring cyclisation. We would expect that after this

H-D Exchange

$$D_2 \Leftrightarrow 2D(a)$$
$$CH_3COCH_3 \Leftrightarrow CH_3COCH_3 \text{ (a)}$$
$$CH_3COCH_3 \text{ (a)} \Leftrightarrow CH_3COCH_2 \text{ (a)} + H \text{ (a)}$$
$$CH_3COCH_2 \text{ (a)} + D \text{ (a)} \Leftrightarrow CH_3COCH_2D \text{ (a)}$$
$$CH_3COCH_2D \text{ (a)} \Leftrightarrow CH_3COCH_2D \text{ (g)}$$

Condensation and Hydrogenation Products

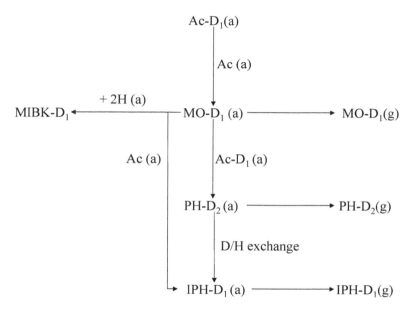

Figure 5 Proposed mechanism for acetone condensation reaction under deuterated conditions.

process a deuterium would be donated to the phorone intermediate to form iso-phorone, resulting in either the isophorone gaining a mass unit or remaining the same. This however, was not been observed, rather a decrease in one mass unit has occurred from the cyclisation of phorone to iso-phorone. This is a puzzling result. One interpretation is that the formation of iso-phorone and phorone occurs by different synthesis routes as originally suggested by Reichle [5] demonstrating that iso-phorone may be formed via three alternative routes under vapour phase conditions. Alternatively, a single D could be selectively lost during the 1,6-Michael addition step.

For the main isotopic species the following mechanism (Figure 5) has been proposed for the production of MO, MIBK, phorone and isophorone from acetone base catalysis under a deuterium stream. A similar mechanism can be invoked for the less common isotopic species.

References

1. K. Tanabe and W. F. Hölderich, *Appl. Catal. A.*, **181**, 399 (1999).
2. F. King, G. J. Kelly, *Catal. Today*, **73**, 75 (2002).
3. G. J. Kelly, F. King, and M. Kett, *Green Chem.* ,**4**, 392 (2002).
4. S. Lippert, W. Baumann, and K. Thomke, *J. Mol. Catal.*, **69**(2), 199 (1991).
5. W. T. Reichle, *J. Catal.*, **63**, 295 (1980).
6. G. J. Kelly and S. D. Jackson, in Catalysis in Application, S. D. Jackson, J. S. J. Hargreaves, and D. Lennon, eds., RSC, Cambridge, 2003, p. 129-135.
7. J. I. Di Cosimo, V. K. Díez, and C. D. Apestegnía, *Appl. Catal. A*, **137**, 149 (1996).
8. J. I. Di Cosimo and C. R. Apesteguía, *J. Mol. Catal. A: Chem.*, **130**, 9177-185 (1998).
9. C. Fiego and C. Perego, *Appl. Catal. A*, **192** , 317 (2000).
10. L. M. Madeira, J. M. Herrmann, F. G. Freire, M. F. Portela, and F. J. Maldonado, *Appl. Cat. A,* **158**, 243 (1997).
11. V. Chikán, A. Molnar, and K. Balazsik, *J. Catal.*, **184**, 134 (1999).

41.   Solid Acid Catalysed Esterification of Amino Acids and Other Bio-Based Acids

N. S. Asthana, Z. Zhang, and Dennis J. Miller

2527 Engineering Building, Department of Chemical Engineering & Material Science, Michigan State University, East Lansing, MI 48824

millerd@egr.msu.edu

Abstract

Esters represent an important class of chemical compounds with applications as solvents, plasticizers, flavors and fragrances, pesticides, medicinals, surfactants, chemical intermediates, and monomers for resins. Recently, esters of amino acids have attracted attention regarding their use as biobased surfactants with excellent adsorption and aggregation properties, low toxicity, and broad biological activity.

Although scattered information is available on amino acid and other acid esterification, detailed kinetic and thermodynamic data for many systems have not been developed to date. We present here some useful data pertaining to kinetics and reaction equilibria for the esterification of various acids using strong cationic exchange resins. Ion exchange resins offer high acidity and ease of separation as advantages over conventional mineral acids for esterification.

Introduction

The traditional source of most organic compounds, including organic acids, is petroleum. In the last a decade or two, there have been concerted efforts to eliminate our dependence on fossil fuels and generate fuel and chemical products from bio-based compounds such as plant carbohydrates. The most successful example of this strategy, at least on a volume basis, is the production of ethanol by fermentation as a fuel additive, although it is recognized that the economic viability of ethanol production is presently dependent on subsidies. Over the last several years our research group has been continuously and aggressively working to develop reaction pathways and separation strategies to produce bio-based chemicals in support of the bio-refinery concept. The higher value addition realized in chemicals production, when integrated across the biorefinery, strengthens overall biorefinery economics and facilitates production of a complete portfolio of biobased fuels and chemicals.

In this paper, we focus on synthesis and application of esters of bio-based organic acids. Organic acid esters are used or have potential for use in many industrial and consumer applications including solvents, paint strippers, surfactants, fragrances, and fuel stabilizers[2]. The chemicals used in these applications need to be ecofriendly in order to comply with growing environmental awareness and concern, and recently bio-based organic acid esters have attracted a deserving attention for further exploration because of their low toxicity. There are several parent organic acids that can be converted to esters with pertinence to the mentioned applications; these include lactic acid, succinic acid, and several amino acids. Lactic acid and succinic acid are both projected to become widely available, low cost acids over the next several years, making them ideal for commodity applications. It is projected that esters of lactic acid and succinic acid can be used as solvents in more than 90% of the solvent applications currently practiced in industry, and thus have potential to capture over 80% of the current market filled by hazardous petroleum-based solvents.[3,4]

In addition to solvent uses, esters of lactic acid can be used to recover pure lactic acid via hydrolysis, which in-turn is used to make optically active dilactide and subsequently polylactic acid used for drug delivery system.[5] This method of recovery for certain lactic acid applications is critical in synthesis of medicinal grade polymer because only optically active polymers with low Tg are useful for drug delivery systems. Lactic acid esters themselves can also be directly converted into polymers, (Figure 1), although the commercial route proceeds via ring-opening polymerization of dilactide.

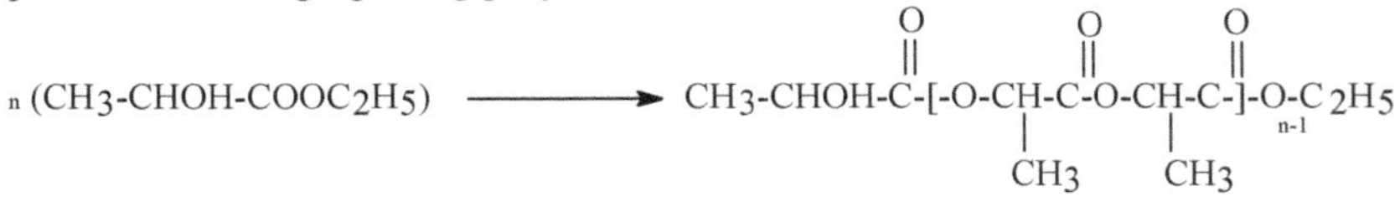

$$n \, (CH_3\text{-}CHOH\text{-}COOC_2H_5) \longrightarrow CH_3\text{-}CHOH\text{-}\overset{O}{\overset{\|}{C}}\text{-}[\text{-}O\text{-}\underset{CH_3}{\underset{|}{CH}}\text{-}\overset{O}{\overset{\|}{C}}\text{-}O\text{-}\underset{CH_3}{\underset{|}{CH}}\text{-}\overset{O}{\overset{\|}{C}}\text{-}]_{n\text{-}1}\text{-}O\text{-}C_2H_5$$

Figure 1

Succinic acid is a dicarboxylic acid and thus has potential to be used in two major ways depending upon the nature of alcohol used in the process. Alcohols containing only one hydroxyl group can be used to synthesis succinate diesters, having excellent solvent properties for a wide range of industrial and consumer use (including commercial paint strippers). On the other hand, poly-esterification with an alcohol having two or more hydroxyl groups can be done to synthesize co-polymers with low Tg value.[6-7] Controlled polymerization also results in low molecular weight compounds having surfactant properties, suitable for use in detergents. The most effective catalyst studied for succinic acid "poly"-esterification was found to be $SnCl_2$.[6] There are also reports on applications of bis-(sulfophenyl) phosphonate-formaldehyde resin for succinic acid esterification, but not much information has been provided on kinetics and thermodynamics of these esterification reaction.

Amino acids are also readily available and relatively inexpensive chemical reagents for industrial use. The recent trend to develop products with multifunctional characteristics, for example as solvents or as fuel additives that act as inhibitors and dispersants – provides a good opportunity for utilization of amino acid esters. The unique multifunctional chemistry, environmental acceptability, and low toxicity properties of amino acids, their esters, and other derivatives make them very interesting compounds for further examination. All amino acids exist in the form of Zwitter-ion in their respective aqueous solution (Figure 2).

$$R\text{-}CH_2\text{-}CH\text{-}COO^-$$
$$\underset{NH_3}{\overset{|}{}_+}$$

Figure 2 Zwitterion structure of amino acids.

Both $-NH_2$ and $-COOH$ groups are hydrophilic; esterifying the $-COOH$ group under acidic conditions would render surfactant properties to the final derivative (Figure 3). If both the $-COOH$ and $-NH_2$ functionalities of the amino acid are derivatized, making it a secondary amino compound, then it can be used as fuel stabilizer, although the exact mechanism of $-NH-$ as free radical stabilizer is still uncertain.

$$R\text{-}CH_2\text{-}CH\text{-}COO\ R_1$$
$$\underset{NH_3}{\overset{|}{}_+}$$

Figure 3 Esterified product of amino-acids in acidic condition.

The esters of organic acids and amino acids are generally produced by acid-catalyzed condensation reactions of alcohols and acids[1]. There is some scattered information available on amino acid esterification using mineral acid or homogeneous catalysts. Even though mineral acids are effective in obtaining good conversion, they require additional separation and purification steps to obtain esters of desired quality. In contrast, the use of solid acid catalysts has gained attention recently, in particular the application of ion exchange resins for reactions conducted in aqueous phase. The use of solid acid catalysts allows better control of acidity and minimizes the number of purification steps required, but further work using these heterogeneous catalysts is still required in detail to facilitate the neat reaction and separation of products from catalysts. In this paper, we have demonstrated the successful application of ion exchange resins as heterogeneous catalysts for esterification of organic and amino acids, and have characterized the kinetics of esterification for these resins.

Results and Discussion

As a model esterification reaction, the formation of ethyl lactate has been studied and its complete kinetic and thermodynamic analysis has been performed. The formation rate of ethyl lactate has been examined as a function of temperature and catalyst loading. In early experiments, it was determined that lactic acid itself catalyzes esterification, so that there is significant conversion even without ion exchange resin present. The Arrhenius plot for both resin-catalyzed and "uncatalyzed" reactions indicates that the uncatalyzed rate is approximately one-twentieth that of the catalyzed rate; however, the activation energies of the two cases are essentially the same at 50 kJ/mol. It is thus clear that the primary role of the ion exchange resin as catalyst is to provide acidic sites for reaction, not to alter the reaction pathways. A second set of experiments was conducted at different ion exchange resin catalyst loadings to ascertain the effect of loading on rate. Increasing the catalyst loading increases the esterification rate linearly, as expected for a kinetically controlled catalytic reaction. However, the plot of initial rate vs. loading does not pass through the origin; if one extrapolates the plot, the intercept is approximately 3.0×10^{-6} $gmol/cm^3/sec$, which implies that the reaction is autocatalytic due the presence of lactic acid and this initiates the reaction on its own. The rate expression for autocatalytic reaction is given by:

$$r_{autocatalytic} = 10.52.e \ (-6186.2/RT)C_{La} \ .\alpha[(C_{La}.C_{EtOH}) - (C_{EtLa}.C_{H2O})/K_{eq})]$$

While in presence of catalyst the reaction is assumed to LHHW type of mechanism and its rate expression is given by:

$$r_{catalytic}=\{11.12e(-6002.2/RT).W_{cat}^2[(K_{La}.K_{EtOH}.C_{La}.C_{EtOH})-$$
$$(K_{Etla}.K_{H2O}.C_{EtLa}.C_{H2O})/K_{eq}]\}/ \ (1+K_{La}.C_{La}+K_{EtOH}.C_{EtOH}+K_{EtLa}.C_{EtLa}+K_{H2O}.C_{H2O})^2$$

Therefore, the overall kinetic rate expression for ethyl lactate conversion from lactic acid and ethanol can be given by:

$$r_{overall} = r_{autocatalyzed} + r_{catalyzed}$$

The equilibrium conversion of lactic acid was measured by allowing reaction to proceed until no further changes were observed. We then used the data to calculate the equilibrium constant for the esterification reaction as a function of temperature. The final form of the equilibrium constant is given by the following equation.

$$K_e = C_{EtLa} \ C_{H2O} / C_{La} \ C_{EtOH} = exp \ (-0.896 + 667.2/T); \ T \ in \ (K)$$

Esterification of succinic acid
A series of experiments was conducted to form monoethyl and diethyl succinate using either sulfuric acid or acidic ion-exchange resin as catalysts. The esterification of succinic acid is modeled as a simple series reaction sequence.

Reaction 1) SA + EtOH $\xrightleftharpoons{Ke_1}$ MES + H_2O

Reaction 2) MES + EtOH $\xrightleftharpoons{Ke_2}$ DES + H_2O

SA = Succinic acid; EtOH = Ethanol; MES = Monoethyl succinate; DES = Diethyl succinate

Reaction 1) $r_1 = k_1$ [SA][EtOH] $- k_{-1}$ [MES][H2O]
Reaction 2) $r_2 = k_2$ [MES][EtOH] $-k_{-2}$ [DES][H2O]

If r_1 = rate of reaction of SA and r_2 = rate of formation of DES, then rate of formation of MES = $r_2 - r_1$. The equilibrium constant for each of the esterification reactions was determined by allowing the reaction to proceed until no further reaction was observed. For the quantities of resin used, the rate of resin-catalyzed esterification is much slower than that for sulfuric acid-catalyzed reaction. All results are summarized in Table 1.

Table 1 Summary of succinic acid esterification results.

Run[a]	Molar ratio[a]	Catalyst Wt%	Time (min)	Temp. °C	Ke_1	Ke_2	k_1[c]	k_2[c]
1	1:20:0	1[b]	183	78	4.65	0.75	1.6	0.51
4	1:16:13	0.48[b]	410	80	5.56	1.29	0.14	0.052
7	1:8:10	1.00[b]	180	80	7.40	1.70	0.29	0.13
11	1:16:10	1.60 Amberlite IR-120	423	79	4.40	0.81	0.074	0.049

[a] SA:EtOH:H$_2$0 (SA: Sucinic acid and EtOH: Ethanol)
[b] H$_2$SO$_4$ (Catalyst)
[c] unit of rate constant (liter.mol^{-1}.min^{-1})

Esterification of amino-acids
Four different amino acids have been selected for esterification to study the effect of R-group substituent of amino acid on rate and ease of esterification. The four acids are alanine, serine, aspartic acid and lysine. Their respective esters were prepared by reported methods to authenticate and compare with those prepared by our method. Alanine was esterified with ethanol to yield the ethyl ester, keeping $-NH_3^+$ group intact. This was also confirmed by acidity of final reaction mixture (pH~ 3.2). There was about 50% conversion of alanine to its ethyl ester. Further work on ester formation, including qualitative and quantitative analysis, is in process.

Experimental Methods and Analysis

Esterification of lactic acid
All reactions involving lactic acids were performed in 300 mL Parr Autoclave batch reactor. All reagents, including the resin catalyst, were charged into the reactor and heated up to the desired reaction temperature. Stirring was commenced once the desired temperature was reached; this was noted as zero reaction time. Reaction sample were withdrawn periodically over the course of reaction and analysed for ester, water and alcohol using a Varian 3700 gas chromatograph with a thermal conductivity detector (TCD) and a stainless steel

column (2m x 3.25 mm) packed with a liquid stationary phase of Porapack -R. Lactic acid concentration was determined by titration using 0.1 N NaOH solution.

Esterification of amino acids and succinic acid
Amino acid and succinic acid esterification were conducted in a three neck round bottom flask on a hot plate with magnetic stirrer. The reactor assembly includes a thermocouple to measure reaction temperature and a reflux condenser to eliminate the loss of volatile reagents due to evaporation. Initially all reagents are charged in the reactor and temperature of reaction mixture was allowed to reach at desired reaction temperature. At that time, the ion exchange catalyst was added to the reaction mixture. Reaction mixtures of succinic acid esterification were analyzed via HPLC using 30 wt% acetonitrile in water with 0.005 N H_2SO_4 as the mobile phase and a Biorad HPX-87H aminex column. The column was operated at $25°C$ at a mobile phase flow rate of 0.5 ml/min. The concentration of all reactants and products were calculated using internal calibration standards.

Acknowledgements
The support of the National Corn Growers Association is greatly appreciated.

References

1. Kirk and Othmer, Encylcopedia of Chemical Technology, 4th Edition, Wiley, New York, **9**, 755-780 (1997).
2. P. Juyal and O. N. Anand, *Petroleum Science and Technology*, **20**(9-10), 1009 (2002).
3. K. Watkins, *Chem. Eng. News*, **80**(2), 15 (2002).
4. G. Fornasari, *Chimica e l'Industria* (Milan), **82**(1), 26 (2000).
5. J. I. Choi and W. H. Hong *J. Chem.l Eng. Jpn*, **32**(2), 184 (1999).
6. C. Y. Zhu, Z. Q. Zhang, Q. P. Liu, Z. P. Wang, and J. Jin. *J. of Appl. Polym. Sci.*, **90**(4), 982 (2003).
7. B. D. Ahn, S. H. Kim, Y. H. Kim, and J. S. Yang. *J. Appl. Polym. Sci.*, **82**(11), 2808 (2001).

42. One Pot Transformation of Geraniol into Citronellol and Menthol over Cu/Al$_2$O$_3$

Nicoletta Ravasio,[a] Achille Fusi,[b] Rinaldo Psaro,[a] and Federica Zaccheria[c]

[a]CNR-ISTM, via Golgi 19, I-20133 Milano, Italy
[b]Dip. Chimica Inorganica, Metallorganica e Analitica, Università di Milano,
Via Venezian 21, I-20133 Milano
[c]Consorzio INSTM, via B. Varchi 59, I-50132 Firenze, Italy

n.ravasio@istm.cnr.it

Abstract

Geraniol can be converted into citronellol and menthol over Cu/Al$_2$O$_3$ under catalytic hydrogenation conditions owing to chemoselective hydrogenation and a three-functional process taking place on the catalyst surface.

Introduction

Terpenes are very cheap precursors to fragrances, flavours, drugs and agrochemicals (1) but their high reactivity often makes it difficult to achieve a selective transformation. However the target is worth pursuing as almost all products can be useful. We have long been involved in catalytic transformation of terpenes and terpenoids studying acid catalysed rearrangements of terpenic epoxides over different amorphous silica-aluminas (2,3), hydrogenation reactions catalysed by supported copper catalysts (4), selective epoxidation of terpenic alcohols catalysed by titanium silicates (5). In particular we are interested in setting up bifunctional processes by exploiting acidic and hydrogenation sites (6) or acidic and epoxidation sites (7) present on the catalyst surface. Thus, the one step transformation of (+)-citronellal into (-)-menthol (pathway *a*, Scheme 1), has been realized with yield ≅ 90% and stereoselectivity up to 80% under mild conditions in the presence of Cu/SiO$_2$ by exploiting the presence of acidic and hydrogenation sites on the catalyst surface the unusual reducibility of an olefinic bond under these conditions and the chemoselectivity of the process (8) while the same substrate can be converted into isopulegol epoxide with an overall 68% yield in a two step process (pathway *b*, Scheme 1) over Ti-MCM41 by exploiting the remarkable acidity of the catalyst in toluene and adding TBHP as an oxidant in CH$_3$CN once the cyclization is completed (7).

Here we report that geraniol **1**, under catalytic hydrogenation conditions in the presence of a Cu/Al$_2$O$_3$ catalyst, gives two valuable products, namely citronellol **2** and menthol **3**.

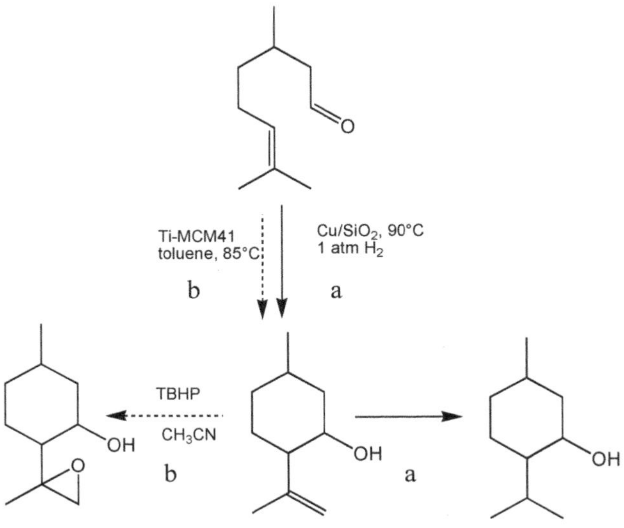

Scheme 1

Experimental Section

The catalysts 8% and 5% Cu/Al$_2$O$_3$ were prepared as already reported (4) by using Al$_2$O$_3$ (BET=300 m^2/g, PV=1.0 ml/g) from Grace Davison. Catalytic tests were carried out in different solvents under 1 atm of H$_2$.

The catalysts were treated at 270 °C for 20 minutes in air, for 20 minutes under vacuo and reduced at the same temperature with H$_2$ at atmospheric pressure. The substrate (0.100 g, 0.65 mmol) was dissolved in the solvent (8 mL) and the solution transferred under N$_2$ into a glass reaction vessel where the catalyst (0.100 g) had been previously treated. Catalytic tests were carried out with magnetic stirring under H$_2$ at 90 °C.

Reaction mixtures were analysed by GC (mesitylene as internal standard) using a SP-2560 (100m) capillary column.

Results and Discussion

The hydrogenation of geraniol over Cu/Al_2O_3 in hydrocarbon solvents gives mixtures of citronellol **2** and menthol **3**.

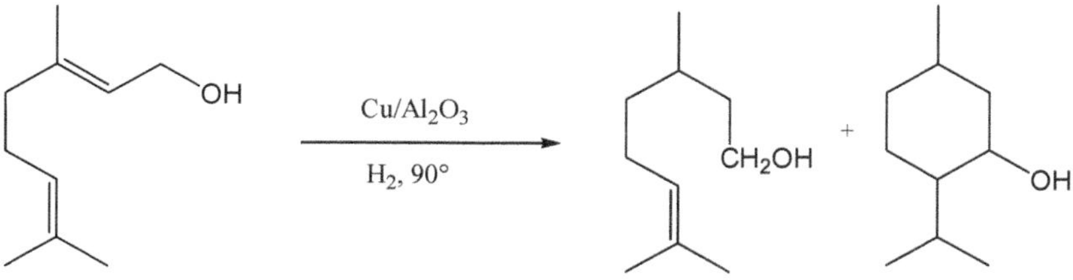

Entry	% Cu	solvent	Conv[a]	t[b]	2	3	Stereo[c]	Dehydr[d]
			(%)	(h)	(%)	(%)	(%)	(%)
1	8	n-heptane	95	2	56	30	77	10
2	5	"	77	7	46	42	78	12
3	5	dioxane	52	6	79	5	82	15
4	5	2-propanol	13	12	>98			

[a] after 30 minutes; [b] Time to reach complete conversion of isopulegol; [c] amount of (-) menthol; [d] amount of dehydration products

Citronellol is formed through selective hydrogenation of the C=C bond activated by the presence of the OH group, whereas menthol **3** is the product of a *three-functional* process involving isomerization of the allylic alcohol **1** to citronellal **4**, *ene* reaction to isopulegol **5** and final hydrogenation (Scheme 2).

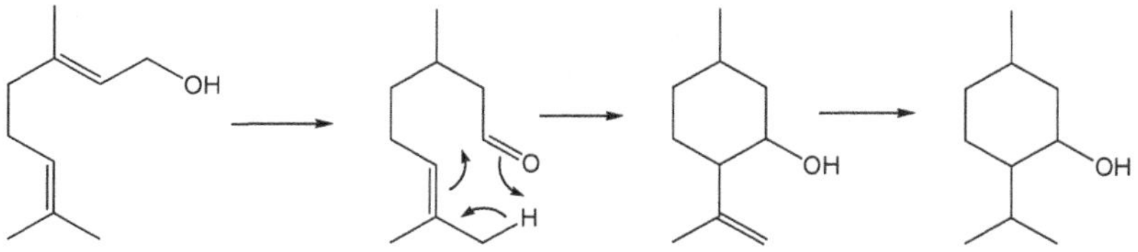

Scheme 2

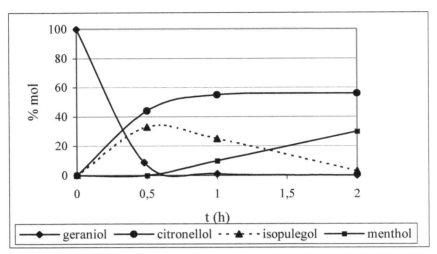

Figure 1 Hydrogenation of geraniol over 8%Cu/Al$_2$O$_3$.

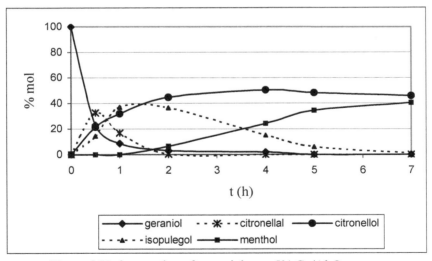

Figure 2 Hydrogenation of geraniol over 5% Cu/Al$_2$O$_3$.

When the reaction is carried out over the 8% catalyst **2** and **5** are soon formed and in 2 hours also the hydrogenation of **5** to **3** is also complete, thus giving a 56% yield in **2** and about 30% in **3**. In order to increase the yield in menthol, we carried out the reaction over a 5% Cu catalyst. In this case the reaction is much slower and the intermediate formation of citronellal **4** can be observed (Figure 2). On the other hand the yield in citronellol can be very easily improved by using an oxygenated solvent like dioxane that inhibits the Lewis acid catalysed reaction or by using 2-propanol. In this way quantitative yield in citronellol could be obtained.

This is a quite remarkable result, as the chemoselective hydrogenation of geraniol over a heterogeneous catalyst has rarely been reported. It can be carried out over platinum containing zeolite (9), over Pt/Al_2O_3 modified with carboxylic acids (10), over Ni/diatomaceous earth and alkali hydroxides or carbonates (11) or NiRaney and alkali or alkaline earth metal hydroxides (12), yields never exceeding 85%.

References

1. K. Bauer, D. Garbe, and H. Surburg, in Common Fragrance and Flavours Materials, Wiley-VCH, New York, 1997.
2. N. Ravasio, M. Finiguerra, and M. Gargano, *Chem. Ind.* (Dekker) (*Catal. Org. React.*) 513 (1998).
3. F. Zaccheria, R. Psaro, N. Ravasio, and L. De Gioia, *Chem. Ind.* (Dekker) (*Catal. Org. React.*) 601 (2000).
4. N. Ravasio, M. Antenori, M. Gargano, and P. Mastrorilli, *Tetrahedron Lett.*, **37**, 3529 (1996).
5. M. Guidotti, N. Ravasio, R. Psaro, G. Ferraris, and G. Moretti, *J. Catal.*, **214**, 242 (2003).
6. N. Ravasio, V. Leo, F. Babudri, and M. Gargano, *Tetrahedron Lett.*, **38**, 7103 (1997).
7. M. Guidotti, G. Moretti, R. Psaro, and N. Ravasio, *Chem. Commun.*, 1789 (2000).
8. N. Ravasio, N. Poli, R. Psaro, M. Saba, and F. Zaccheria, *Topics Catal.*, **13**, 195 (2000).
9. D. Tas, R. F. Parton, K. Vercruysse, and P. A. Jacobs, *St. Surf. Sci. Catal.* **105B**, 1261 (1997).
10 H. Kuno, K. Takahashi, M. Shibagaki, and H. Matsushita, *Bull. Chem. Soc. Jpn*, **62**, 3779 (1989).
11. Jpn. Kokai Tokyo Koho (1982) JP 57024320.
12 V. Paul, Indian Patent (1979) IN 147167.

43. Investigation of Chlorine Substitution Effects in the Claisen-Schmidt Condensation of 2′-Hydroxyacetophenone with Chlorobenzaldehydes Over MgO

Sirena C. Hargrove-Leak, Janine Lichtenberger, and Michael D. Amiridis

University of South Carolina, Department of Chemical Engineering, Swearingen Engineering Center, Columbia, SC 29208

amiridis@engr.sc.edu

Abstract

The Claisen-Schmidt condensation of 2′-hydroxyacetophenone and different chlorinated benzaldehydes over MgO has been investigated through kinetic and FTIR spectroscopic studies. The results indicate that the position of the chlorine atom on the aromatic ring of the benzaldehyde substantially affects the rate of this reaction. In particular, the rate increases in the following order: *p*-chlorobenzaldehyde < *m*-chlorobenzaldehyde < *o*-chlorobenzaldehyde. The difference between the meta and para-substituted benzaldehyde can be attributed to electronic effects due to the difference in the Hammett constants for these two positions. Steric effects were found to be responsible for the higher rate observed with the *o*-chlorobenzaldehyde.

Introduction

The synthesis of fine chemicals and pharmaceuticals has traditionally been achieved via homogeneous catalytic methods. Heterogeneous catalysis is an attractive alternative to these processes because of its inherent waste minimization and the potential to consolidate steps in a process. The synthesis of flavanones (i.e., condensation of 2′-hydroxyacetophenone and benzaldehydes to form a 2′-hydroxychalkone followed by an isomerization leading to the corresponding flavanone) represents an example of an important, homogeneously-catalyzed reaction that can be successfully heterogenized over solid catalysts [1]. This reaction has been previously investigated in our group in an effort to understand the reactant-solvent-catalyst interactions occurring when it is carried out over MgO in various solvents [2 - 4].

Substitution effects have been observed for both the homogeneously and heterogeneously base-catalyzed Claisen-Schmidt condensation of ketones and aldehydes with functional groups substituted in the para-position [5 - 12]. In this study, we extend our investigation of observed substitution effects in the reaction between 2′-hydroxyacetophenone and benzaldehydes to include meta- and ortho-substitution. It is anticipated that both steric and electronic effects may play a role in these cases. In order to explore these effects we have combined kinetic and infrared studies and focused on the reaction of 2′-hydroxyacetophenone with substituted chlorobenzaldehydes.

Results and Discussion

Normalized 2′-hydroxyacetophenone concentration versus time data obtained during the reaction of equimolar amounts of 2′-hydroxyacetophenone with different chlorobenzaldehydes are shown in Figure 1. These data indicate substantial differences in the reactivity of the different chlorobenzaldehydes, depending on the position of the Cl atom. In particular, the initial reaction rate (equal to the slope of the concentration versus time curve) nearly doubles in going from the para- to the ortho-chlorobenzaldehyde (i.e., para-chlorobenzaldehyde: 160 kmol/g-cat/s, meta-chlorobenzaldehyde: 191 kmol/g-cat/s and ortho-chlorobenzaldehyde 268 kmol/g-cat/s). A similar high initial rate was also observed with 2,3-dichloro-benzaldehyde (298 kmol/g-cat/s). The difference in reactivity between meta- and para-chlorobenzaldehyde is consistent with anticipated electronic effects and can be explained by means of the Hammett constants for Cl substitution in these two positions (i.e., 0.37 for 3-chlorobenzaldehyde versus 0.23 for 4-chlorobenzaldehyde). Steric effects have been previously reported to affect the reactivity of ortho-substituted molecules, due to the close proximity in this case of the additional functional group to the reaction center. Consequently, the orientation of the functional group in the ortho position may sterically affect the ability of the active center to participate in the reaction [13 - 14].

The reactivity of the different substituted benzaldehydes on the MgO surface was further investigated with the use of *in situ* FTIR spectroscopy. Spectra obtained following the adsorption of meta-chlorobenzaldehyde on MgO at different temperatures in the 70 to 160°C range are shown in Figure 2. Following adsorption of *m*-chlorobenzaldehyde at 70°C (Figures 2a and 2b) strong bands at 1652, 1620, and 1325 cm^{-1} as well as a weaker one at 1710 cm^{-1} are present in the spectra. These bands can be assigned to the aldehydic group vibrations of different adsorbed benzaldehyde species [15]. The band at 1710 cm^{-1} is only visible after the first few minutes on stream and can be assigned to physisorbed *m*-chlorobenzaldehyde as indicated by the small shift from the frequency of the C = O vibration in the gas phase, previously observed at 1720 cm^{-1}. The bands at 1652 and 1325 cm^{-1} show the same changes in intensity at the different adsorption temperatures. In particular, they are growing in intensities at temperatures up to 100°C, then slowly decrease, and finally disappear at 160°C. These bands can be assigned to a benzaldehyde species

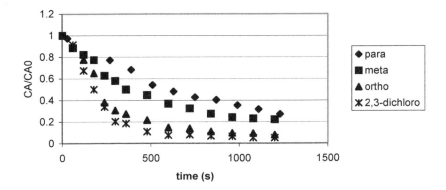

Figure 1 Normalized 2'-hydroxyacetophenone concentration versus time data obtained during the reaction of 2'-hydroxyacetophenone with different chlorobenzaldehydes over MgO: ◆ para-chlorobenzaldehyde, ■ meta-chlorobenzaldehyde, ▲ ortho-chlorobenzaldehydes and * 2,3-dichlorobenz-aldehyde. (Conditions: Initial concentrations: 1.5 mole/L 2'-hydroxyacetophenone, 1.5 mole/L benzaldehyde in DMSO at 160°C with 0.1 wt% MgO)

bonded to the MgO surface through both the carbonyl carbon and hydrogen, as shown in Figure 3a [15]. In particular the band at 1652 cm^{-1} can be assigned to a weakened C = O bond [15], while the band at 1325 cm^{-1} can be assigned to a weakened aldehydic C – H bond. In the gas phase the aldehydic C – H bending vibration is observed at 1388 cm^{-1} for benzaldehyde [16] and at 1372 cm^{-1} for para-chlorobenzaldehyde [17]. A similar band at 1370 cm^{-1} was observed in the gas phase spectrum of *m*-chlorobenzaldehyde, but upon adsorption only a very weak feature is present at 1368 cm^{-1} during the first few minutes (Figure 2a). Subsequently, this band shifts to the 1325 cm^{-1} position, presumably due to a weakening of the aldehydic C – H bond. The 1325 cm^{-1} bond eventually disappears at elevated temperatures, suggesting complete rupture of the aldehydic C – H bond. This transformation is depicted in the species shown in Figure 3b. The band at 1620 cm^{-1} can be assigned to the C = O bond of the same species formed after the abstraction of the aldehydic hydrogen [15]. Since this band is present in the spectra of Figure 2 from the onset of adsorption it is possible that initially the species shown in Figure 3a and 3b co-exist on the catalyst surface with the former been slowly converted to the later. This process is accelerated at higher temperatures.

The two strong bands at 1595 and 1472 cm^{-1} can be assigned to ring vibrations [16 - 18] and a mixed mode containing some additional aromatic CH [16] or aldehydic CH bending [18], respectively.

Finally, the spectra of adsorbed *m*-chlorobenzaldehyde also contain two strong bands at 1560 and 1427 cm^{-1}. Neither of these bands is present in the spectrum of gas phase *m*-chlorobenzaldehyde. These bands can be assigned to the symmetric and asymmetric C – O stretching vibrations of a surface bridging benzoate species, as the one shown in Figure 3c [15, 19]. These species are present on the surface even at 70°C, in agreement with van Hengstum et al. [15], who observed benzoate formation at 50°C during the adsorption/ oxidation of an aromatic benzaldehyde on V$_2$O$_5$/TiO$_2$. Two additional weaker bands, observed at 1502 and 1403 cm^{-1}, can also be assigned to a different surface benzoate as shown in Figure 3d [15, 20]. These surface benzoates are undesired by-products, which are not involved in the Claisen-Schmidt condensation reaction [21]. Since no formation of benzoic acid was observed in the bulk, we believe that they are only formed in small quantities on the catalyst surface as a result of the interaction between benzaldehyde and surface oxygen.

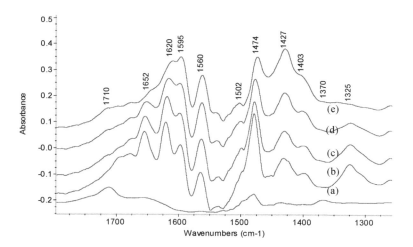

Figure 2 FTIR spectra of *m*-chlorobenzaldehyde adsorbed on MgO: (a) After 1min exposure of *m*-chlorobenzaldehyde at 70°C; (b) after 30 min exposure of *m*-chlorobenzaldehyde at 70°C; (c) after 5 min exposure at 110°C; (d) after 5 min exposure at 130°C and (e) after 5 min exposure at 160°C.

Additional adsorption studies were also carried out using para- and ortho-chlorobenzaldehydes (Figure 4). The results obtained with *p*-chlorobenzaldehyde are very similar to the ones obtained with *m*-chlorobenzaldehyde and were omitted from this paper for brevity. In contrast, substantial differences are observed in the spectra of adsorbed *o*-chlorobenzaldehyde. In particular, in this case the 1620 cm^{-1} band characteristic of the dissociated surface benzaldehyde species shown in Figure 3b is very strong even after only 5 min on stream at 70°C. An additional strong band is present in the low temperature spectra at 1697 cm^{-1} and can be assigned to the C = O double bond of a physisorbed benzaldehyde species. This band decreases in intensity with increasing temperature, while a shoulder at 1672 cm^{-1} grows in intensity at the same time. According to van Hengstum et al. [15] the 1672 cm^{-1} band can be assigned to a benzaldehydes species attached to the catalyst surface through the carbonyl oxygen, as shown in Figure 3a. A band at 1340 cm^{-1} increases in intensity together with the shoulder at 1672 cm^{-1} and can be assigned to a weakened aldehydic C – H bond (similar to the 1325 cm^{-1} band observed with *m*-chlorobenzaldehyde). No benzoate formation was observed in the spectra of adsorbed *o*-chlorobenzaldehyde, even at elevated temperatures, although the type of dissociated surface benzaldehyde species observed are capable of forming benzoates when no functional groups are present in the ortho-position.

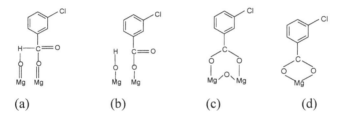

(a)   (b)   (c)   (d)

Figure 3 Schematic representations of the proposed surface species formed following the adsorption of meta-chlorobenzaldehyde on MgO.

In summary, the FTIR results indicate that the different chlorobenzaldehyde isomers interact with the MgO surface in four distinct modes as: intact physisorbed species, species chemisorbed through both the carbonyl carbon and hydrogen atoms, a species in which the aldehydic bond has been ruptured, and finally benzoate species. Benzoate formation was observed with both the para- and meta-chlorobenzaldehydes, but not with ortho-chlorobenzaldehyde. This behavior can be attributed to the steric effect of the Cl atom in the ortho position. In particular, in this position the Cl atom is coplanar with the benzene ring and the aldehydic group. Therefore, it has to bend out of this plane for benzoates to be formed and hence, benzoate formation is restricted. Since benzoates are not active in the Claisen-Schmidt condensation reaction [21], their presence in the case of para- and meta-chlorobenzaldehydes may slow the progression of the desired Claisen-Schmidt condensation reaction by blocking active catalyst sites. Such a hypothesis is consistent with the observed kinetic results, which suggests a substantially higher reaction rate with *o*-chlorobenzaldehyde, the isomer that does not form benzoates.

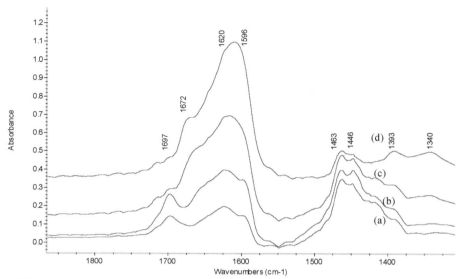

Figure 4 FTIR spectra of *o*-chlorobenzaldehyde adsorbed on MgO: (a) After 1min exposure of *o*-chlorobenzaldehyde at 70°C; (b) after 20 min exposure of *o*-chlorobenzaldehyde at 70°C; (c) after 5 min exposure at 110°C; and (d) after 5 min exposure at 160°C.

Experimental Section

Activity measurements were conducted in a dimethyl sulfoxide (DMSO) solvent (Aldrich, 99.9% purity) at constant temperature (160°C), atmospheric pressure, and under total reflux using a batch reactor with a nitrogen purge stream (to prevent oxidation of the benzaldehydes to benzoic acid). A 0.1 wt% catalyst loading, 60/80 mesh catalyst particle size, and a stirring rate of 500 rpm were used. Equimolar amounts of 2'-hydroxyacetophenone (Aldrich, 99.9% purity) and the corresponding benzaldehyde examined (i.e., para-, meta-, ortho-, and 2,3 chlorobenzaldehyde; Aldrich 97–99% purity), were used, in agreement with the reaction stoichiometry. The samples were analyzed offline with an SRI 8610C gas chromatograph (GC) equipped with a Supelco SPB-5 fused silica capillary column. Initial reaction rates were calculated using the concentration versus time data collected in the first 5 minutes of the reaction.

FTIR spectra were collected with a Nicolet 740 spectrometer and a custom built *in situ* gas flow cell. The spectrometer was equipped with a MCT-B detector cooled by liquid nitrogen. Approximately 15 mg of the MgO catalyst sample was pressed into a self-supported disc and placed in a sample holder located at the center of the cell. The temperature in the cell was measured with a thermocouple placed close to the catalyst sample. Transmission spectra were collected in a single beam mode with a resolution of 2 cm^{-1}. Prior to introduction of the chlorobenzaldehydes to the catalyst, the sample was pretreated in flowing nitrogen for 2hr at the reaction temperature to remove any water or other impurities adsorbed on the surface. Spectra were then collected at different time intervals following exposure of the catalyst to a flowing nitrogen stream carrying a small concentration of the different chlorobenzaldehydes, obtained through passage through a saturator maintained at room temperature.

References

1. M. J. Climent, A. Corma, S. Iborra, and J. Primo, *J. Catal.* **151**, 60 (1995).
2. M. T. Drexler, M. D. and Amiridis, "Heterogeneous Synthesis of Flavanone over MgO," Chemical Industries (Dekker) (*Catal. Org. React.*) 451-457 (2000).
3. M. T. Drexler M. D. and Amiridis, *Catal. Lett.* **79**, 175 (2002).
4. M. T. Drexler, and Amiridis, *J. Catal.,* **214**, 136 (2003).
5. E. Coombs, and D. P. Evans, *J. Chem. Soc.* 1295 (1940).
6. D. S. Noyce, W. A. and Pryor, *J. Am. Chem. Soc.* **77**, 1397 (1955).
7. A. T. Nielsen and W. J. Houlihan, "The Aldol Condensation," in Organic Reactions, A.C. Cope, Ed., Wiley , New York, 1968.
8. S. Kandlikar, B. Sethuram, and T. N. Rao, *Zeitschrift für Physikalische Chemie,* **95**, 87 (1975).
9. M. J. Climent, H. Garcia, and J. Primo, *Catal. Lett.,* **4**, 85 (1990).
10. D. Tichit, M. H. Lhouty, A. Guida, B. H. Chiche, F. Figueras, A. Auroux, D. Bartalini, E. and Garrone, E., *J. Catal.,* **151**, 50 (1995).
11. G. W. Haas, and M. L. Gross, *Am. Soc. Mass Spectr.* **7**, 82 (1996).
12. S. Hargrove and M. Amiridis
13. A. A. Jameel and M. K. Pillay, *J. Ind. Chem. Soc.* **76**, 101 (1999).
14. H. Gangwani, P. K. Sharma, and K. K. Banerji, *Int. J. Chem. Kin.,* **32**, 615 (2000).
15. A. J. van Hengstum, J. Pranger, S. M. van Henstum-Nijhuis, J. G. van Ommen, P. J. and Gellings, *J. Catal.,* **101**, 323 (1986).
16. H. Lampert, W. Mikenda, and A. Karpfen, *J. Phys. Chem.,* **101**, 2254 (1997).
17. S. H. W. Hankin, O. S. Khalil, L. Goodman, *J. Mol. Spec.,* **72**, 383 (1978).
18. J. H. S. Green and D. J. Harrison, *Spectrochim. Acta,* **32A** (1976) 1265.
19. S. Besselmann, E. Löffler, and M. Muhler, *J. Mol. Catal. A,* **162**, 401 (2000).
20. K.-H. Kung, K.-H. and M. B. McBride, *Soil Sci. Soc. Am. J.* **53**, 1673 (1989).
21. J. Lichtenberger, S. Hargrove, and M. Amiridis, paper in prep.

IV. Symposium on Catalytic Oxidation

44. The Effect of Single Metal Variation in Mo-V-Nb-W/Al₂O₃ System on Propane Selective Oxidation

Giacomo Grasso and Julian R.H. Ross

Centre of Environmental Research, University of Limerick, Limerick, Ireland

giacomo.grasso@ul.ie

Abstract

A catalytic system Mo-V-Nb-W supported on alumina was prepared by impregnation and investigated for the selective oxidation of propane. The effects of the variation of each metal and of the catalyst preparation were analysed. The results show that Mo and V species supported on alumina can lead to catalysts with high selectivity to propene and reasonable selectivity to acrolein. The presence of Nb and W seems to have little effect. The catalyst can be affected by the method of impregnation.

Introduction

As consequence of an increased demand from the market for propene and C3 oxygenated products over the last fifteen years, the oxidative dehydrogenation (ODH) and the selective oxidation of propane have become reactions attracting much research into both possible catalyst formulations and reaction mechanisms. Previous work [1] from our laboratory has shown that a combination of Mo, V and other oxides supported on a low-area alumina calcined at 1150°C (α-alumina) have interesting properties as catalysts for the ODH of propane; when water vapour was added to the reactant stream, they also gave significant yields of acrolein, particularly for a formulation containing the oxides of Mo, V, Nb and W. In order to gain a better understanding of this system and to optimise the methods of catalyst composition and preparation, a series of libraries of such catalysts have been prepared and tested, using both traditional and high throughput screening techniques. This paper summarises some of the results obtained with catalysts made using manual methods which illustrate the effects of catalyst composition as well as of preparation method on the properties of the final catalyst. The paper also emphasises the importance of the testing conditions used in determining the measured properties.

Experimental Section

Catalyst preparation. One library of catalysts, indicated below with the name Library 1, was prepared by wet impregnation to have a range of concentrations of Mo, V, Nb and W on α-alumina. Table 1 shows the compositions of the standard material (S) and of a series containing variable quantities of Mo and V. Equivalent series with variable Nb and W contents were prepared but are not shown. These samples were prepared according to a 'standard' *wet impregnation* method as follows: the alumina support precursor (Alcan AA400) was first calcined at 1150°C; a sample of 2g of this was then treated sequentially by aqueous solutions of the salts of each metal (ammonium molybdate, AnalaR; ammonium metavanadate, AnalaR; niobium oxalate, CBMM; ammonium metatungstate hydrate, Aldrich), the amount of each salt added corresponding to the desired final catalyst composition (assuming that all the salts added are subsequently taken up by the support). After leaving the system to equilibrate in an excess of solution overnight (about 15h), the samples were then dried using a rotary evaporator at 80°C and were then finally calcined for 6h at 650°C. The actual amount of metal incorporated for each sample was not analysed, it being assumed that all was deposited (see above); hence, if this assumption is not fully justified, the data presented show only self-consistent trends, giving an indication of the range of compositions over which the best performance can be obtained by this catalytic system.

Table 1 Compositions of catalysts of Library 1.

Series	Sample Code	Wt% added per g of alumina			
		Mo	V	Nb	W
Standard	S	6	0.4	1	1
Mo series	3Mo	3	0.4	1	1
	4.5Mo	4.5	0.4	1	1
	7.5Mo	7.5	0.4	1	1
	9Mo	9	0.4	1	1
V series	0.2V	6	0.2	1	1
	0.3V	6	0.3	1	1
	0.8V	6	0.8	1	1
	1V	6	1	1	1
	2V	6	2	1	1

A second library of six catalysts, shown in Table 2, codes F1 to F6, was prepared to examine the effect of varying the method and sequence used for the impregnation process; in this series, the samples contained only Mo and V at nominally identical concentrations (with a ratio Mo/V of ca. 6), with V approx 1% (but not comparable with previous series). Two additional samples, F7 and F8, were prepared to show any effects of including Nb and W (each in amount of ca. 1%) in this formulation.

The samples F1, F2, F3, F7 and F8 were prepared by *wet impregnation* using excess solutions as above; the preparation of F1, F7 and F8 involved feeding the support with each solution in turn before finally drying and calcining (as for the samples of Table 1) whereas F2 and F3 involved drying and calcining before addition of the second component. In contrast, samples F4, F5 and F6 were prepared by *incipient wetness (or 'dry') impregnation*: each solution was fed to 2g of the calcined alumina in aliquots of 800μL - this being the volume required just to fill the pores without wetting the powder - with drying between each impregnation (ca. 15min at 130°C). The preparation of sample F4 involved a sequence of impregnation steps involving the addition of both species simultaneously whereas F5 and F6 involved impregnation by a single species per sequence (Mo first for the former, V first for the latter) and calcining after each sequence. Catalysts F7 and F8 were prepared by wet impregnation to verify whether or not the presence of Nb and W had any effect on this system.

Table 2   Catalysts Composition and preparation methods for Library 2.

Sample Code	Metal oxides (nominal w/w%)			Impregnation method	Sequence
F1	6Mo	1V		Wet	Simultaneous/cal
F2	6Mo	1V		Wet	Mo/calc/V/calc
F3	1V	6Mo		Wet	V/calc/Mo/calc
F4	6Mo	1V		Dry	Simultaneous/cal
F5	6Mo	1V		Dry	Mo/calc/V/calc
F6	1V	6Mo		Dry	V/calc/Mo/calc
F7	6Mo	1V	1Nb	Wet	Simultaneous/cal
F8	6Mo	1V	1W	Wet	Simultaneous/cal

Catalyst testing. The materials prepared were tested in a quartz microreactor using 300mg of each sample (particle size 212-425μm) diluted with 100mg of quartz grains of the same size. The feed to the bed contained 15vol% each of propane and oxygen; water vapour was added to the feed by a saturator kept at a constant temperature of 47°C, this giving a concentration of 10vol%; nitrogen was added (7.5vol%) as an internal standard to allow corrections to be made for changes occurring in the volume flow with the formation of products; and He was used as inert diluent to make up the concentration to 100%. The WHSV (weight hourly space velocity) based on the weight of catalyst alone was 8000 ml g^{-1} h^{-1}. The catalytic experiments were carried out in the temperature range from 400 to 650°C.

The effluent from the reactor was analysed by a Varian 3400 Gas Chromatograph using Molsieve 5A and Poraplot Q columns to separate respectively the permanent gases and the C1-C3 hydrocarbons and oxygenated compounds. They were connected in series using a 10-way valve system in order to invert the order of the columns immediately after the exit of CO from the Poraplot Q to avoid CO_2, hydrocarbons and oxygenates entering and deactivating the Molsieve column.

Conversions of propane and oxygen are defined as moles of propane/oxygen reacted per mole of propane/oxygen fed. The yield of the C3 products was calculated as moles of compound formed per mole of propane reacted; for the C2 products, the number of moles formed was multiplied by 2/3 and for the C1 products, by 1/3.

Results and Discussion

Effect of catalyst composition. Figures 1 and 2 show the effect on the behaviour of the different catalysts of varying the concentrations of Mo (Figure 1) and V (Figure 2), the concentrations of all the other components being kept constant; the results obtained for the standard catalyst (Sample S) are common to both figures. There is a wide variation in both conversion and product selectivity over both the Mo and V series; however, there was a much smaller range of change for the Nb and W series and so the results for these series are not shown here.

For the V series (Figure 2), the propane conversion increased steadily with V content at lower temperatures and the same steady increase applied to the oxygen conversions over the whole temperature range. The propane conversion flattened off with increasing temperature for the more active catalysts once the oxygen conversion had reached approximately 100%, as for the 4.5wt%Mo samples above. The yields of propene again paralleled the propane conversion data and the maxima achieved once more seemed to depend on the achievement of 100% oxygen conversion when the exit of the bed becomes reducing in character. (Some ethane and methane were found under these conditions, these molecules seemingly being formed in stoichiometric amounts from a proportion of the propane feed.) A further series of experiments (not shown) demonstrated that the behaviour of the catalysts was reproducible on cycling through the temperature range several times, there being no evidence for irreversible changes. Hence, we believe that the surface of the catalyst is not affected by the reductive conditions.

The acrolein yield also went through a maximum for each sample of the V series, this moving to lower temperatures and decreasing in magnitude with increasing V content. These experiments also showed that propanal was formed in significant proportions at temperatures when the oxygen was fully consumed (above the maximum in acrolein production) and this also seems to be a function of the existence of reducing conditions at the exit of the bed.

Figure 1 Effect of Mo variation in Library 1 (Table 1).

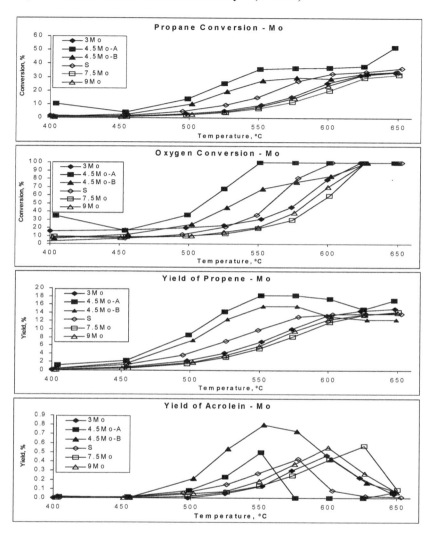

Figure 2 Effect of V variation in Library 1 (Table 1).

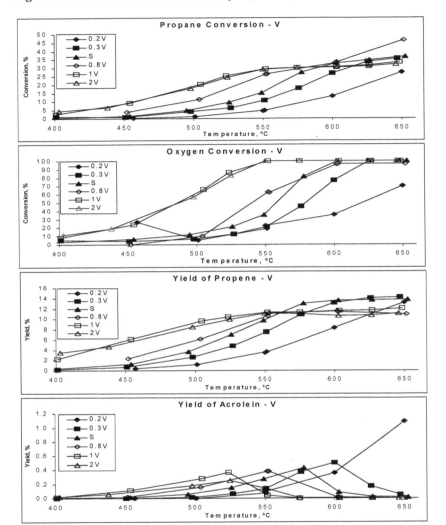

(This was also the case for the 4.5Mo samples discussed above.) These observations seem to indicate that the acrolein yield might be increased if the reductive conditions could be avoided for example by increasing the O_2/C_3H_8 ratio in the feed or by staged admission. Additional experiments carried out in this regime (to be reported in detail elsewhere) demonstrated that a slight increase in the acrolein yield/selectivity occurred when the propane content of the feed was decreased but that there was no allover improvement because CO_2 formation was increased to a greater extent, at the expense of propylene formation.

Effect of W/F variation. In order to examine in more detail the reaction sequence and the importance of oxidising and reducing conditions in the formation of various products, a series of experiments was carried out using the most active (and least selective) sample of the V series, that containing 2wt% V, to examine the effect of the residence time on the selectivity of the catalyst at various reaction temperatures. In this series, a range of different weights of the V catalyst was used with various dilutions with quartz granules, the total bed weight being kept constant; the total reactant flow rate was also kept constant. Figure 3a-d show the conversions of propane and oxygen and the selectivities to propene and acrolein as a function of W/F for different reaction temperatures. The selectivity to propene decreased with increasing catalyst weight over the whole range studied; the value at low weights (i.e. pseudo residence times) decreased from a value close to 100%, indicating that propene is probably the primary product of the reaction. The irregular behaviour shown by data for propene formation at low temperature (400 and 450°C) could be due to the lower precision of data at low propane conversions. However, Figure 3e shows that the value of the selectivity to CO_2 at low weights (pseudo residence times) for the T=400-450°C decreases from a value around 45% and so it is probable that the behaviour of selectivities to propene and CO_2 at low temperature is at least partially due to (partial) reduction of the catalyst. The selectivity to acrolein, on the other hand, increased and then decreased, the maximum occurring at pseudo residence times lower than those giving 100% conversion of oxygen, i.e. when there was still oxygen present in the catalyst bed. The other significant products (CO and CO_2) are accounted for most of the balance. The behaviour of selectivity to CO (Figure 3f) seems different from that to CO_2: it decreases at higher temperatures, especially at pseudo residence times higher than those giving 100% conversion of oxygen. As the decreasing of CO is thus associated with a more reductive environment, it is probable that the production of CO takes place on the catalyst surface through a route different from that for CO_2. At higher temperatures, cracking to give ethylene and methane was also observed (not shown). Small quantities of propanal and acetone (not shown) were also found. The propanal appeared only under conditions when the oxygen was fully reacted.

Bettahar et al. [4], in a review on the selective oxidation of propane and propene on mixed metal oxide catalysts, have presented a scheme showing the relative enthalpies for various possible intermediate species and products. Our results appear most consistent with the route to acrolein passing through propene (dehydrogenation of propane and subsequent O-insertion on the terminal C). There is the possibility that the alternative parallel propanal route (O-insertion and then dehydrogenation of the neighbouring C-C bond) also occurs to some extent since propanal only appears in the gas phase when there is insufficient oxygen present to bring about the final dehydrogenation step in the transition of propanal to acrolein.

Figure 3 Effect of variation of W/F on the conversions and selectivities at various temperatures for catalyst 2V (Library 1).

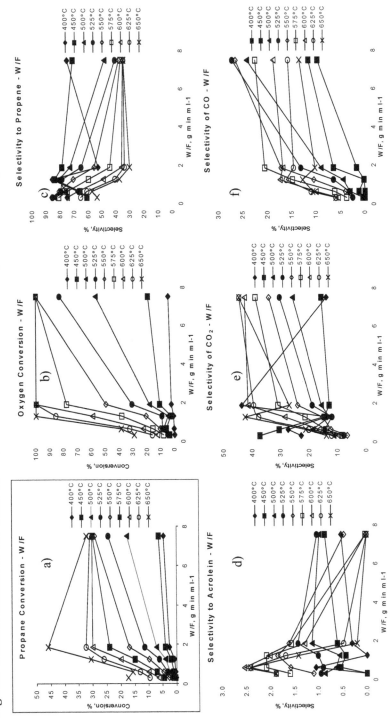

This will be investigated in future work. The acetone which is also formed shows a behaviour independent of the acrolein; as the intermediate of reaction forming this is probably propan-2-ol (not detected in our experiments) instead of propene, this lack of relationship is not unexpected.

Figure 4 Results obtained with the catalysts of the F series (Table 2).

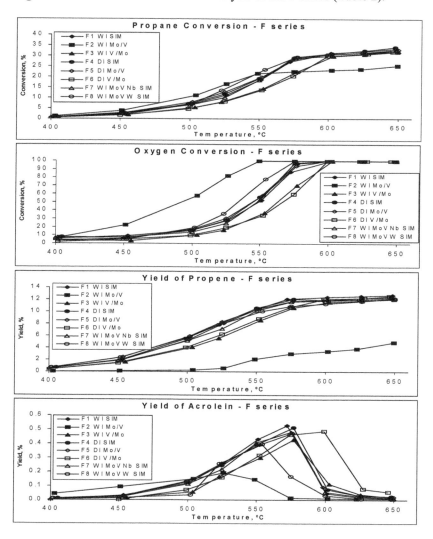

Effect of preparation. The aim of testing this batch of catalysts was to observe any differences in the behaviours of MoV/α-Al$_2$O$_3$ catalysts prepared by wet and dry impregnation and to observe any influence caused by intermediate calcination between two impregnation steps. The only significant difference seemed to arise with sample F2 in which Mo was added to the support and calcined before the addition of vanadium, the impregnation occurring by the so-called wet method involving excess solution. This catalyst had higher activity than the others and showed significantly lower selectivities towards propene and acrolein. We conclude that the wet impregnation procedure used with this sample gave rise to crystallites of V^{5+} outside the pores and that these were much less selective than crystallites in which the Mo and V species were in close proximity to one another. With dry impregnation, the Mo and V species were more effectively confined to the pores and so the interaction was more pronounced. Samples F7 and F8 gave results similar to those for all but the sample F2 and so we conclude once more that neither Nb nor W has any significant effect on the behaviour of this type of catalyst, at least when the compositions are in the range studied here.

The results presented in this paper therefore show that V and Mo species supported on alumina can give rise to a catalyst which has a high selectivity for the oxidation of propane to propene and a reasonable selectivity to acrolein and that both species are essential to give the optimal behaviour. Contrary to our previous observations and what observed for bulk catalysts [5], the presence of Nb and W seem to have little effect, perhaps because the methods used here restrict the active phase to a monolayer whereas previously prepared materials may have contained multilayer oxidic species.

Acknowledgements

The authors acknowledge financial support from the EU under project number GRD12000-25262 (KEMiCC).

References

1. J. R. H. Ross, E. M. O'Keeffe and G. Grasso, in Proceedings of ACS Spring Meeting, Personal Communication, Orlando, USA, April 2002.
2. B. P. Barbero and L. E. Cadús, *Appl. Catal. A:Gen.*, **252**(1) 133 (2003).
3. P. Michorczyk and J. Ogonowski, *Appl. Catal. A:Gen.* **251**, 425 (2003).
4. M. M. Bettahar, G. Costentin, L. Savary and J. C. Lavalley, *Appl. Catal. A: Gen.*, **145**, 1 (1996).
5. R. K. Grasselli, D. J. Buttrey, P. De Santo Jr., J, D. Burrington, C. G. Lugmair, A.F. Volpe Jr. and T. Weingand, in Proceedings of Europacat VI Congress, Innsbruck, Austria, September 2003.

45. Partial Oxidation of Propylene Over Supported Rhodium Catalysts

Rajesh A. Khatri, Rahul Singh, Steven S. C. Chuang, and Robert W. Stevens Jr.

Department of Chemical Engineering, The University of Akron, Akron, OH 44325

schuang@uakron.edu

Abstract

Transient infrared studies of partial oxidation over 2 wt% Rh/Al_2O_3 showed that the total oxidation of propylene to CO_2 occurred at a higher rate than the partial oxidation to propylene oxide; total and partial oxidation occurred in parallel pathways at 250 °C. Temperature-programmed desorption revealed that adsorbed propylene oxide can convert to other C_3 oxygenates such as acetone and propanal. The parallel reaction pathways for partial and total oxidation on Rh/Al_2O_3 suggest that selective poisoning of the total oxidation sites could be a promising approach to obtain high selectivity toward propylene oxide under high propylene conversion.

Introduction

Propylene oxide (PO) is an important intermediate in the manufacture of a wide range of valuable products: propylene glycol, ethers, isopropanolamines, and various propoxylated products for polyurethanes (1). The current processes for the large scale synthesis of PO include (i) the chlorohydrin process and (ii) the peroxide process (1, 2).

Chlorohydrin process:

$$2H_3C-\underset{H}{C}=CH_2 + 2\,HOCl \longrightarrow H_3C-\underset{H}{\overset{OH}{C}}-CH_2Cl + H_3C-\underset{H}{\overset{Cl}{C}}-CH_2OH$$

$$\downarrow Ca(OH)_2$$

$$2H_2O + CaCl_2 + 2\,H_3C-\underset{H}{C}\overset{O}{\underset{}{\diagup\!\diagdown}}CH_2$$

Peroxide process:

$$H_3C-\underset{H}{C}=CH_2 + \underset{\underset{O}{\overset{\parallel}{}}}{\underset{R'COOH}{ROOH}} \xrightarrow{cat} H_3C-\underset{H}{\overset{O}{\overset{\triangle}{C}}}-CH_2 + \underset{R'COOH}{ROH}$$

The disadvantage of the chlorohydrin process is the use of toxic, corrosive, and expensive chlorine; the major drawback of the peroxide process is the formation of co-oxidates in larger amounts than the desired PO. The direct epoxidation of propylene using O_2 (i.e., partial oxidation of propylene) from air has been recognized as a promising route.

$$H_3C-\underset{H}{C}=CH_2 \quad + \quad 1/2\ O_2 \quad \longrightarrow \quad H_3C-\underset{H}{\overset{O}{\overset{\triangle}{C}}}-CH_2$$

The catalysts which have been tested for the direct epoxidation include: (i) supported metal catalysts, (ii) supported metal oxide catalysts (iii) lithium nitrate salt, and (iv) metal complexes (1-5). Rh/Al_2O_3 has been identified to be one of the most active supported metal catalysts for epoxidation (2). Although epoxidation over supported metal catalysts provides a desirable and simple approach for PO synthesis, PO selectivity generally decreases with propylene conversion and yield is generally below 50%. Further improvement of supported metal catalysts for propylene epoxidation relies not only on catalyst screening but also fundamental understanding of the epoxidation mechanism.

The objective of this study is to investigate the mechanism of propylene oxidation by a transient infrared spectroscopic technique over Rh/Al_2O_3. This technique allows simultaneous measurement of the dynamics of adsorbed species by in situ infrared spectroscopy and the product formation profile by mass spectrometry.

Results and Discussion

Pulse Studies of 1 cm^3 O_2 into the propylene/He flow

Figure 1 shows the series of infrared spectra collected during O_2 pulse studies into flowing He/propylene at 250 °C. The initial exposure of the catalyst to the He/propylene flow produced bands at 1982 cm^{-1} and 1810 cm^{-1}. The band at 1982 cm^{-1}, in the range of adsorbed CO, is a result of interaction of propylene with surface OH. This is evidenced by a decrease in the OH intensity which is accompanied by an increase in the intensity of the 1982 cm^{-1} band. This 1982 cm^{-1} band can also be produced from adsorption of PO. The band at 1810 cm^{-1} is due to CH_2 wagging of propylene; the band at 1590 and 1465 cm^{-1} can be assigned to an acetate species (6). The acetate species can be produced from interaction of C_3H_6 with Al_2O_3-supprted Pt and Cu catalyst in the presence and absence of O_2 (6,7).

Pulsing O_2 led to the growth of the C-O-C band at 978 cm^{-1} for PO (8) and a decrease in the 1982 cm^{-1} band. The MS profiles in Figure 2 shows the increase in the m/e intensity at 28 and 44. The latter can be assigned to CO_2 of which increase and decrease in intensity coincided with that of CO_2 IR intensity in Figure 1. The m/e=28 can be assigned to PO whose intensity increases and decreases in parallel with C-O-C IR intensity band at 978 cm^{-1}. The assignment of m/e=28 to CO can be ruled out because of the absence of CO as evidenced by infrared spectra in Figure 1. The intensity of the 1982 cm^{-1} band rapidly disappeared upon introduction of the O_2 pulse indicating the species is an active precursor for oxidation. MS profiles in Figure 2 also revealed that CO_2 profile led that of PO indicating that the total oxidation of propylene to CO_2 occurred at a higher rate than the partial oxidation to PO; total and partial oxidation occurred in a parallel pathway.

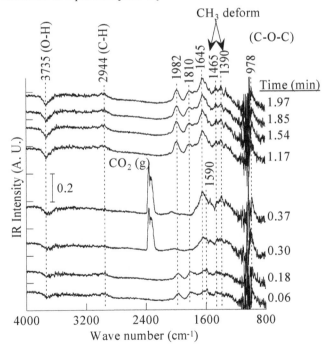

Figure 1 Infrared spectra of pulsing 1 ml O_2 into the propylene/He flow.

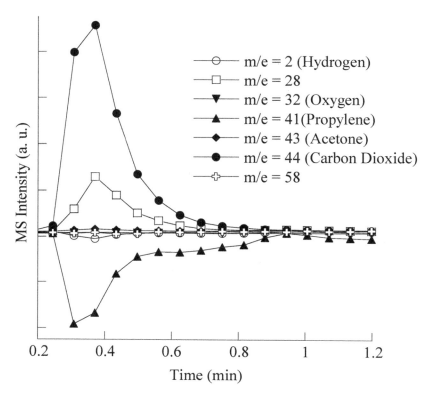

Figure 2 Mass spectrometry profiles of pulsing 1 ml O_2 into the propylene/He flow.

Step switch

Figure 3 shows IR spectra taken during the step switch from steady state He/C_3H_6 to $He/O_2/C_3H_6$. The exposure of the catalyst to propylene prior to the O_2 step switch produced the bands similar to those in Figure 1. Introduction of O_2 into the reaction by switching from He/C_3H_6 to $He/O_2/C_3H_6$ flow led to immediate formation of a CO_2 band followed by the PO C-O-C band at 977 cm^{-1}.   The bands at 977, 1390, and 1640 cm^{-1} are consistent with those in adsorbed PO which is shown as the upper-most spectrum in Figure 3.

The MS analysis shows that the CO_2 profile led that of the PO profile (results not shown). The step switch results further confirm that CO_2 formation is faster than PO formation and that both reactions take place in parallel. GC analysis of the steady state effluent stream from the reactor revealed that propylene conversion was 10.5% at 250 °C; product formation rates were determined to be 1.33, 0.12, and 34.3 µmol/min, respectively, for acetone, PO, and CO_2. The yield calculated on basis of total conversion was 10%, 1%, and 89% for acetone, PO, and CO_2, respectively.

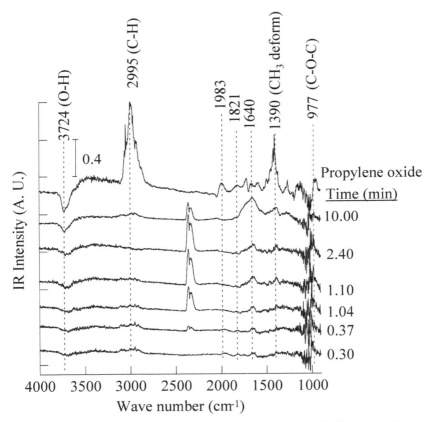

Figure 3 Infrared spectra taken during the step switch from steady state He/C_3H_6 to $He/O_2/C_3H_6$ flow.

Although acetone was a major product, it was not observed by infrared spectroscopy. Flowing helium/acetone over the catalyst at room temperature gave a prominent carbonyl band at 1723 cm^{-1} (not show here). In this study, a DRIFTS (diffuse reflectance infrared Fourier transform spectroscopy) cell was placed in front of a fixed reactor; DRIFTS only monitored the adsorbed and gaseous species in the front end of the catalyst bed. The absence of acetone's carbonyl IR band in Figure 3 and its presence in the reactor effluent suggest the following possibilities: (i) acetone formation from partial oxidation is slower than epoxidation to form PO and/or (ii) acetone is produced from a secondary reaction of PO.

Temperature-programmed desorption (TPD)

Figure 4 and Figure 5 show the IR and MS results during the TPD studies, respectively. Figure 4 shows that exposure of the catalyst to PO/He led to formation of the C-H bands at 2872 cm^{-1} and 2973 cm^{-1}, CH$_3$ deformation band at about 1400 cm^{-1}, and the C-O-C band at 977 cm^{-1} and decrease in the OH band, suggesting the interaction of the surface OH with adsorbed PO. The decrease in the surface OH intensity can be clearly observed by comparing the spectra at different temperatures with the spectrum prior to introduction of PO in the 3300 – 3960 cm^{-1} region. The interaction of PO with the surface also led to a shift in C-H stretching from 3000 cm^{-1} in gaseous PO to 2872-2973 cm^{-1}.

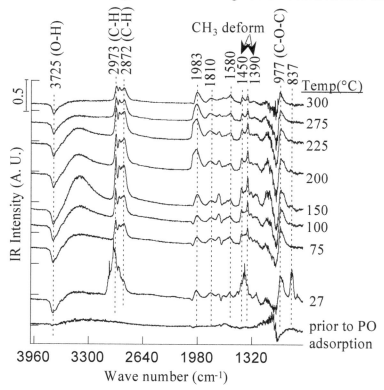

Figure 4 IR spectra taken during TPD at 10 °C/min in flowing 30 ml/min.

The bands in 1580 – 1983 cm^{-1}, which are not observed in the gaseous PO, could be due to the adsorbed C$_3$ oxygenates produced from adsorbed PO. The spectra of adsorbed PO observed in the 800 – 2000 cm^{-1} during TPD resembled those of the pulse reaction studies in Fig. 1, suggesting that adsorbed PO from gaseous PO may have the same structure as the adsorbed epoxidation intermediate.

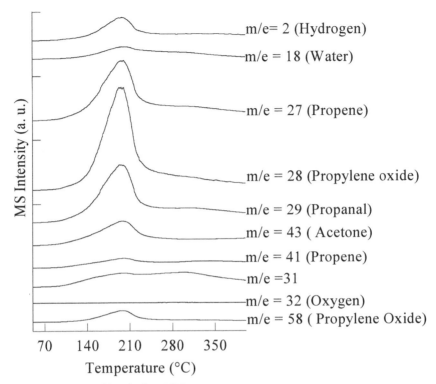

m/e= 2 (Hydrogen)

m/e = 18 (Water)

m/e = 27 (Propene)

m/e = 28 (Propylene oxide)

m/e = 29 (Propanal)

m/e = 43 (Acetone)

m/e = 41 (Propene)

m/e =31

m/e = 32 (Oxygen)

m/e = 58 (Propylene Oxide)

Figure 5 MS profiles during TPD.

TPD of adsorbed PO in Figure 5 shows that PO desorbed at the peak temperature of 180 °C. The profile for m/e=29 is assigned to propanal since it is the only C_3 oxygenated hydrocarbon that gives the highest m/e intensity at 29; The profile for m/e=43 is assigned to acetone since acetone is the only C_3 oxygenated hydrocarbon gives a highest m/e intensity at 43 (8). The broad m/e=31 desorption peak (100 - 360 °C) can be attributed to a C_3 oxygenated compound such as allyl alcohol. The specific assignment of the m/e=31 species is not possible because a number of C_3 oxygenates give the m/e=31. Although PO is the major product during TPD, the observation of acetone and propanal shows that adsorbed PO can convert to other C_3 oxygenates during the TPD on the Rh/Al_2O_3 catalyst.

In summary, the total oxidation of propylene to CO_2 occurred at a higher rate than partial oxidation to propylene oxide and acetone; total and partial oxidations occurred in parallel pathways. The existence of the parallel reaction pathways over Rh/Al_2O_3 suggest that the selective poisoning of total oxidation sites could be a promising approach to obtain high selectivity toward PO under high propylene conversion.

Experimental Section

A 2 wt% Rh/Al_2O_3 catalyst was prepared by incipient wetness impregnation using an aqueous solution of $RhCl_3 \cdot 2H_2O$ (Alfa chemicals) onto γ-alumina support (Alfa Chemicals, 100 m^2/g). The catalyst was dried overnight in air at room temperature and calcined by flowing air at 450 °C for 6 hours. It was then reduced in flowing hydrogen at 450 °C for an additional 6 hours.

The experimental system consists of three sections: (i) a gas metering section with interconnected 4-port and 6-port valves, (ii) a reactor section including an in-situ diffused reflectance infrared Fourier transform spectroscopy reactor (DRIFTS) connected to tubular quartz reactor, (iii) an effluent gas analysis section including a mass spectrometer or a gas chromatograph (9).

The gas metering section is designed to deliver controlled gas flows of C_3H_6 (Praxair, 99.99%), O_2 (Praxair, 99.998%) and He (Praxair, 99.999%) to the reactor system via Brooks 5850 mass flow controllers at a total flow rate 40 ml/min. and 1atm pressure. Feed gas compositions are C_3H_6 (40%), O_2 (10%) and He (50%) for the steady state reaction. Prior to each experiment, the catalyst was reduced in pure flowing H_2 at 34 ml/min for 2 hours at 400 °C.

The tubular reactor consists of a stainless steel tube (3/8" OD) in which approximately 300 mg of 2% Rh/Al_2O_3 is held in place with glass wool and 22 mg of catalyst is loaded in DRIFTS cell. The temperatures are monitored with a K type thermocouple connected to an omega temperature controller. Both pulse and step reaction studies were carried out at 250 °C.

Pulsing O_2 (1 ml) into flowing C_3H_6/He (3/37 ml/min, respectively) over Rh/Al_2O_3 at 250 °C was made by a six-port switching valve with a 1 ml loop. The step switch was made from He/C_3H_6 (36/4 ml/min) to He/C_3H_6/O_2 (35/4/1ml/min) by a four-port valve. During steady state reaction with He/C_3H_6/O_2 (35/4/1 ml/min), the composition of the reactor effluent was determined by SRI 8610 GC with a Porapak R column (6' x 1/8" x 0.085").

Temperature programmed desorption (TPD) was carried out by (i) exposure of the catalyst to a helium/PO flow at room temperature and (ii) heating adsorbed species on the catalyst at a rate of 10 °C/min in flowing helium at 30 ml/min.

Acknowledgements

We would like to thank Ohio Board of Regents for their partial financial support.

References

1. K. Weissermel, and H.–J. Arpe, Industrial Organic Chemistry, VCH Inc., New York, 1997, p. 266 - 271.
2. T. Miyazaki, S. Ozturk, I. Onal, and S. Senkan, *Catal. Today*, **81**, 473 (2003).
3. B. S. Uphade, M. Okumura, S. Tsubota, and M. Haruta, *Appl. Catal. A: Gen*, **190**, 43 (2000).
4. J. Lu, M. Luo, H. Lei, X. Bao, and C. Li, *J. Catal.*, **211**, 552 (2002).
5. T. Miyaji, P. Wu, and T. Tatsumi, *Catal. Today*, **71**, 169 (2001).
6. Y. W. Chi and S. S. C. Chuang, *J. Catal.,* **190**, 75 (2000).
7. D. K. Caption and M. D. Amiridis, *J. Catal.* **184**, 377 (1999).
8. http://webbook.nist.gov/chemistry/ March, 2003.
9. R. W. Stevens, S. S. C. Chuang, and B. H. Davis, *Appl. Catal. A: Gen.*, **252**, 57 (2003).

46. The Effect of Bismuth Promotion on the Selective Oxidation of Alcohols Using Platinum Catalysts

J. R. Anderson[1], G. A. Attard[2], K. G. Griffin[1], and P. Johnston[1]

[1]*Johnson Matthey plc, Catalyst Development, Orchard Road, Royston, Hertfordshire, SG8 5HE, UK*
griffk@matthey.com

[2]*Chemistry Department, University of Cardiff, P.O. Box 912, Cardiff, CF10 3TB, UK*

Abstract

The promotional effect of bismuth on heterogeneous platinum catalysts has been studied for the selective oxidation of alcohols to the carbonyl and carboxylic acid functionality. Using either air or hydrogen peroxide as the oxidant it has been shown that effective promotion from bismuth is attained from precipitation onto the catalyst surface and not as an external reaction component. Catalyst activity tests for the selective oxidation of 1 and 2-octanol indicate that the optimal Bi loading for a 5%Pt/carbon catalyst is approximately 1wt.%. Cyclic voltammetry conducted on equivalent 5%Pt/graphite catalysts demonstrate that Bi is preferentially adsorbed onto Pt step and kink sites and that loadings >1wt.% are present mainly as a bulk Bi phase on the catalyst surface. The results of activity tests indicate that the bulk Bi species is an inactive component of the catalyst. Detailed experimental testing using conditions representative of those used in industry reveal that a 5%Pt,1%Bi/C is active for the selective oxidation of a variety of alcohols. The catalyst can be successfully reused for a number of cycles with minimal loss in activity.

Introduction

The development of clean selective oxidation routes is a challenge to the fine chemical and pharmaceutical industries (1,2). The use of heterogeneous catalysts in conjunction with air or hydrogen peroxide can provide a clean alternative to the use of stoichiometric oxidants (3,4). A potential drawback for this route is premature catalyst deactivation due to either poisoning from the strong adsorption of by-products (5) or the formation of an inactive surface oxide layer (6). An objective in catalyst development is to prepare catalysts having both high activity and stability thereby enabling successful reuse of the catalyst.

The aim of this work is to investigate the changes in activity and morphology of a 5%Pt/C catalyst upon Bi addition. The effect of Bi addition has been studied when Bi has been added as an external component in the reaction mixture and during the catalyst preparation. The impact of variation of the Bi loading has been evaluated by the probing the catalyst surface in a series of electrochemical experiments. Based upon these findings the activity of selected catalysts has been determined under a set of realistic conditions for the selective oxidation reactions shown in Figure 1, using both air and hydrogen peroxide as the oxidant. These reactions have been chosen since the oxidation products are important materials in the fine chemicals industry.

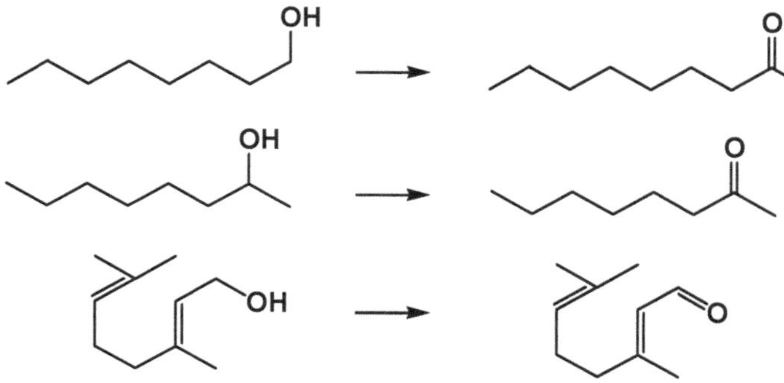

Figure 1 Selective oxidation of 1-octanol, 2-octanol and geraniol to the carbonyl.

Results and Discussion

The activity of 5%Pt/C before and after incorporation of Bi has been evaluated for the selective oxidation reactions above. The effect of incorporation of Bi within the catalyst as well as its use as an external promoter by direct addition of Bi_2O_3 within the reaction mixture has been determined. The results obtained for the selective oxidation of geraniol are shown in Table 1 and indicate that the use of unpromoted 5%Pt/C leads to a low geraniol conversion (17%) after 6h. Addition of Bi_2O_3 into the reaction mixture has a negligible impact on the catalyst activity and conversion is limited to 20%. However, incorporation of Bi by co-precipitation with the Pt precursor salt during catalyst preparation results in a Pt-Bi/C catalyst displaying significantly greater activity than an equivalent non promoted 5%Pt/C. These results clearly demonstrate the requirement of obtaining the correct Bi location to achieve effective promotion of this reaction.

The effect of Bi promotion for the selective oxidation of 1-octanol using H_2O_2 as oxidant is reported in Table 2. Since decomposition of H_2O_2 by Platinum Group Metals is rapid, H_2O_2 is fed continuously into the reactor over 2 hours. The results obtained demonstrate that the presence of Bi_2O_3 as an additive within the reaction mixture displays no significant influence on catalyst activity. However, Bi promoted Pt/C catalysts, prepared by co-precipitation of

Bi and Pt precursor salts, show significantly higher activities in this reaction. These results indicate the promotional effect of Bi is not affected by the oxidant used, but is associated with Bi in close interaction with the active metal (Pt).

Table 1 Selective oxidation of geraniol using air as oxidant. Conditions: 0.085mol geraniol in toluene, 5wt.% loading of catalyst (dry weight), 60°C, 3bar air, 600rpm, 6h.

Catalyst	Additive	Conversion (%)[a]
5%Pt/C	None	17
5%Pt/C	Bi_2O_3[b]	20
5%Pt,1%Bi/C	None	56

[a] 100% selectivity to cis-citral observed.
[b] 0.03g Bi_2O_3 powder added (equivalent to 3wt.% Bi).

Table 2 Selective oxidation of 1-octanol using H_2O_2 as oxidant. Conditions: 0.015mol 1-octanol in toluene, reactant:metal molar ratio 80:1, 60°C, 600rpm, 30%H_2O_2 fed at rate of 1molar equivalent/h, atmospheric pressure.

Catalyst	Additive	Conversion (%)[a]	Octanal Yield (%)	Octanoic Acid Yield (%)
5%Pt/C	None	19	16	3
5%Pt/C	Bi_2O_3[b]	20	17	3
5%Pt,1%Bi/C	None	93	63	30

[a] Values taken after 2 equivalents of H_2O_2 added.
[b] 0.03g Bi_2O_3 powder added (equivalent to 3wt.% Bi).

The interaction of promoter and active metal has been studied further in a series of experiments in which the Bi loading of the catalyst has been varied in the range from 0 to 5wt.%. Catalysts were prepared by co-precipitation of Pt and Bi precursor salts onto the carbon support. The relationship between conversion and Bi loading for the selective oxidation of 1-octanol and 2-octanol is shown in Figure 2. The profiles indicate that the conversion is increased as the Bi loading is increased from 0 to 1wt.%, thus a Bi loading of 0.5wt.% is below the optimal level. At 1wt.% the conversion reaches a plateau and subsequent Bi addition has a negligible impact on activity. These results suggest that the 5%Pt/C catalyst requires at least 1wt.% loading to obtain an optimal promotional effect. It is interesting to note that the further addition of Bi into the system beyond 1wt.% has neither a promotional nor poisoning effect.

In order to understand the Pt-Bi interaction in more detail, the morphology of the catalyst has been studied using cyclic voltammetry. Since this technique requires the use of a conducting material, equivalent Pt-Bi catalysts were prepared on a graphite support for these experiments.

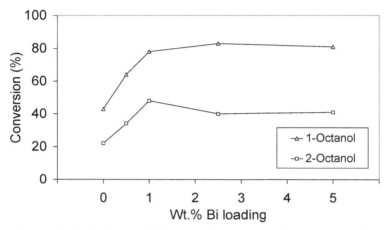

Figure 2 Selective oxidation of 1-octanol and 2-octanol using air as oxidant. Conditions: 0.015mol 1-octanol in toluene, reactant:metal molar ratio 80:1, 60°C, 600rpm, 3 bar air, 6h.

The cyclic voltammetry profiles contain electroadsoprtion peaks that reveal information on the adsorption sites available on the 5%Pt/G catalyst before and after treatment with Bi. Progressing down the series from Figure 3(a) to 3(d) represents an increase in Bi loading from to 0 to 5wt.%. In Figure 3(a) the broad peaks between 0 to 0.2V are attributed to Pt step and kink sites (7-10). These peaks are progressively reduced in (b) and (c) as the Bi atoms preferentially occupy these sites (10-13). In Figure 3(c) there is a broad peak at 0.6V arising from the occupation of Bi on the Pt (111) terraces (12). In Figure 3(d) the Bi loading has increased to 5wt.% and the peak at 0.6V is seen to be more pronounced indicating complete filling of all (111) sites together with at least some Bi forming local, high coverage compression structures as signified by the growth in Bi features at potentials < 0.6V (14-15). It may also be that some second layer Bi is formed in this potential range when the redox charge for this peak is calculated (15). In Figure 3(d) the large peak at 0.2V is assigned to a bulk Bi phase. This observation is in agreement with Attard et al. who claim that the order of site occupation in a heterogeneous Pt catalyst follows the trend: step and kink sites > terrace sites > multi-layers on terrace sites > bulk phase formation (11,14-15).

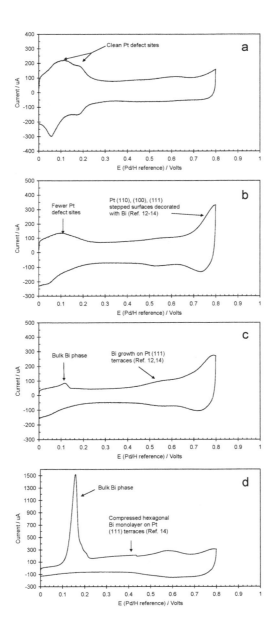

Figure 3 Cyclic voltammetry profiles of (a) 5%Pt/G, (b) 5%Pt,0.5%Bi/G, (c) 5%Pt,1%Bi/G, (d) 5%Pt,5%Bi/G.

It is concluded that the occupation of the step and kink sites plays a crucial role in the promotion of the Pt catalyst. The cyclic voltammetry results can be used to explain the conversion trends observed in Figure 2. For unpromoted 5%Pt/C the Pt step and kink sites are unoccupied and available for adsorption of reactant and oxidant species. During reaction these sites facilitate premature catalyst deactivation due to poisoning by strongly adsorbed by-products (5) and (or) the formation of a surface oxide layer (6). The 5%Pt,0.5%Bi/C catalyst has a portion of these Pt step and kink sites occupied and the result is a partial reduction in the catalyst deactivation and a consequent increase in alcohol conversion. As the Bi level is increased to 1wt.% almost all of the Pt step and kink sites are occupied and the result is a catalyst with high activity. As more Bi is introduced onto the catalyst surface a bulk Bi phase is formed. Since the catalyst activity is maintained it is speculated that the bulk Bi phase is not involved in the catalytic cycle.

The stability of the Bi species has been investigated in a Bi stripping cyclic voltammetry experiment. Figure 4 (a) and (b) show the results of progressing into the oxide region of 5%Pt,5%Bi/G from 0.7-1.2V at 0.1V intervals. These conditions simulate the effect on catalyst morphology as it is subjected to an increasing oxidative atmosphere. Figure 4 (a) shows the five cycles grouped together and Figure 4 (b) has the same profile with cycle 1 removed allowing for closer inspection of the electroadsorption peaks. In Figure 4 (a) the first cycle clearly shows the bulk Bi phase present at 0.2V. In the second cycle this phase has been stripped from the catalyst surface along with the 3rd and successive layers of Bi on the Pt (111) terraces (0.4V). The results show that Bi atoms separated from the Pt surface are the least stable under oxidising conditions. The second cycle also indicates that some of the first and second layers of Bi on Pt (111) terraces are still present (0.6V) and the Pt step and kink sites are still decorated as demonstrated by the lack of any peaks between 0 and 0.2V. In the third cycle the peak at 0.6V has been removed indicating that the Bi on the Pt (111) terraces has been stripped off. The third to fifth cycles are characterised by the progressive increase of a broad peak at 0-0.2V and subsequent decrease of the feature at 0.8V indicative of the removal of Bi atoms from the Pt kink and step sites. It has been shown that the cycling deeper into the oxide region of a Pt catalyst will lead to some re-structuring of Pt (110) defect sites at 0.1V (8). This may partly explain the strong growth of the peak in that region. Nonetheless Bi stripping appears to follow a trend opposite to the order of occupation, namely: bulk phase > multi-layers on terrace sites > terrace sites > step and kink sites. It is therefore suggested that the similar catalyst activity observed between 5%Pt/C catalysts doped with Bi loadings from 1-5wt.% is due to the excess Bi forming a bulk-like phase on the catalyst surface which is stripped off under reaction conditions. A metal leaching study of 5%Pt,2.5%Bi/C during the selective oxidation of 1-octanol revealed that in the first reaction cycle 57% of the Bi leached into the reactant liquor. A cyclic voltammetry study of the used catalyst revealed the disappearance of the peak at 0.2V indicating the Bi had leached from the bulk phase.

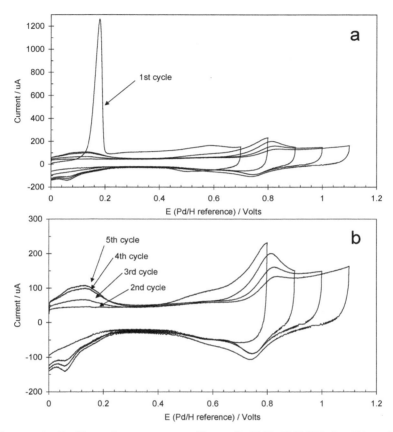

Figure 4 Cyclic voltammetry profiles of 5%Pt,5%Bi/G in Bi stripping experiment, (a) cycles 1-5 and (b) cycles 2-5.

Based upon the activity and electrochemical experimental results the 5%Pt,1%Bi/C catalyst was chosen for further detailed evaluation. For the catalyst to be effective in industrial applications it is desirable that it should remain active for a number of reaction cycles. The recycle capability of 5%Pt,1%Bi/C was evaluated under realistic conditions for a number of selective oxidation reactions, see Table 3.

Table 3 indicates that 5%Pt,1%Bi/C is active for three reaction cycles in the selective oxidation of the chosen alcohols. For primary alcohols the use of water as solvent can promote the aldehyde to carboxylic acid reaction (3). This effect is observed in the selective oxidation of 1-octanol where octanoic acid is formed with 97% selectivity in the first cycle dropping to 81% in the third. In the selective oxidation of geraniol only citral is observed as the oxidation product. The presence of the double bond stabilises the aldehyde even in the presence of water. For 2-octanol the reaction is 100% selective to 2-octanone and the catalyst undergoes negligible deactivation after three cycles.

Table 3 Recycle testing of 5%Pt,1%Bi/C for the selective oxidation of various alcohols. Conditions: 15ml substrate / 15ml water, 60°C, 3 bar air.

Alcohol	Solvent	Conversion (%)		
		Fresh	1st Recycle	2nd Recycle
1-Octanol[a]	Water	96	93	81
2-Octanol[b]	Water	92	87	92
Geraniol[c]	Water	47	41	40

[a] 24h, 5wt.% catalyst loading, 1200rpm.
[b] 6h, 5wt.% catalyst loading, 1200rpm.
[c] 6h, 10wt.% catalyst loading, 600rpm.

Conclusions

The work demonstrates that the incorporation of Bi is effective in promoting catalytic activity in the selective oxidation of various alcohols using both air and hydrogen peroxide as oxidant. Effective promotion requires intimate contact between Bi and Pt. Addition of Bi as an external reaction component to the reaction mixture is ineffective as promoter. Catalyst testing and cyclic voltammetry experiments have highlighted the need to balance the loading of Bi into the catalyst. Insufficient promoter addition leads to premature catalyst deactivation whereas overloading results in Bi loss during reaction (Bi stripping). For a 5%Pt/C catalyst the optimal Bi loading was found to be approximately 1wt.% Electrochemical experiments show that at this loading there is sufficient Bi present to preferentially occupy the highly active Pt step, kink and terrace sites without forming large quantities of a bulk Bi. Using 5%Pt,1%Bi/C catalysts, the selective oxidation of 1-octanol, 2-octanol and geraniol can be conducted with little loss in activity for three reaction cycles.

Experimental Section

Bi promoted Pt catalysts were prepared using JM proprietary methods. Aqueous solutions of Pt and Bi salts were co-precipitated onto the catalyst support and reduced using chemical reducing agents. Thereafter the materials were washed, filtered and retained as pastes. Graphite supported materials were dried prior to use. Similar preparation methods were used for all Pt and Pt-Bi catalysts.

Catalyst testing was carried out in two different reactors. The reactions using air as the oxidant were performed in a 50ml glass lined autoclave (working volume 30ml). A pressure of 3 bar air was used and the O_2 partial pressure was kept constant by maintaining a 5ml/min air flow through the reactor. The reactions with H_2O_2 were carried out in a 50ml stainless steel reactor (working volume 30ml). A $30\%H_2O_2$ solution was pumped into the reactor at a rate of 1equivalent/hour at atmospheric pressure. Unless stated otherwise all reactions were performed at 60°C and 600rpm stirring speed using toluene as solvent. Mesitylene was used as internal standard and reaction products were analyzed by GC.

Table 4 Chemisorption characterisation of fresh Pt and Pt-Bi catalysts supported on carbon and graphite. H_2 chemisorption data calculated from the charge in the hydrogen under-potential deposition (H-UPD) region of the cyclic voltammetry profiles. Values given as m^2/g (total catalyst).

Catalyst	CO Chemisorption (m^2/g)	H_2 Chemisorption (m^2/g)
5%Pt/C	8.1	-
5%Pt,0.5%Bi/C	5.4	-
5%Pt,1%Bi/C	2.8	-
5%Pt,2.5%Bi/C	1.6	-
5%Pt,5%Bi/C	1.7	-
5%Pt/G	4	4.8
5%Pt,0.5%Bi/G	3.2	2.8
5%Pt,1%Bi/G	1.7	1.7
5%Pt,5%Bi/G	0	0.9

Cyclic voltammetry experiments were obtained under a nitrogen atmosphere at a sweep rate of 1mV/s using a gold mesh electrode (0.064mm diameter) and 3mg catalyst. The electrolyte solution of 0.5M sulfuric acid was made with ultrapure water. The electrolyte was degassed for 30min prior to experimentation to remove any dissolved oxygen. The electrochemical potentials are quoted with reference to a saturated palladium-hydrogen electrode.

Acknowledgements

We gratefully acknowledge the support provided by the E.U. GROWTH programme (GRD-2000-25053) Catalytic Oxidation as a Tool for Sustainable Fine Chemicals Manufacture (SUSTOX). We thank Andy Smith (Johnson Matthey, Royston) for his assistance with the analytical experiments.

References

1. R. A. Sheldon and J. Dakka, *Catal. Today,* **19**, 215 (1994).
2. R. A. Sheldon, I. C. E. Arends, and A. Dijksman, *Catal. Today,* **57**, 157 (2000).

3. K. G. Griffin, P. Johnston, S. Bennett, and S. Kaliq, Chemical Industries (Dekker), **89**, (*Catal. Org. React.*), 169 (2002).
4. R. Anderson, K. G. Griffin, P. Johnston, and P. L. Alsters, *Adv. Synth. Catal.*, **345**, 517 (2003).
5. T. Mallat and A. Baiker, *Appl. Catal.*, **79**, 41 (1991).
6. M. Besson, F. Lahmer, P. Gallezot, P. Feurtes, and G. Flèche, *J. Catal.*, **152**, 116 (1995).
7. A. Rodes, K. El Achi, M. A. Zamakhchari, and J. Clavilier, *J. Electroanal. Chem. and Interfacial Electrochem.*, **284**, 245 (1990).
8. J. Clavilier, K. El Achi, and A. Rodes, *Chem. Phys.*, **141**, 1 (1990).
9. J. Clavilier, *ACS Symp. Ser.* (*Electrochem. Surf. Sci.: Mol. Phenom. Electrode Surf.*), **378**, 202 (1988).
10. G. A. Attard, A. Ahmadi, D. J. Jenkins, O. A. Hazzazi, P.B. Wells, K. G. Griffin, P. Johnston, and J. E. Gillies, *Chem. Phys. Chem.*, **4**, 123 (2003).
11. E. Herrero, V. Climent, and J. M. Feliu, *Electrochem. Commun.*, **2**, 636 (2000).
12. J. Clavilier, J. M. Feliu, and A. Aldaz, *J. Electroanal. Chem. and Interfacial Electrochem.*, **243**, 419 (1988).
13. J. Clavilier, J. M. Feliu, A. Fernandez-Vega, and A. Aldaz, *J. Electroanal. Chem. and Interfacial Electrochem.*, **269**, 175 (1989).
14. R. W. Evans and G. A. Attard, *J. Electroanal. Chem.*, **345**, 337 (1993).
15. M. Price, M. Phil. Thesis, University of Cardiff (2000).

47. Preparation and Activity of a V-Ti/Silica Catalyst for Olefin Epoxidation

Susannah L. Scott,[a] Azfar Hassan[b]

[a] *Department of Chemical Engineering, University of California—Santa Barbara*
[b] *Department of Chemistry, University of Ottawa, Canada*

sscott@engineering.ucsb.edu

Abstract

A heterogeneous olefin epoxidation catalyst containing both V and Ti in the active site was prepared by sequential non-hydrolytic grafting. The silica was exposed first to $VO(O^iPr)_3$ vapor followed by $Ti(O^iPr)_4$ vapor. Formation of propene is evidence for the creation of Ti-O-V linkages on the surface. Upon metathesis of the 2-propoxide ligands with ^tBuOOH, the catalyst becomes active for the gas phase epoxidation of cyclohexene. The kinetics of epoxidation are biphasic, indicating the presence of two reactive sites whose activity differs by approximately one order of magnitude.

Introduction

Synergy between different metal sites in bimetallic catalysts is a well-known phenomenon. For example, $V-Ti-SiO_2$ is a more active and more selective catalyst than any of $V-SiO_2$, $Ti-SiO_2$ or $V-TiO_2$ for the oxidation of *o*-xylene to phthalic anhydride (1) as well as the selective catalytic reduction of NO_x (2). Both V-based and Ti-based epoxidation catalysts are well-known, in homogeneous and heterogeneous forms (3). A mixed V-Ti catalyst for epoxidation has not previously been reported. Such a system may be designed to catalyze the epoxidation of functionalized olefins.

One route to heterogeneous bimetallic catalysts with well-defined active sites involves deposition of bimetallic molecular precursors. A more general route is sequential grafting of monometallic precursors, which are more readily available. We previously reported low temperature, non-hydrolytic syntheses of $V-SiO_2$ (4) and $Ti-SiO_2$ (5) by grafting of the corresponding metal alkoxides on the surface hydroxyl sites. We have also prepared a family of heterobimetallic materials $V-Ti-SiO_2$ by sequential grafting of the respective metal complexes (6). In this report, we examine how one of these bimetallic systems performs in epoxidation catalysis compared to the monometallic materials.

The stoichiometry and kinetics of gas phase epoxidation of cyclohexene by silica-supported $Ti(O^iPr)_4$ upon treatment with *tert*-butylhydroperoxide were the subject of an earlier report in this series (7). Silica-supported $VO(O^iPr)_3$ is inactive towards cyclohexene under the same conditions.

Experimental Section

Aerosil 200 silica, with a surface area of *ca.* 200 m^2/g, was a gift of Degussa Corp. It was partially dehydroxylated at 500°C for at least 4 h prior to the grafting of metal complexes at room temperature. $Ti(O^iPr)_4$ and $VO(O^iPr)_3$ (Aldrich) were stored in Schlenk tubes under an inert atmosphere. The liquid metal complexes were subjected to several freeze-pump-thaw cycles to remove dissolved gases before use. Grafting was performed using breakseal techniques, as described previously (2,5). Anhydrous *tert*-butylhydroperoxide (Aldrich, 10 M in decane) was dried over $MgSO_4$ and stored in a glass bulb under vacuum. Cyclohexene was dried, vacuum-distilled and stored over activated molecular sieves in a glass bulb. Liquid reagents were introduced into the reactor by vapor phase transfer via a high vacuum manifold.

Metal-modified silicas were exposed to excess ^tBuOOH vapor in order to generate the supported *tert*-butylperoxide complexes, followed by evacuation to remove ^iPrOH and unreacted ^tBuOOH. Reaction kinetics were monitored as the uptake of cyclohexene from the gas phase, using a ThermoNicolet Nexus FTIR spectrometer to measure the intensity of the $\upsilon(C=C)$ mode. *In situ* spectra were recorded in custom-made glass reactors under vacuum. Formation of cyclohexene oxide was confirmed by GC/MS on an HP 6890 equipped with a DBI capillary column (J&W Scientific).

Metal loadings on silica were determined at the end of each experiment. V and Ti were extracted by boiling about 15 mg of the catalyst in 1 M H_2SO_4 for 1 h, followed by addition of 0.1 mL 30% H_2O_2 and dilution with 1 M H_2SO_4 to 25 mL. After filtering, UV-visible spectra were recorded on a Cary 100 spectrophotometer. The spectra of peroxo complexes of Ti and V have λ_{max} at 405 and 462 nm, respectively. In the case of bimetallic materials, the absorbance at each wavelength was measured, and the concentration of each metal was calculated by solving two simultaneous equations with two unknowns.

Results and Discussion

The room temperature reaction of Aerosil 200 silica pretreated at 500°C with excess $VO(O^iPr)_3$ results in complete reaction of the surface hydroxyls and quantitative formation of the mononuclear surface complex **1**, eq 1 (4).

$$\equiv SiOH + VO(O^iPr)_3 \rightarrow \equiv SiOVO(O^iPr)_2 + {}^iPrOH \qquad (1)$$

$$\mathbf{1}$$

The only volatile product is ^{i}PrOH, as was confirmed by FTIR and GC/MS. Although **1** contains no unreacted hydroxyl groups, and does not react further with VO(O^{i}Pr)$_3$, it reacts with Ti(O^{i}Pr)$_4$ according to eq 2.

$$\equiv\text{SiOVO(O}^i\text{Pr)}_2 + 2\ \text{Ti(O}^i\text{Pr)}_4 \rightarrow \equiv\text{SiOVO[OTi(O}^i\text{Pr)}_3]_2 + 2\ ^i\text{PrOH} + 2\ C_3H_6 \quad (2)$$
$$\textbf{1} \qquad\qquad\qquad\qquad\qquad\qquad \textbf{2}$$

Formation of 2-propanol and propene was confirmed by GC/MS. Chemisorption of Ti(O^{i}Pr)$_4$ is therefore suggested to be a consequence of nonhydrolytic condensation of the 2-propoxide ligands, which generates V-O-Ti bridges in the absence of surface hydroxyl groups. After removal of volatiles and desorption of physisorbed metal complexes, analysis reveals the presence of two equiv. Ti per V site on the surface, Table 1 (experiments 3 and 4).

Table 1 Catalyst composition and rate constants for cyclohexene epoxidation.

Expt	wt.%Ti	wt.% V	Ti/V	k_1 / (mol Ti)$^{-1}$ s^{-1}	k_2 / (mol Ti)$^{-1}$ s^{-1}
1	0.0	2.0	-	no activity	no activity
2	2.5	0.0	-	1371	124
3	3.8	2.0	1.9	1038	84
4a	3.1	1.6	1.9	1104	92
4b*				1051	79

* Catalyst regenerated by addition of ^{t}BuOOH followed by evacuation of excess reagent.

The 2-propoxide ligands of **2** undergo quantitative exchange in the presence of excess *tert*-butylhydroperoxide, eq 3:

$$\equiv\text{SiOVO[OTi(O}^i\text{Pr)}_3]_2 + \text{xs}\ ^t\text{BuOOH} \rightarrow \equiv\text{SiOVO[OTi(OO}^t\text{Bu)}_3]_2 + 6\ ^i\text{PrOH} \quad (3)$$
$$\textbf{2} \qquad\qquad\qquad\qquad\qquad\qquad \textbf{3}$$

The resulting material is active for the gas phase epoxidation of simple olefins. Addition of cyclohexene resulted in the formation of cyclohexene oxide as the sole volatile product, detected by GC/MS.

Rate of Epoxidation

The kinetics of epoxidation were measured *in situ* as the rate of uptake of olefin by the catalyst, since the epoxide product remains adsorbed on the catalyst surface in the absence of solvent. Addition of 2 Torr cyclohexene vapor (30 μmol) to a 14.2 mg sample of **3** (9.0 μmol Ti, 4.5 μmol V) at room temperature resulted in the loss of the υ(C=C) mode in the gas phase IR spectrum over the course of one hour, Figure 1a. The form of the curve is biexponential (8); the curve fit represents the contributions of two consecutive reactions with pseudo-first-order rate constants differing by one order of magnitude. Pseudo-first-order rate constants were converted to second-order rate constants by dividing by n_{Ti}, Table 1.

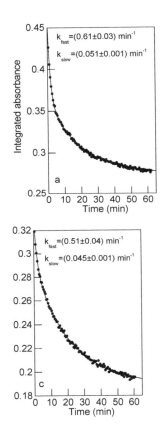

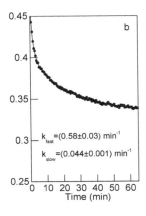

Figure 1 Rate of disappearance of cyclohexene (30 μmol) catalyzed by (a) **3** (9.0 μmol Ti; 4.5 μmol V), upon first addition of cyclohexene; (b) upon second addition of cyclohexene after regeneration of **3** with 'BuOOH; (c) **4** (6.2 μmol Ti) treated with 'BuOOH.

The catalyst can be regenerated by evacuation to remove cyclohexene oxide and addition of fresh 'BuOOH at the end of the first kinetic run. When a second dose of cyclohexene vapor was introduced to **3**, very similar kinetic behavior was observed, Figure 1b. However, the smaller absorbance change implies that less cyclohexene was epoxidized, likely because of incomplete removal of the epoxide which blocks the active sites.

As mentioned above, V(O^{i}Pr)$_3$-modified silica treated with ^{t}BuOOH shows no activity towards cyclohexene at room temperature. However, the material obtained by grafting Ti(O^{i}Pr)$_4$ directly on unmodified silica heated at 500°C, eq 4, does convert olefins to their epoxides upon treatment with ^{t}BuOOH. The catalyst **4** also shows biphasic kinetic behavior towards cyclohexene, Figure 1c.

$$\equiv SiOH + 2\ Ti(O^iPr)_4 \rightarrow \equiv SiOTi_2O(O^iPr)_5 + 2\ ^iPrOH + C_3H_6 \qquad (4)$$
$$\mathbf{4}$$

Biphasic behavior suggests the presence of two active sites. The magnitude of the difference in their activities suggests these may correspond to diperoxo and monoperoxometal sites. Under our experimental conditions, the monoperoxo site would be formed upon reaction of the diperoxo site with olefin in the absence of excess ^{t}BuOOH.

Conclusions

The presence of V does not diminish the activity of a grafted Ti-SiO$_2$ catalyst for olefin epoxidation. However, activity towards simple olefins such as cyclohexene is not enhanced. Since homogeneous V catalysts are known to catalyze the epoxidation of functionalized olefins (e.g., allylic alcohols), the ability of a mixed V-Ti/SiO$_2$ catalyst to achieve such transformations will be the next focus of our investigations.

Acknowledgement

This work was supported by a Strategic Grant from the Natural Sciences and Engineering Research Council of Canada.

References

1. C. R. Dias, M. F. Portela, M. Galan-Fereres, M. A. Banares, M. L. Granados, M. A. Pena, and J. L. G. Fierro, *Catal. Lett.*, **43**,117-121 (1997).
2. R. A. Rajadhyaksha, G. Hausinger, H. Zeilinger, A. Ramstetter, H. Schmelz, and H. Knozinger, *Appl. Catal.*, **51**, 67-79 (1989).
3. D. E. De Vos, B. F. Sels, and P. A. Jacobs, *Adv. Synth. Catal.* **345**, 457-473 (2003).
4. G. L. Rice and S. L. Scott, *Langmuir*, **13**, 1545-1551 (1997).
5. A. O. Bouh, G. L. Rice, and S. L. Scott, *J. Am. Chem.Soc* **121**, 7201-7210 (1999).
6. G. L. Rice and S. L. Scott, *Chem. Mater.*, **10**, 620-625 (1998).
7. A. O. Bouh, A. Hassan, and S. L. Scott, Chemical Industries (Dekker), **89**, (*Catal. Org. React.*), 537-543 (2002).
8. J. H. Espenson, Chemical Kinetics and Reaction Mechanisms, 2nd ed., McGraw-Hill, New York, 1995, p. 71 -75.

48. Aerobic Formaldehyde Oxidation Under Mild Conditions Mediated by Ce-Containing Polyoxometalates

Oxana A. Kholdeeva,[†,*] Maria N. Timofeeva,[†] Gennadii M. Maksimov,[†] Raisa I. Maksimovskaya,[†] Anastasia A. Rodionova,[†] and Craig L. Hill [‡,*]

[†]*Boreskov Institute of Catalysis, Lavrentieva 5, Novosibirsk 630090, Russia*
[‡]*Department of Chemistry, Emory University, 1515 Pierce Drive, Atlanta, GA 30322*

Abstract

Aerobic oxidation of formaldehyde in water under mild conditions (20-40 °C, 1 atm of air or O_2) in the presence of Ce-substituted POMs affords formic acid with high selectivity.

Introduction

The development of catalysts for the oxidation of organic compounds by air under ambient conditions is of both academic and practical importance (1). Formaldehyde is an important intermediate in synthetic chemistry as well as one of the major pollutants in the human environment (2). While high temperature (> 120 °C) catalytic oxidations are well known (3), low temperature aerobic oxidations under mild conditions have yet to be reported. Polyoxometalates (POMs) are attractive oxidation catalysts because these extensively modifiable metal oxide-like structures have high thermal and hydrolytic stability, tunable acid and redox properties, solubility in various media, etc. (4). Moreover, they can be deposited on fabrics and porous materials to render these materials catalytically decontaminating (5). Here we report the aerobic oxidation of formaldehyde in water under mild conditions (20-40 °C, 1 atm of air or O_2) in the presence of Ce-substituted POMs (Ce-POMs).

Results and Discussion

Various Ce-POMs have been synthesized and evaluated as catalysts for the title reaction at ambient conditions along with simple Ce-compounds (Figure 1). The most active catalyst was the monosodium acid salt of the Ce(IV)-monosubstituted silicotungstate, $NaH_3SiW_{11}CeO_{39} \cdot 7H_2O$ (**1**), especially when combined with a 4-fold molar excess of $AgNO_3$ (Figure 1). Surprisingly, no precipitation of the silver salt of the Ce-POM occurred, contrary to many other POMs, including $PMo_{10}V_2O_{40}^{5-}$ (6). The accelerating effect of $AgNO_3$, compound known to co-catalyze oxidations but be minimally active itself, was considerably less pronounced when the reaction temperature was increased.

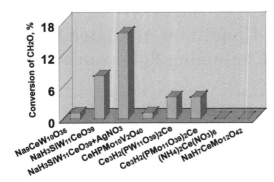

Figure 1 Oxidation of CH_2O (4.8 mmol) by air (1 atm) in the presence of Ce-compounds (0.02 mmol) in H_2O (5 ml) at 20°C for 45 h.

The existence of $SiW_{11}CeO_{39}^{4-}$ was previously suggested (7); however, contrary to $PW_{11}CeO_{39}^{3-}$ (8), it had never been isolated. We have synthesized **1** by using the previously described electrodialysis method (8,9) and characterized it by using elemental analysis, potentiometric titration, IR, UV-vis, ^{183}W and ^{17}O NMR, and CV. All these data are consistent with a monomeric Keggin structure of **1**, containing one Ce(IV) ion per one Si atom. Significantly, dimeric Ce-POMs, $Na_{10}(NH_4)_2[(SiW_{11}O_{39})_2Ce]$ and $H_{10}[(PW_{11}O_{39})_2Ce]$, show very low activity (Table 1), most likely as a consequence of coordinate saturation and the non-lability of the Ce centers.

Table 1 Aerobic oxidation of formaldehyde in water in the presence of Ce-containing compounds.[a]

Catalyst	CH_2O conversion (%)	Yield of $HCOOH^b$ (%)	TON[c]
without catalyst	0	-[d]	-
$NaH_3SiW_{11}CeO_{39}$	15	14	3.4
$Na_{2.4}H_{1.6}SiW_{11}CeO_{39}$	2	-	-
$Na_3PW_{11}CeO_{39}$	0	-	-
$H_{10}[(PW_{11}O_{39})_2Ce]$	1	-	-
$Na_{10}(NH_4)_2[(SiW_{11}O_{39})_2Ce]$	4	4	0.9
$Ce(NO_3)_3$	0	-	-
$Ce(SO_4)_2^{e}$	9	9	2.0

[a]Reaction conditions: CH_2O, 0.59 mmol; catalyst, 0.026 mmol; air, 1 atm; H_2O, 5 ml; 40 °C; 5 h; [b]GC yield based on initial CH_2O; [c]TON = (moles of CH_2O consumed/moles of catalyst) x 100; [d]not determined; [e]H_2SO_4, 0.125 M.

It is noteworthy that $Na_{2.4}H_{1.6}SiW_{11}CeO_{39}$ and $Na_3PW_{11}CeO_{39}$ were practically inactive compared to $NaH_3SiW_{11}CeO_{39}$ (Table 1), indicating that the number of protons in the Ce-POM is a crucial factor in catalytic activity. The stoichiometric oxidation of formaldehyde with Ce(IV) in aqueous acid solutions was reported in the early 1970's (10). We have found that $Ce(SO_4)_2$ shows considerable catalytic activity in aerobic formaldehyde oxidation; however, it requires the use of a large excess of H_2SO_4 (Table 1).

The oxidation of CH_2O in the presence of **1** affords formic acid in nearly quantitative yields (eq 1). Only traces (<0.5%) of CO and CO_2 were detected by GC in the reactions performed at 20-40 °C. No methylformate was found.

$$CH_2O + 1/2\ O_2 \Rightarrow HCOOH \qquad (1)$$

Formic acid is stable under the reaction conditions. Optimization of the reaction conditions led to a system that afforded 30 turnovers of **1** after 5 h (Table 2). Based on both IR and UV-vis, the Ce-POM retains its structure under the turnover conditions. The formic acid product inhibits the reaction, so procedures to remove it during reaction are needed to increase CH_2O conversions.

Table 2 Aerobic oxidation of formaldehyde in water in the presence of **1**.[a]

[**1**] x 10^3 (M)	[CH_2O] (M)	CH_2O conversion (%)	Yield of $HCOOH^b$ (%)	TON^c
12.4	0.117	9	8	0.8
12.4[d]	0.117	10	9	0.9
6.2	0.117	20	20	3.8
6.2[d]	0.117	21	20	4.0
3.1	0.117	15	14	5.7
1.6	0.117	11	10	8.0
3.1	0.233	18	17	13.5
3.1	0.467	20	19	30.1
12.4	0.467	25	23	9.4

[a]Reaction conditions: 1 atm of air, 40°C, 5 h; [b]GC yield based on initial CH_2O; [c]turnovers = (moles of CH_2O consumed/moles of catalyst) x 100; [d]1 atm of O_2 was used instead of air.

The reaction is retarded by the addition of the radical chain scavengers 2,6-di-*t*-butyl-4-methylphenol and hydroquinone. The oxidation rate is not affected by the oxygen pressure (Table 2), indicating that re-oxidation of Ce(III) in **1** is not rate limiting. The oxidation rate is first order in formaldehyde and goes through a maximum with increasing concentration of **1**. All these phenomena are consistent with a chain radical mechanism of oxidation (11,12).

Importantly, **1** can be heterogenized on NH_2-functionalized supports such as xerogels, fibers, etc., and in this form used repeatedly without significant loss of catalytic activity (Figure 2). The supported Ce-POM can be separated from the reaction mixture by simple filtration.

Experimental Section

The formaldehyde solutions were prepared by diluting 35.5% aqueous formalin with water. Synthesis of **1** was performed in a two-chamber electrodialyzer described in (8,9). To 30 ml of an aqueous solution of 0.2 M $Na_8SiW_{11}O_{39}$

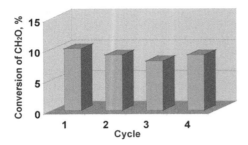

Figure 2 Oxidation of CH₂O (1.17 mmol) by air (1 atm) in the presence of **1** supported on NH₂-xerogel (500 mg; for composition, see Experimental) in H₂O (5 ml) at 40°C for 5 h.

2.604 g of $Ce(NO_3)_3$ $6H_2O$ were added. The solution over the solid was passed through the anodic chamber of the electrodialyzer. Turning on the current (density 0.1 A/cm²) causes dissolution of the solid and lightening of the solution. Ce(III) is oxidized to Ce(IV), and Na^+ is partially replaced by H^+ in the first 0.5h. The total time of the dialysis was 5h. Concentration of the solution followed by drying (1 h, 100 °C) afforded yellow plates of **1** (100% yield). E.A. (wt.%, found/calc.): Na 0.73/0.78, Si 1.1/0.95, W 67/68, Ce 4.4/4.7. Weight loss upon calcination at 600 °C: 5.2 wt%. IR (KBr, cm⁻¹): 1005 (wk), 965, 910, 780, 730 (wk), 525. ¹⁸³W NMR (-δ): 111.0(1), 116.0(2), 129.5(2), 151.4(2), 152(2), 178.5(2). ¹⁷O NMR (δ): 24 (Si-O); 376, 393, 401, 407, 417 (W-O-W); 658 (W-O-Ce); 706, 713, 718, 730 (W=O). When the total time of the electrodialysis is only 0.5h, a less acidic salt, $Na_{2.4}H_{1.6}SiW_{11}CeO_{39}\cdot15H_2O$, is formed. IR (KBr, cm⁻¹): 1010, 960, 900, 785, 730 (wk), 515. ¹⁸³W NMR (-δ): 109.7(1), 115.8(2), 129.5(2), 150.6(2), 151.2(2), 77.6(2). The NH₂-functionalized xerogel (NH₂-X) was prepared using 10% 3-(aminopropyl) triethoxysilane by a procedure similar to that described in the literature (13). The supported Ce-POM was prepared by dissolving **1** (365 mg) in water (5 ml), adding 500 mg of NH₂-X, storing overnight at RT, filtering, washing with water and drying in air.

Catalytic experiments were performed in 100-ml thermostated glass reactors filled with O_2 or air (1 atm). An aqueous solution (5 ml), containing formaldehyde and catalyst, was vigorously stirred at 20-40 °C. Aliquots were analyzed by both GC (ethanol int. std.) and titrimetric methods. GC analyses of liquid phase were performed using a "Tsvet-500" gas chromatograph equipped with a flame ionization detector and a 2m x 3mm Poropak-Q column (N_2, carrier gas, 150-170 °C, 25° min⁻¹). The gas phase was analyzed using a "Tsvet-500" gas chromatograph equipped with a catharometer (He, carrier gas). Poropak-T (98 °C) and Na-X (70 °C) columns were used to quantify CO_2 and CO, respectively. ¹⁸³W and ¹⁷O NMR spectra were recorded at 16.67 and 54.24 MHz, respectively, on an MSL-400 Brüker spectrometer.

Acknowledgments

CRDF (grant RC1-2371-NO-02) funded the research. We thank also J. Mrowiec-Bialon and N. Trukhan for the preparation of NH_2-X and the supported Ce-POM, respectively.

References

1. (a) The Activation of Dioxygen and Homogeneous Catalytic Oxidation, D. H. R. Barton, A. E. Martell, and D. T. Sawyer, eds., Plenum, New York, 1993; (b) G.-J. Brink, I. W. C. E. Arends, and R. A. Sheldon, *Science,* **287**, 1636 (2000).
2. R. M. Harrison, in Indoor Air Pollution and Health, R. E. Hester, R. M. Harrison, eds., Royal Society of Chemistry: Cambridge, **10**, 101 (1996).
3. (a) M. Ai, *J. Catal.,* **83**, 141 (1983); (b) G. Y. Popova and T. V. Andrushkevich, *Kinet. Catal. Engl. Tr.,* **38**, 258 (1997).
4. (a) Polyoxometalate Chemistry From Topology via Self-Assembly to Applications, M. T. Pope and A. Müller, eds., Kluwer, Dordrecht, 2001; (b) C. L. Hill and C. M. Prosser-McCartha, *Coord. Chem. Rev.,* **143**, 407 (1995); (c) R. Neumann, *Prog. Inorg. Chem.,* **47**, 317 (1998).
5. (a) R. D. Gall, C. L. Hill, and J. E. Walker, *J. Catal.,* **159**, 473 (1996); (b) L. Xu, E. Boring, and C. L. Hill, *J. Catal.,* **195**, 394 (2000).
6. J. T. Rhule, W. A. Neiwert, K. I. Hardcastle, B. T. Do, and C. L. Hill, *J. Am. Chem. Soc.,* **123**, 12101 (2001).
7. N. Haraguchi, Y. Okaue, T. Isobe, and Y. Matsuda, *Inorg. Chem.,* **33**, 1015 (1994).
8. G. M. Maksimov, R. I. Maksimovskaya, and I. V. Kozhevnikov, *Zh. Neorgan. Khim.,* **37**, 2279 (1992).
9. O. M. Kulikova, R. I. Maksimovskaya, S. M. Kulikov, and I. V. Kozhevnikov, *Izv. Akad. Nauk SSSR, Ser. Khim.,* 1726 (1991) (in Russian).
10. (a) P. S. Sakhla and R. N. Mehrotra, *J. Inorg. Nucl. Chem.,* **34**, 3781 (1972); (b) *Indian J. Chem.,* **10,** 1081 (1972); (c) A. K. Wadhawan, *Indian J. Chem.,* **11**, 567 (1973).
11. (a) N. M. Emanuel, E. T. Denisov, and Z. K. Maizus, Chain Reactions of Hydrocarbon Oxidation in Liquid Phase, Nauka, Moscow, 1965; (b) R. A. Sheldon and J. K. Kochi, Metal-Catalyzed Oxidations of Organic Compounds; Academic Press: New York, 1981.
12. O. A. Kholdeeva, V. A. Grigoriev, G. M. Maksimov, M. A. Fedotov, A. V. Golovin, and K. I. Zamaraev, *J. Mol. Catal. A, Chemical,* **114**, 123 (1996).
13. C. Alie, R. Pirard, A. J. Lecloux, and J.-P. Pirard, *J. Non-Cryst. Solid,* **246**, 216 (1999).

V. Symposium on Catalysis in Organic Synthesis

a. General Papers

49.

2004 Paul N. Rylander Award: Palladium-Catalyzed Annulation and Migration Reactions

Richard C. Larock

Department of Chemistry, Iowa State University, Ames, Iowa 50011

larock@iastate.edu

Abstract

The palladium-catalyzed annulation of alkynes by functionally-substituted aryl and vinylic halides or triflates provides a very convenient and efficient approach to a wide variety of heterocycles and carbocycles. This chemistry has lead to the discovery of a number of novel palladium-catalyzed processes in which the palladium migrates from one carbon to another within the molecule providing a unique way to form carbon-carbon bonds in remote locations within the same molecule.

Introduction

Some years ago we began a program to explore the scope of the palladium-catalyzed annulation of alkenes, dienes and alkynes by functionally-substituted aryl and vinylic halides or triflates as a convenient approach to a wide variety of heterocycles and carbocycles. We subsequently reported annulations involving 1,2-, 1,3- and 1,4-dienes; unsaturated cyclopropanes and cyclobutanes; cyclic and bicyclic alkenes; and alkynes, much of which was reviewed in 1999 (Scheme 1).[1] In recent days our work has concentrated on the annulation of alkynes. Recent developments in this area will be reviewed and some novel palladium migration processes that have been discovered during the course of this work will be discussed.

Palladium-Catalyzed Annulation of Alkynes

A wide variety of heterocycles can be readily prepared by the heteroannulation of alkynes. For example, the palladium-catalyzed annulation of internal alkynes by 2-iodoanilines provides easy access to 2,3-disubstituted indoles by a process that involves initial reduction of $Pd(OAc)_2$ to $Pd(0)$, oxidative addition of the aryl halide to $Pd(0)$, *cis*-addition of the arylpalladium

Scheme 1

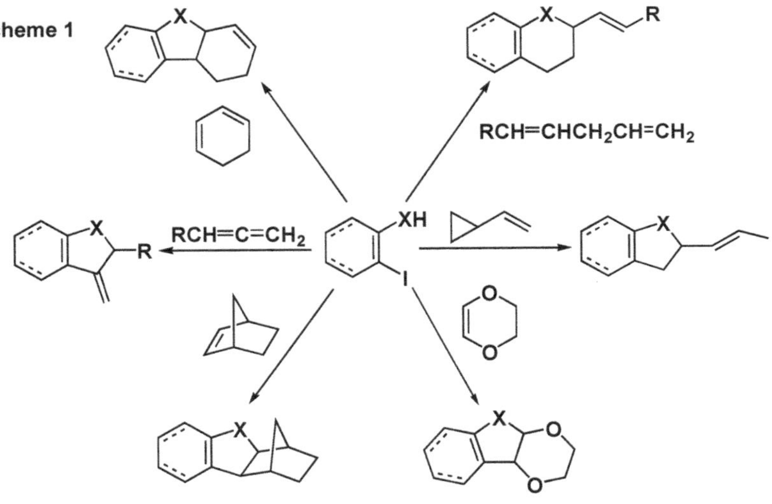

species to the alkyne, and nucleophilic displacement of the Pd moiety by the nitrogen, which in turn regenerates the Pd(0) catalyst (Scheme 2).[2] This chemistry works on a wide variety of alkynes and the regiochemistry of alkyne insertion is generally quite high with the more hindered group on the alkyne ending up next to the nitrogen. This chemistry has afforded a wide range of indoles and has been carried out on solid supports to afford indole libraries.

Scheme 2

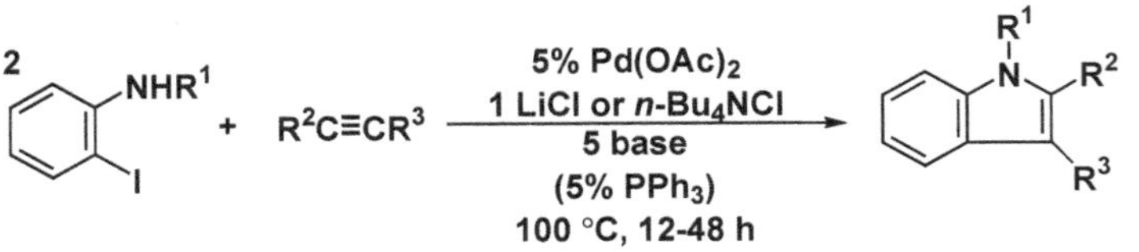

A wide variety of other functional groups can be employed in these annulations. For example, this chemistry has been extended to the synthesis of 1,2-dihydroisoquinolines, benzofurans and benzopyrans (Scheme 3).[3] One can also employ vinylic halides and triflates in this process.[4]

Scheme 3

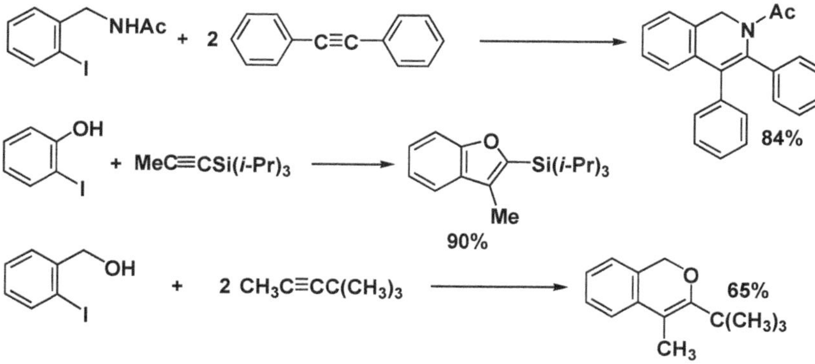

84%

90%

65%

More recently, this chemistry has been extended to the synthesis of isoquinolines by employing the *t*-butylimines of *o*-halobenzaldehydes (Scheme 4).[5] By employing vinylic halides, one obtains highly substituted pyridines.

Scheme 4

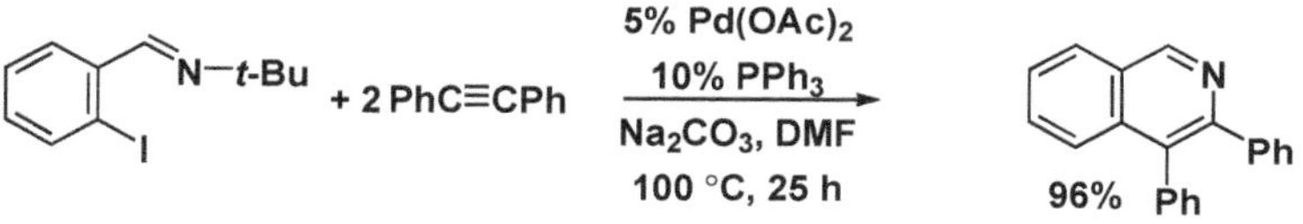

5% Pd(OAc)$_2$
10% PPh$_3$
Na$_2$CO$_3$, DMF
100 °C, 25 h

96%

This chemistry has been extended to terminal alkynes by first carrying out the cross-coupling of the alkyne and aryl halide using catalytic amounts of Pd and Cu salts and then employing catalytic amounts of CuI to affect the cyclization (Scheme 5).[6]

Scheme 5

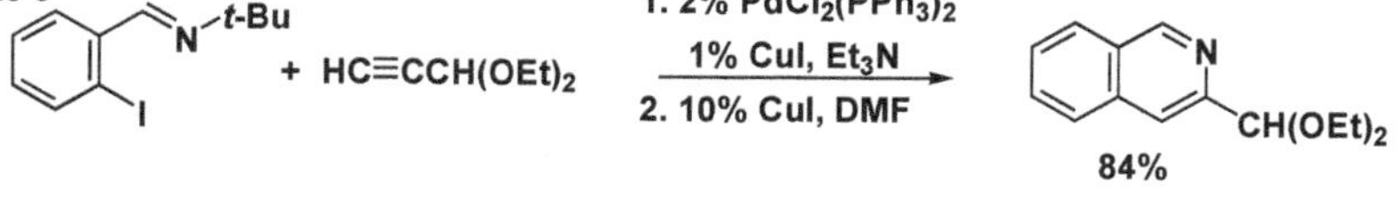

1. 2% PdCl$_2$(PPh$_3$)$_2$
1% CuI, Et$_3$N
2. 10% CuI, DMF

84%

By employing imines derived from indoles, one can easily prepare β- and γ-carbolines (Scheme 6).[7] In this chemistry, terminal alkynes can also be

Scheme 6

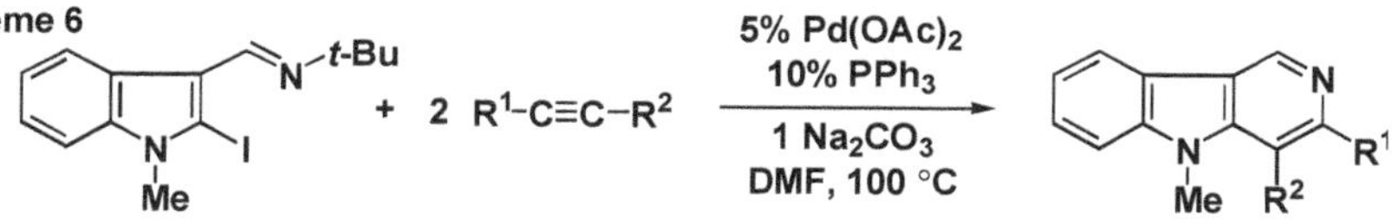

5% Pd(OAc)$_2$
10% PPh$_3$
1 Na$_2$CO$_3$
DMF, 100 °C

employed in a one step process. The alkyne can also be tethered to the nitrogen to affect intramolecular annulation (Scheme 7).[8]

Scheme 7

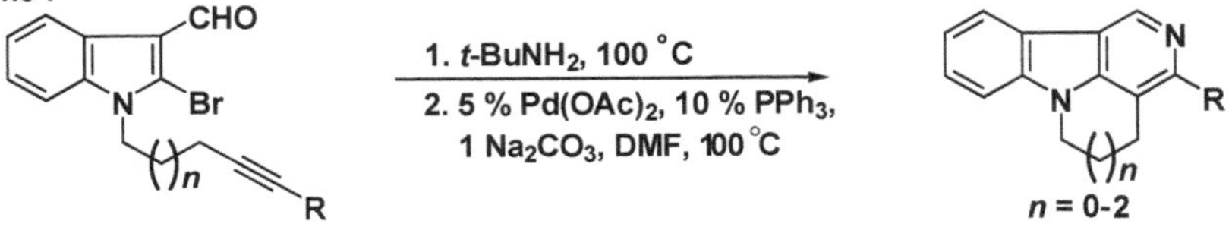
CHO
N
Br
n
R
1. t-BuNH2, 100 °C
2. 5 % Pd(OAc)2, 10 % PPh3,
1 Na2CO3, DMF, 100 °C
N
N
R
n
n = 0-2

One drawback to this alkyne annulation chemistry is that it requires either symmetrical alkynes or unsymmetrical alkynes in which the two substitutents on the internal alkyne are sterically quite different or else one obtains mixtures of regioisomers. One way to overcome this problem is to prepare the corresponding arylalkyne through catalytic Pd/Cu chemistry and then effect electrophilic cyclization using organic halides and a Pd catalyst (Scheme 8).[9]

Scheme 8

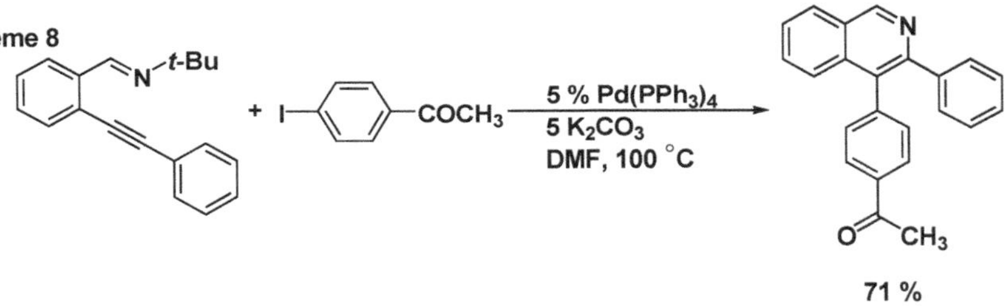
N
t-Bu
+ I
COCH3
5 % Pd(PPh3)4
5 K2CO3
DMF, 100 °C
N
O
CH3
71 %

This reaction works with aryl, allylic and 1-alkynyl halides, but not vinylic halides. A vinylic group can be introduced into the 4 position of an isoquinoline by using $PdBr_2$ as the electrophile and an alkene (Scheme 9).[10] This process

Scheme 9

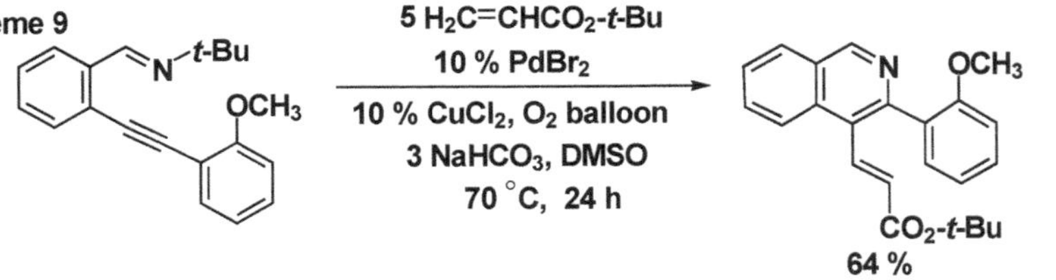
N
t-Bu
OCH3
5 H2C=CHCO2-t-Bu
10 % PdBr2
10 % CuCl2, O2 balloon
3 NaHCO3, DMSO
70 °C, 24 h
N
OCH3
CO2-t-Bu
64 %

most likely involves electrophilic cyclization by Pd(II) to afford the isoquinoline with a Pd moiety in the 4 position. This proposed species subsequently reacts with the alkene in a Heck reaction to produce the final product and Pd(0), which is reoxidized to Pd(II) by Cu salts and oxygen.

If the arylation chemistry is carried out under 1 atm of CO, an isoquinoline with a ketone moiety in the 4 position is produced in good yields (Scheme 10).[11]

Scheme 10

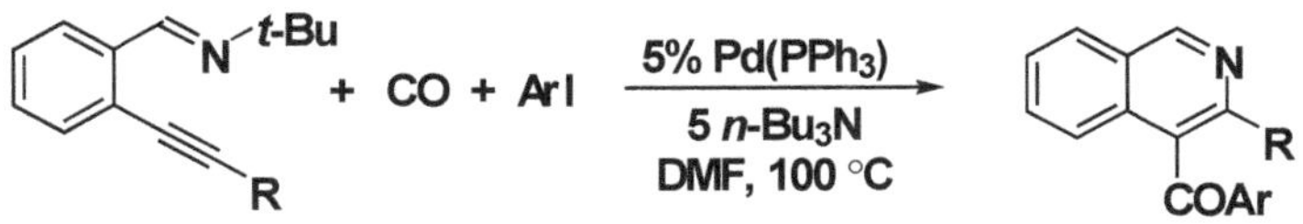

Imines derived from *o*-iodoaniline and arenecarboxaldehydes react with internal arylalkynes and catalytic Pd(0) to afford isoindolo[2,1-*a*]indoles by a process that involves alkyne insertion, addition across the C=N double bond and substitution of the aromatic ring (Scheme 11).[12] This process exhibits very

Scheme 11

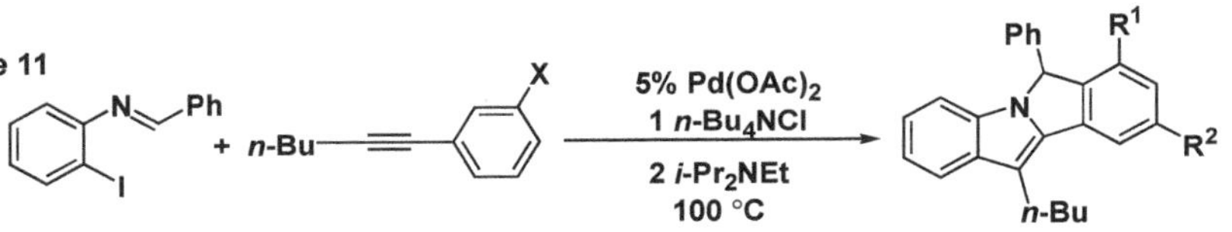

X	R^1	R^2	%
CH$_3$	-	CH$_3$	81
CH$_3$O	CH$_3$O	-	78
CF$_3$	-	CF$_3$	95
CO$_2$Et	CO$_2$Et	-	74

unusual regiochemistry. It appears that substitution of the arene is directed by chelation of the Pd to the oxygen in the functional groups of the substituents.

One of the most attractive features of organopalladium chemistry is its ability to accommodate essentially all important functional groups commonly used in organic chemistry. Yet, when certain functional groups are placed in close proximity to the vinylic palladium intermediates generated during our annulation processes, a number of unexpected reactions have been observed. For example, esters do not normally react with organopalladium compounds, but when one employs *o*-haloarenecarboxylates and vinylic analogues in our alkyne annulation chemistry, they participate to afford good yields of isocoumarins and pyrones (Scheme 12).[13]

Scheme 12

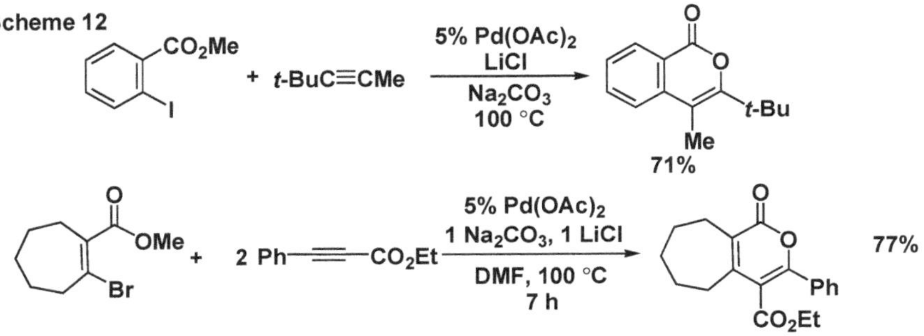

Recently we have been able to produce coumarins by alkyne annulation in the presence of CO (Scheme 13).[14] Production of the coumarin ring system is

Scheme 13

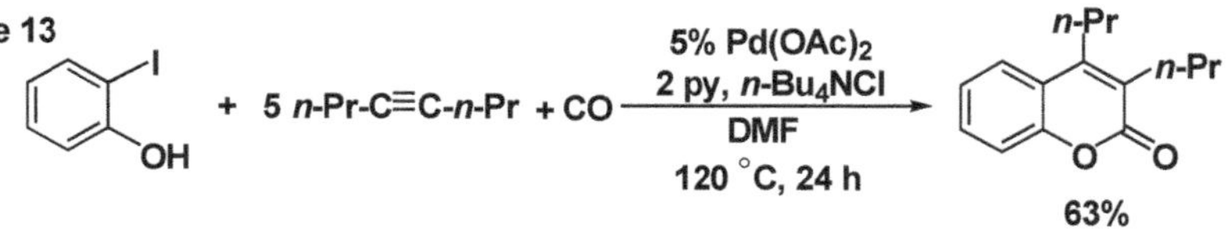
I
OH
+ 5 n-Pr-C≡C-n-Pr + CO
5% Pd(OAc)2
2 py, n-Bu4NCl
DMF
120 °C, 24 h
n-Pr
n-Pr
O
O
63%

quite unusual, since the literature would have you believe that CO should insert into the carbon-palladium bond more readily, which would afford isomeric chromones. This process appears to involve reversible CO insertion, which does not lead to product, and irreversible alkyne insertion, which does. The major drawback to this chemistry at present is the fact that the insertion of unsymmetrical alkynes is not very regioselective.

If one employs *o*-iodoaniline derivatives in this double insertion chemistry, 2-quinolones are generated in good yield after a basic work-up to remove the nitrogen protecting group (Scheme 14).[15]

Scheme 14

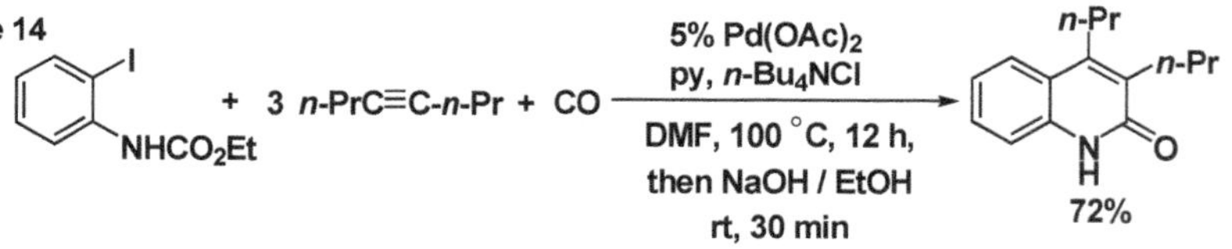
I
NHCO2Et
+ 3 n-PrC≡C-n-Pr + CO
5% Pd(OAc)2
py, n-Bu4NCl
DMF, 100 °C, 12 h,
then NaOH / EtOH
rt, 30 min
n-Pr
n-Pr
N
H
O
72%

We have also been able to generate carbocycles by annulation processes. For example, the reaction of halobiaryls and 1 atm of CO in the presence of a Pd catalyst affords excellent yields of fluorenones (Scheme 15).[16]

Scheme 15

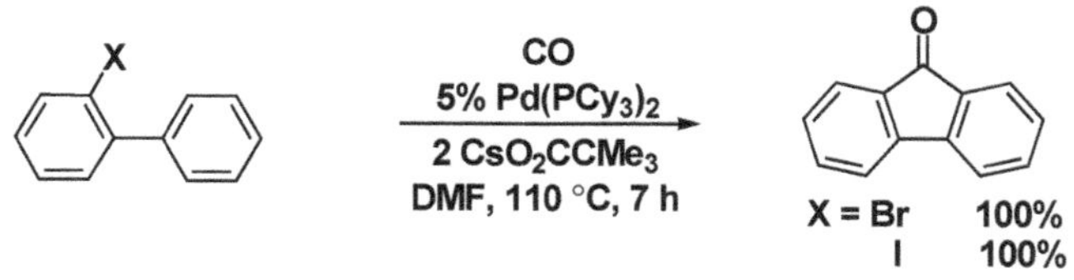
X
CO
5% Pd(PCy3)2
2 CsO2CCMe3
DMF, 110 °C, 7 h
O
X = Br 100%
I 100%

Carbocycles can also be synthesized by the annulation of alkynes by arene-containing vinylic halides or triflates (Scheme 16).[17] This process involves

Scheme 16

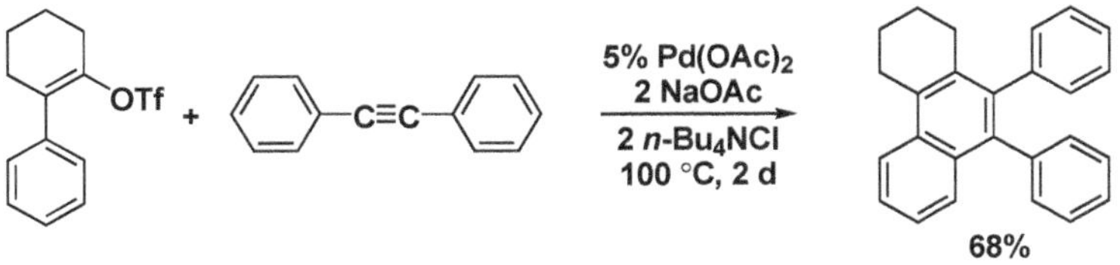

68%

alkyne insertion, followed by substitution of the arene. It can also be carried out using halobiaryls (Scheme 17).[18] Surprisingly, this process exhibits

Scheme 17

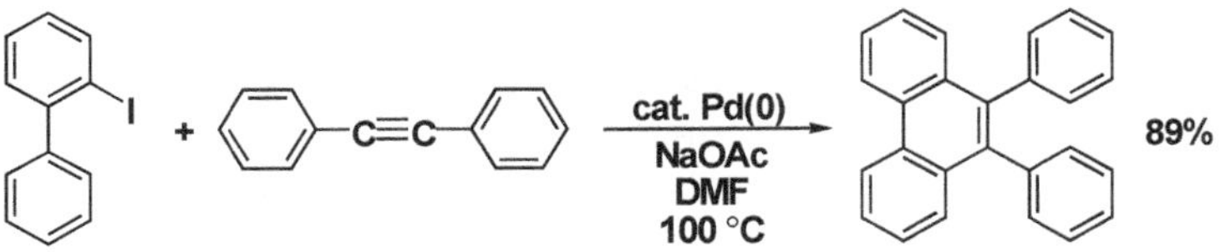

89%

very poor regioselectivity when unsymmetrical alkynes are utilized. Even alkynes, such as silylalkynes, which usually afford high selectivity in this annulation methodology, give mixtures of regioisomers. This observation has lead to the discovery of a whole new class of very interesting Pd migration processes.

Palladium Migration Chemistry

The poor regioselectivity of alkyne insertion in our polycyclic aromatic hydrocarbon synthesis (Scheme 17) suggested to us that perhaps the palladium intermediate in that process was actually undergoing migration from one aromatic ring to the other, perhaps by a Pd(IV) hydride intermediate, to establish an equilibrium mixture of two regioisomeric arylpalladium intermediates under our reaction conditions (Scheme 18). This, indeed, appears to be true as

Scheme 18

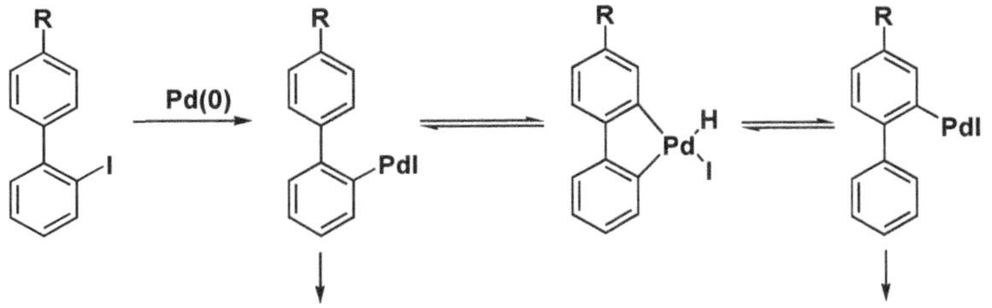

evidenced by the formation of the same 50:50 mixture of Heck products from two isomeric iodobiphenyls when the reaction is run in the presence of ethyl acrylate (Scheme 19).[19] When one employs a biaryl with two very different

Scheme 19

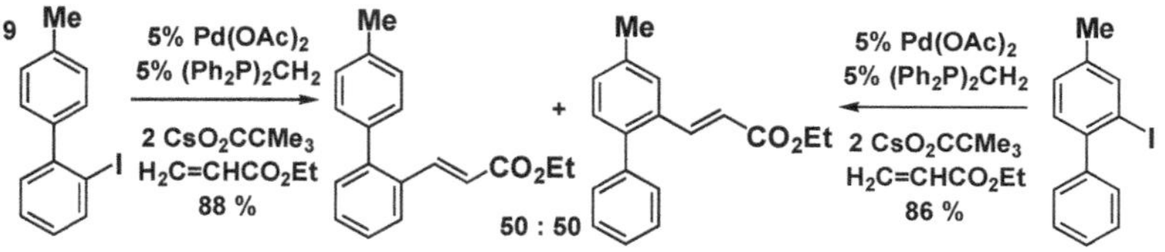

aromatic rings, one obtains the anticipated olefin product using a standard Heck procedure and a regioisomeric olefin when using our palladium migration conditions (Scheme 20).

Scheme 20

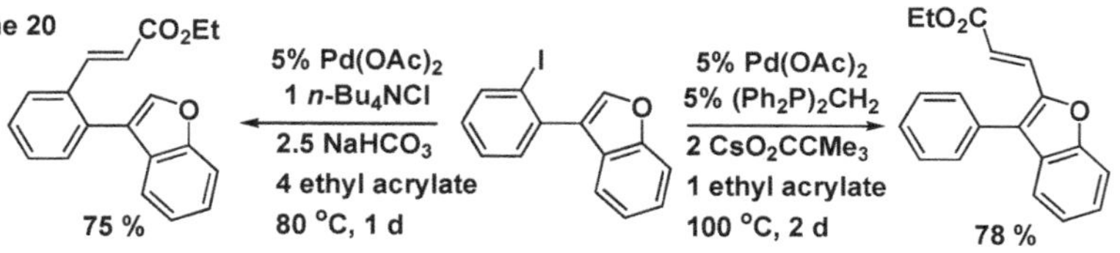

Similar Pd migration products have also been observed from alkyne insertion chemistry using our standard migration conditions (Scheme 21).[19]

Scheme 21

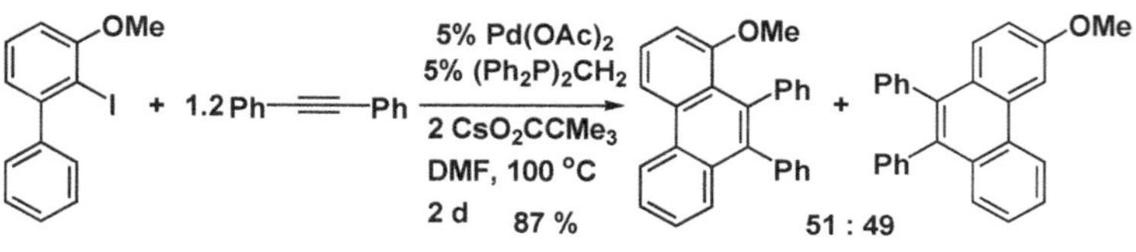

Similar observations have been made when employing a modification of the Suzuki reaction (Scheme 22).[20] Thus, isomeric iodobiphenyls, when cross-

Scheme 22

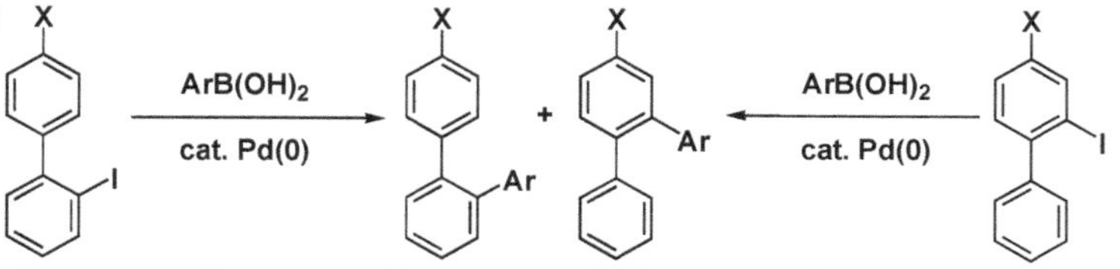

X	Ar	%	Ratio	%
Me	p-MeO$_2$CC$_6$H$_4$	78	51:49 / 49:51	83
Me	C$_6$H$_5$	79	52:48 / 50:50	69
Me	p-MeOC$_6$H$_4$	93	52:48 / 49:51	90
OMe	p-MeOC$_6$H$_4$	85	42:58 / 39:61	75
CO$_2$Et	p-MeOC$_6$H$_4$	84	40:60 / 34:66	68

coupled under modified Suzuki conditions, afford mixtures of regiosomeric Suzuki products. The electronic effects of simple substitutents on the biaryl system are generally minimal. However, biaryls with substantially different aromatic rings once again afford predominantly the product where Suzuki coupling takes place almost exclusively on the heterocyclic ring (Scheme 23).

Scheme 23

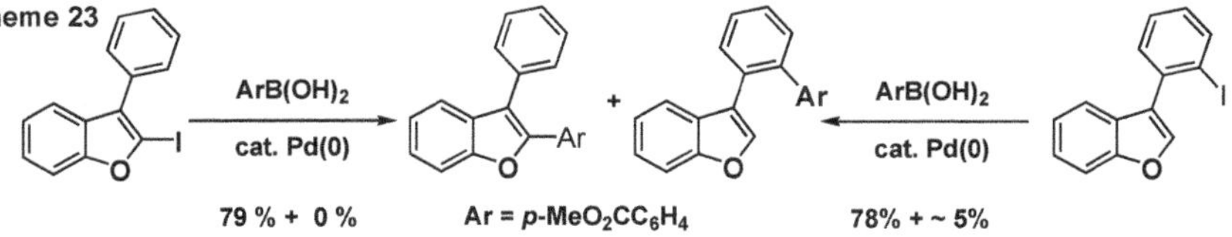
ArB(OH)2
cat. Pd(0)
Ar
ArB(OH)2
cat. Pd(0)
79 % + 0 %
Ar = p-MeO2CC6H4
78% + ~ 5%

This chemistry becomes synthetically useful when one of the isomeric palladium intermediates can react with a neighboring substituent and the other isomer cannot. Thus, we have taken advantage of this effect to synthesize a range of polycyclic aromatic hydrocarbons by Pd migration and subsequent arylation (Scheme 24).[21] This provides a unique way to form new carbon-carbon bonds in a location remote from the original functionality.

Scheme 24

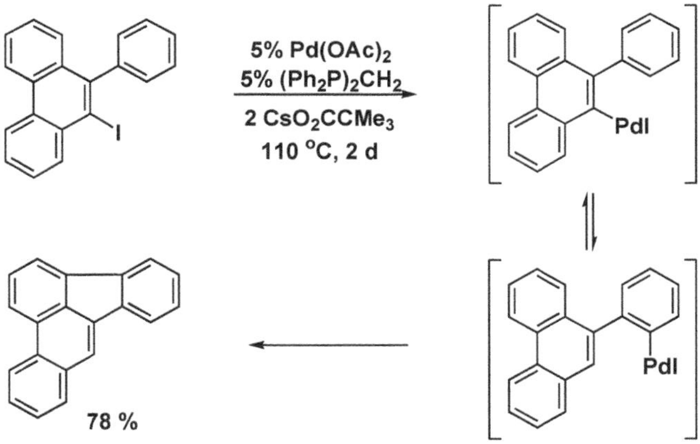
5% Pd(OAc)2
5% (Ph2P)2CH2
2 CsO2CCMe3
110 °C, 2 d
PdI
PdI
78 %

About the same time that we discovered this aryl to aryl palladium migration chemistry, we observed a similar vinylic to aryl palladium migration process, which readily affords alkylidene fluorenes (Scheme 25).[22] In this case,

Scheme 25

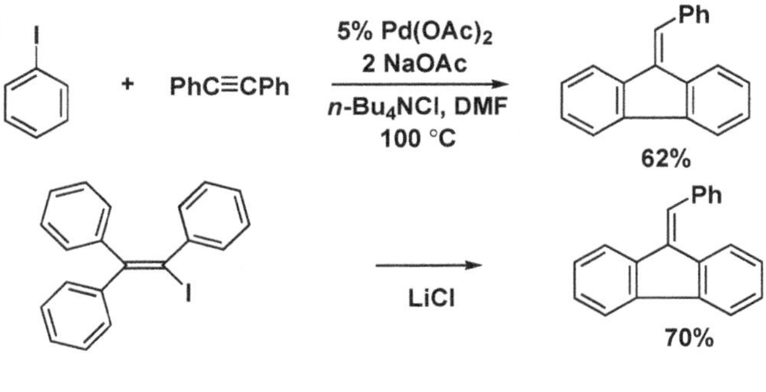
PhC≡CPh
5% Pd(OAc)2
2 NaOAc
n-Bu4NCl, DMF
100 °C
Ph
62%
LiCl
Ph
70%

the initially generated vinylic palladium species apparently undergoes clean migration to the neighboring arene and subsequent intramolecular arylation to produce the observed products (Scheme 26).

Scheme 26

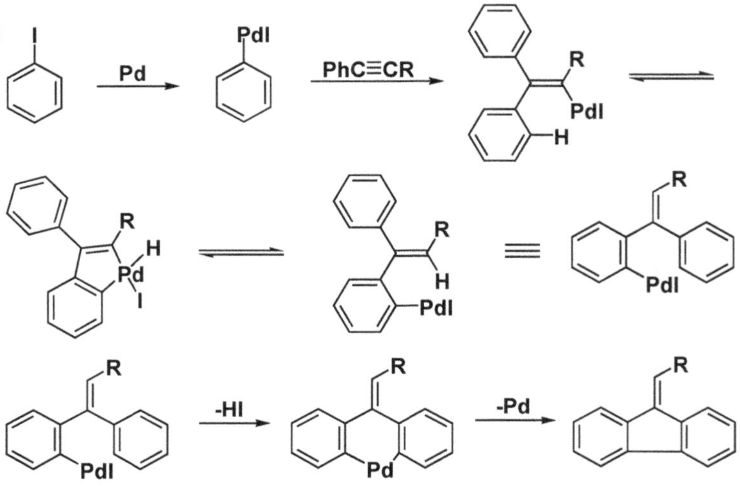

Scheme 27

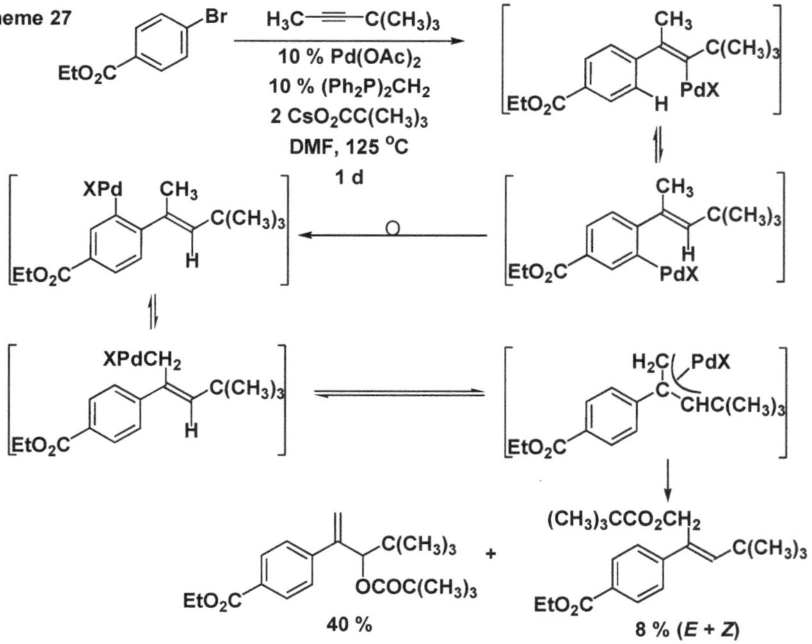

Recently, we have observed a process where the palladium apparently migrates from a vinylic to an aryl to an allylic position all in one reaction (Scheme 27).[23] This provides a unique new way to produce π-allylpalladium intermediates, which have proven very valuable as intermediates in organic synthesis.

Several different alkyl to aryl palladium migration processes have been also recently discovered using our standard palladium migration conditions (Scheme 28).[24] Thus, both intermolecular Heck and intramolecular arylation

Scheme 28

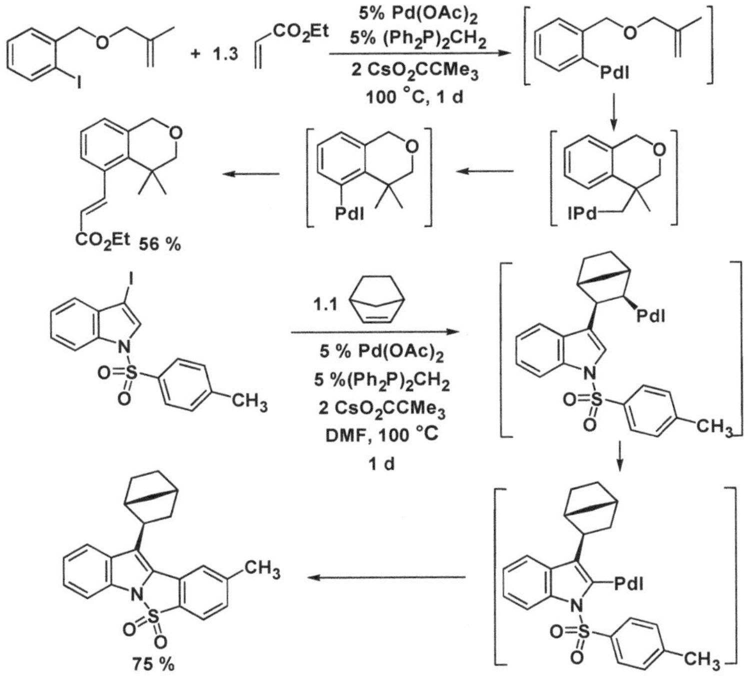

processes may be used to terminate the reaction sequence. If the starting aryl halide contains an aromatic ring in the side chain, alkyl to aryl palladium migration is followed by intramolecular arylation (Scheme 29).[25]

Scheme 29

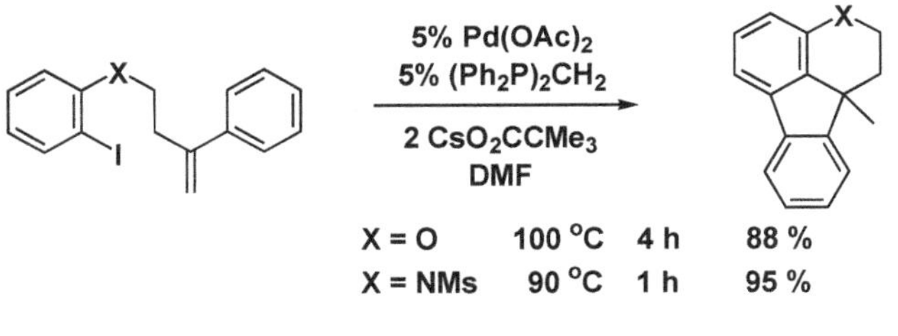

X = O	100 °C	4 h	88 %
X = NMs	90 °C	1 h	95 %

We have also recently observed a unique example of an aryl to acyl migration (Scheme 30).[26] In this case, the intermediate is easily trapped by running the reaction in an alcohol solvent to form the carbamate.

Scheme 30

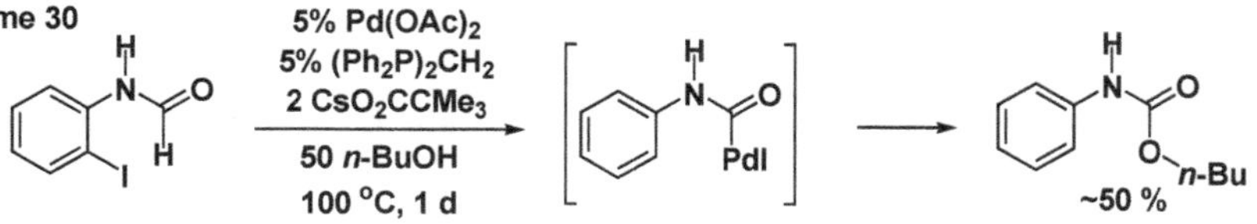

Arylpalladium intermediates bearing neighboring imine functionality have also been observed to rearrange to imidoyl palladium species, which are easily trapped by reaction with water (Scheme 31).[26] In a novel application of

Scheme 31

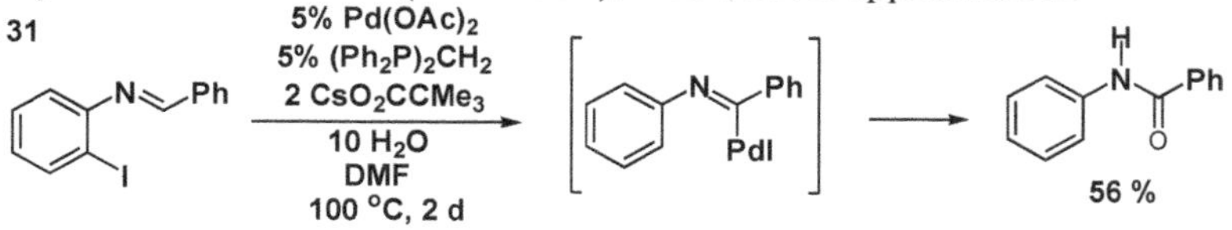

this chemistry, we have prepared the imine of a biarylcarboxaldehyde and observed its quantitative rearrangement to a fluorenone imine, which can be easily hydrolyzed to the corresponding ketone (Scheme 32).[27]

Scheme 32

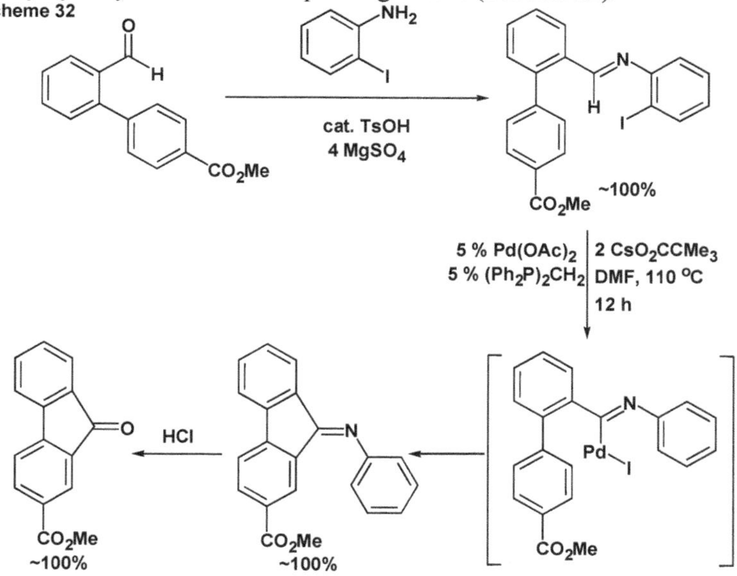

We believe that this novel new palladium migration chemistry provides a unique new way to generate a wide variety of organopalladium intermediates that will find considerable synthetic utility in the future.

Acknowledgements

I am indebted to the many hardworking students who have contributed so much both experimentally and intellectually to the chemistry described in this article. Many of their names appear in the references that follow. I also wish to acknowledge the Petroleum Research Fund, the National Institutes of Health, and the National Science Foundation for their financial support; Kawaken Fine Chemicals Co. and Johnson Matthey Inc. for donating the palladium salts; and Frontier Scientific for providing the arylboronic acids used in this work.

References

1. (a) R. C. Larock, *J. Organometal. Chem.,* **576**, 111 (1999). (b) R. C. Larock, Perspectives in Organopalladium Chemistry for the XXI Century, Ed. J. Tsuji, Elsevier Press, Lausanne, Switzerland, 1999, p. 111-124. (c) R. C. Larock, *Pure Appl. Chem.*, **71**, 1435 (1999).
2. R. C. Larock, E. K. Yum and M. D. Refvik, *J. Org. Chem.,* **63**, 7652 (1998).
3. R. C. Larock, E. K. Yum, M. J. Doty and K. K. C. Sham, *J. Org. Chem.*, **60**, 3270 (1995).
4. R. C. Larock, M. J. Doty and X. Han, *Tetrahedron Lett.,* **39**, 5143 (1998).
5. K. R. Roesch, H. Zhang and R. C. Larock, *J. Org. Chem.,* **66**, 8042 (2001).
6. K. R. Roesch and R. C. Larock, *J. Org. Chem.*, **67**, 86 (2002).
7. H. Zhang and R. C. Larock, *J. Org. Chem.,* **67**, 9318 (2002).
8. H. Zhang and R. C. Larock, *J. Org. Chem.,* **68**, 5132 (2003).
9. G. Dai and R. C. Larock, *J. Org. Chem.,* **68**, 920 (2003).
10. Q. Huang and R. C. Larock, *J. Org. Chem.,* **68**, 980 (2003).
11. G. Dai and R. C. Larock, *J. Org. Chem.*, **67**, 7042 (2002).
12. K. R. Roesch and R. C. Larock, *J. Org. Chem.,* **66**, 412 (2001).
13. R. C. Larock, M. J. Doty and X. Han, *J. Org. Chem.,* **64**, 8770 (1999).
14. D. V. Kadnikov and R. C. Larock, *Org. Lett.,* **2**, 3643 (2000).
15. D. Kadnikov and R. C. Larock, work submitted for publication.
16. M. A. Campo and R. C. Larock, *J. Org. Chem.,* **67**, 5616 (2002).
17. R. C. Larock and Q. Tian, *J. Org. Chem.,* **63**, 2002 (1998).
18. R. C. Larock, M. J. Doty, Q. Tian and J. M. Zenner, *J. Org. Chem.,* **62**, 7536 (1997).
19. M. A. Campo and R. C. Larock, *J. Am. Chem. Soc.,* **124**, 14326 (2002).
20. M. A. Marino, H. Zhang, Q. Huang and R. C. Larock, work in progress.
21. M. A. Campo, Q. Huang, T. Yao, Q. Tian, R. C. Larock, *J. Am. Chem. Soc.,* **125**, 11506 (2003).
22. R. C. Larock and Q. Tian, *J. Org. Chem.*, **66**, 7372 (2001).
23. J. Zhao, M. A. Campo and R. C. Larock, work in progress.
24. M. A. Campo , G. Dai and R. C. Larock, work in progress.
25. Q. Huang, A. Fazio and R. C. Larock, work in progress.
26. M. A. Campo, D. E. Emrich and R. C. Larock, work in progress.
27. D. Yue and R. C. Larock, work in progress.

50.   New Palladium-Catalyzed Processes for the Synthesis of Selectively Substituted Aromatics

Marta Catellani, Elena Motti, Sara Deledda, Fiorenza Faccini

Dipartimento di Chimica Organica e Industriale dell'Università, Parco Area delle Scienze, 17/A, 43100 Parma, Italy

marta.catellani@unipr.it

Abstract

Alkylaromatic palladacycles, formed in situ by reaction of palladium acetate with a strained and rigid cycloolefin such as norbornene and an *ortho*-substituted aryl iodide, allow the selective arylation of the aromatic ring by a second molecule of aryl iodide with subsequent norbornene expulsion. Coupling the resulting complex with appropriate reagents liberates the organic product thus regenerating the catalyst.

Introduction

Selective aromatic functionalization has been a permanent object of research since the nineteenth century. Catalysis has offered a powerful tool to achieve this goal. Over the years we have worked out a complex catalytic system consisting of an inorganic compound such as a palladium salt and an organic molecule containing a strained double bond such as norbornene (1, 2). We have seen that these two catalysts cooperatively react with an aromatic iodide, an alkyl iodide and a terminal olefin. The following equation reports an example (L = solvent and/or olefin) (3).

As reported in Scheme 1 the process involves a series of steps. The alkylpalladium species **1** forms through oxidative addition of the aromatic iodide to palladium(0) followed by norbornene insertion (4-7). The ready generation of complex **2** (8-11) from **1** is due to the unfavourable stereochemistry preventing β-hydrogen elimination from **1** (12). Complex **2** further reacts with alkyl halides RX to form palladium(IV) complex **3** (13-15). Migration of the R group to the metallacycle aryl site gives the palladium(II) complex **4** (14). Repetition of steps from **1** to **4** leads to the doubly alkylated species **7**. Norbornene expulsion from the latter gives **8** (16).

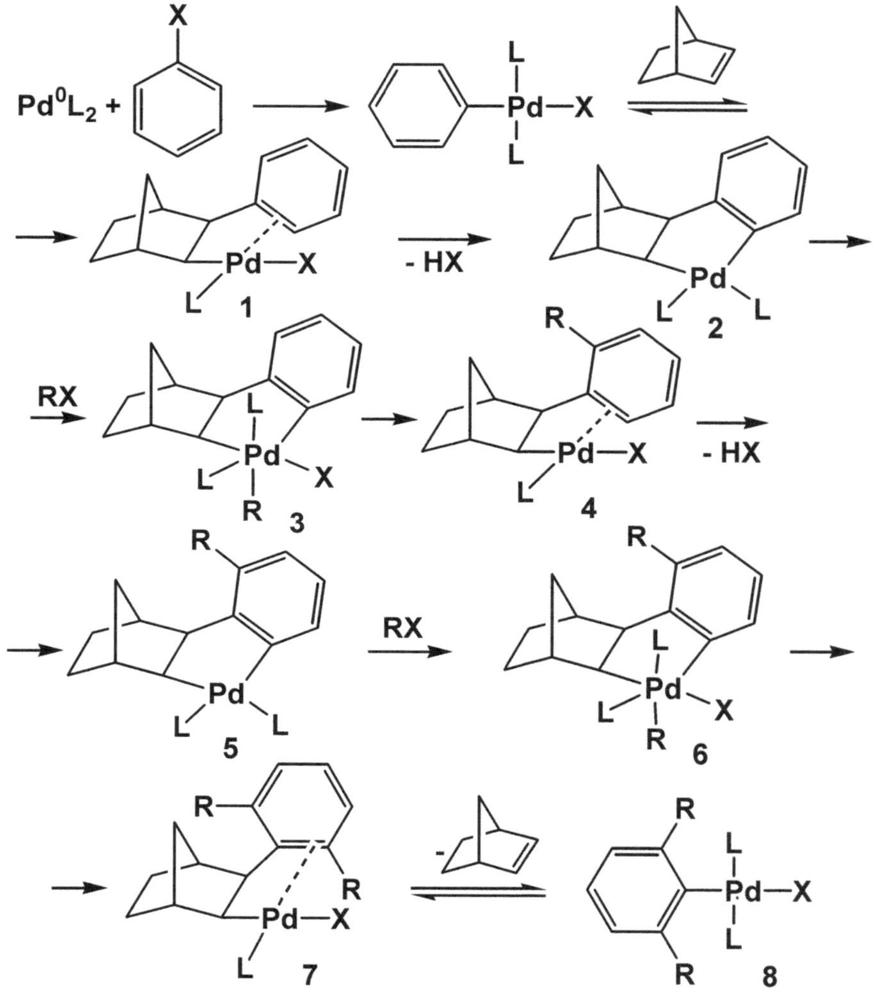

Scheme 1

Relevant complexes have been isolated and fully characterized (4-10, 13-16). The resulting arylpalladium complex **8** is able to react with various compounds such as terminal alkenes, alkynes, aryl boronic acids or hydrogen-transfer agents to give an organic molecule and palladium(0) (3, 17, 18).

We have found that palladacycles **2** and **5** also react with aryl halides but the reaction pathway follows a different course in the absence (complex **2**) or in the presence of an *ortho*-substituent in the aromatic ring as in complex **5**. We could establish that aryl iodides selectively react with the norbornyl site of **2** to give **10** (19, 20) through **9** (21). With palladacycles of type **5**, containing an *ortho*-substituent in the aromatic ring, however, the aryl iodide reacts exclusively at the aryl site of **5** to afford **11**. Owing to steric hindrance by the *ortho*-substituents, norbornene deinserts to afford the biphenylyl structure **12** (Scheme 2, L = solvent) (21).

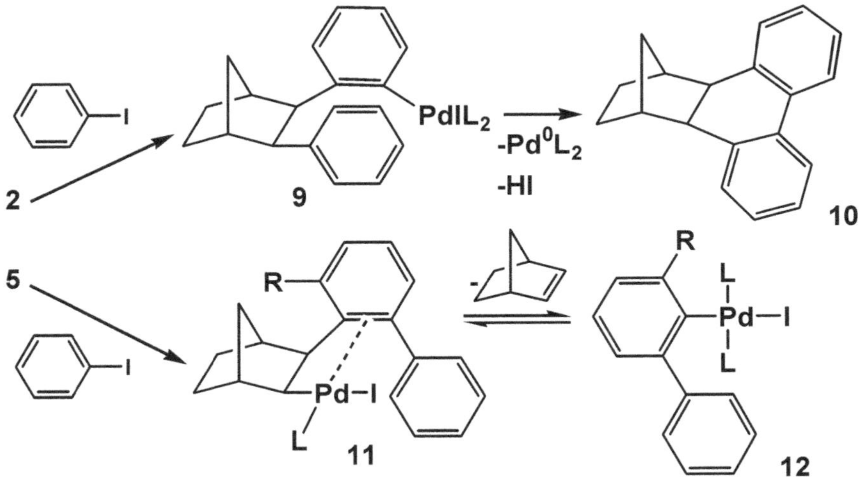

Scheme 2

Results and Discussion

Starting from the stoichiometric sequence leading to **12** we worked out appropriate reactions able to terminate the sequence by giving rise to organic products with concomitant liberation of palladium(0), as previously mentioned for the alkylation reactions (1, 2). Solvent of choice proved to be the coordinating ones such as dimethylformamide (DMF) and N-methylpyrrolidinone (NMP) while the best base turned out to be potassium carbonate.

The presence of an *ortho*-substituent in the aryl iodide required the use of a higher temperature, which also was needed to cause the aryl iodide to react with palladacycle **5**, with the consequence that secondary reactions became important.

The main difficulty we had to overcome, however, was the propensity of the terminating agent to interact with any of the preceding steps of the sequence, thus causing the formation of a complex mixture.

Controlling secondary reactions required using appropriate ratios among reagents. Special attention was devoted to norbornene: although it acts as a catalyst its concentration must be kept sufficiently high to enable it to form the palladacycle by insertion, but not so high as to prevent its liberation at the end of the sequence.

Under these conditions activated olefins such as methyl acrylate, styrene and vinyl phenylsulfoxide were found to be effective in reacting with **12** (22).

Table 1 Reaction of an *ortho*-substituted aryl iodide with a terminal olefin in the presence of K_2CO_3, $Pd(OAc)_2$ and norbornene.[a]

Run[a]	R	Y	Compound **14** yield (%)[b]
1	Me	CO_2Me	79
2	Me	SOPh	81
3	i-Pr	CO_2Me	84
4	Et	Ph	87
5	CO_2Me[c]	CO_2Me	98

[a] Under nitrogen at 105 °C for 16 h in DMF. Molar ratio of the reagents as indicated in the title: 40:24:40:1:12. $Pd(OAc)_2$ concentration: 0.2×10^{-2} M.
[b] Isolated yield based on the aryl iodide. Conversion is over 93%.
[c] Complete conversion after 6 h.

Different results were obtained using allylic alcohols as terminating agents. In spite of the expected reactivity of the alcoholic function as hydrogen donor only products deriving from the reaction of the double bond were obtained (23) according to the general pattern previously described for Heck-type reactions with allylic alcohols (24).

Table 2 Reaction of an *ortho*-substituted aryl iodide with an allylic alcohol in the presence of K$_2$CO$_3$, Pd(OAc)$_2$ and norbornene.[a]

Run[a]	R	Y	Conversion of 13 (%)[b]	Compound 15 yield (%)[c]
1	Me	H	75	56
2	Me	Me	84	60
3	*n*-Bu	Me	71	54
4	-(CH)$_4$-	Me	100	75
5	*i*-Pr	Ph	92	62

[a] Under nitrogen at 105 °C for 24 h in DMF. Molar ratio of the reagents as indicated in the title: 80:45:80:1:20. Pd(OAc)$_2$ concentration: 0.4 X 10^{-2} M.
[b] Determined by GC
[c] Isolated yield based on the aryl iodide.

Passing from olefins to acetylenic compounds as terminating agents we observed that terminally unsubstituted alkynes reacted quite unselectively, so we decided to use disubstituted alkynes. Among the latter diarylacetylenes gave phenanthrenes according to the following equation, while alkylacetylenes gave allenes (2-biphenylpropadienes selectively substituted by alkyl groups in the 1,3 positions of propadiene and 3,2' of biphenyl) (25-27).

Table 3 Reaction of an *ortho*-substituted aryl iodide with a diarylacetylene in the presence of K_2CO_3, *n*-Bu_4NBr, $Pd(OAc)_2$ and norbornene.[a]

Run[a]	R	R_1	Compound **16** yield (%)[b]
1	Me	C_6H_5	82
2	Et	C_6H_5	85
3	*i*-Pr	C_6H_5	93
4	OMe	C_6H_5	64
5	*i*-Pr	4-$MeOC_6H_4$	80

[a] Under nitrogen at 105 °C for 24 h in DMF. Molar ratio of the reagents as indicated in the title: 40:20:60:120:1:10. $Pd(OAc)_2$ concentration: 0.3×10^{-2} M.
[b] Isolated yield based on the aryl iodide.

So far we have considered olefins and alkynes as terminating agents. Another reaction, however, was available to effect C-C coupling in the last step, the Suzuki reaction. Results were satisfactory under the conditions shown in the Table (28).

Table 4 Reaction of an *ortho*-substituted aryl iodide with an arylboronic acid in the presence of K_2CO_3, $Pd(OAc)_2$ and norbornene.[a]

Run[a]	R	R_1	Compound **17** (%)[b]
1	*n*-Bu	H	73
2	OMe	H	82

3	CO$_2$Me	H	89
4	*n*-Bu	4-Me	72
5	*n*-Pr	2-Me	73

[a] Under nitrogen at 105 °C for 90 h in DMF. Molar ratio of the reagents as indicated in the title:200:120:400:1:100. Pd(OAc)$_2$ concentration: 0.1 X 10^{-2} M.
[b] Isolated yield based on the aryl iodide. Conversion is over 91%.

The last reaction we consider here, hydrogenolysis, is the most simple and straightforward but at the same time it is the most difficult to control, because the high hydrogen transfer rate adversely affects every step of the sequence. Although many hydrogen donors are available the one that led to the most satisfactory results was benzyl alcohol (29).

Table 5 Reaction of an *ortho*-substituted aryl iodide with benzyl alcohol in the presence of K$_2$CO$_3$, Pd(OAc)$_2$ and norbornene.[a]

Run[a]	R	Compound **18** yield (%)[b]
1	Et	84
2	*i*-Pr	86
3	-(CH)$_4$-	78
4	OMe	48
5	CO$_2$Me	87

[a] Under nitrogen at 105 °C for 24 h in NMP. Molar ratio of the reagents as indicated in the title: 80:80:160:1:20. Pd(OAc)$_2$ concentration: 0.2 X 10^{-2} M.
[b] Isolated yield based on the aryl iodide. Conversion is over 95%.

The reactions reported so far involve the same complex **5** described in Scheme 1, which is now obtained directly by oxidative addition of the *o*-substituted aryl iodide to palladium(0) followed by norbornene insertion and cyclization. Scheme 3 shows how the sequence proceeds by reaction of a second molecule of the aryl iodide with complex **5** to form an aryl-aryl bond selectively, possibly through a palladium(IV) intermediate (15, 30). The complete regioselectivity is due to the presence of the *ortho*-substituent in the aromatic ring of palladacycle **5**. The resulting norbornylpalladium species undergoes C-C bond cleavage with norbornene deinsertion, thus affording the biphenylylpalladium structure **20**. Only at this stage the terminating agent reacts with the arylpalladium complex. So all the previous steps involved in the construction of the biphenylylpalladium species must be faster than the reaction with a terminating agent as reported in the Tables. To this end an accurate balance of the various factors is required.

Although many other reaction pathways are possible only a few by-products such as **21**, resulting from a competitive ring closure step, are present to a small extent.

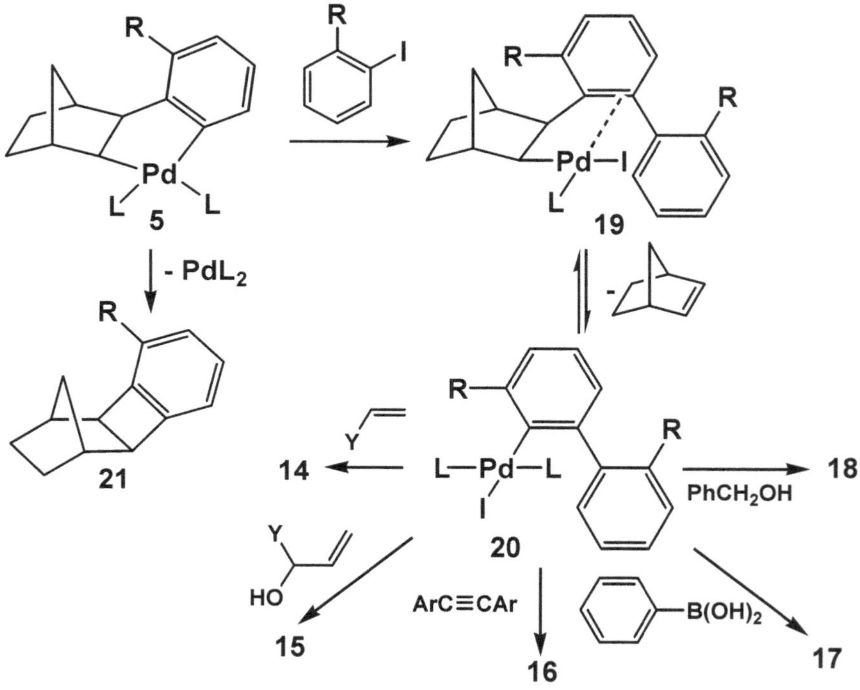

Scheme 3

Experimental Section

General procedure for the reaction of an o-substituted aryl iodide and a terminal olefin. Synthesis of vinylbiphenyls.
A DMF solution (10 mL) of the desired *o*-substituted aryl iodide (0.88 mmol), norbornene (26 mg, 0.27 mmol) and the appropriate olefin (0.53 mmol) was introduced under nitrogen into a Schlenck-type flask containing Pd(OAc)$_2$ (5 mg, 0.022 mmol) and K$_2$CO$_3$ (124 mg, 0.90 mmol). The reaction mixture was stirred at 105 °C for 16 h. After cooling the organic layer was diluted with CH$_2$Cl$_2$ (20 mL) and extracted twice with a 5% solution of H$_2$SO$_4$ (15 mL). The organic layer was washed with water (20 mL) and dried over Na$_2$SO$_4$. The crude

was analysed by GC, GC-mass and TLC. The product was isolated by flash cromatography on silica gel using mixtures of hexane-EtOAc as eluent.

General procedure for the reaction of an o-substituted aryl iodide with an allylic alcohol. Synthesis of biphenyls containing an oxoalkyl chain.
The reaction was carried out as reported above using $Pd(OAc)_2$ (5 mg, 0.022 mmol), K_2CO_3 (250 mg, 1.80 mmol), the desired *o*-substituted aryl iodide (1.80 mmol), norbornene (43 mg, 0.45 mmol) and the appropriate allylic alcohol (0.88 mmol) in DMF (5 mL) under nitrogen at 105 °C for 24 h.

General procedure for the reaction of an o-substituted aryl iodide with a diarylacetylene. Synthesis of phenanthrene derivatives.
The reaction was carried out as reported above using $Pd(OAc)_2$ (5 mg, 0.022 mmol), K_2CO_3 (182 mg, 1.32 mmol), Bu_4NBr (851 mg, 2.64 mmol), the desired *o*-substituted aryl iodide (0.88 mmol), norbornene (21 mg, 0.22 mmol) and the appropriate diarylacetylene (0.44 mmol) in DMF (7.5 mL) under nitrogen at 105 °C for 24 h.

General procedure for the reaction of an o-substituted aryl iodide with an arylboronic acid. Synthesis of terphenyl derivatives.
The reaction was carried out as reported above using $Pd(OAc)_2$ (1.8 mg, 0.008 mmol), K_2CO_3 (420 mg, 3.2 mmol), the desired *o*-substituted aryl iodide (1.6 mmol), norbornene (76 mg, 0.8 mmol) and the appropriate arylboronic acid (0.96 mmol) in DMF (8 mL) under nitrogen at 105 °C for 90 h.

General procedure for the reaction of an o-substituted aryl iodide in the presence of benzyl alcohol. Synthesis of biphenyl derivatives.
The reaction was carried out as reported above using $Pd(OAc)_2$ (5 mg, 0.022 mmol), K_2CO_3 (484 mg, 3.50 mmol), the desired *o*-substituted aryl iodide (1.80 mmol), norbornene (43 mg, 0.45 mmol) and benzyl alcohol (190 mg, 1.80 mmol) in NMP (10 mL) under nitrogen at 105 °C for 24 h.

Acknowledgements

This work was supported by the University of Parma and the Ministero dell'Università e della Ricerca Scientifica e Tecnologica (Cofinanziamento MIUR, project n. 2003039774).

References

1. M. Catellani, *Synlett*, 298 (2003).
2. M. Catellani, Handbook of Organopalladium Chemistry for Organic Synthesis, E.-I. Negishi Ed., Wiley, New York, 2002, pp. 1479-1489.
3. M. Catellani, F. Frignani and A. Rangoni, *Angew. Chem. Int. Ed.*, **36**, 119 (1997).
4. H. Horino, M. Arai and M. Inoue, *Tetrahedron Lett.*, 647 (1974).

5.   M. Portnoy, Y. Ben-David, I. Rousso and D. Milstein, *Organometallics*, **13**, 3465 (1994).
6.   C.-S. Li, C.-H. Cheng, F.-L. Liao and F.-L. Wang, *Chem. Commun.*, 710 (1991).
7.   M. Catellani, C. Mealli, E. Motti, P. Paoli, E. Perez-Carreno and P. S. Pregosin, *J. Am. Chem. Soc.*, **124**, 4336 (2002).
8.   M. Catellani and G. P. Chiusoli, *J. Organometal. Chem.*, **346**, C27 (1988).
9.   C.-H. Liu, C.-S. Li and C.-H. Cheng, *Organometallics*, **13**, 18 (1994).
10. M. Catellani, B. Marmiroli, M. C. Fagnola and D. Acquotti, *J. Organometal. Chem.*, **507**, 157 (1996).
11. J. Cámpora, P. Palma and E. Carmona, *Coord. Chem. Rev.*, **193-195**, 207 (1999).
12. J. Sicher, *Angew. Chem. Int. Ed.*, **11**, 200 (1972).
13. M. Catellani and B. E. Mann, *J. Organometal. Chem.*, **390**, 251 (1990).
14. G. Bocelli, M. Catellani and S. Ghelli, *J. Organometal. Chem.*, **458**, C12 (1993).
15. A. J. Canty, *Acc. Chem. Res.*, **25**, 83 (1992).
16. M. Catellani and M. C. Fagnola, *Angew. Chem. Int. Ed.*, **33**, 2421 (1994).
17. M. Catellani and F. Cugini, *Tetrahedron*, **55**, 6595 (1999).
18. M. Catellani, E. Motti and M. Minari, *Chem. Commun.*, 157 (2000).
19. M. Catellani and G.P. Chiusoli, *J. Organometal. Chem.*, **286**, C13 (1985).
20. O. Reiser, M. Weber, A. de Meijere, *Angew.Chem. Int. Ed.*, **28**, 1037 (1989).
21. M. Catellani and E. Motti, *New J. Chem.*, **22**, 759 (1998).
22. E. Motti, G. Ippomei, S. Deledda and M. Catellani, *Synthesis*, in press.
23. M. Catellani, S. Deledda, B. Ganchegui, F. Hénin, E. Motti and J. Muzart, *J. Organomealt. Chem.*, **687**, 473, 2003.
24. W. Smadja, S. Czernecki, G. Ville, C. Georgoulis, *Organometallics,* **6**, 166 (1987).
25. M. Catellani, E. Motti and S. Baratta, *Org. Lett.*, **3**, 3611 (2001).
26. R. C. Larock, M. J. Doty, Q. Tian and J. M. Zenner, *J. Org. Chem.*, **62**, 7536 (1997).
27. G. Dyker, *J. Org. Chem.*, **58**, 1993 (1993).
28. E. Motti, A. Mignozzi and M. Catellani, *J. Mol. Catal. A: Chem.*, **204-205**, 115 (2003).
29. S. Deledda, E. Motti and M. Catellani, unpublished results.
30. A. Bayler, A. J. Canty, J. H. Ryan, B. W. Skelton, A. H. White, *Inorg. Chem. Commun.*, **3**, 575 (2000).

51.

O-Acylphosphites: New and Promising Ligands for Isomerizing Hydroformylation

Detlef Selent,[a] Klaus-Diether Wiese,[b] and Armin Börner[a]

[a]*Leibniz-Institut für Organische Katalyse, Buchbinderstr. 5-6, D-18055 Rostock, Germany*
[b]*Oxeno C-4 Chemie, Paul-Baumann-Str. 1, D-45772 Marl; Germany*

detlef.selent@ifok.uni-rostock.de

Abstract

Bidentate phosphorus ligands bearing an *O*-acyl phosphite moiety show superior modifying properties to the rhodium catalyst used in the hydroformylation of internal olefins. Results obtained for the hydroformylation of internal octenes and 2-pentene, respectively, are presented. The new ligands do markedly enhance the isomerization activity of the rhodium center. Internal hydroformylation is clearly disfavoured. At 120 °C/ 20 bar CO/H_2, a predominant terminal reaction is achieved. Thus, a 0.65...0.8 molar fraction of the desired terminal product is obtained with an aldehyde chemoselectivity exceeding 99.7%. Depending on the ligand structure and the olefinic substrate used, excellent turn over frequencies between 3000 and 7000 h^{-1} have been estimated. Further results concerning the coordination behaviour of the new ligands towards the precatalyst [acacRh(COD)] itself, as well as high pressure NMR investigations in the formation of *O*-acylphosphite-phosphite hydrido rhodium complexes, are presented.

Introduction

The regioselective synthesis of *n*-aldehydes is an important goal of the rhodium catalyzed alkene hydroformylation reaction. Considerable progress has been achieved within this field by the development of bidentate phosphines and phosphites as modifying ligands for the hydroformylation of terminal olefins.[1] However, for industry the use of less reactive internal olefins as substrates for the production of terminal aldehydes is of interest also. Internal olefins are available on an a technical scale by low-cost routes and are hydroformylated mainly with the aid of modified/unmodified cobalt catalysts.[2,3] It is still a greater challenge to develop rhodium catalysts able to convert these substrates with the desired good *n*-regioselectivity and high activity. Thus, diphosphine modified rhodium(I) allows for selectivities of ~90% to *n*-nonanal starting from 2-octene but turn over frequencies are below 400 h^{-1}, better rates are obtained with the shorter 2-butene as a substrate.[4,5] Interesting even from the more practical point of view were results obtained for a hydroxy phosphonite which in the presence of rhodium forms a ligand mixture and finally leads to ~50% nonanal from internal octenes at a $TOF = 2840$ h^{-1}, comparable to that of the unmodified cobalt counterpart used in industry.[6,7] High activity for 1-olefin hydroformylation and isomerization is known for several phosphite derived catalysts, therefore encouraging to conduct further research in this field.[8,9] Herein we report on the outstanding performance of *O*-acylphosphite modified rhodium catalysts used in the hydroformylation of internal octenes and 2-pentene.

Results and Discussion

A catalyst used for the *n*-regioselective hydroformylation of internal olefins has to combine a set of properties, which include high olefin isomerization activity, see reaction *b* in Scheme 1 outlined for 4-octene. Thus the olefin migratory insertion step into the rhodium hydride bond must be highly reversible, a feature which is undesired in the hydroformylation of 1-alkenes. Additionally, β-hydride elimination should be favoured over migratory insertion of carbon monoxide of the secondary alkyl rhodium, otherwise *iso*-aldehydes are formed (reactions *a*, *c*). Then, the fast regioselective terminal hydroformylation of the 1-olefin present in a low equilibrium concentration only, will lead to enhanced formation of *n*-aldehyde (reaction *d*) as result of a dynamic kinetic control.

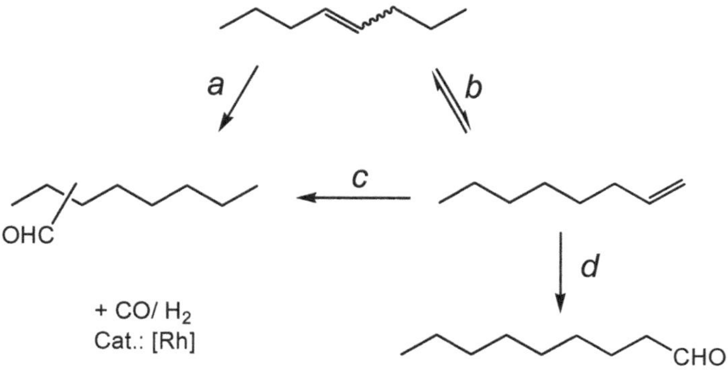

Scheme 1

At the present, it is difficult to predict a distinct rhodium catalyst showing the appropriate properties. Furthermore, the reaction conditions applied will influence the outcome of the reaction also. Low carbon monoxide pressure favours β-hydride elimination by enhanced CO dissociation which allows for the formation of vacant sites at the metal center. Increased reaction temperatures will lead to enhanced isomerization rates also.[1] For a proper modification of the rhodium catalyst it will be desirable to introduce highly π-acidic phosphorus ligands able to enhance the tendency of carbon monoxide dissoziation. We therefore decided to structurally modify phosphites by incorporation of a keto function, thus forming *O*-acyl derivatives. Table 1 summarizes spectroscopic data obtained for model [acacRh(CO)*L*] type complexes with different monodentate ligands *L*, including the *O*-acylphosphite complex **4**. This complex exhibits an ν(CO) of 2019 cm^{-1} in IR and a $^1J_{PRh}$ of 302.4 Hz in ^{31}P-NMR spectroscopy, respectively, therefore even exceeding the corresponding values obtained for the phosphite complex **3**. From the data a dependency of π-acidity of the ligands may be derived, following the order phosphonite~*O*-acylphosphonite<phosphite<*O*-acylphosphite.

Table 1 Spectroscopic data of complexes [acacRh(CO)(*L*)] (**1-4**), *L*=

| | (1) | (2) | (3) | (4) |

complex	^{13}C-NMR CO ligand δ [ppm] J(C-Rh) / J(C-P), [Hz]	^{31}P-NMR δ [ppm], J(P-Rh) [Hz]	IR ν(CO) [cm^{-1}]
1	188.0(dd), 187.6(dd) 80.1, 74.4 / 28.6, 27.7	146.9(d), 246.9 137.9(d), 245.5	2004
2	188.4(dd) 76.3 / 29.5	163.6(d), 233.1	2001
3	189.7(dd) 74.4 / 34.3	117.1(d), 289.9	2007
4	186.8(dd) 71.5 / 39.1	123.7(d), 302.4	2019

We have used the *O*-acylphosphite functional group to redesign the sterically demanding diphenol diphosphites first described in the patent literature.[10,11] Thus, reacting a hydroxy phosphite with the appropriate P-Cl derivative, *e.g.* commercial available racemic 2-chloro-4H-1,3,2-benzodioxaphosphorin-4-one in the presence of base gives the new, unsymmetrical bidentate hybrid ligands **5** and **8**, respectively, as a mixture of two diastereoisomers, compare chart 1.[12] Attempts to use these modifiers in rhodium catalyzed hydroformylation of internal octenes gave surprising results. High catalyst activities together with selectivities towards nonanal exceeding 60% were obtained at 130 °C/ 20 bar CO/H$_2$, see Table 2. Nearly complete conversion is obtained after 3 h with all the new catalytic systems applied. Turn-over-frequencies at 20% conversion exceed 3000 h^{-1}. However, a significant influence of the ligand substitution pattern on activity and also selectivity is obvious. Ligands with methoxy substituents

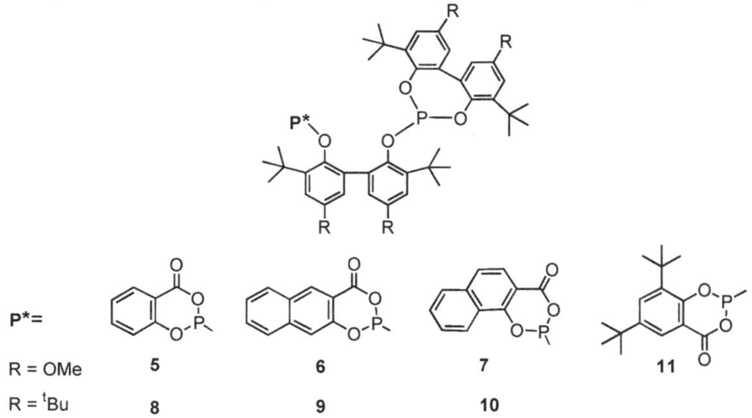

P*=				
R = OMe	**5**	**6**	**7**	**11**
R = ^{t}Bu	**8**	**9**	**10**	

Chart 1 Ligands used for isomerizing hydroformylation of internal octenes and 2-pentene.

at the ligand periphery cause higher activity, but lower regioselectivities than their *tert.*-butyl counterparts. Additionally, compounds derived from sterically more demanding hydroxy naphthoic acids tend to induce higher selectivities. This effect is more pronounced for the O-Me than for the *tert*-butyl ligand series.

Table 2 Hydroformylation ^{a} of isomeric *n*-octenes ^{b} with ligands **5-10**.

L	yield [%] ^{c}	n-nonanal [%] ^{d}	ROH [%] ^{e}	TOF [h^{-1}] ^{f}
5	94	63.9	0.2	>4600 ^{g}
5 ^{h}	83	64.1	0.2	4751
6	95	64.2	0.2	4310
7	96	69.0	0.3	4448
8	95	67.3	0.1	3119
9	93	67.9	0.1	3324
10	95	68.5	0.2	3336

a) Runs were performed with [n-Octene]$_0$ = 1.68 *M* at *T*=130 °C, *p*= 20 bar CO/H$_2$ (1:1), *t*=3 h, solvent toluene. Precatalyst: [acacRh (COD-1.5)]. [Rh]= 1.07 m*M*, Rh:P:olefin=1:10:1570. *b*) technical substrate, 3.3% 1-octene; 48.4% *Z/E*-2-octene; 29.2% *Z/E*-3-octene; 16.4% *Z/E*-4-octene; 2.1% sceletal isomeric octenes; 0.6% octane. *c*) Overall aldehyde yield. *d*) Percentage based on overall aldehyde yield. *e*) Overall alcohol yield, based on olefin supply. *f*) mol aldehyd x mol^{-1} catalyst h^{-1} at 20% conversion, *g*) detection limit of gas flow meter. *h*) [Rh]= 0.107 m*M*.

The ligands synthesized were also applied to the isomerizing hydroformylation of more reactive 2-pentene. At 120 °C/ 20 bar quantitative conversion of olefin to aldehydes was achieved within 40 min. Trends similar to those described for internal octene hydroformylation were found. The regioselectivity obtained for the individual ligands tends to be ~5% higher compared to that for the octenes. Thus, in the presence of **10** 75% of n-hexanal were determined, compare Table 3. Obviously, 2-pentene is able to react more smoothly to the terminal isomer compared to olefins having the double bond in an more internal position. Illustrative for this effect are also literature results obtained for 2- and 4-octene.[4,5]

Table 3 Hydroformylation[a] of 2-pentene with ligands **5-10**.

L	5	6	7	8	9	10
yield[b] [%]	100	100	100	100	100	100
n-hexanal [%]	69.4	69.8	74.1	73.8	74.1	75.0

a) Runs were performed with $[2\text{-pentene}]_0$ = 0.70 M at T=120 °C, p= 20 bar, t= 3 h; precatalyst: [acacRh(cyclooctadiene-1,5)]; [Rh]= 1.07 mM, Rh/L = 5, solvent toluene. b) Overall aldehyde yield.

Additional results obtained for **5** clearly showed the influence discussed above for changes of pressure and temperature, respectively. At 120 °C, 10 bar, regioselectivity raises to 70%, whereas it drops to 66% at 110 °C/ 20 bar. The dependency on overall pressure and temperature was studied also including propylene carbonate as a solvent using ligand **6**. All reactions were performed under constant syngas pressure (CO/H_2=1:1) by simultaneously monitoring the gas flow. A first order dependency of the decay of gas consumption versus time was observed for all runs. The pseudo constants $k_{obs.}$ which allow for a comparison of catalytic activities of the different catalytic systems applied are also given in Table 4.

Table 4 Hydroformylation of 2-pentene with **6**: Variation of T, p, solvent.[a]

solvent	T [°C]	p [bar]	yield [%]	n-hexanal [%]	$k_{obs.}$ [min^{-1}]	TOF_0 [h^{-1}]
toluene	120	20	100	69.8	0.1142	4064
pc[b]	120	20	100	71.7	0.1572	6750
pc	110	20	100	67.1	0.0869	3686
pc	110	15	100	72.8	0.1095	4136
pc	110	10	100	75.6	0.0895	3909
pc	100	10	100	75.3	0.0767	3095

a) Other experimental conditions see Table 3, except t= 1.5 h for runs with propylene carbonate applied as a solvent. b) pc= propylene carbonate.

The polar solvent induces higher catalytic activity with an initial TOF= 6750 min⁻¹. The influence of solvent on the microkinetices of the reaction is quite unclear, though an similar positive effect of polar solvents on reaction rates is known for rhodium catalyzed hydroformylation.[13] In propylene carbonate at 20 bar a decrease of temperature by 10 K is accompanied by a selectivity drop of 4.6%. The advantage of propylene carbonate is shown at 110 °C/ 15 bar where the rate observed equals that in toluene at 120 °C, but a significantly higher regioselectivity to *n*-hexanal is obtained. Moreover, the polar solvent allows for high activity even at 100 °C/10 bar. As a consequence of reduced pressure, a maximum of regioselectivity is observed at >75%. An even higher regioselectivity of 81.4% is obtained already at 20 bar, when the 3,5-di(*tert.*-butyl) salicylic acid derived ligand **11** is used. With $k_{obs.}= 0.0456$ min⁻¹ reaction proceeds with a relatively low rate. Obviously, not only the internal to terminal olefin isomerization but also hydroformylation of internal double bonds is slowed down in presence of **11**. The same effect causes high selectivities when diphosphites are used for the reaction of 1-olefins. Besides regioselective hydroformylation, these catalysts do only effectively isomerize the terminal olefin to the internal one resulting in reduced *chemoselectivity* of the overall reaction.[1,14]

We have chosen compound **5** to get insight to the coordination behavior of the new *O*-aclyphosphite ligands. Attempts to isolate pure diastereoisomers for further investigations were found possible but rather tedious by preparative HPLC, therefore we proceeded with the mixture of diastereomers. Treatment of [acacRh(COD-1.5)] with one equivalent of **5** in toluene at room temperature affords spontaneous liberation of the diolefin and quantitative formation of the new complex [acacRh(**5**)] (**12**), which was isolated as an orange microcrystalline solid. The phosphorus NMR spectrum of complex **12** is characterized by two doubletts of doubletts for each diastereomer, resonating at δ 132.2 ($J_{PRh}= 317.7$ Hz, *O*-acyl-P *major*), 135.4 ($J_{PRh}= 292,6$ Hz, phosphite-P, *major*), 132.2 ($J_{PRh}= 215.0$ Hz, *O*-acyl-P *minor*), 134.1 ($J_{PRh}= 188.7$ Hz, phosphite-P, *minor*) ppm, with almost equal coupling constants $^2J_{PP}= 111$ Hz for the major and 110 Hz for the minor diastereomer, respectively. The newly formed compound **12** was reacted with syngas in an autoclave in toluene solution. It was proved by NMR that complete conversion occured at 20 bar CO/H₂ (1:1) at 40 °C within 4 h. Workup under argon afforded an orange-brown solid containing a mixture of two diastereomeric hydrido complexes with the hydride signals centered at 10.0 for the major and 9.95 ppm for the minor diastereomer, respectively, with equal values of ¹H-¹⁰³Rh couplings of 5 Hz, see Figure 1, d). These hydrido complexes are accompanied by a 20% amount (fraction of overall signal intensity in ³¹P-NMR) of an impurity showing no ³¹P-¹H(hydride) coupling and a complex ³¹P-NMR spectrum. With respect to the known instability of the [HRh(CO)L₂] type complexes to hydrogen elimination this impurity is likely to be the Rh(0)-dimeric complex [Rh(CO)₂(**5**)]₂ which may form eight diasteromers. We could indirectly support this assignment by high pressure NMR. Thus, the crude hydrido complex was dissolved in toluene-D_8 and pressurized with 20 bar syngas in an sapphire high pressure NMR tube at 25 °C. Within a few minutes the initial color of solution turned from orange-brown to yellow. ^1H- and ^{31}P-NMR spectroscopy verified the formation of pure [HRh(CO)$_2$(**5**)] (**13**), representing 97% of overall signal intensity in the ^{31}P- NMR spectrum. The equilibrium to occur between [HRh(CO)$_2$(**5**)] and the suggested dimer was proofed by evaporating the solvent of the probe solution to dryness and redissolving the solid obtained under argon, which again gave the known complex mixture. Detailed analysis of the NMR spectra of **13** with

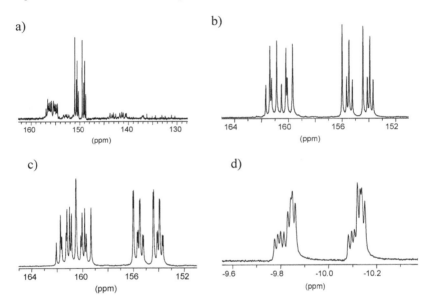

Figure 1 ^{31}P-NMR (toluene-D_8) of [HRh(CO)$_2$(**5**)] (**13**): a) crude, argon; b) pure, 20 bar CO/H$_2$ (1:1), proton decoupled; c) pure, proton coupled; d) ^1H-NMR, hydride region.

255.3 Hz again gave higher values for P$_{o\text{-acyl}}$ rhodium couplings compared to 191.4 for the P$_{phosphite}$ in the major diastereomer which holds qualitatively also for the minor one. P-H couplings of 115.0 and 123.4 Hz for the phosphite unit and much lower values of 8.4 and 9.8 Hz for the O-acyl bound phosphorus, respectively, were observed. That indicates the bidentate ligand coordinating in a preferably axial-equatorial fashion under the assumption of an idealized trigonal pyramidal complex structure. The bulky phosphite unit is placed in the axial position *trans* to the hydrido ligand in both diastereomers which is consistent for a ligand exhibiting a bite angle significantly lower than 120 °C. We tried to crystallize **13** in a diastereomerically pure form to get more reliable structural data by X-ray structural analysis. But, any attempts made were not successful. Therefore a series of precursors were treated with **5**, including the π–allyl complex [(C₃H₅)Rh(COD-1,5)]. The latter reacted sponteanously with the bidentate ligand by substitution of the diolefin comparable to the reaction of [acacRh(COD-1,5)] described above. The straightforward reaction formed [(C₃H₅)Rh(**5**)] (**14**) which was isolated as a yellow solid. From **14** crystalline material suitable for diffraction studies could obtained. The X-ray analysis revealed a diasteromer exhibiting the same relative configuration of both stereocenters, see Figure 2. The Rh-P$_{phosphite}$ distance of 220.2 pm was found to be substantially longer compared to 216.0 pm for Rh-P$_{o\text{-acyl}}$. A P-Rh-P bite angle of 103.7° was determined which fits to the

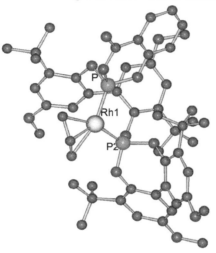

Figure 2 SCHAKAL drawing of [[(C₃H₅)Rh(**5**)]. P1= P$_{o\text{-acyl}}$, P2= P$_{phosphite}$. For distances Rh1-P1/P2 and angle P-Rh-P see text. Space group $P\bar{1}$, triclinic, a = 10.382(2), b = 11.219(2), c = 12.817(3) Å, α = 85.00(3), β = 79.14(3), γ = 83.65(3)°, V= 1453.8(5) Å³, Z= 1, ρ$_{calcd}$ = 1.151 g cm⁻³.

spectroscopic data obtained for hydrido complex **13**. Interestingely, [(C₃H₅)Rh(**5**)] under syngas is easily transformed to [HRh(CO)₂(**5**)] quantitatively provided a temperature of 60 °C is applied. This shows the allyl complex a suitable precursor for hydroformylation. The allyl to hydrido complex transformation may proceed by hydrogenolysis of the Rh-allyl bonding with elimination of propene, a reaction common for a series of allyl rhodium phosphite complexes.[15]

    In conclusion, bidentate *O*-acylphosphite ligands were shown to be very useful modifyers for the rhodium catalyzed isomerizing hydroformylation of internal olefins. First insight to the possible structure of the catalyst's resting state was gained by HP-NMR. Detailed mechanistic studies as well as attempts to even improve the catalytic performance of this new ligand class, and their use in other catalytical applications, respectively, are underway.

Experimental Section

All reactions were carried out using standard Schlenk techniques under an atmosphere of argon (Linde AG). Chemicals were purchased from Aldrich Chemical Co.. Ligands **5 – 11** were prepared according to described methodology.[12] Hydroformylation experiments were performed under constant pressure in an 200 ml autoclave (Fa. Buddeberg) equipped with a Bronkhorst hitec gas flow meter and a Bronkhorst pressure controler. Reaction solutions were analyzed by GLC using solvent toluene as an internal standard. NMR spectra were obtained on a Bruker ARX 400 spectrometer (400.1 MHz for ^{1}H-NMR, 100.6 MHz for ^{13}C-NMR and 162.0 MHz for ^{31}P-NMR); shifts are given relative to TMS (^{1}H, ^{13}C). For HPNMR experiments, 0.2 mmol of complexes were dissolved in 2 ml toluene-D$_8$, the solution filtered, transfered to a 10 mm high pressure sapphire NMR tube and pressurized with syngas (CO/H$_2$=1:1) to 20 bar. Mass spectra were measured on a Intectra AMD 402/3 facility. Elemental analyses data were detected on a Leco CHNS-932 apparatus. IR-spectra were recorded on a Nicolet Magna-IR 550 spectrometer.

X-ray structural analysis. Suitable crystals of compound **14** were obtained from toluene/ether solutions. X-ray data were collected on a STOE-IPDS diffractometer using graphite monochromated Mo-Kα radiation. The structure was solved by direct methods (SHELXS-86)[16] and refined by full-matrix-least-squares techniques against F^2 (SHELXL-93).[17] Crystal dimensions: 0.3 · 0.2 · 0.1 mm, yellow-orange prisms, 3612 reflections measured, 3612 were independent of symmetry and 1624 were observed ($I > 2\sigma(I)$), R1 = 0.048, wR^2 (all data) = 0.151, 295 parameters.

Syntheses of complexes 1 – 4. To a stirred suspension of [acacRh(CO)$_2$] (0.131 g, 0.507 mmol) in CH$_2$Cl$_2$ (1 ml) the appropriate ligand (0.507 mmol) was added as a solution in toluene (2.5 ml) at room temperature. After stirring for 1 h the solvent was removed in vacuo to give yellow residues in nearly quantitative yields, which were shown to be spectroscopically pure. For details of spectroscopic characterisation see text, table 1. Elemental analyses (calc.): **1** (C$_{41}$H$_{46}$O$_8$RhP, 800.69 g/mol), C 61.51 (61.50); H 5.98 (5.79)%. **2** (C$_{33}$H$_{38}$O$_6$RhP, 664.54 g/mol), C 59.49 (59.65); H 5.76 (5.72)%. **3** (C$_{48}$H$_{70}$O$_6$RhP, 876,96 g/mol), C 65.37 (65.74); H 8.10 (8.05)%. **4** (C$_{27}$H$_{32}$O$_7$RhP, 602.43 g/mol), C 54.22 (53.83); H 5.76 (5.35)%.

[acacRh(5)] (12). A stirred solution of [acacRh(COD-1,5)] (620.4 mg, 2 mmol) in toluene (10 ml) was treated with a solution of ligand **5** (1.822 g, 2 mmol) in the same solvent (10 ml). After stirring for 1 h, the solvent was removed in vacuo to give a yellow oil. Addition of hexane (15 ml), after virgorous stirring, gave a suspension from which the yellow solid was filtered off and dried for 2 h at 70 °C/ 0.1 mbar. Yield: 2.086 g (93 %). ^{31}P-$\{^1H\}$-NMR (toluene-D$_8$): δ 132.2 (J_{PRh}= 317.7 Hz, *O*-acyl-P *major*), 135.4 (J_{PRh}= 292,6 Hz, phosphite-P, *major*), 132.2 (J_{PRh}= 215.0 Hz, *O*-acyl-P *minor*), 134.1 (J_{PRh}= 188.7 Hz, phosphite-P, *minor*) ppm, $^2J_{PP}$= 111 Hz (*major*), 110 Hz (*minor*). Elemental analysis (calc. for C$_{56}$H$_{67}$O$_{12}$P$_2$Rh, 1096.91 g/mol), C 61.56 (61.31); H 6.32 (6.16).

[C$_3$H$_5$Rh(5)] (14). A solution of ligand **5** (2.684 g, 2.946 mmol) in toluene (35 ml) was added to a stirred solution of [C$_3$H$_5$Rh(COD-1.5)] (0.743 g, 2.946 mmol) at room temperature. After stirring for 1 h and filtration the filtrate was evaporated to dryness. The residue was suspended in pentane (2x30 ml) and the solvent removed in vavuo. The yellow solid obtained was dried for 2 h at 0.1 mbar and then recrystallized from toluene/pentane to obtain the pure compound. Yield: 1.707 g (55 %). Elemental analysis (calc. for C$_{54}$H$_{65}$O$_{11}$P$_2$Rh, 1054.95 g/mol): C 61.55 (61.48); H 5.84 (5.87), P 5.69 (5.87)%. EI-MS (70 eV): *m/e* 1054 (100%, *M*$^+$), 1013 (25%, *M*$^+$-C$_3$H$_5$). ^{31}P-NMR (toluene-D$_8$): δ 154.1 (^1J(P-Rh)= 348.2 Hz, ^2J(P-P)= 62.5 Hz), 167.3 (^1J(P-Rh)= 327.3 Hz, ^2J(P-P)= 62.5 Hz), *minor*; 155.1 (^1J(P-Rh)= 349.5 Hz, ^2J(P-P)= 61.1 Hz), 167.4 (^1J(P-Rh)= 330.0 Hz, ^2J(P-P)= 61.1 Hz), *major*.

References

1. Rhodium Catalyzed Hydroformylation, P. W. N. M. van Leeuwen, and C. Claver, Eds., Kluwer, Dordrecht, 2000.
2. C. D. Frohning, C. W. Kohlpaintner, and H.-W. Bohnen, in Applied Homogeneous Catalysis with Organometallic Compounds, Vol. 1, B. Cornils, W. A. Herrmann, Eds., VCH, Weinheim, 2002, p. 30.
3. H.-W. Bohnen and B. Cornils, *Adv. Catal.,* **47**, 1 (2002).
4. H. Klein, R. Jackstell, K.-D. Wiese; C. Borgmann, and M. Beller, *Angew. Chem.,* **113**, 3505 (2001).
5. L. A. van der Veen, P. C. J. Kamer, P. W. N. M. van Leeuwen, *Angew. Chem.,* **111**, 349 (1999).
6. D. Selent, K.-D. Wiese, D. Röttger, A. Börner, *Angew. Chem.,* **112**, 1694 (2000).
7. D. Selent, W. Baumann, R. Kempe, A. Spannenberg, D. Röttger, K.-D. Wiese, and A. Börner, *Organometallics,* **22**, 4265 (2003).
8. P. W. N. M. van Leeuwen and C. F. Roobeek, *J. Organometal. Chem.,* **258**, 343 (1983).
9. A. van Rooy, E. N. Orij, P. C. J. Kamer, and P. W. N. M. van Leeuwen, *Organometallics,* **14**, 34 (1995).

10. E. Billig, A. G. Abatjoglou, and D. R. Bryant, U.S. Patent 4748261, 1988.
11. E. Billig, A. G. Abatjoglou, and D. R. Bryant, U.S. Patent 4769498, 1988.
12. D. Selent, D. Hess, K.-D. Wiese, D. Röttger, C. Kunze, and A. Börner, *Angew. Chem.,* **113**, 1739 (2001).
13. R. M. Deshpande, B. M. Bhanage, S. S. Divekar, R. V. Chaudhari, *J. Mol. Catal.,* **78**, L37 (1993).
14. W. L. Gladfelter, B. Moasser, and C. R. Roe, *Organometallics,* **14**, 3832 (1995).
15. A. J. Sivak and E. L. Muetterties, *J. Am. Chem. Soc.,* **101**, 4878 (1979).
16. G. M. Sheldrick, *Acta Crystallogr.,* *A46*, 467 (1990).
17. G. M. Sheldrick: SHELXL-93, University of Göttingen, Germany, 1993.

52. The Synthesis of Carbamate from the Reductive Carbonylation of Nitrobenzene over Pd-based Catalysts

Bei Chen and Steven S. C. Chuang

Department of Chemical Engineering, The University of Akron, Akron OH 44325-3906

schuang@uakron.edu

Abstract

Pd/Al_2O_3-$FeCl_3$, and Ce-Pd/Al_2O_3-$FeCl_3$ catalysts exhibit activity for the synthesis of ethylphenylcarbamate from the reductive carbonylation of nitrobenzene with ethanol at 453 K and 2.07 – 2.93 MPa. The advantage of the use of Al_2O_3-supported Pd catalyst is the easy of catalyst recovery form the reactants/product mixture.

Introduction

Isocyanates are important chemical feedstock for the manufacture of fertilizers, pesticides, and various forms of polyurethane (1-4). The current isocyanate synthesis process involves phosgenation of amines.

$$RNH_2 + COCl_2 \rightarrow RNCO + HCl \qquad [1]$$

where R is an alkyl or aryl group. The phosgenation process suffers from the difficulty in handling hazardous phosgene as well as the highly corrosive by-product HCl. To minimize the environmental impact of $COCl_2$, extensive research has been directed toward the development of effective catalysts for oxidative and reductive carbonylation (1-16).

Oxidative carbonylation:
$$RNH_2 + R'OH + CO + 1/2O_2 \rightarrow RNHCOOR' + H_2O \qquad [2]$$
Reductive carbonylation:
$$RNO_2 + 3CO + R'OH \rightarrow RNHCOOR' + 2CO_2 \qquad [3]$$
Thermal decomposition:
$$RNHCOOR' \rightarrow RNCO \text{ (isocyanate)} + R'OH \qquad [4]$$
where R and R' are either alkyl or phenyl groups.

Carbamate, RNHCOOR', produced from carbonylation pathways can be selectively converted to isocyanate (1.4). Carbonylation pathways offer a number of advantages: (i) the environmentally benign nature of the reactants, (ii) the high selectivity of the reaction processes, (iii) the stability and low toxicity of carbamate products and (iv) the wide range of applications of carbamate as chemical feedstock.

The starting reactant is aniline for the oxidative carbonylation and nitrobenzene for the reductive carbonylation. The major advantage of the oxidative carbonylation is that the oxidative carbonylation is more thermodynamically favorable than the reductive carbonylation. The former can occur at a significantly milder condition than the latter (11-16). However, nitrobenzene is the feedstock for the production of aniline:

$$RNO_2 + 3H_2 \rightarrow RNH_2 + 2H_2O \hspace{3cm} [5]$$

The reductive carbonylation has an advantage of low feedstock cost. A wide range of homogenous metal complexes have been tested for both reactions (1-16). The major drawback of the use of metal complex catalysts is the difficulty of catalyst recovery and purification of the reaction products (12). In addition, the gaseous reactants have to be dissolved in the alcohol/amine mixture in order to have an access to the catalyst. The reaction is limited by the solubility of the gaseous CO and O_2 reactants in the liquid alcohol reactant (17).

We have demonstrated that supported Pd and Cu catalysts are effective in catalyzing the oxidative carbonylation at low pressure reaction condition and the supported metal catalysts can be easily separated from the product mixture in both fixed bed and slurry phase reactors (12,17). The objective of this study is to investigate the feasibility of using Al_2O_3-supported Pd catalysts for catalyzing the reductive carbonylation of nitrobenzene with ethanol.

Results and Discussion

Figure 1 shows the infrared spectra of the reactant and product mixture taken as a function of time over $PdCl_2(PPh_3)_2$. The characteristic bands of carbamate at 1713-1731 cm^{-1} and 1607-1620 cm^{-1} increased with the reaction time. The former is due to the C=O stretching vibration; the latter is due to the N-H bending vibration in carbamate. The ^{13}C NMR spectrum of the product sample gave a carbamate's C=O peak at 154.5 ppm. Both IR and NMR results confirmed the formation of ethyl-N-phenyl carbamate over $PdCl_2(PPh_3)_2$, Pd/Al_2O_3-$FeCl_3$, and Ce-Pd/Al_2O_3-$FeCl_3$.

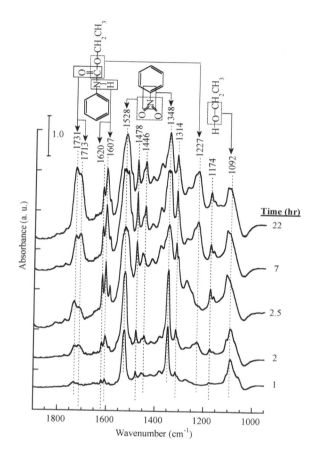

Figure 1 Transmission IR spectra of reductive carbonylation over PdCl$_2$(PPh$_3$)$_2$ at 2.93 MPa and 453 K.

Figure 2 shows the carbamate yields as a function of time over PdCl$_2$(PPh$_3$)$_2$, Pd/Al$_2$O$_3$-FeCl$_3$, and Ce-Pd/Al$_2$O$_3$-FeCl$_3$. The carbamate yield is defined as the ratio of moles of carbamate to the initial moles of aniline. The initial TOF for carbamate yield (the number of mole of carbamate molecule produced per hour per Pd atom on the catalyst) can be determined from the slope of the carbamate yield at t=0 in Figure 2.

The results of TOF and carbamate yields are summarized in Table 1. Although PdCl$_2$(PPh$_3$)$_2$ exhibited the highest activity, supported Pd exhibits good activity for the carbamate synthesis for the reductive carbonylation.

Addition of ceria caused a decrease in both TOF and yields. Decreasing the reaction pressure caused both TOF and yields to decrease.

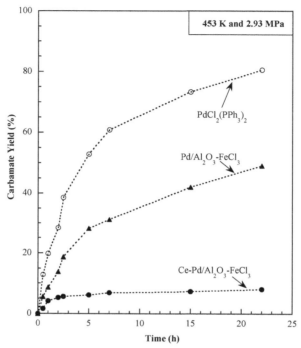

Figure 2 Carbamate yield (%) versus reaction time over PdCl$_2$(PPh$_3$)$_2$, Pd/Al$_2$O$_3$-FeCl$_3$, and Ce-Pd/Al$_2$O$_3$-FeCl$_3$ at 2.93 MPa and 453 K.

Table 1 Carbamate yield of final solution and initial TOF over PdCl$_2$(PPh$_3$)$_2$, Pd/Al$_2$O$_3$-FeCl$_3$, and Ce-Pd/Al$_2$O$_3$-FeCl$_3$ at 453 K.

P (MPa)	Catalyst	Carbamate Yield (%)	TOF (h^{-1})
2.93	PdCl$_2$(PPh$_3$)$_2$	80.5	92.3
2.93	Pd/Al$_2$O$_3$-FeCl$_3$	49.1	47.7
3	Ce-Pd/Al$_2$O$_3$-FeCl$_3$	8.0	15.2
2.07	Pd/Al$_2$O$_3$-FeCl$_3$	40.6	33.2

Conditions: Nitrobenzene (0.10 mol); ethanol (0.17 mol); 0.046 g PdCl$_2$(PPh$_3$)$_2$ or 0.35 g Pd/Al$_2$O$_3$/Ce-Pd/Al$_2$O$_3$ with 1.0 mmol FeCl$_3$; reaction time (22 h).

The results of this study demonstrate that Pd/Al$_2$O$_3$-FeCl$_3$, exhibited activity at about 60% activity of PdCl$_2$(PPh$_3$)$_2$ for the synthesis of ethylphenylcarbamate from the reductive carbonylation of nitrobenzene with ethanol at 453 K and 2.07 – 2.93 MPa. The major advantage of the use of Al_2O_3-supported metal catalyst is the easy of catalyst recovery form the reactant and product mixture. Fine-tuning catalyst composition could lead to further enhancement of catalyst activity. Combining the high activity with the intrinsic merit of ease of catalyst recovery, the reductive carbonylation over supported Pd-based catalyst could provide an effective pathway for carbamate syntheses.

Experimental Section

Materials

Palladium (II) chloride and ferric (III) chloride were purchased from Sigma Chemicals Co.; Cerium(III) nitrate from Pfaltz & Bauer, Inc.; γ-Al_2O_3 (100 m^2/g) from Alfa Chemicals; ethanol (HPLC grade) and nitrobenzene (Certified ACS) from Fisher Scientific. Bis(triphenylphosphine)palladium(II) dichloride was generously donated by Dr. Ivan J.B. Lin (National Dong Hwa University, Taiwan). All compounds were used without further treatment. Carbon monoxide (99.0%) was obtained from Praxair.

Catalyst Preparation and Characterization

Pd/Al_2O_3 catalyst was prepared by incipient wetness impregnation of γ-Al_2O_3 support with a $PdCl_2$ solution at a pH of 2.8 and a temperature of 333 K. The catalyst was calcined in flowing air at 673 K for 8 h and then reduced in flowing H_2 at 673 K for 2 h. The resulting catalyst is in the form of fine powder. Pd loading on the Pd/Al_2O_3 catalyst was determined to be 1.75 % by inductively coupled plasma (ICP) analysis (Galbraith Laboratories, Inc.). X-ray diffraction (XRD) line-broadening gave an average Pd crystallite size of 6.4 nm, corresponding to a dispersion of 10%. Ce-Pd/Al_2O_3 (1.75 wt% and 20 wt% Ce) was prepared by coimpregnation with a $Ce(NO_3)_3$ and $PdCl_2$ solution. XRD data yielded an average Pd crystallite size of 5.2 nm, corresponding to a dispersion of 12% Pd for Ce-Pd/Al_2O_3.

Reductive Carbonylation

The batch reductive carbonylation reaction was carried out in a 500-cm^3 autoclave (Pressure Products Industries, Inc.) with 0.17 mol ethanol, 0.10 mol nitrobenzene, 0.046 g $PdCl_2(PPh_3)_2$ (or 0.35 g Pd/Al_2O_3 and Ce-Pd/Al_2O_3 with 1.0 mmol $FeCl_3$ as promoter) at 453 K, 2.93 MPa, and a stirring speed of 700 r.p.m.

One cm^3 of the reactant/product/catalyst mixture was sampled periodically during the reaction for the transmission infrared analysis (Nicolet Magna 550 Series II infrared spectrometer with a MCT detector). The concentrations of reactants and products were obtained by multiplying integrated absorbance of each species by its molar extinction coefficient. The molar extinction coefficient was determined from the slope of a calibration curve, a plot of the peak area versus the number of moles of the reagent in the IR cell. The reaction on each catalyst was repeated and the relative error for the carbamate yield measured by IR is within ±5%.

The final reactant/product mixture was analyzed by a Gemini 300 ^{13}C nuclear magnetic resonance (NMR) spectrometer using CDCl$_3$ as a solvent.

Acknowledgements

This work was supported by the NSF Grant CTS 9816954 and the Ohio Board of Regents Grant R5538. We greatly appreciate Prof. Ivan J. B. Lin at National Dong Hwa University, Taiwan, for providing PdCl$_2$(PPh$_3$)$_2$.

References

1. S. Cenini, M. Pizzotti, and C. Crotti, Metal Catalyzed Deoxygenation Reactions by Carbon Monoxide of Nitroso and Nitro Compounds (D. Reidel Publishing), **6**, (*Aspects of Homogeneous Catalysis: A Series of Advances*), 97-198 (1988).
2. G. W. Parshall and J. E Lyons, Advanced Heterogeneous Catalysts For Energy Applications; U. S. Department of Energy, 1994; Chapter 6.
3. G. W. Parshall and S. D. Ittel, Homogeneous Catalysis—The Applications and Chemistry of Catalysis by Soluble Transition Metal Complexes, 2nd ed., Wiley, New York, 1992.
4. K. Weissermel and H.-J. Arpe, Industrial Organic Chemistry, 2nd ed., VCH, Weinheim, 1993.
5. J. G. Zajacek, J. J. McCoy and K. E. Fuger, U. S. Patent 3,919,279 (1975) and U. S. Patent 3,962,302 (1976) to Arco.
6. I. J. B Lin and C.-S. Chang, *J. Mol. Catal.,* **73**, 167 (1992).
7. Y. Ono, M. Shibata, and T. Inui, *J. Mol. Catal. A,* **153**, 53 (2000).
8. F. W. Hartstock, D. G. Herrington, and L. B. McMahon, *Tetrahedron Lett.,* **35**, 8761 (1994).
9. K. V. Prasad and R. V. Chaudhari, *J. Catal.,* **145**, 204 (1994).
10. W. McGhee, D. Riley, K. Christ, Y. Pan, and B. Parnas, *J. Org. Chem.,* **60**, 2820 (1995).
11. B. Chen and S. Chuang, *J. Mol. Catal.,* **195**, 37 (2003).

12. S. S. C. Chuang, Y. Chi, B. Chen, and P. Toochinda, U. S. Patent 6,541,543, 2003.
13. B. Chen and S. S. C. Chuang, *Green Chem.*, **5(4),** 484 (2003).
14. S. S. C. Chuang, P. Toochinda, and M. V. Konduru, Green Engineering-ACS Symposium Series 766, P. T. Anastas, L. G. Heine, and T. C. Willisamson, Eds. ACS, Washington, D.C., 2001, p.136-148.
15. P. Toochinda, S. S. C. Chuang, and Yawu Chi, Chemical Industries (Dekker) **89** (*Catal. Org. React.*) 369-378 (2003).
16. F. Ragani and S. Cenini, *J. Mol. Catal.*, **109**, 1 (1996).
17. P. Toochinda and S. S. C. Chuang, *Ind. Eng. & Chem. Res.* (in press).

53.   Microwave Synthesis in the Suzuki-Miyaura Cross Coupling Reaction with Pd/C

John R. Sowa Jr.,* Thomas Carrell, Megha Kandhari,
Joanna Manansala, Nicole Heuschkel, Katharine Hodock,
Amanda Thiel and Lubabalo T. Bululu

Department of Chemistry and Biochemistry, 400 South Orange Ave., South Orange, NJ 07079

sowajohn@shu.edu

Abstract

Using a conventional "dorm-room quality" microwave oven, we have successfully performed Suzuki-Miyaura cross-coupling reactions catalyzed by Pd/C. Shorter reaction times are obtained (13 min of irradiation) using microwave irradiation compared to conventional methods of heating (> 1 h). Yields with relatively non-volatile aryl bromides range from 65 to 83%. Lower yields (15 – 27%) are obtained with relatively volatile aryl bromides substrates which may be evaporating during the course of the reaction. Ease of reaction set-up, rate enhancement from the microwave irradiation and facile work-up provided by the use of Pd/C makes this a very efficient procedure to run.

Introduction

Suzuki-Miyaura cross-coupling has evolved into an extremely useful reaction for small- and large-scale preparations of biphenyl and styrene derivatives. (1) There is considerable interest in the use of heterogeneous catalysts such as Pd/C and Ni/C for this reaction because heterogeneous catalysts are more easily separated from the reaction products than are organic soluble, homogeneous catalysts. (2, 3) This ease of processing facilitates product isolation and results in greatly reduced palladium and ligand contamination in the crude reaction product. At a recent ORCS meeting, we reported that the Pd/C catalyst can be recycled in batch mode and we also demonstrated potential for semi-continuous flow systems. (3b) More recently, we demonstrated a large scale application of this reaction for preparation of an investigative pharmaceutical intermediate which resulted in low palladium contamination of the initial crude product. (3c)

$$X-\text{C}_6\text{H}_4-B(OH)_2 \ + \ Br-\text{C}_6\text{H}_4-Y \ \xrightarrow[\substack{\text{base} \\ \text{EtOH/H}_2\text{O}}]{\text{Pd/C}} \ X-\text{C}_6\text{H}_4-\text{C}_6\text{H}_4-Y$$

The use of microwave ovens in organic synthesis including catalytic organic reactions has also been of considerable recent interest. (4) Primarily this interest stems from greatly reduced reaction times for reactions conducted with microwave radiation compared to conventional thermal methods of heating. Coupling reactions such as Suzuki-Miyaura, Heck, Sonogashira have been demonstrated using microwave irradiation using homogeneous, nano-particle and heterogeneous catalysts. (5) Pd/C has been used in a microwave oven for transfer hydrogenation catalysis. (6a) However, the purpose of this study was to ascertain the suitability of Pd/C as a catalyst for microwave synthesis in the Suzuki-Miyaura coupling reaction.

Results and Discussion

In this study we explore the heterogeneous catalyst, Pd/C, for the Suzuki-Miyaura coupling reaction using a microwave oven as an alternative method for "heating" the reaction. Following a design of Prof. A. K. Bose, the reaction flask is loosely capped (*see below for cautionary statement*) and surrounded by beakers of water acting as a heat-sink. (6) This design allows irradiation (600 W) of the reaction mixture using a microwave oven that is not equipped with electronic power control. The irradiated mixture vigorously boils within a few minutes which promotes mixing of the reaction. However, solvent evaporation also occurs. A general protocol was developed to minimize evaporation in which the reaction mixture is initially irradiated for 5 min and then irradiated for four additional 2 min intervals for a total of 13 min of irradiation. Between each interval, the reaction mixture is allowed to cool for one min for an overall total of 18 min reaction time.

The Pd/C catalyst chosen for this study was in the unreduced form. The results in Table 1 clearly indicate that Suzuki-Miyaura coupling can be performed with an unreduced Pd/C catalyst using microwave irradiation. Each reaction is carried out to apparent completion as indicated by the disappearance of the aryl bromide starting material as measured by HPLC. However, the variation in product yields is quite remarkable. The highest yields are obtained with products **1, 2** and **6** which range from 65 to 83%. These yields are comparable to those obtained with thermally driven coupling reactions. The purities of the isolated crude products are excellent as indicated by HPLC analyses showing > 99% area percent purity. Only trace amounts (< 1%) of homo-coupled product from the aryl boronic acid (e.g., biphenyl) starting are detected. In contrast, low yields (15 – 27%) are obtained for products **3, 4** and **5**. In these reactions complete disappearance of the aryl bromide starting material is also observed. However, since the aryl bromide starting materials are relatively volatile, the low yields are likely caused by evaporation of the aryl bromide starting materials.

To minimize evaporation of the solvent and evaporation of the aryl bromide starting materials we attempted two reactions in a sealed system with a rubber septum fastened tightly with a plastic clamp to the neck of the Erlenmeyer flask. *Caution! Significant pressure developed which in one case blew off the rubber septum and in another case shattered the reaction flask within 3 min of reaction time.* Currently we are performing studies using a microwave oven with power modulation such that reducing the irradiation power may minimize evaporation and improve reaction yields. We note that recently commercial microwave ovens have become available that are designed with reflux condensers. Although quite expensive ($8 - $20 K) these design improvements may improve safety as well as reaction yields. (5) In addition, we have not investigated solventless Pd/C catalyzed reactions similar to those reported for Pd/alumina and Pd/KF/alumina. (5a,b)

Table 1 Yields of microwave assisted, Pd/C catalyzed, Suzuki-Miyaura cross-coupling reactions with 100% consumption of the aryl bromide.

	Product[a]	total irradiation time[b]	% yield (HPLC)	% isolated yield
1	—OMe	13	94	79
2	—CN	13	97	65
3	—F	13	38	15
4		13	49	27
5	—CF$_3$	13	55	26
6	MeO— —CN	13	95	83

[a] The left half of the product comes from the aryl boronic acid and the right half comes from the aryl bromide. Thus, products **1 - 5** use $PhB(OH)_2$ as the aryl boronic acid and product **6** utilizes 4-MeOPhB(OH)$_2$.
[b] The reaction mixture was irradiated for 5 min then 4 x 2 min intervals at full power (600 W) with a one min cool between each interval.

Generally we observe that microwave reactions are complete in shorter times and overall reaction set-up, work-up and clean-up is more efficient than using conventional heating methods. Roughly 3 – 4 microwave reactions can be completed in an 8 h day compared to 1 – 2 thermally driven reactions. However, it has been well noted that microwave induced reactions are considerably faster than the thermal counterparts. (4,5) The microwave method that we developed causes rapid boiling of the solvent mixture and the boiling is evident during most of the 13 min irradiation time, thus, we might assume this is equivalent to refluxing the reaction mixture for 13 min at 80 °C. Since our overall reaction time includes 5 min of cooling intervals, we decided that a conservative comparison to thermal method would be to heat in an oil bath the exact same reaction mixture for 18 min at 80 °C. Under these conditions, the reaction between phenyl boronic acid and p-bromoanisole only went to 63% conversion in 18 min and the reaction was still incomplete after 1 h. This result which was successfully duplicated indicates that the thermally driven reaction is substantially slower than the microwave induced reaction.

In this study we show that the Pd/C catalyzed Suzuki-Miyaura coupling reaction can be performed in a microwave oven. Overall the microwave synthesis is faster than comparable thermal methods and the combination of the ease of use of the microwave oven and the facile work-up with Pd/C makes this a very efficient method for performing coupling reactions.

Experimental Section

Reagents were purchased from commercial sources and used without further purification. Product identities were confirmed by comparison of HPLC retention times with authentic samples. The Pd/C catalyst was purchased as a dry, edge-coated, unreduced catalyst with 5 wt% in Pd. HPLC analyses were performed on a HP 1090 (stationary phase: C18, 25 cm x 0.46 cm; mobile phase: CH_3CN/H_2O 1% H_3PO_4, 90% H_2O to 100% CH_3CN gradient over 25 min, 1 ml/min flow, UV-Vis detection). For safety aspects of handling Pd/C, see ref. 7.[7]

General microwave assisted Suzuki-Miyaura cross-coupling procedure
    Caution! A reviewer expressed concern about using volatile organic solvents in a conventional microwave due to the potential for fire. Although we did not experience any problems of this sort, as a precaution, the entire microwave was placed in and experiments were operated in a fume hood. To a 125-mL Erlenmeyer is added Pd/C (30 mg, 0.014 mmol, 1 mol %), aryl boronic acid (3.2 mmol), K_2CO_3 (0.44 g, 3.2 mmol), aryl bromide (2.7 mmol). The flask and contents are purged with argon or nitrogen for 2 min. A mixture of ethanol/water (5:1, 20 mL) is then added and the suspension is sparged for an addition 2 min with argon or nitrogen. The Erlenmeyer flask is loosely capped and placed in a microwave oven. The flask is surrounded by four 250 mL beakers each containing 150 mL of water. The reaction mixture is initially irradiated at full power (600 W) for 5 min and then irradiated for four additional 2 min intervals for a total of 13 min of irradiation. Between each interval, the reaction mixture is allowed to cool for one min. When complete, the reaction solution is cooled, diluted with acetone (10 mL) and gravity filtered (Whatman #1 qualitative filter paper) into 250 mL of water. The catalyst cake is washed with acetone (10 mL) into the same 250 mL of water. The filtrate containing a white precipitate is cooled to 0 °C and collected on a Buchner funnel, washed with water (25 mL) and dried by suction in air. If necessary, the product is

further dried by standing in air, vacuum pump (ca. 1 Torr) or placing in a desiccator.

Acknowledgements

This project was initiated by JM and NH and continued by KH, and AT as part of the CHEM 1108 Freshman Chemistry Research Rotation taught by Prof. W. R. Murphy, Jr. at Seton Hall University. We thank Merck & Co., Inc. for partial support of this research through an unrestricted grant. We also thank the Camille and Henry Dreyfus Foundation for support of TC through the Partners-in-Science Program and the ACS Project SEED program for support of MK.

References

1. (a) J. Hassan, M. Sevignon, C. Gozzi, E. Schulz, M. Lemaire, *Chem. Rev.*, **102**, 1359-1469 (2002). (b) A. Suzuki, *J. Organometal. Chem.*, **576**, 147-168 (1999).
2. (a) G. Marck, A. Villiger, R. Buchecker, *Tetrahedron Lett.*, *35*, 3277-3280 (1994). (b) B. H. Lipshutz, *Adv. Synth. Catal.*, **343**, 313-326 (2001).
3. (a) C. R. LeBlond, A. T. Andrews, Y.-K. Sun, J. R. Sowa, Jr. *Org. Lett.*, **3** 1555-1557 (2001). (b) F. P. Gortsema, C. LeBlond, L. Semere, A. T. Andrews, Y.-K. Sun, J. R. Sowa, Jr., Chemical Industries (Dekker), **89**, (*Catal. Org. React.*), 643-652 (2003). (c) D. A. Conlon, B. Pipik, S. Ferdinand, C. R. LeBlond, J. R. Sowa, Jr., B. Izzo, P. Collins, G.-J. Ho, J. M. Williams, Y.-J. Shi, Y.-K. Sun, *Adv. Synth. Catal.*, **345**, 931-935 (2003).
4. (a) B. L. Hayes, Microwave Synthesis, Chemistry at the Speed of Light, CEM Publishing, Matthews, NC, 2002. (b) C. Schoenfeld, M. Loechner, L. Favretto, *Amer. Lab.*, **35**(22), 22-27 (2003).
5. (a) P. Villemin, F. Caillot, *Tetrahedron Lett.*, **42**, 639-642 (2001). (b) B. Basu, P. Das, Md. M. H. Bhuiyan, S. Jha, *Tetrahedron Lett.*, **44**, 3817-3820 (2003). (c) B. M. Choudary, S. Madhi, N. N. Chowdari, M. L. Kantam, B. Sreedhar, *J. Am. Chem. Soc.*, **124**, 14127-17136 (2002). (d) J. W. Han, J. C. Castro, K. Burgess, *Tetrahedron Lett.*, **44**, 9359-9362 (2003). (e) U. S. Sorensen, E. Pombo-Villar, *Helv. Chim. Acta*, **87**, 82-89 (2004).
6. (a) B. K. Banik, K. J. Barakat, D. R. Wagle, M. S. Manhas, A. K. Bose, *J. Org. Chem.*, **64**, 5746-5753 (1999). (b) A. K. Bose, B. K. Banik, N. Lavlinskaia, M. Jayaraman, M. S. Manhas, *Chemtech, 27*(9), 18-24 (1997).
7. M. P. Reynolds, H. Greenfield, Chemical Industries (Dekker), **68**, (*Catal. Org. React.*) 371-376 (1996).

b. Deprotection

54. Improved Catalyst for the Removal of Carbobenzyloxy Protective Groups

Kimberly B. Humphries, Konrad Möbus, Tracy D. Dunn, and Baoshu Chen

Degussa Corporation, 5150 Gilbertsville Hwy, Calvert City, KY 42029

Abstract

The carbobenzyloxy group (Cbz or Z) is a very useful protecting group in organic synthesis, in particular for the synthesis of proteins and peptides, which are used as drug intermediates. The Cbz group can be easily removed via catalytic hydrogenation. Typically, Pd supported on activated carbon powder is the catalyst of choice for such reaction (1). In this work, we report a more active 5%Pd on activated carbon catalyst for the deprotection of Cbz-phenylalanine. This new Degussa Cbz catalyst is highly active and shows a shorter reaction time compared to the state of the art catalyst. The protected amino acid Cbz-phenylalanine was dissolved in alcoholic solvent and stirred under H_2 in the presence of Pd/C. The initial reaction between hydrogen and the Cbz-phenylalanine produced toluene and a carbamic acid intermediate. This unstable intermediate rapidly decomposes to yield CO_2 and the unprotected amino acid. The progress of the reaction was monitored by the infrared detection of CO_2 in the off-gas.

Introduction

The chemical pathways to the complex molecules produced by the fine chemical, pharmaceutical and agrichemical industries often involve multi-step synthesis. Protective groups are commonly used to block reactive sites that would otherwise react in an undesired fashion. A variety of protective groups have been successfully introduced and are now well established in organic synthesis. A common group for the protection of amines is the carbobenzyloxy group, also called Cbz or Z group. This group can be easily removed in the presence of a heterogeneous catalyst and hydrogen under ambient conditions (2). For the development of heterogeneous catalysts there is a need for tests that can easily distinguish between different catalysts, thus providing a tool for fast catalyst development. For hydrogenation reactions most often the consumed hydrogen can be measured directly, and this correlates to the catalyst's activity. This is also true for the hydrogenolysis of benzyl ethers. However, this technique cannot be applied to the hydrogenolysis of benzyl carbamates, since one equivalent of CO_2 is produced for each hydrogen consumed, keeping the pressure in the reactor constant. Taking samples at certain time intervals during the course of the reaction and determining the product via HPLC analysis is very time consuming and therefore not very practical for the determination of the performance of a catalyst. This contribution describes an online test method for determining the catalyst activity for the cleavage of benzyl carbamates and the development of an improved catalyst using this method.

Results and Discussion

The deprotection of the Cbz protected amino acid proceeds via a two step mechanism (Figure 1). The first step comprises the catalytic hydrogenolysis of the benzyloxy group of the Cbz-protected amino acid (**1**). Toluene (**3**) is formed from the O-benzyl group as well as an unstable carbamic acid intermediate (**2**). This intermediate decomposes to form the unprotected amino acid (**4**) and carbon dioxide (**5**).

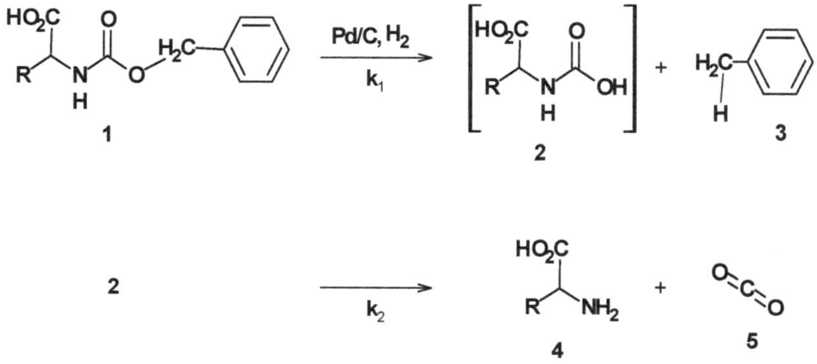

Figure 1 Reaction scheme for the hydrogenolysis of a Cbz-protected amino acid.

In our investigations, a solution of Cbz-protected amino acid in ethanol/water (1:1) was prepared. The starting material is not water-soluble but is quite soluble in alcoholic solvents; conversely, the water-soluble deprotected amino acid is not very soluble in alcoholic solvents. Thus, the two solvent mixture was used in order to keep the entire reaction in solution. This solution was then stirred under hydrogen in the presence of a heterogeneous catalyst, typically Pd/C. The off-gas was channeled through a non-dispersive infrared gas analyzer to measure the amount of carbon dioxide in the off-gas (Figure 2). This method gave a quick and easy determination of the differences in catalytic activity among different heterogeneous catalyst samples.

In order to investigate the catalytic activity of different heterogeneous catalysts for the removal of Cbz-protecting groups, it is important to consider the kinetics of the overall reaction. If the decomposition of the carbamic acid intermediate **2** is very fast compared to the hydrogenation of the benzyloxy group, ($k_2 \gg k_1$), the evolution of CO_2 would be in direct relation to the catalyst activity. If the decomposition of the carbamic acid **2** is the rate-determining step of the overall reaction ($k_1 \gg k_2$), measuring the CO_2 evolution would not correlate to the catalyst performance. Thus, we investigated the kinetics to determine which step was rate-determining.

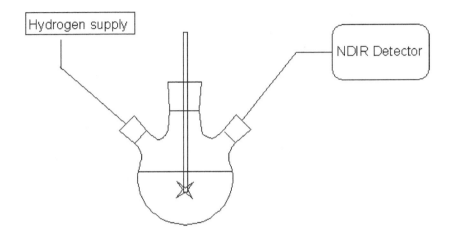

Figure 2 Schematic drawing of reaction apparatus.

In order to examine the kinetics of the hydrogenolysis in more detail, a sequence of reactions were conducted. Cbz-protected glycine (10mmol) was hydrogenated, and the reaction was quenched after 2, 4, 6, 8 and 10 minutes. The concentration of the unprotected glycine was determined by GC after derivatization. Figure 3 shows the correlation between the unprotected amino acid (**4**, R=-H) that was produced during the reaction and the carbon dioxide that was detected in the off-gas. The amount of glycine detected after 2 and 4 minutes closely follows the recorded amount of CO_2 in the off-gas. The detected glycine at subsequent time intervals is lower than what would be expected based upon the detection of CO_2 in the off-gas. One theory for this may be that the glycine over time becomes absorbed onto the surface of the carbon and is not easily removed with washing. In any case, this data suggests that at any time during the reaction there is only a very small amount of the carbamic acid intermediate present. This is an indication that the carbamic acid intermediate rapidly decomposes, $k_2 \gg k_1$, and the hydrogenation of the benzyloxy group is the rate-determining step. Since the catalyst affects only the step involving the hydrogenation of the benzyloxy group, this indicates that indeed the catalytic activity can be directly measured via CO_2 formation. A similar carbamic acid intermediate is formed during the water-toluene diisocyanate reaction; kinetic studies on this reaction have shown that the decomposition of carbamic acid into the amine and carbon dioxide proceeds more rapidly than the initial reaction between the isocyanate functional group and water (3).

Next, we investigated the experimental parameters for hydrogenolysis of Cbz-protected amino acids. It is important to carefully select the experimental parameters so that the reactions are not limited by diffusion of hydrogen to the catalytically active sites. The diffusion of hydrogen can be affected by temperature, agitation speed, as well as the number of catalytically active sites

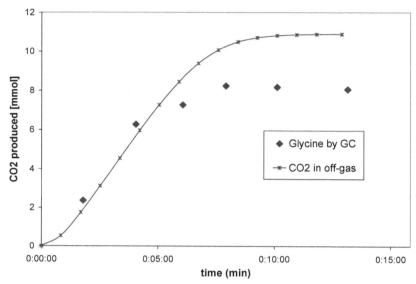

Figure 3 Correlation between carbon dioxide and glycine at different stages of the reaction.

available in the reaction (amount of catalyst present). For this study, we varied agitation speed using a catalyst loading of 200 mg 5% Pd/C dry in the deprotection of 20 mmol Cbz-glycine (**1**, R=-H) operating at 25 °C. Figure 4 shows that the reaction rate did not change when operating within the range of 1500-2000 rpm. An agitation speed of 600 rpm did have a negative effect on the reaction rate as this slower agitation limits the diffusion of hydrogen. For subsequent experiments, 2000 rpm was used as the standard agitation speed and the temperature was held constant at 25 °C. Unless otherwise noted, the catalyst loading was 200 mg of a 5% Pd/C dry catalyst.

Once the reaction kinetics were determined and the appropriate reaction conditions were set, the analyses of different catalysts was launched. A variety of different heterogeneous catalysts were evaluated for the deprotection of 20 mmol Cbz-glycine (**1**, R=-H). Catalysts composed of different platinum group metals supported on activated carbon were evaluated as well as palladium supported on alumina. Figure 5 confirms that Pd supported on activated carbon is indeed the catalyst of choice for this type of hydrogenolysis reaction.

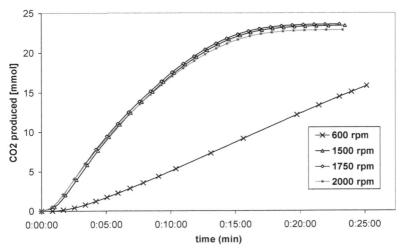

Figure 4 Effect of agitation speed on reaction rate.

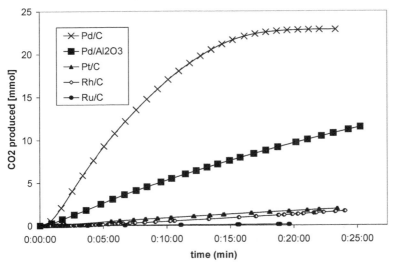

Figure 5 Comparison of different heterogeneous catalysts
[a]Catalyst amount 200 mg.
[b]Precious metal nominal loading 5%.

Palladium gave the highest activity of all the platinum group metals evaluated; platinum, rhodium and ruthenium exhibited very poor activity. The choice of support was also demonstrated to be very important; the activated carbon supported Pd catalyst showed a nearly fourfold increase in activity than did Pd supported on alumina.

An improved Pd supported on activated carbon catalyst for removal of Cbz-protecting groups was developed during the course of this work. Figure 6 shows the improved activity of Catalyst B over our former state-of-the-art catalyst for Cbz-removal, Catalyst A, during the hydrogenolysis of 10 mmol Cbz-phenylalanine (**1**, R=-CH$_2$C$_6$H$_5$). These catalysts were also compared for the hydrogenolysis of Cbz-glycine (**1**, R=-H), and Cbz-tryptophan (**1**, R=-CH$_2$-C=CHNHC$_6$H$_5$). In order to dissolve Cbz-glycine in the ethanol/water mixture, the addition of a small amount of HCl was necessary. Since it is known that this hydrogenolysis reaction is promoted by acidic conditions, the addition of HCl may have superseded the effect of the catalyst in Figure 7.

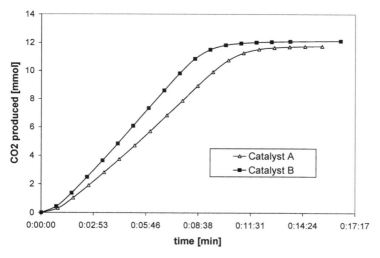

Figure 6 Improved catalyst for the deprotection of Cbz-phenylalanine.

Experimental Section

Commercially supplied carbobenzyloxy-amino acids were used in this study (Carbobenzyloxyglycine, 99%, N-(Carbobenzyloxy)-L-phenylalanine, 99% and N-Carbobenzyloxy-L-tryptophan, 99%) and were purchased from Aldrich. Deionized water and commercial absolute ethanol were used as solvents.

The deprotection of the carbobenzyloxy protected amino acids was carried out in a low-pressure test unit (V= 200 ml) equipped with a stirrer, hydrogen inlet

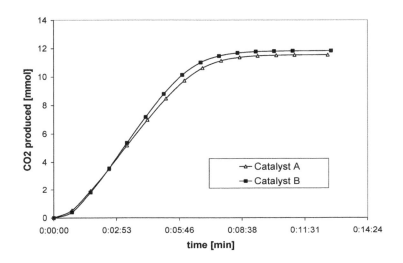

Figure 7 Improved catalyst for the deprotection of Cbz-glycine.

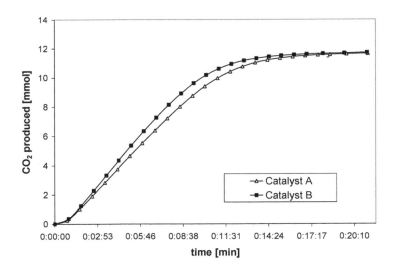

Figure 8 Improved catalyst for the deprotection of Cbz-tryptophan.

and gas outlet. The gas outlet was attached to a Non Dispersive InfraRed (NDIR) detector to measure the carbon dioxide. During the reaction the temperature was kept at 25 °C at a constant agitation speed of 2000 rpm, except as noted otherwise. In a typical reaction run, 10 mmol of Cbz protected amino acid and 200 mg of 5%Pd/C catalyst dry were stirred in a mixture of 70 ml ethanol/water (1:1) [Note: For the complete dissolution of Cbz-glycine, the addition of 5mL of 0.1M hydrochloric acid was necessary.] The hydrogen feed was started and the flow into the reactor was kept constant at 500 ml/minute. The progress of the reaction was monitored by the infrared detection of CO_2 in the off-gas.

Acknowledgements

Thanks to Degussa Corporation Analytical Services for derivatization and GC analysis for selected samples.

References

1. L. Seif, K. Partyka and J Hengeveld, Chemical Industries (Dekker), **40**, (*Catal. Org. React.*), 197-216 (1990).
2. S. Nishimura, Handbook of Heterogeneous Catalytic Hydrogenation for Organic Synthesis, Wiley, New York, 2001, pg. 589-590.
3. J. Stoffer, D. Ihms, D. Schneider, and C. McClain, Proceedings of the Water-Borne and Higher-Solids Coatings Symposium (1987), 14th 428-442.

55. The Effect of N-containing Modifiers on the Deprotection of a Carbobenzyloxy Protected Amino Acid

Kimberly B. Humphries, Tracy D. Dunn, Konrad Möbus, and Baoshu Chen

Degussa Corporation, 5150 Gilbertsville Hwy, Calvert City, KY 42029

Abstract

The removal of carbobenzyloxy (Cbz or Z) groups from amines or alcohols is of high interest in the fine chemicals, agricultural and pharmaceutical industry. Palladium on activated carbon is the catalyst of choice for these deprotection reactions. Nitrogen containing modifiers are known to influence the selectivity for certain deprotection reactions. In this paper we show the rate accelerating effect of certain N-containing modifiers on the deprotection of carbobenzyloxy protected amino acids in the presence of palladium on activated carbon catalysts. The experiments show that certain modifiers like pyridine and ethylenediamine increase the reaction rate and therefore shorten the reaction times compared to non-modified palladium catalysts. Triethylamine does not have an influence on the rate of deprotection.

Introduction

Carbobenzyloxy groups are widely used as protective groups in organic synthesis (1). The removal of carbobenzyloxy groups is easily achieved via hydrogenolysis in the presence of heterogeneous catalysts at low hydrogen pressure and ambient temperature (2-4). Other examples for hydrogenolysis reactions where a σ bond is directly cleaved are debenzylation and hydro-dehalogenation reactions. The effect of N-containing modifiers on the selectivity of debenzylation reactions has been previously reported. It was shown that the presence of a wide variety of mono and bifunctional aliphatic amines and pyridines selectively inhibit the cleavage of aliphatic benzyl ethers, whereas the reactivity of other functional groups was not effected. (5-6). In another work, the addition of triethylamine to the reaction mixture containing the palladium on carbon catalyst leads to the complete hydrogenolysis of an intermediate for the production of ABT-866, which otherwise would have stalled out (7).

More hindered nitrogen bases were not effective modifiers, suggesting an interaction of the N-containing base with the palladium metal surface of the catalyst (8). Ethylenediamine forms a complex with palladium on carbon catalyst, which was isolated and investigated by X-ray Photoelectron Spectroscopy (XPS). The spectroscopic data indicated an interaction of palladium with the nitrogen atoms of ethylenediamine (9). Depending on the solvent the Cbz protective group can be retained (10). In this paper we report the influence of various N-containing bases on the rate of hydrogenolysis of carbobenzyloxy groups.

Results and Discussion

The removal of a carbobenzyloxy group can be separated into two steps (Figure 1). The first step comprises the hydrogenolysis of the benzyl oxygen bond of the Cbz-protected amino acid **1** to form a carbamic acid intermediate **2** and toluene **3**. The carbamic acid intermediate decarboxylates to give the deprotected amino acid **4** and one equivalent of carbon dioxide **5**.

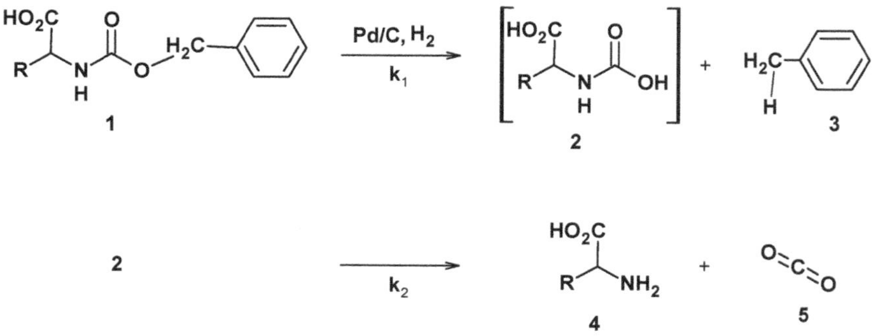

Figure 1 Removal of carbobenzyloxy (Cbz) groups from protected amino acids.

The decomposition of the carbamic acid intermediate **2** proceeds very rapidly, leaving the hydrogenolysis as the rate-determining step of the overall reaction $k_2 \gg k_1$ (11). Thus the formation of carbon dioxide can be directly correlated to the activity of a catalyst. This is illustrated in Figure 2, which shows the rate of formation of CO_2 during the deprotection of CBz-phenylalanine in the presence of a palladium on activated carbon catalyst. When the reaction was repeated with added ethylenediamine the rate of CO_2 formation increased. This clearly demonstrates the rate accelerating effect of ethylenediamine on the hydrogenolysis reaction. The use of ethylenediamine as a modifier can shorten the reaction time significantly.

To explain why this modifier inhibits debenzylation and accelerates a very similar reaction (Cbz removal), the interaction of the nitrogen containing base and the substrate with the palladium surface have to be considered. The addition of a nitrogen containing base to a catalyst slurry leads to adsorption of the compound on the surface of the catalyst. The nitrogen-containing base occupies the active sites on the palladium crystallites where typically the hydrogenation takes place. The interaction with the palladium metal is comparably strong.

Before hydrogenation can take place there must be an adsorption of the substrate on the palladium surface. Obviously this is blocked by certain modifiers, depending on the strength and the coordination number of the ligand. The benzylethers that were investigated earlier do not adsorb very strong on the palladium surface and are not able to displace the modifier, thus the hydrogenolysis reaction is inhibited. In the case of Cbz groups the presence of the carbamate group leads to stronger adsorption followed by hydrogenolysis. The accelerating effect of the N-containing modifier can be explained by the role of the N-containing base in increasing the basic character of the metal surface, thus making it more attractive for adsorption of acidic species, e.g. other protected amino acids.

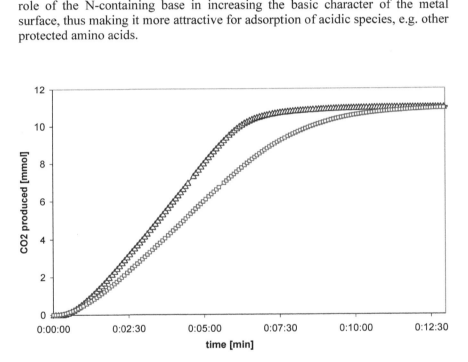

Figure 2 Formation of carbon dioxide during the deprotection of CBz protected phenylalanine (with and without ethylenediamine) in the presence of 5%Pd/C.

Figure 3 demonstrates the influence of different modifiers on the reaction rate. Compared to the unmodified 5%Pd catalyst, the addition of 20 µl of pyridine or 20 µl of ethylenediamine increased the reaction rate from 1.5 to 2 mmol / (min∗gram) catalyst. Adding twice the amount of modifier (40 µl pyridine) does not increase the reaction rate any further. Obviously all the available surface palladium is already effectively modified. In contrast triethylamine does not exhibit any influence on the reaction rate.

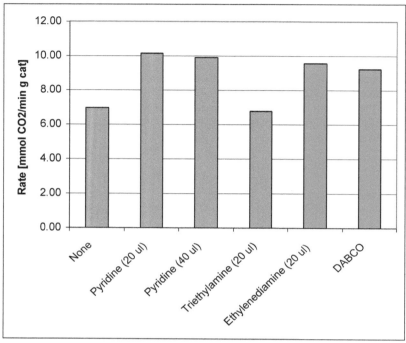

Figure 3 Influence of different N-containing bases on the reaction rate of Cbz deprotection in the presence of 5%Pd on activated carbon catalyst.

Conclusions

In addition to the already known inhibiting effect of N-containing bases on debenzylation reactions we have shown that similar modifiers can increase the rate of hydrogenolysis of a CBz protected amino acid. As these reactions are carried out on industrial scale the addition of certain modifiers can increase the reaction rate, thus leading to shorter reaction times and higher productivity.

Experimental Section

A commercially available 5% palladium on activated carbon catalyst from Degussa was used for the investigation. Commercially supplied N-(Carbobenzyloxy)-L-phenylalanine (99%) was purchased from Aldrich. Modifiers such as pyridine, triethylamine, ethylenediamine and DABCO (Diazabicyclooctane) with a purity >99 % are also available commercially and were used as received.

The deprotection of carbobenzyloxy protected phenylalanine was carried out in a low-pressure test unit (V= 200 ml) equipped with a stirrer, hydrogen inlet and gas outlet. The gas outlet was attached to a Non Dispersive InfraRed (NDIR) detector to measure the carbon dioxide. During the reaction the temperature was kept at 25 °C at a constant agitation speed of 2000 rpm. In a typical reaction run, 10 mmol of Cbz protected phenylalanine and 200 mg of 5%Pd/C catalyst were stirred in a mixture of 70 ml ethanol/water (1:1). The Cbz protected phenylalanine is not water-soluble but is quite soluble in alcoholic solvents; conversely, the water-soluble deprotected phenylalanine is not very soluble in alcoholic solvents. Thus, the two solvent mixture was used in order to keep the entire reaction in the solution phase. Twenty μl of the corresponding modifier was added to the reaction mixture, and hydrogen feed was started. The hydrogen flow into the reactor was kept constant at 500 ml/minute and the progress of the reaction was monitored by the infrared detection of CO_2 in the off-gas.

References

1. T. W. Green, P. G. M. Wuts, Protective Groups in Organic Synthesis, 2nd ed., Wiley, New York, 1991.
2. S. Nishimura, Handbook of Heterogeneous Catalytic Hydrogenation for Organic Synthesis, Wiley, New York, 2001, p. 582.
3. F. Notheisz, Mihaly Bartok, in Fine Chemicals through Heterogeneous Catalysis, A. Sheldon, H. Van Bekkum, ed., Wiley-VCH, Weinheim, Germany, 2001, pp. 415-426.
4. R. L. Augustine, Heterogeneous Catalysis for the Synthetic Chemist, Dekker, New York, (1995), p. 524.
5. L. S. Seif, K. M. Partyka, J. D. Hengeveld, Chemical Industries (Dekker), **40**, (*Catal. Org. React.*), 197-216 (1990).
6. H. Sajiki, *Tetrahedron Lett.*, **36**, 3465 (1995).
7. L. S. Seif, S. M. Hannick, D. J. Plata, H. E. Morton, P. N. Sharma, Chemical Industries (Dekker), **89**, (Catal. Org. React.), 633-641 (2003).
8. H. Sajiki, K. Hirota, *Tetrahedron*, **54**, 13981 (1998).
9. H. Sajiki, K. Hattori, K. Hirota, *J. Org. Chem.*, **63**, 7990 (1998).
10. K. Hattori, H. Sajiki, and K. Hirota, *Tetrahedron*, **56**, 8433 (2000).
11. See K. Humphries, T. Dunn, K. Moebus, B. Chen, Improved Catalyst for the Removal of Carbobenzyloxy Protecting Groups, this issue, Ch. 54.

56.   Selective Debenzylation in the Presence of Aromatic Chlorine on Pd/C Catalysts Effects of Catalyst Types and Reaction Kinetics

Steve S. Y Wang[a], Jun Li[a], Karen TenHuisen[a], Jale Muslehiddinoglu[a], Srinivas Tummala,[a] San Kiang[a], and Jianping J. P. Chen[b]

[a]*Bristol-Myers Squibb Company, Pharmaceutical Research Institute,*
New Brunswick, NJ 08903
[b]*Engelhard Corporation, Beachwood R&D Center*
Beachwood, OH 44122

Abstract

Reductive debenzylation has been widely used in the fine chemical and pharmaceutical industries. When an aromatic halogen is present, selective debenzylation verses dehalogenation has been a challenging task. In our earlier studies, transfer hydrogenolysis under different solvent, pH and salt conditions was investigated. For gaseous hydrogenolysis, a variety of palladium catalysts with different metal loadings, oxidation states and support materials were evaluated. A newly developed DeLink[TM] 10%Pd/CPS4 catalyst had the desired selectivity to convert (S) to product (A) in this reaction. In contrast, the low metal loading 3%Pd/C (Engelhard nonstandard catalyst) readily over-hydrogenated the debenzylated product (A), generating the dechlorinated by-product (B).

   With the consecutive reaction path identified, our objective in this work was to better understand the reaction mechanism and the surface catalytic behavior under optimized conditions. Neither debenzylation nor dechlorination reactions were structure-sensitive over supported palladium catalysts. A sequence of elementary steps based on the dual nature of Pd sites, dissociative hydrogen adsorption and noncompetative adsorption between (S) and hydrogen atoms were proposed. Using the quasi-equilibrium and two-step reaction concepts in the catalytic cycle, we have derived a Langmuir-Hinshelwood model which reasonably describes the kinetics of the debenzylation reaction. The Madon-Boudart test and the Weisz-Prater criterion show that the kinetic data is free of mass transfer limitations. The secondary sequential dechlorination from the product A to B was a slow reaction. The equilibrium conversion varies significantly with the palladium loading. Catalysts with lower metal loading tend to have higher metal dispersion and possibly provide more edge and corner sites to further hydrogenate the debenzylated product.

Introduction

An important goal in pharmaceutical process development is the design of highly selective chemical transformations leading to cost-effective processes in which separation or purification steps are minimized. The hydrogenolysis of benzyl protecting groups using supported palladium catalysts are well documented (1,2,3). In our efforts to develop a selective debenzylation process, catalytic transfer hydrogenolysis of (S) was studied using either formic acid, ammonium formate or sodium hypophosphite as hydrogen donors. The reactions afforded mainly debenzylated product (A), but approximately 5% of the deschloro by-product (B) was observed. The dehalogenation proceeds quickly under basic and neutral conditions but slowly in the presence of acid (4). In an earlier report, H_3PO_3 was used as a catalyst inhibitor to increase the selective debenzylation of a tertiary amine in the presence of aromatic halogen (5).

For gaseous hydrogenolysis, dehalogenation was found to be highly dependent on the nature of the solvent. In protic media such as methanol, deschloro (B) levels were very high (23-35%). In aprotic media such as ethyl acetate or methylene chloride, the level of deschloro (B) was 1-2%. Recently, a variety of palladium catalysts was evaluated using ethyl acetate as the solvent to optimize the catalytic process. The aims of this work were a) to examine the catalytic parameters which influence the activity and selectivity of the hydrogenolysis, b) to identify suitable catalysts and process conditions to minimize the dechlorinated by-product formation and c) to elucidate the surface reaction mechanism and kinetics using the concept of quasi-equilibrium and two- step reaction in the catalytic cycle.

Experimental Section

The hydrogenation reactions were carried out using the Endeavor reactor system (Argonaut technologies), which allows multiple hydrogenations in parallel on a small scale with individual control of pressure and temperature. Experimental conditions were 0.5 g of starting material (S) dissolved in 5 mL of ethyl acetate operating at 25 °C, 45 psia and an agitation speed of 700 rpm. Hydrogen uptake values and HPLC analyses were used to evaluate the reaction rate. In our development work, monitoring the reaction in situ has been used to evaluate the reaction kinetics and to perform the parametric studies. In-line probes such as FTIR and Raman as well as hydrogen uptake and calorimetry were integrated with high pressure reactors. For this debenzylation reaction system, no distinct spectroscopic peak was identified to monitor the progress of the reaction. Since the reaction was not significantly exothermic, calorimetric measurement was not useful. The typical experiment was conducted in a 100 mL autoclave using 6 g of starting material (S) dissolving in 60 mL ethyl acetate. Besides monitoring the hydrogen uptake, samples were periodically withdrawn from the reactor for HPLC analysis. In this work a series of supported Pd catalysts were used. Carbon supported catalysts, including ESCAT and the newly developed DeLink^{TM} series, are commercially available Engelhard catalysts. Highly dispersed Al_2O_3, SiO_2 and TiO_2 supported catalysts were prepared by Penn State University. Carbon monoxide chemisorption was performed in a Micromeretics 2910 system to determine the metal dispersion.

Results and Discussion

Table 1 shows the comparison of the catalytic activities and selectivities between different supported palladium catalysts. All catalysts were prepared by

Table 1 Catalytic performance of the debenzylation reaction on different types of supported palladium catalyst.

Catalyst	Condition/ Metal Location	Dispersion, (%) / metal crystallite Size(nm)	Reaction Time (hours)	Product Distribution%		
				Starting material	Desired product	De-chlorinated by-product
5%Pd/C	Prereduced Uniform	19.5/ 5.8	2	6.12	92.45	1.43
			4	3.14	94.96	1.90
			18	0.39	96.61	3.00
1.7% Pd/Al$_2$O$_3$	Unreduced Uniform	48.5/ 2.3	2	72.8	27.11	0.09
			4	46.3	53.37	0.33
			18	1.98	95.7	2.32
0.93% Pd/SiO$_2$	Unreduced Uniform	>90/ 1.25	2	63.36	36.34	0.30
			4	17.15	80.26	2.59
			18	0.49	94.89	4.62
0.62% Pd/TiO$_2$	Unreduced Uniform	75.0 / 1.51	4	31.06	68.87	0.07
			18	0.89	98.85	0.26
0.62% Pd/TiO$_2$	Reduced* Uniform	MSI effects	4	1.92	94.78	3.30
			14	nd	94.54	5.46

* Prereduced at 500 °C, Metal-Support Interaction (MSI) suppressed CO adsorption.

an impregnation method with a variety of treatment procedures. Charged quantities were adjusted so that that the number of exposed palladium atoms was equal despite differences in percent dispersions and palladium loadings. Among the four different supports, Pd/C showed favored catalytic activity with shorter reaction time. Highly dispersed catalysts did not have strong catalytic activities. Variation of the palladium crystallite or aggregate size on the catalyst surface showed that the debenzylation reaction rate was independent of the structure. The activity of Pd/TiO$_2$ was significantly improved after a high temperature

(500 °C) reduction, however the dehalogenation product level also increased from very low levels in the unreduced case to 3.3% after 4 hr reaction.

In order to select a suitable catalyst, a variety of Pd/C catalysts were screened. Table 2 shows that the dechlorinated by-product was minimized to the level less than 1% as the catalyst metal loading increased from 3% to 10%. The catalytic activity for the debenzylation reaction was examined and compared based on the rate of hydrogen uptake (Figures 1 and 2).

Table 2 Catalytic performance of the debenzylation reaction on carbon supported palladium catalyst.

Catalyst	Condition/ Metal Location	Dispersion (%), by CO Chemisorption	Reaction Time (hours)	Product Distribution%		
				Starting material	Desired product	De-chlorinated by-product
3% Pd/C	Unreduced Eggshell	36.7	3	nd	96.54	3.46
			12	nd	93.59	6.41
3% Pd/C	Prereduced Mixed	28.6	2	5.16	93.62	1.22
			4	0.16	97.67	2.17
			18	nd	95.69	4.31
5% Pd/C (Escat 147)	Unreduced Eggshell	29.3	2	0.55	98.43	1.02
			4	nd	98.29	1.71
5%Pd/C (Escat 142)	Pre-reduced Eggshell	26.0	2	3.16	96.06	0.78
			4	nd	98.69	1.31
5% Pd/C (Escat 168)	Prereduced Uniform	19.5	2	6.12	92.45	1.43
			4	3.14	94.96	1.90
			18	0.39	96.61	3.00
5% Pd/C (Escat 148)	Prereduced Uniform	13.6	2	32.45	66.77	0.78
			4	10.24	88.6	1.16
			8	0.24	97.6	2.16
DeLink™ 10% Pd/CPS4	Unreduced Eggshell	29.5	4	3.26	96.33	0.41
			19	nd	99.24	0.76
10%Pd/CP L-6805-24A	Unreduced Eggshell	15.8	3	11.40	88.6	nd
			6	6.50	93.5	nd
			9	nd	99.5	0.50

It is clear that ESCAT 147 (5% edge coated, unreduced) gives the highest activity, followed by ESCAT 142 (5% edge coated, reduced) and ESCAT 168 (5%Pd/C uniform, reduced) under the same reaction conditions. Based on these results, catalyst screening was continued with other edge-coated, unreduced catalysts. Figure 2 shows the hydrogen uptake curves for three different Pd loading levels. When the same amount of exposed palladium atoms was used, DeLink™ 10%Pd/C gave the best activity and selectivity. The 3% Pd/C had similar activity to ESCAT 147 , but showed poor selectivity. With this catalyst, the hydrogen uptake was 30% more than the theoretical value of 1.2 mmole for debenzylation. HPLC analysis showed over reduction of the product (A) to

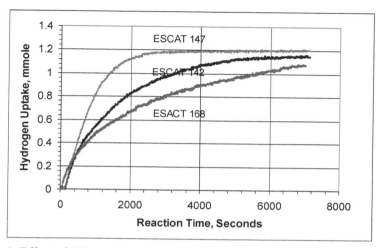

Figure 1 Effect of Pd metal location and oxidation state on reaction rate.

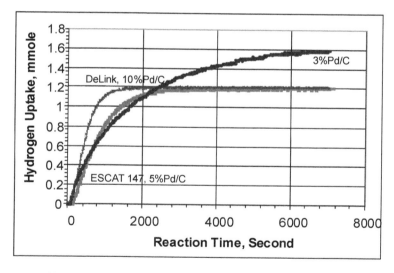

Figure 2 Effect of metal content on reaction rate.

generate the dechlorinated by-product (B). To further examine the high-palladium metal loading catalyst DeLink™ 10%Pd/C, the reaction was conducted in a Parr reactor with evaluation of reaction kinetics and the product distribution. Figure 3 shows the concentration profile of reactant and product during hydrogenation. Under the given reaction conditions, the dechlorinated by-product content was less than 0.8% even after 19 hours, thus establishing that dehalogenation occurs primarily, if not exclusively from the desired product (A). Under the same conditions, the 3%Pd/C catalyst generated greater than 8% of the undesired deschloro by-product.

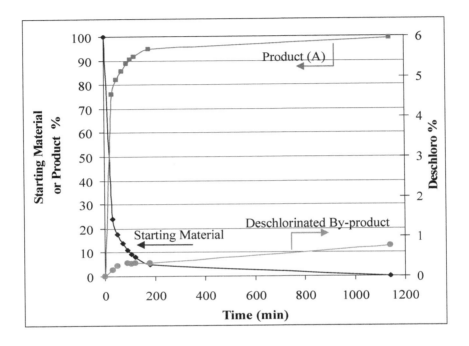

Figure 3 Product distribution as a function of time for debenzylation of SM(S) on DeLink™ 10%Pd/CPS4. Parr Reactor, 25 °C, 45 psig hydrogen.

Based on the above experimental evidence, the consecutive reaction order S → A → B was considered as the reaction path for this catalytic hydrogenolysis reaction on supported palladium catalyst.

In aromatic catalytic hydrogenation studies (6,7), aromatic species are considered strongly adsorbed on the Pd surface. Dissociative adsorption of H_2 on group VIII metals has been well documented in the literature (8,9). M. Salmeron (10) recently reported using scanning tunneling microscopy to watch the process of dissociative hydrogen adsorption. Hydrogen adsorption required active sites containing three or more vacancies, in which case, hydrogen atoms quickly filled two vacancies and left the rest open. With these assumptions, the following sequence of elementary steps was proposed.

$$\text{SM} + \text{S1} \overset{K_{SM}}{\Leftrightarrow} \text{SM-S1}$$

$$\text{H}_2(\text{L}) + \text{S2} \overset{K_{H_2}}{\Leftrightarrow} \text{H}_2\text{-S2}$$

$$\text{SM-S1} + \text{H}_2\text{-S2} \overset{k}{\Rightarrow} \text{SM-H-S1} + \text{S2} + \text{C}_6\text{H}_5\text{CH}_3$$
$$\text{(RDS)}$$

$$\text{SM-H-S1} \rightarrow \text{A-S1}$$

$$\text{A-S1} \overset{1/K_A}{\Leftrightarrow} \text{A} + \text{S1}$$

$$\text{A-S1} + \text{H}_2\text{-S2} \rightarrow \text{B} + \text{S1} + \text{S2} + \text{HCl}$$
$$\text{(slow)}$$

Gomez-Sainero et al. (11) reported X-ray photoelectron spectroscopy results on their Pd/C catalysts prepared by an incipient wetness method. XPS showed that Pd^0 (metallic) and Pd^{n+} (electron-deficient) species are present on the catalyst surface and the properties depend on the reduction temperature and nature of the palladium precursor. With this understanding of the dual sites nature of Pd, it is believed that organic species S and A are chemisorbed on to Pd^{n+} (**S1**) and H_2 is chemisorbed dissociatively on to Pd^0(**S2**) in a non-competitive manner. In the catalytic cycle, quasi-equilibrium ($\Leftrightarrow$) was assumed for adsorption of reactants, SM and hydrogen in liquid phase and the product A (12). Applying Horiuti's concept of rate determining step (13,14), the surface reaction between the adsorbed SM on site S1 and adsorbed hydrogen on S2 is the key step in the rate equation.

$$rate = k\theta_{SM}\theta_{H_2}$$

Expressions for θ_{SM} and θ_{H_2} can be derived and related to rate (k) and equilibrium constants (**K**). The S1 and S2 site balances are $\theta_{SM} + \theta_A + \theta_{S_1} = 1$ and $\theta_{H_2} + \theta_{S_2} = 1$ respectively (θ_{S_1}, θ_{S_2} are empty sites). Based on Henry's law, the gas-phase hydrogen pressure and the liquid-phase hydrogen concentration may be used interchangeably. The rate expression can be written as follows:

$$rate = \frac{k\,K_{SM}C_{SM}}{[1 + K_{SM}C_{SM} + K_A C_A]} \frac{\left(K_{H_2}P_{H_2}\right)^{\frac{1}{2}}}{[1 + \left(K_{H_2}P_{H_2}\right)^{\frac{1}{2}}]}$$

Based on the assumption that hydrogen atoms are adsorbed in pairs, instead of the expression of dissociative hydrogen atom adsorption, it's adequate to use H_2 adsorption as the fractional surface coverage for the rate equation.

$$rate = \frac{k\, K_{SM} C_{SM}}{[1 + K_{SM} C_{SM} + K_A C_A]} \frac{K_{H_2} C_{H_2}}{[1 + K_{H_2} C_{H_2}]}$$

Since both hydrogen in the solution and the product A are weakly adsorbed species, equilibrium constants K_A and K_{H_2} are very small, which leads to $K_A C_A \ll 1$ and $K_{H_2} C_{H_2} \ll 1$. Thus, the rate expression for the debenzylation can be simplified as a conventional Langmuir-Hinshelwood model.

$$rate = \frac{k K_{SM} C_{SM} K_{H_2} C_{H_2}}{1 + K_{SM} C_{SM}} \approx \frac{k' K_{SM} C_{SM} K_{H_2} P_{H_2}}{1 + K_{SM} C_{SM}}$$

The dependence of the initial turnover frequency (hydrogen consumption rate) on starting material concentration (0.12 – 0.5 M) and hydrogen pressure (2.5 – 5.5 atm) with DeLink™ 10%Pd/C at 298 K in ethyl acetate is shown in Figure 4. The reaction order with respect to SM is approximately zero. The catalyst surface probably becomes more highly covered with adsorbed SM species, approaching saturation coverage on **S1**, however dihydrogen molecules may still be adsorbed on **S2** and not blocked by SM adsorption. The initial rate for the debenzylation reaction exhibits a near first-order dependency on hydrogen pressure, which is consistent with this being the RDS. Our experimental results show that DeLink™ 10%Pd/C is less temperature (298-318K) dependent with respect to its debenzylation activity. The activation energies are 6.5 and 15.5 kcal/mol for DeLink™ 10%Pd/C and 3%Pd/C respectively.

The weakly adsorbed product A can be rapidly desorbed or react further with adsorbed hydrogen depending on the availability of certain Pd sites, **S2**. The secondary hydrogenation is a slow reaction which is structure-insensitive over Pd/C. It is believed that the dechlorination generates HCl which may inhibit further reduction and eventually reach equilibrium conversion. A plausible explanation for the higher level of dechlorinated by-product B formation with lower loading catalysts may be the higher percentage of exposed atoms in either edge or corner sites contributing to the further reduction of the debenzylation product A.

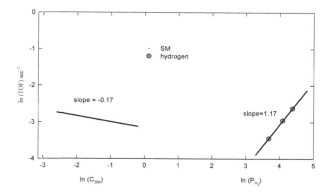

Figure 4 Fit of kinetic model to the initial turnover frequency for debenzylation
at 298 K, 2.5-5.5 atm H_2 and 0.12- 0.5 M S in ethyl acetate.

The kinetic data obtained in this study are free of all transport limitations including external and internal mass transfer effects as well as the rate of hydrogen transfer from the gas to the liquid phase. Under these conditions, mixing is good enough so that the rate of H_2 transport (k_La) is no longer rate controlling and the reaction occurs in the kinetic regime. One of the definitive criteria for evaluation of transport limitations with supported metal catalysts is the Madon-Boudart test (15). The test requires at least two catalysts with similar metal dispersion but widely varying concentrations of surface atoms. In the absence of transport limitations, a ln-ln plot of the activity verses the surface metal atoms concentration will exhibit a linear correlation with a slope of unity. In our study, a family of Pd/C with very close dispersion value was chosen. The slope of M-B testing results is about 1.2. Further verification of the absence of internal diffusion limitations was shown by using the Weisz-Prater criterion (16,17). The dimensionless W-P parameter was evaluated for DeLink™ 10%Pd/C at the reaction conditions, and it was in the order of 10^{-3} that is significantly less than –0.3, as shown in C_{wp} equation representing the ratio of the reaction rate to the rate of internal diffusion.

Table 3 Madon-Boudart test for the absence of mass transfer effects, reaction conditions: 45 psig hydrogen / 298 K / 700 RPM.

Catalyst	Dispersion (%) (CO Chemisorption)	Amount of Catalyst Charged (mg)	Pd Concentration (μmol/g cat.)	Initial Activity (μmol/g cat.$\bullet$sec)	Initial TOF (sec^{-1})
3% Pd/C	31.0	50.65	280.46	4.08	0.047
5% Pd/C	29.3	20.40	465.3	10.12	0.074
10% Pd/C	29.7	45.40	938.4	19.75	0.071
10% Pd/C	29.7	16.10	938.4	19.05	0.068

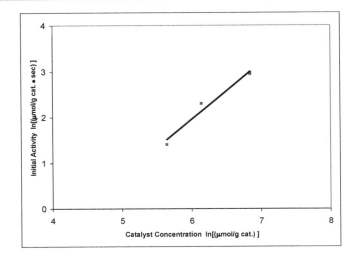

Figure 5 Madon-Boudart test for mass transfer limitation over Pd/C catalysts -a plot of ln (activity) versus ln (Pd surface concentration).

Weisz-Prater criterion uses measured values of the rate of reaction to determine if internal diffusion is limiting the reaction.

$$C_{wp} = \text{actual reaction rate / diffusion rate}$$
$$= r_{obs}\, \rho R^2 / C_s\, D_e = 0.0051 << 0.3$$

Where
D_e : effective diffusivity for organic molecule in dilute solution, 1.87×10^{-5} cm^2/sec
C_s : surface concentration, 2.47×10^{-4} mole/cm^3
R: radius of the catalyst particle, 0.005 cm
ρ: density of the catalyst, 0.5 g/cm^3
r: rate of the reaction, 1.88×10^{-6} mole/g cat. sec

Summary

The hydrogenolysis of benzyl protecting groups in the presence of aromatic chlorine was studied at concentrations of 0.12 − 0.5 M of pharmaceutical intermediate (S) in ethyl acetate at 298 K and between 2.5 −5.5 atm hydrogen. Engelhard unreduced and surface loaded (eggshell structure) palladium catalysts showed reasonable activity in the selective debenzylation process. A newly developed DeLink™ 10%Pd/CPS4 was identified that had the desired activity and selectivity to convert (S) to product (A) in contrast to the low metal loading 3%Pd/C (Engelhard nonstandard catalyst), which could readily over-hydrogenate the debenzylated product (A) and generate the dechlorinated by-product (B).

A consecutive reaction was identified. The kinetic behavior on Pd/C was modeled by a conventional Langmuir-Hinshelwood expression invoking noncompetative adsorption on Pd sites of a dual nature and the addition of H_2 adsorption as the rate determining step. The kinetics exhibited a near zero-order dependence on starting material concentration , and a first-order dependence on hydrogen pressure, which has a reasonable RDS consistency within the model. The desired product could be further slowly hydrodechlorinated to the undesired by-product. The equilibrium conversions from product to deschloro were significantly affected by the metal loading. The edge or corner sites generated from Pd catalysts with low loading may contribute to dechlorination.

Debenzylation is an industrially important process. The current study might furnish an example of using Horiuti's concept of rate determining step to elucidate a hydrogenolysis reaction for high molecular weight systems used in pharmaceutical industry. Understanding the mass transfer roles is also a critical factor for successfully scaling up and developing a hydrogenation process.

Acknowledgements

The authors thanks Dr. M. Futran , Dr. M. A. Vannice and Dr. W. L. Parker for fruitful discussions.

References

1. P. N. Rylander, Catalytic Hydrogenation Over Platinum Metals, Academic Press, NY (1967).
2. J. P. Chen, et al., Chemical Industries (Dekker) **89**, (*Catal. Org. React.*) 313-328 (2002).
3. M. Studer, H.-U. Blaser, *J. Mol. Catal. A: Chem,* **112**, 437-445 (1996).
4. J. Li and S. Y.Wang, et al., *Tetrahedron Lett.*, **44**(21), 4041-4043 (2003).
5. M. G. Scaros, et al., Catalysis of Organic Reactions, Marcel Dekker, Inc, New York, p. 457 (1995).
6. A. Stanislaus and B. H. Cooper, *Catal. Rev. Sci. Eng.*, **36**(1) 75-123 (1994).

7. P. Chou and M. A.Vannice, *J. Catal.*, **107**, 140 (1987).
8. R. Z. C. van Meerten, T. F. M. de Graaf, and J. W. E. Coenen, *J. Catal.*, **46**, 1 (1977).
9. U. K. Singh and M. A. Vannice, *J. AIChE.*, **45**, 5 (1999).
10. Chemical Engineering Progress, p.14 July 2003.
11. L. M. Gomez-Sainero, X. L. Seoane, J. L. G. Fierro, and A. Arcoya., *J. Catal.*, **209**, 279 (2002).
12. M. Boudart and G. Djega-Mariadassou, *Catal. Lett.*, **29**,7 (1994).
13. M. Boudart and K. Tamaru., *Catal. Lett.*, **9**,15 (1991).
14. M. Boudart, *J. AIChE*, **18**, 3 (1972).
15. R. J. Madon and M. Boudart., *Ind. Eng. Chem. Fundam.*, **21**, 438 (1982).
16. H. S. Fogler, Elements of Chemical Reaction Engineering, p.625 (1992).
17. C. Wilke and P. Chang, *J. AIChE*, **1**, 264 (1955).

c. Asymmetric Catalysis

57. A Comparison Between the Homogeneous and the Anchored Rhodium Skewphos Catalyst

Nagendranath Mahata, Setrak Tanielyan and Robert Augustine

Center for Applied Catalysis, Seton Hall University, South Orange, NJ 07079
USA

augustro@shu.edu

Abstract

The anchored catalyst, $Rh(COD)(\underline{S}kewphos)/PTA/Al_2O_3$, and the homogeneous analog, $Rh(COD)(Skewphos)^+BF_4$, have been used for the hydrogenation of dimethyl itaconate at several moderate to high substrate/catalyst ratios. While the rates of hydrogenation using the homogeneous catalyst are faster than the those obtained using the anchored species at these turnover numbers, the product enantioselectivities (%ee's) observed with the homogeneously promoted reactions are much lower than those seen with the anchored catalysts. In fact, with the homogeneous catalyst the product ee's decrease from 70% to 62% as the TON increases from 1000 to 20,000. With the anchored catalyst, the product ee's remain constant at 79-81% regardless of the reaction TON. Details of these reactions will be presented along with some pressure effect data and the extent of metal loss, where appropriate.

Introduction

Previous studies on the use of Anchored Homogeneous Catalysts (AHC's) have been concerned with studying the effect which different reaction variables had on the activity, selectivity and stability of these catalysts (1-9). These reactions were typically run at relatively low substrate/catalyst ratios (turnover numbers-TON's), usually between 50 and 100. While these low TON reactions made it possible to obtain a great deal of information concerning the AHC's, in order to establish that these catalysts could be used in commercial applications it was necessary to apply them to reactions at much higher TON's and, also, to make direct comparisons with the corresponding homogeneous catalyst under the same reaction conditions.

The standard method used to prepare these AHC's was by anchoring a pre-formed complex onto an alumina support which had been treated with a heteropoly acid such as phosphotungstic acid (PTA). Alternately, the AHC can be prepared by treating an anchored $Rh(COD)_2$ precursor with an appropriate ligand (8). We report here the use of AHC's which have been prepared by this ligand exchange procedure to promote high TON hydrogenations. The ligand used in the exchange was the readily available, Skewphos (DBPP). The resulting catalyst was used in high TON hydrogenations of dimethyl itaconate (DMIT).

Results and Discussion

The anchored catalyst, Rh(COD)(Skewphos)/PTA/Al$_2$O$_3$, (AHC-Skew) was prepared by treating the anchored precursor, Rh(COD)$_2$/PTA/Al$_2$O$_3$ with a solution containing an equivalent of the ligand, Skewphos (BDPP). After washing, the catalyst was used for the hydrogenation of dimethyl itaconate (DMIT), normally at ambient temperature and 50 psig of H$_2$. Corresponding hydrogenations were also run using 20 μmoles of the homogeneous, Rh(COD)(Skewphos)$^+$BF$_4$, catalyst under the same reaction conditions. For direct comparisons, the AHC-Skew catalyst used contained 20 μmoles of the active complex. In contrast to what was found in our initial low TON work, we found that at high TON applications the homogeneous catalysts were more active than the AHC-Skew. Table 1 lists comparison data for the hydrogenation of DMIT over AHC-Skew and the homogeneous Rh(Skewphos) at various TON's.

Table 1 Comparison of reaction rate and product ee in the hydrogenation of DMIT over AHC-Skew and the homogeneous Rh(Skewphos) catalyst.

		AHC-Skew			Homogeneous		
TON	Use #	TOF hr^{-1}	%ee	% Conv	TOF hr^{-1}	%ee	% Conv.
1000	1	5300	79	98	9900	70	98
	2	5500	80	98	8100	70	100
	3	5200	79	100	6100	69	99
	4	5300	81	100			
10,000	1	5000	80	100	1560	62	100
	2	4500	80	100	-^{a}	-	-
	3	4000	80	100			
20,000	1	5200	77	100	24,000	61	100
	2	3700	79	100	2500	52	60^b
50,000	1	3900	79	100	7500	63	90^c

^{a} 20% completion after 10 hours.
^{b} After 5 hrs; 70% at 10 hrs; 80% at 20 hrs.
^{c} After 6 hrs; 100 % at 16 hrs (TOF = 2800 hr-1).

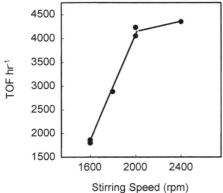

Stirring Speed (rpm)

Figure 1 Effect of stirring speed on the reaction rate in the 1000 TON hydrogenation of DMIT over AHC-Skew containing 20 μmoles of active complex. Run at R.T. and 50 psig with 20 mmoles of DMIT in 15 mL of EtOH.

Interestingly, it should be noted that even though the homogeneous catalyst was more active than AHC-Skew, the product ee's obtained from the homogeneously catalyzed reactions were consistently lower than those from the reactions promoted by the anchored catalyst, with the difference increasing as the TON increased. In addition, as we have seen with most homogeneous catalysts, once the Rh(Skewphos) had been used in a high TON application, it was deactivated for the hydrogenation of more substrate which was subsequently added to the reaction mixture. There was even a considerable deactivation during the very high TON hydrogenations. The AHC-Skew, however, was more stable under these conditions and could be re-used several times with only a little loss of activity. Analysis of the AHC-Skew catalyzed reaction mixtures showed that less than 1 ppm of Rh was present in each one.

One of the more obvious reasons for the slower reaction rates observed with the anchored catalyst as compared to the homogeneous catalysts is the fact that hydrogen and substrate diffusion are more important in heterogeneously catalyzed reactions than in those promoted by homogeneous catalysts. Fig. 1 shows the relationship between the turnover frequency (TOF hr-1) and the stirring rate for 1000 TON hydrogenations of DMIT over AHC-Skew. There is possibly some degree of kinetic control in reactions run at 2000 rpm but complete elimination of diffusion control has probably not been attained even at 2400 rpm because of the splashing of solid catalyst particles onto the reactor wall which was observed at the higher speeds. Because of this we believe that with the reactor system used, 2000 rpm is the maximum value reasonably attainable. The product ee's remained constant at 78-80% and all reactions went to completion.

These reactions were run using an EtOH solution of DMIT but hydrogenation of neat DMIT at 9000 TON was also accomplished with the AHC-Skew which was re-used three times giving consistent product ee's of 80-82% and with only a slight loss of activity. The homogeneously catalyzed hydrogenation, on the other hand, proceeded only slightly faster for the first injection of the substrate and gave product ee 20% lower than that obtained with

Table 2. Stirring rate effect at 200 psig in DMIT hydrogenation over AHC-Skew.

rpm	TOF hr^{-1}	%ee
1000	9700	79
1200	9700	79
1400	7500	80
1600	7000	78

Table 3. Pressure effect in DMIT hydrogenation over AHC-Skew.

psig	TOF hr^{-1}	%ee
200	7600	78
500	15,000	76
1000	21,500	75
200	7100	76

the AHC-Skew. After the first amount of DMIT had been hydrogenated, the homogeneous catalyst was almost completely deactivated since after addition of more DMIT it took ten hours to go to 20% completion.

Table 4. Multiple high TON DMIT hydrogenations over AHC-Skew at 1000 psig.

	20,000 TON			100,000 TON	
Use #	TOF hr^{-1}	%ee	Use #	TOF hr^{-1}	%ee
1	27,000	73	1	40,000	76
2	25,000	75	2	37,500	75
3	21,500`	77			
4	18,500	78			
5	14,000	77			

As was mentioned above, the design of our suspended stirrer reactors does not provide adequate stirring to keep the very reactive heterogeneous catalysts from operating in the diffusion regime, especially with high TON reactions. On the other hand, the stirring used in most autoclaves is more efficient than what we are able to attain with our low pressure suspended stirrer reactor system. The effect of stirring on 5,000 TON DMIT hydrogenations over AHC Skew run at 200 psig is given in Table 2. The apparent deactivation was the result of catalyst splashing on the sides of the reactor. Because of this, all high pressure reactions were run at 1000 rpm. Table 3 describes the effect of pressure on this reaction. Table 4 lists the reaction data obtained for five 20,000 TON and two 100,000 TON hydrogenations of DMIT run at 1000 psig. In the 20,000 TON hydrogenations the catalyst contained 10 μmoles of active complex and in the 100,000 TON reactions it contained 4 μmoles of the complex. The corresponding homogeneously catalyzed reactions run at these higher pressures all gave products with ee's in the 60 – 65% range.

Experimental Section

The AHC catalyst (Rh(COD)(Skewphos)/PTA/Al$_2$O$_3$) (AHC-Skewphos) was prepared by treating the anchored precursor Rh(COD)$_2$/PTA/Al$_2$O$_3$ with an alcoholic solution of the ligand, Skewphos, using the general procedure described previously (1-3). The catalyst contained 0.5% Rh which corresponds to about a 4.5% load of the anchored complex. The hydrogenations were run using the low and high pressure apparatus previously described (10) under the conditions listed in the discussion.

Acknowledgements

Financial support of this work from Johnson Matthey is gratefully acknowledged.

References

1. S. K. Tanielyan, R. L. Augustine, US Patents 6,005,148 (1999) and 6,025,295 (2000), to Seton Hall University.
2. S. K. Tanielyan, R. L. Augustine, R.L. PCT Int. Appl., WO-9828074: *Chem. Abstr..* **129**. 109217 (1998).
3. R. Augustine, S. Tanielyan, S. Anderson and H. Yang, *Chem. Commun. (Cambridge),* 1257 (1999).
4. S. K. Tanielyan, R. L. Augustine, Chemical Industries (Dekker), **75** *(Catal. Org. React.),* 101 (1998).
5. R. L. Augustine, S. K. Tanielyan, S. Anderson, H. Yang, Y. Gao, Chemical Industries (Dekker), **82** *(Catal. Org. React.),* 497 (2000).
6. P. Goel, S. Anderson, J. Nair, C. Reyes, G. Gao, S. Tanielyan, R. Augustine Chemical Industries (Dekker), **89** *(Catal. Org. React.),* 523 (2003).
7. C. Reyes, Y. Gao, A. Zsigmond, P. Goel, N. Mahata, S. K. Tanielyan and R.L. Augustine, Chemical Industries (Dekker), **82** *(Catal. Org. React.),* 627 (2000).
8. S. Anderson, H. Yang, S. K. Tanielyan and R. L. Augustine, Chemical Industries (Dekker), **82** *(Catal. Org. React.),* 557 (2000).
9. R. L. Augustine, S. K. Tanielyan, N. Mahata, Y. Gao, A. Zsigmond and H. Yang, *Appl. Catal. A*, **256**, 69 (2003.
10. R. L. Augustine, S. K. Tanielyan, Chemical Industries (Dekker), **89** *(Catal. Org. React.)*, 73 (2003).

58. Asymmetric Heck Reactions Using Dipeptido Phosphinite Metal Complexes

Dushmanthi Jayasinghe and Heinz-Bernhard Kraatz

Department of Chemistry, University of Saskatchewan, 110 Science Place, Saskatoon, SK, Canada, S7N 5C9

kraatz@skyway.usask.ca

Abstract

A novel phosphinito dipeptido ligand series were prepared, and fully characterized. These ligands readily form metal complexes with Pd(II) and Pt(II) precursors. The Pd(II) complexes were investigated for their suitability in asymmetric Heck reaction using 3,4-dihydrofuran as a substrate.

Introduction

Amino phosphinites and amidophosphine-phosphinito (AMPP) ligands based on naturally occurring chiral entities such as simple hydroxyl containing amino acids and sugar based ligands have been extensively explored recently for the development of chiral and catalytically active P(III) ligands for transition metal complexes.[1-9] Hydroxyproline based AMPP ligands have been successfully used in asymmetric hydrogenation and hydroformylation of olefins and ketones.[8] However most ligands were composed of only a single amino acid entity. Gilbertson reported the reactivity of a helical phosphino peptide.[6,7]

The Heck reaction has proven to be an extremely useful method for the formation of C-C bond at a vinyl carbon centre. There are numerous reported examples of enantioselctive Pd catalyzed C-C bond forming reactions.[10-13] Surprisingly, reports of Heck transformations using amino acid based phosphine, phosphinite ligands are rare. Recently Gilbertson reported a proline derived phosphine-oxozoline ligand in a catalytic asymmetric Heck reaction.[5] In this paper we present some novel amino acids derived ligands as part of a catalytic system for use in asymmetric Heck reactions.

Results and Discussion

Scheme 1 summarizes our synthetic approach. By protecting the carboxyl groups using a suitable protecting group, the three hydroxy amino acids, serine, threonine and tyrosine were conveniently coupled with Boc-Phe-OH to obtain the corresponding peptides (**1-4**) in good yields.

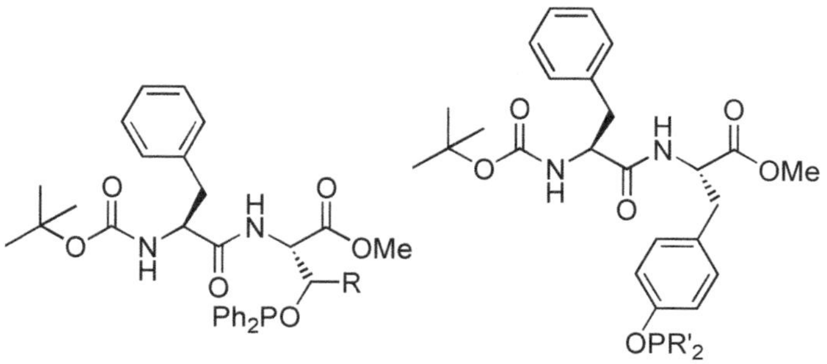

Scheme 1 Phosphinite derivatives of serine, threonine and tyrosine containing dipeptides: **1**   R = H, Boc-Phe-Ser(OPPh$_2$)OMe; **2**   R = Me, Boc-Phe-Thr(OPPh$_2$)OMe; **3** R' = Ph, Boc-Phe-Tyr(OPPh$_2$)OMe; **4** R' = cyhex, Boc-Phe-Tyr(OPcyhex$_2$)OMe.

Reactions of the hydroxy peptides with selected phosphine chlorides in the presence of a base and catalytic amounts of DMAP resulted the corresponding phosphinites, which readily form Pd and Pt complexes (Scheme 2).

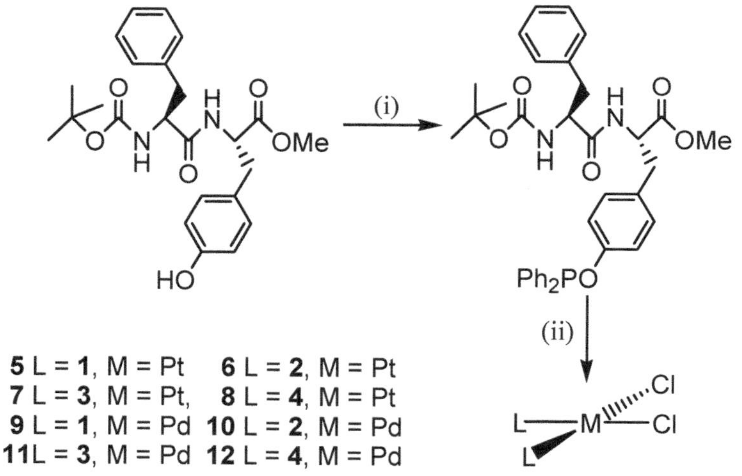

5 L = 1, M = Pt   6 L = 2, M = Pt
7 L = 3, M = Pt,   8 L = 4, M = Pt
9 L = 1, M = Pd   10 L = 2, M = Pd
11 L = 3, M = Pd   12 L = 4, M = Pd

Scheme 2 Synthesis of dipeptido phosphinite ligands **1**–**4** and corresponding metal complexes, **5-12** (i) ClPPh$_2$, CH$_2$Cl$_2$, Et$_3$N, DMAP; (ii) (BzCN)$_2$MCl$_2$, CH$_2$Cl$_2$, M = Pt, Pd.

All ligands and metal complexes were characterized spectroscopically. The ^{31}P-NMR spectra of ligands **1-4** appear in a region of δ 111.6–140.6. Upon metal complexation an upfield shift is observed. Table 1 summarizes the selected spectroscopic properties of the new phosphinite ligands (**1-4**) and corresponding metal complexes (**5-12**). With exception of **8** and **12** all complexes have a *cis* coordination geometry as judged from J$_{P-Pt}$ and the IR spectra (341 cm^{-1} and 355 cm^{-1} for **8** and **12**, respectively). Due to the steric pressure imposed on the complexes by ligand **4**, complexes **8** and **12** adopt trans configuration. However a minor *cis* component is present in the reaction solution. Phosphinite ligands and complexes are generally considered as not very stable. However, these ligands and complexes exhibit a considerable stability as evidenced by the mass spectroscopy (ESI). Even in gaseous state ligands and the complexes are fairly stable with minimum ligand dissociation.

Table 1 Summary of the spectroscopic features of ligands **1 – 4** and their respective Pd and Pt complexes **5-12**. αH^1, αH^2, βH^1, βH^2 are defined in the scheme given below (δ in ppm, J in Hz).

Compound	'H-NMR (CDCl₃)					^{31}P-NMR (J$_{P-Pt}$) (CDCl₃)
	NH	αH^1	αH^1	βH^1	βH^1	
1	6.73	4.78	4.29	3.05	3.90	119.5
2	6.63	4.65	4.48	3.07	4.71	113.3
3	6.13	4.71	4.32	3.03-2.86		111.6
4	6.28	4.72	4.33	3.11-2.88		148.9
5	6.76	4.73	4.29	2.95	3.95	86.0 (4470)
6	6.75	4.72	4.25	2.92	3.88	78.0 (4220)
7	6.73	4.75	4.39	3.02	4.71	87.0 (4080)
8	6.78	4.63	4.28	2.97	4.97	118.2 (2611)
9	6.33	4.62	4.24	3.02-2.86		111.5
10	6.53	4.63	4.22	2.97-2.85		105.6
11	6.52	4.71	4.37	3.15-3.00		112.4
12	6.53	4.69	4.38	3.04-2.97		131.2

Pd complexes **9 – 12** were tested for their catalytic behavior in the asymmetric Heck reaction involving the phenylation of 2,3-dihydrofuran (Scheme 3). The results are summarized in Table 2. The two isomeric products of 2-phenyl-2,5-dihydrofuran are formed with varying yields from 80% to 0%. The obtained ee's are high. Complex **12** is shown to be catalytically inactive. The lack of catalysis in complex **12** is rationalized by differences in the steric requirements between the diphenylphosphinites **1-3** (cone angle >140°) and the more sterically hindered cyclohexyl-phosphinite **4** (cone angle >170°) and the resulting stereochemistry on the Pd center. The ligands in complex **12** adopt a *trans* geometry. The bulky ligands effectively shield the metal and prevent the approach of the olefin, thereby preventing catalysis.

Table 2 Phenylation of 2,3-dihydrofuran using the Pd complexes **9-12**.[a,b]

Run	Complex	Base	Yield in %	ee in %
1	9	*i*-Pr$_2$NEt	60	95 (*R*)
2	10	*i*-Pr$_2$NEt	80	97 (*R*)
3	11	*i*-Pr$_2$NEt	60	87 (*R*)
4	12	*i*-Pr$_2$NEt	0	0

[a]Temperature 70°C; all in benzene.
[b]Products were identified by GC-MS and compared to literature values. Yields and ee's were calculated from the GC (Varian ProStar # 1, 15 psi He, Chirasil-D-Val capillary column, 25m x 0.25mm). Product: 2-phenyl-2,3-dihydrofuran; t_R = 9.9 min, (*R*) 2-phenyl-2,5-dihydrofuran t_R = 10.6 min.

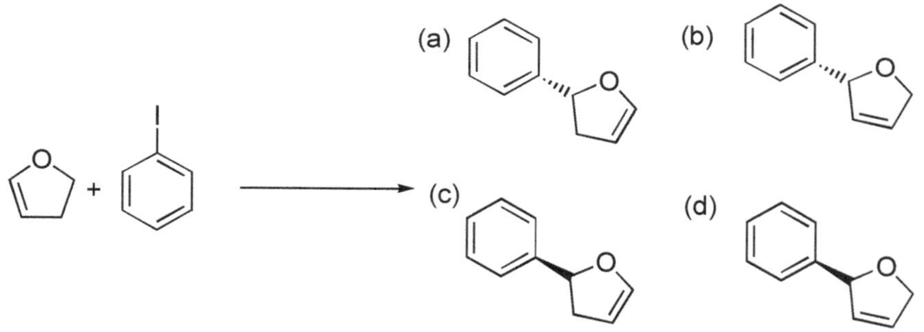

Scheme 3 Asymmetric Heck coupling of 3,4-dihydrofuran and iodobenzene in the presence of Pd complexes **9-12** giving a mixture of regio and stereo isomers: a) 2(R)-phenyl-2,3-dihydrofuran, b) 2(R)-phenyl-2,5-dihydrofuran c) 2(S)-phenyl-2,3-dihydrofuran d) 2(S)-phenyl-2,5-dihydrofuran.

Our future work towards the investigation of the catalytic potential of our novel catalysts will involve more inert substrates such as aryl chlorides and bromides.

Experimental Section

General procedure for phosphinito ligands. To a flask containing the N,C-protected hydroxy dipeptides and CH$_2$Cl$_2$ (20 ml), 3eq. NEt$_3$ and DMAP (10 mol%) were added. An equimolar amount of alkyl or aryl substituted phosphine chloride was added dropwise using an 1-ml Hamilton syringe. The mixture was stirred overnight, filtered through basic alumina, and the solvent was removed in vacuo, giving the desired phosphinites 1-4 in quantitative yields.

General procedure for phosphinite-metal complexes. To the suspension of metal precursor ($Cl_2Pt(COD)$, $Cl_2Pt(PhCN)_2$ or $Cl_2Pd(PhCN)_2$) in CH_2Cl_2 (10ml) a ligand solution was added (2 eq. of 1-4 in 10 mL CH_2Cl_2). The mixture was stirred for 24 h at room temperature and evaporated, giving the metal complex as colourless to yellow solids in quantitative yields.

Example synthesis of [Boc-Phe-Thr(OPPh$_2$)-OMe]$_2$PdCl$_2$ (10). Boc-Phe-Thr(OPPh$_2$)-OMe (0.5 mmol), cis-(BzCN)PdCl$_2$ (0.25 mmol). Product: yellow solid; 100 % yield. HRMS of $C_{62}Cl_2H_{74}N_4O_{12}P_2Pd$ Calc.$[M - Cl]^+$ 1271.091 Observed $[M - Cl]^+$ 1271.362.

General procedure for the Heck reaction. A mixture of iodobenzene (5 mol equ.), diisopropylethylamine (3 mol equ.), 2,3-dihydrofuran (1 mol eq.), and the complexes **9-12** (3 mol %) in degassed benzene was stirred at 70 °C. The progress of the reaction was monitored by GC. Upon completion, the mixture was filtered through basic alumina (58 mesh, Aldrich) to remove Pd and the products were identified using GCMS.

Acknowledgement

We wish to thank NSERC, ORCS and the University of Saskatchewan for financial support. H.B.K is the Canada Research Chair in Biomaterials.

References

1. S. R. Gilbertson and G. W. Starkey, *J. Org. Chem.*, **61**, 2922-2923 (1996).
2. T. V. RajanBabu, A. T. Ayers, G. A. Halliday, K. Y. You, and J. C. Calabrese, *J. Org. Chem.*, **62**, 6012-6028 (1997).
3. Y. Yan and T. V. RajanBabu, *J. Org. Chem.*, **65**, 900-906 (2000).
4. T. V. RajanBabu, Y. Yan, and S. Shin, *J. Org. Chem.*, **123**, 10207-10213 (2000).
5. S. R. Gilbertson, D. Xie, and Z. Fu,. *J. Org. Chem.*, **66**, 7240-7246 (2001).
6. S. R. Gilbertson, G. Chen, J. Kao, A. Beatty, and C. F. Campana, *J. Org. Chem.*, **62**, 5557-5566 (1997).
7. S. R. Gilbertson and L. Ping, *Tetrahedron Lett.*, **43**, 6961-6965 (2002).
8. F. Agbossou, J. Carpentier, and A. Mortreux, *Chem. Rev.*, **95**, 2485-2506 (1995).
9. P. V. Galka, and H. B. Kraatz, *Chem. Ind. (Dekker)*, **82** (Catal. *Org. React.*), 589-594 (2001).
10. P. V. Galka and H. B. Kraatz, *J. Organometal. Chem.*, **674**, 24-31 (2003).
11. T. G. Kilory, A. J. Hennessy, D. J. Connolly, Y. M. Malone, A. Farrell, and P. J. Guiry, *J. Mol. Cat.*, **196**, 63-65 (2003).
12. A. J. Hennessy, D. J. Connolly, Y. M. Malone, and P. J. Guiry, *Tetrahedron Lett.*, **41**, 7757-7761 (2000).
13. M. Casey, J. Lawless, and C. Shirran, *Polyhedron*, **19**, 517-520 (2000).

59.   Enantioselective Hydrogenations of Exo- and Endocyclic C=C Double Bond with Highly Mesoporous Carbon Supported Pd

É. Sípos[a], G. Fogassy[a], P.V. Samant[b], J.L. Figueiredo[b], and A. Tungler[a]

[a]Department of Chemical Technology, Technical University of Budapest, H-1521
[b]Laboratório de Catálise e Materiais, Chemical Engineering Department, Faculty of Engineering, University of Porto, 4200-465 Porto, Portugal

Abstract

Highly mesoporous carbon supported Pd catalysts were prepared using sodium formate and hydrogen for the reduction of the catalyst precursors. These catalysts were tested in the enantioselective hydrogenation of isophorone and of 2-benzylidene-1-benzosuberone. The support and the catalysts were characterized by different methods such as nitrogen adsorption, hydrogen chemisorption, SEM, XPS and TPD.

Introduction

The type of catalyst strongly influences the enantioselectivity of heterogeneous catalytic hydrogenations (1). In the enantioselective saturation of the C=C bond of isophorone over (-)-dihydroapovincaminic acid ethyl ester ((-)-DHVIN) modified Pd catalysts (scheme 1) the optical purity strongly depended on the type and properties of the support used (2, 3, 4).

In the most effective, chirally modified catalytic systems, Pt/cinchonidine and Raney-Ni/tartaric acid, the enantioselectivity was also sensitive to the method of catalyst preparation and on support properties (5, 6).

In the enantioselective hydrogenation of isophorone in the presence of (-)-DHVIN modifier the best optical purity was afforded by small dispersion (<0,05) Pd black catalyst (up to 55%) (7). The influence of the preparation method of Pd black on the optical yield was reported (8). A correlation was found between the oxidation state of the metal surface and the enantioselectivity, the catalyst having more oxidised species on its surface giving higher enantiomeric excess, while the Pd black with lower surface area was more enantioselective.

The Pd black catalyst was found to be the most effective in the hydrogenation of isophorone (modifier (S)-α,α-diphenyl-2-pyrrolidinemethanol, DPPM, Scheme 1, e.e. up to 42%) (9) and that of 2-benzyl-1-benzosuberone too (modifier cinchonidine, CD, Scheme 2, e.e. up to 54%) (10).

The object of the present study was to use in the above mentioned hydrogenations improved carbon supported catalysts, which could compete with the Pd black catalyst. Carbon materials are common supports, their surface properties can be modified easily and it is possible to prepare carbons with different proportion of micro-, meso- and macropores, which can be key factors influencing their performances. A highly mesoporous carbon was synthesised and used as support of Pd catalysts in the enantioselective hydrogenations. To our knowledge this is the first report on the use of highly mesoporous carbon for the preparation of Pd catalysts for liquid-phase hydrogenation.

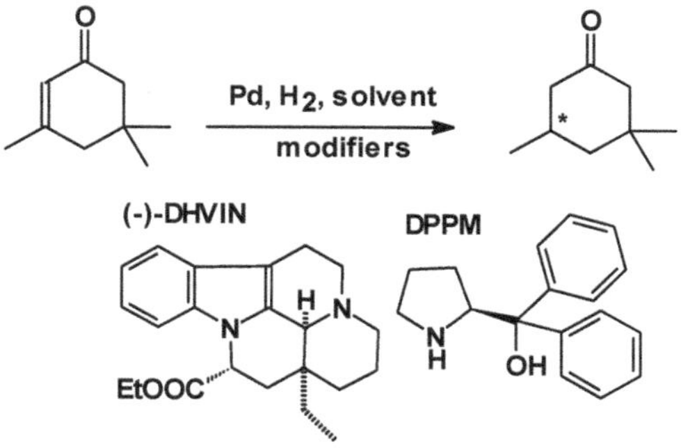

Scheme 1 Hydrogenation of isophorone.

Experimental Section

Materials

A highly mesoporous carbon xerogel was synthesized by the condensation of resorcinol and formaldehyde in molar ratio of 1:2. An aqueous solution of sodium carbonate was used as a base catalyst, in a molar ratio 1:200 with resorcinol. The clear solution obtained was cured by employing [1-1-1] cycle keeping at room temperature, 50 °C and 90 °C for one day each, respectively. The gel was exchanged with acetic acid, and it was further exposed to successive extractions of acetone and cyclohexane, to remove the solvent from the pores. The gel was first dried at ambient temperature and then it was further exposed to warm air at 50 °C for 2-3 h. Carbonisation of the gel was carried out in nitrogen atmosphere by heating it to 800 °C at a heating rate of 10 °C/min and further keeping it at this temperature for 6 h. The surface of the carbon material was modified by treatment with 5% oxygen in nitrogen at 400 °C for 10 h to reach a burn-off of 14% (referred as CX-14). Prior to surface activation the carbon was subjected to extraction with 2M HCl in a Soxhlet apparatus for 6 h, in order to remove sodium and washed copiously with water until neutral pH.

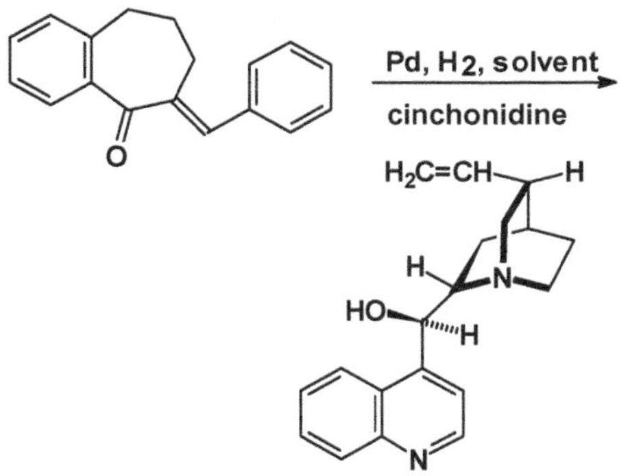

Scheme 2 Hydrogenation of 2-benzylidene-1-benzosuberone.

Catalysts with 10 wt% Pd on highly mesoporous carbon xerogel CX-14 were prepared as follows. The calculated amount of the K_2PdCl_4 precursor was added to the aqueous suspension of the support. The pH value was adjusted to 11 by addition of KOH. The suspension was boiled for 1 h, then the reducing agent (HCOONa, three times the stoichiometric amount) (type 1) was added to the boiling mixture. After half an hour the suspension was cooled, the catalyst was filtered and washed with distilled water until neutral pH. The other method comprised, after similar preparation steps, the reduction of the $Pd(OH)_2$ precursor with H_2 gas in water at atmospheric pressure (type 2). Selcat Pd/C contains 10wt% Pd on a high surface area activated carbon ($S_{BET} = 1200$ m^2/g), the manufacturer is Fine Chemical Comp. Budapest, Hungary.

Apovincaminic acid ethylester was supplied by Richter Gedeon Co., (-)-dihydroapovincaminic acid ethyl ester was prepared according to the procedure described in (11). *(S)-α,α*-diphenyl-2-pyrrolidinemethanol was synthesised as described (12). 2-benzylidene-1-benzosuberone was prepared as described in (13). Isophorone was supplied by Merck. Cinchonidine was purchased from Fluka.

Catalysts characterization

The support and the catalysts were characterised by means of nitrogen adsorption, XPS, TPD and SEM. The nitrogen adsorption isotherms were determined at 77 K in a Coulter Omnisorp 1000 CX equipment, and were analysed by the BET equation (S_{BET}), and by the t-plot for mesopore surface area (S_{meso}) and micropore and mesopore volume (V_{micro}, V_{meso}), using the standard isotherm for carbon materials. The catalyst samples were previously outgassed at 120 °C.

The dispersion of the catalyst was measured by hydrogen chemisorption at 27 °C in the Coulter Omnisorp 100 CX equipment. The catalysts were subjected to an *in situ* reduction under hydrogen at 350 °C for 1 h before chemisorption analysis. The palladium dispersion was calculated by assuming a stoichiometry of 1:1 (H: Pd). As Pd hydride formation can not be excluded, the dispersion data are only for comparison of the two catalysts with the same metal loading. The dispersion of the commercial Selcat type Pd/C catalysts, which was found to be D>70%, was determined by hydrogen and carbonmonoxide adsorption. The dispersion of Pd black (D<1%)was determined by successive H_2/O_2 titration (8).

The amount and type of surface functional groups of the carbon support were studied by temperature programmed desorption (TPD) in helium atmosphere. The analysis was carried out in a custom built set-up consisting of an U- shaped quartz tubular micro reactor cell, placed inside an electrical furnace, with helium flow of 25 ml /min. The samples were subjected to a 5 °C/min linear temperature rise up to 1100 °C. A SPECTRAMASS Dataquad quadrupole mass spectrometer was used to monitor the desorbed CO (m /z =28) and CO_2 (m /z =44) products from the samples. XPS analysis was carried out in a VG Scientific ESCALAB 200A spectrometer using non-monochromatized Mg Kα radiation (1253.6 eV) to determine the surface chemical composition of Pd catalyst on carbon. Basically, the Pd $3d^5$, $3d^3$ high- resolution spectrum was analysed for Pd oxidation states by deconvoluting the peaks with mixed Lorentzian-Gaussain functions.

The surface analysis for morphology and average particle size was carried out with JEOL JSM 6301 F scanning electron microscope (SEM). The micrographs of the samples were observed at different magnifications under different detection modes (secondary or back-scattered electrons).

Hydrogenation

The hydrogenations were carried out in a stainless steel autoclave with magnetic stirrer at room temperature and 50 bar hydrogen pressure. Before the hydrogenation, the reaction mixtures were stirred under nitrogen (2-3 bar) for 10 minutes in the reaction vessel.

The working-up procedure of the reaction mixtures of the 2-benzyl-1-benzosuberone was the catalyst filtration and the removal of the solvent in vacuum. The residue was dissolved in dichloromethane and extracted with 5 % HCl and distilled water. The organic phase was separated and dried over Na_2SO_4. After filtration, the solvent was removed in vacuum.

Analysis

The reaction mixtures of isophorone were analysed with a gas chromatograph. The GC analyses were carried out with gas chromatograph equipped with a β-cyclodextrine capillary column (analysis temperature: dihydroisophorone at 110 °C) and FID. The chromatograms were recorded and peak areas were calculated with Chromatography Station for Windows CSW32® v.1.2 (DataApex Ltd. 2001, Prague).

The HPLC analyses of the reaction mixtures of the 2-benzyl-1-benzosuberone were carried out on a Chiracel OJ column (0.46×25 cm). The column contains silica-gel as packing material coated with a cellulose derivative. The eluent was hexan/2-propanol 90:10 v/v, the flow rate was 1.0 ml/min and the column pressure was 50 kg/cm². The UV-absorbance was measured at 249 nm. Enantiomeric excesses were calculated according to the following equation:

$$e.e. \ (\%) = \frac{[A]-[B]}{[A]+[B]} \cdot 100$$

where [A] is the concentration of major enantiomer and [B] is the concentration of minor enantiomer.

Results and Discussion

Characterization of the carbon and the Pd/C catalysts

Table 1 gives the textural properties of the support and catalyst samples. As expected the pore volumes and the surface areas of the catalysts are lower than those of the support. This indicates that the palladium blocks some part of the surface of carbon.

The adsorptive behaviour of a carbon, as well as its catalytic and electrical properties, strongly depend on the nature and concentration of the oxygen containing surface complexes (14). The main characteristic of the TPD profiles (Figure 1) of the catalyst can be a decrease or increase of the amount of surface oxygen containing groups of the support oxidised in gas phase, expressed by ΔCO and ΔCO_2, the difference between the CO and CO_2 values of the support and those of corresponding catalysts (Table 2). The TPD profiles indicate that the nature of the CO_2- and CO-generating groups are the same on the support and on the catalysts. All spectra exhibit a broad peak between 550-850 °C, assigned to anhydride and lactone type linkages generating CO_2, above 800 °C in case of CO assigned to quinone and ether groups.

Table 1 Textural properties of the samples.

Sample	S_{BET}[a] (m^2/g)	S_{meso}[b] (m^2/g)	V_{micro}[b] (cm^3/g)	V_{meso}[c] (cm^3/g)
CX-14	954	601	0.14	0.57
Pd/C type 1	797	496	0.12	0.49
Pd/C type 2	805	510	0.11	0.52

[a]Calculated from the BET equation.
[b]Calculated from the t-plot.
[c]Calculated from the adsorption isotherm.

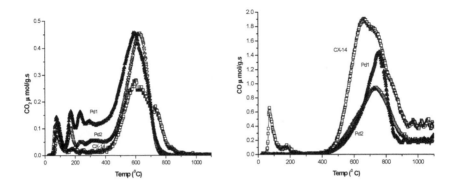

Figure 1 Gas evolution profiles (CO_2 and CO) of support and catalysts (Pd1 is for Pd/C Type 1, Pd2 is for Pd/C Type 2).

Table 2 Surface properties of the samples.

Sample	CO ($\mu molg^{-1}$)	CO_2 ($\mu molg^{-1}$)	ΔCO	ΔCO_2
CX-14	6784	1023		
Pd/C type 1	3126	1606	-3658	+583
Pd/C type 2	2953	1155	-3821	+132

The amount of the CO-yielding complexes decreased on the catalyst surface compared to the carbon support. It indicates that the palladium is attached to CO generating groups. The increased concentration of CO_2 containing complexes on the catalysts surfaces can be explained by the preparation method.

The anchoring and the reduction methods of precious metal precursors influence the particle size, the dispersion and the chemical composition of the catalyst. The results of SEM and H_2 chemisorption measurements are summarised in Table 3. The XPS measurements indicate that the catalysts have only metallic Pd phase on their surface. The reduction of catalyst precursor with sodium formate resulted in a catalyst with lower dispersion than the one prepared by hydrogen reduction. The mesoporous carbon supported catalysts were prepared without anchoring agent, this explains why they have much lower dispersion than the commercial catalyst which was prepared in the presence of a spacing and anchoring agent (15).

The micrographs of the samples Pd/C type 1 and Pd/C type 2 are shown in Figure 2 and Figure 3, respectively. The structure of the catalysts is rather uniform. The surface of the support is covered evenly by the palladium. The elementary palladium particles are stuck together to a big grain, their size is different, as can be seen in the figures, in the case of catalyst type 1 the particle size is approximately 10× bigger than that of type 2.

Table 3 Characterization of the catalysts by SEM and H_2 chemisorption.

Pd/C catalyst	Reducing agent	Average particle size (μm)	Dispersion (%)
Type 1	HCOONa	>1	8.40
Type 2	H_2	>0.1	30.58
Selcat[a]	H_2	<100nm	>70

[a]Dispersion measured by hydrogen and carbonmonoxide adsorption.

Catalytic tests

The Pd catalysts on the highly mesoporous carbon support were tested in the enantioselective hydrogenation of isophorone (Scheme 1) and of 2-benzylidene-1-benzosuberone (Scheme 2). The enantioselectivities are summarised in Table 4. The reaction conditions were optimised earlier (10, 11, 16). The results obtained with a high dispersion (D>0.7) Pd/C Selcat commercial catalyst and with small dispersion (D<0.01) Pd black are included for comparison. The applied Pd metal:substrate weight ratio is different for the mesoporous carbon supported catalysts (1/175), for the commercial Pd/C (1/560) and for the Pd black (1/19). The reason for the different catalyst loading is that the activity of these catalysts is extremely different (Pd/C>>Pd/mesoC>>black), we tried to compensate this with the catalyst amount in order to complete the reactions during nearly the same time. According to our expectations the highly mesoporous carbon supported Pd catalysts afforded higher e.e. than the commercial Pd/C catalyst (2, 9), but only about one half of the optical purities obtained with Pd black catalysts (4, 15). The activity, enantioselectivity and the dispersion of the tested catalysts seem to be in correlation, the higher is the dispersion, the higher is the activity and the lower is the enantioselectivity. But this is not the only property which influences the enantiodifferentiating ability, in the case of supported catalysts the nature of the support, moreover the structure and surface properties of it can have an effect on the chiral modification.

In the hydrogenation of isophorone the catalyst type 1 of smaller dispersion resulted in higher enantiomeric excesses especially with DPPM modifier. In the hydrogenation of 2-benzylidene-1-benzosuberone the catalyst type 2 of higher dispersion was more enantioselective. These reverse tendencies or smaller relative difference in e.e. for the latter reaction can be attributed to the use of modifiers with totally different structure and working mode.

a)   b)

Figure 2   The SEM pictures of Pd/C type 1 catalyst: a)secondary mode, b) back scattered electron.

a)   b)

Figure 3   The SEM pictures of Pd/C type 2 catalyst: a)secondary mode, b) backscattered electron.

Table 4 Enantiomeric excesses in the enantioselective hydrogenation of isophorone and 2-benzyl-1-benzosuberone on highly mesoporous carbon supported Pd catalysts.

Substrate	Modifier	Catalyst	React. time (h)	Conv. (%)	e.e. (%)
Isophorone[a]	(-)-DHVIN	Type 1	6	100	14 (R)
		Type 2	2	89	10 (R)
[b]		Pd/C Selcat	7.5	87	10 (R) (2)
[c]		Pd black	8	100	54 (R)
Isophorone[d]	DPPM	Type 1	6	100	33 (S)
		Type 2	6	100	14 (S)
		Pd/C Selcat	2	100	10 (S) (9)
[e]		Pd black	4	100	41
2-benzyl-1-benzosuberone[f]	CD	Type 1	5	100	20 (S)
		Type 2	5	82	27 (S)
[g]		Pd/C Selcat	5	100	0
		Pd black	8	100	54 (S)

Reaction conditions: [a]0.35 g isophorone, 0.02 g catalyst, 0.005 g AcOH, 0.001 g (-)-DHVIN, 3 ml MeOH, [b]5.6 g isophorone, 0.1g catalyst, 0.05 g AcOH, 0.01 g, 50 ml MeOH, [c]5.6 g isophorone, Pd black 0.3 g, 0.02 g (-)-DHVIN, 0.2 g AcOH, 50 ml methanol, 50 bar, 25 °C, [d]0.35 g isophorone, 0.02 g catalyst, 0.01 g DPPM, 1.5 ml MeOH, 1.5 ml water, [e]1.15 g isophorone, 0.037 g DPPM, 0.05 g Pd black, 5 ml MeOH, 5 ml water, 50 bar, 25 °C, [f]0.25 g 2-benzyl-1-benzosuberone, 0.025 g catalyst, 0.00125 g cinchonidine, 10 ml toluene, [g]2.5 g 2-benzyl-1-benzosuberone, 0.25 g catalyst, 0.0125 g cinchonidine, 100 ml toluene.

Conclusion

The smaller surface area values of the catalysts indicate that the palladium blocks some part of the surface of this mesoporous carbon. The decreased amount of the CO-yielding complexes on the catalyst surface compared to that of the carbon support indicates that the palladium is attached to CO generating groups. The increased concentration of CO_2 containing complexes on the catalysts surfaces can be due to the steps of the preparation. To clarify this it needs further experiments.

The reduction of the catalyst precursor with sodium formate resulted in a lower Pd dispersion than the catalyst prepared by hydrogen reduction, the particle size is much larger in the former catalyst. The mesoporous carbon supported Pd catalysts are near to those of Pd on titania with respect to their enantiodifferentiating ability. Besides the metal dispersion, the availability of the Pd surface in the pores for the large modifier molecules seems to be the determining factor of the enantioselectivity.

The interesting properties of the mesoporous carbon supported Pd need further studies.

Acknowledgement

The authors acknowledge the financial support of the Hungarian OTKA Foundation under the contract number T 043153, from the Portuguese Foundation for Science and Technology (FCT) and FEDER (contract number POCTI/1181) and of the TéT Portugese-Hungarian Foundation (GRICES-OMFB).

References

1. A. Baiker, *J. Mol. Catal. A*, **163**, 205 (2000).
2. T. Tarnai, A. Tungler, T. Máthé, J. Petró, R. A. Sheldon, and G. Tóth, *J. Mol. Catal. A*, **102**, 41 (1995).
3. G. Farkas, L. Hegedûs, A. Tungler, T. Máthé, J. L. Figueiredo, and M. Freitas, *J. Mol. Catal. A*, **153**, 215 (2000).
4. É. Sípos, G. Farkas, A. Tungler, and J. L. Figueiredo, *J. Mol. Catal. A*, **179**, 107 (2002).
5. Y. Nitta, F. Sekine, T. Imanaka, and S. Teranishi, *J. Catal.*, **74**, 382 (1982).
6. H. U. Blaser, H. P. Jalett, M. Müller, and M. Studer, *Catal. Today*, **37**, 441 (1997).
7. A. Tungler, T. Máthé, K. Fodor, R. A. Sheldon, and P. Gallezot, *J. Mol. Catal. A*, **108**, 145 (1996).
8. G. Farkas, É. Sípos, A. Tungler, A. Sárkány, and J. L. Figueiredo, *J. Mol. Catal. A*, **170**, 101 (2001).
9. É. Sípos, A. Tungler, and I. Bitter, *React. Kinet. Catal. Lett.*, **79**, 101 (2003).
10. G. Fogassy, A. Tungler, and A. Lévai, *J. Mol. Catal. A*, **192**, 189 (2003).
11. A. Tungler, T. Máthé, T. Tarnai, K. Fodor, J. Kajtár, K. I., B. Herényi, and R. A. Sheldon, *Tetrahedron Asymm.*, **6**, 2395 (1995).
12. D. J. Mathre, T. K. Jones, L. C. Xavier, T. J. Blacklock, R. A. Reaner, J. J. Mohan, E. T. T. Jones, K. Hoogsteen, M. W. Baum, and E. J. J. Grabowski, *J. Org. Chem.*, **56 (2)**, 751 (1991).
13. N. R. El-Rayyes and N. H. Bahtiti, *J. Heterocyclic Chem.*, **26**, 209 (1989).
14. L. R. Radovic and F. Rodríguez-Reinoso, in *Chemistry and Physics of Carbon 25*, P. A. Thrower, Ed., Dekker, New York, 1997, p. 243.
15. T. Máthé, A. Tungler, J. Petró, Hung. Pat. 177860, US Pat. 4,361,500.
16. É. Sípos, A. Tungler, I. Bitter, and M. Kubinyi, *J. Mol. Catal. A*, **186**, 187 (2002).

60.   Influence of Achiral Tertiary Amines on the Enantioselective Hydrogenation of α,β-diketones Over Cinchonidine-Pt/Al₂O₃ Catalyst

J. L. Margitfalvi and E. Tálas

Chemical Research Center, Hungarian Academy of Sciences, 1025 Budapest, Pusztaszeri út 59-67 Hungary; joemarg@chemres.hu

N. Marín-Astorga and T. Marzialetti

Departamento de Fisicoquímica, Facultad de Ciencias, Universidad de Concepción, Casilla 160-C, Concepción, Chile

Abstract

Enantioselective hydrogenation of 2,3-butanedione and 3,4-hexanedione has been studied over cinchonidine - Pt/Al₂O₃ catalyst system in the presence or absence of achiral tertiary amines (quinuclidine, DABCO) using solvents such as toluene and ethanol. Kinetic results confirmed that (i) added achiral tertiary amines increase both the reaction rate and the enantioselectivity, (ii) both substrates have a strong poisoning effect, (iii) an accurate purification of the substrates is needed to get adequate kinetic data. The observed poisoning effect is attributed to the oligomers formed from diketones.

Introduction

Pt/Al₂O₃-cinchona alkaloid catalyst system is widely used for enantioselective hydrogenation of different prochiral substrates, such as α-ketoesters [1-2], α,β-diketones, etc. [3-5]. It has been shown that in the enantioselective hydrogenation of ethyl pyruvate (Etpy) under certain reaction conditions (low cinchonidine concentration, using toluene as a solvent) achiral tertiary amines (ATAs: triethylamine, quinuclidine (Q) and DABCO) as additives increase not only the reaction rate, but the enantioselectivity [6]. This observation has been explained by a virtual increase of chiral modifier concentration as a result of the shift in cinchonidine monomer - dimer equilibrium by ATAs [7].

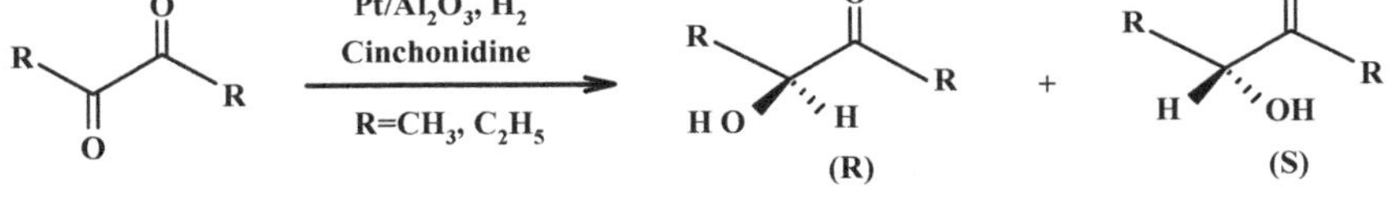

Scheme 1 Reaction scheme of hydrogenation of α,β-diketones.

In this study enantioselective hydrogenation of diketones (2,3-butanedione (BD), 3,4-hexanedione (HD)) (see Scheme 1) was investigated in the presence or absence of ATAs using solvents, such as toluene and ethanol. In addition, the importance of the purification of these substrates will be discussed. The main goal of this study is to get further information about the effect of ATAs in case of substrates, such as of α,β–diketones.

Experimental Section

Enantioselective hydrogenation of 2,3-butanedione and 3,4-hexanedione has been investigated using cinchonidine (CD) as chiral modifier. The injection technique was used to introduce CD into the reactor [7]. The reaction temperature and the hydrogen pressure were 20 °C and 50 bar in all experiments. Commercial 5% Pt/Al_2O_3 (E4759) catalyst was used. The diketones (from Fluka) were used as received or after purification using either activated carbon or distillation (at 36 °C at 10 torr). The solvents (toluene and ethanol) were Reanal products. The ATAs (quinuclidine and DABCO purchased from Fluka) were injected together with CD (coinjection). Samples were taken at different reaction time and analysed by a gas chromatograph equipped with a capillary column Chiraldex B-TA (0,25 mm; 30 m) and flame ionisation detector. The highest enantiomeric excess value measured in a given reaction, ee_{max}, was used for comparison. The conversion and enantioselectivity at 4-hr of each reaction, $conv_{240}$ and ee_{240}, were measured as well for comparison. The kinetic curves of the enantioselective hydrogenation had the characteristic two step behaviour described earlier [7]. First order rate constants measured in the first hour of reaction and between 1 and 4 hours (k_1 and k_2, respectively) were used to compare the activity of catalysts under various experimental condition.

Results

The reproducibility of the enantioselective hydrogenation of diketones was investigated first. It has been mentioned by other authors that the rate of hydrogenation of diketones is less then that of the α-keto esters [3]. The difference in the rate of hydrogenation may be attributed to the increased strength of adsorption of diketones to the platinum surface compared to that of the α-keto esters [3,5]. In addition, we consider that the poisoning effect of oligomers formed from diones cannot also be neglected. These high molecular weight compounds can be formed either during the storage of the diones or in the hydrogenation reaction. According to mass spectrometry measurements, these compounds appeared in different amounts in the various batches of HD. Oligomers were enriched especially in the distillation residue [8].

The results of hydrogenation obtained using different batches of 3,4-hexanedione and different purification methods are given in Table 1. The data presented in Table 1 show that different batches of 3,4-hexanedione behave differently (see runs 1,6,9 and 2,3,4). In this series of experiments, very low CD concentrations were used. The introduction of additional amount of CD did not improve the activity (see runs 4,5). This result suggests that poisoning of Pt rather then the undesired transformation of CD is involved in the loss of activity. This is supported by that active carbon treatment of the starting material resulted in an increase in the reaction rate and the highest activity was found with the material purified via distillation. No poisoning effect was found in the latter case (i.e., $k_1 \cong k_2$). The results obtained in this series of experiment show also that upon increasing the concentration of CD both the k_1 and ee values increased, but k_2 was not affected. In all experiments except of N° 11, the rate constant of 1st kinetic regime is significantly greater than that of the 2nd (i.e., $k_1 >> k_2$). In some experiments (see exps. N° 4 and 5) the reaction practically stopped after 2-3 hours. The similar is shown by run N° 8. Addition of doubled amount of catalyst increased k_1, but not k_2. All these observations can be explained by the strong poisoning effect.

Table 1 Effect of treatment of 3,4-hexanedione.

No	Substrate batch	Treat-ment	[CD] 10^{-4} x, M	k_1 min^{-1}	k_2 min^{-1}	ee$_{max}$	ee$_{240}$	conv$_{240}$
1	HD1	no	0.12	0.0014	0.0003	0.192	0.085	0.136
2	HD2	no	0.24	0.0079	0.0004	0.553	0.550	0.423
3	HD3	no	0.24	0.0020	0.0001	0.314	0.238	0.143
4	HD4	no	0.24	0.0038	>0.0001	0.223	0.211	0.204
5[a]	HD4	no	0.24	0.0044	>0.0001	0.229	0.216	0.150
6	HD4	no	0.12	0.0022	0.0002	0.185	0.103	0.131
7	HD4	Carbon	0.12	0.0047	0.0006	0.327	0.269	0.423
8[b]	HD4	Carbon	0.12	0.0071	0.0007	0.289	0.277	0.455
9	HD5	no	0.12	0.0018	0.0002	0.109	0.107	0.153
10	HD5	Carbon	0.12	0.0028	0.0003	0.192	0.181	0.204
11	HD5	Distillation	0.12	0.0066	0.0058	0.215	0.110	0.788

[a]Further amount of cinchonidine (4.8×10^{-5} M) injected at 60 min, [b]double amount of Pt/Al$_2$O$_3$; CD injection technique, T$_r$:20 °C; p$_{H2}$: 50 bar, catalyst: 0.125g 5% Pt/Al$_2$O$_3$ (E4759), in all experiments after 240 min the yield of hexanediols is less than 5 %.

Figure 1A shows the conversion-ee dependencies in the enantioselective hydrogenation of 2,3-butanedione (BD). As shown that the introduction of quinuclidine (Q), as an ATA, significantly increased the ee values in the whole conversion range. Upon Q addition, first order rate constant k_1 increased from 0.0102 to 0.0158 while k_2 remained almost unchanged (0.0008 and 0.0005, respectively). After 4 hours reaction time, measurable amount of butanediols was found. The yields of butanediols were as follows: in the absence of quinuclidine: R,R=2.0%, S,S=1.7%, R,S=5.4%, in its presence: R,R=2.3%; S,S=1.2%; R,S=3.7%. Figure 1B shows the ee-conversion dependencies obtained in the enantioselective hydrogenation of 3,4-hexanedione. Further experimental data obtained upon investigating the ATA-effect in the enantioselective hydrogenation of 3,4-hexanedione are given in Table 2 and Figure 2.

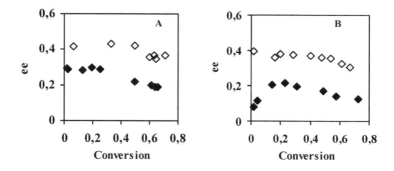

Figure 1A Enantioselective hydrogenation of 2,3–butanedione.
Figure 1B Enantioselective hydrogenation of 3,4–hexanedione.
[CD] = 1.2×10^{-5} M; [substrate]$_0$ = 1 M; T_r = 20 °C; p_{H2} = 50 bar; catalyst = 0.125g; solvent = toluene, purification: Figure 1A = as received; Figure 1B = distillation.
♦ - in the absence of quinuclidine, ◇ - in the presence of 6×10^{-5} M quinuclidine.

Table 2. Hydrogenation of 3,4 hexanedione to 4-hydroxy-3-hexanone

No	Solvent	[CD] 10^{-4} x M	[QN] 10^{-4} x M	k_1 min^{-1}	k_2 min^{-1}	ee_{max}	ee_{240}	$conv_{240}$
1	toluene	1	0	0.0094	0.0004	0.512	0.459	0.511
2	toluene	0.12	0	0.0047	0.0006	0.327	0.269	0.423
3	toluene	0.12	0.6	0.0067	0.0003	0.451	0.325	0.437
4[a]	toluene	0.12	0.6	0.0058	0.0004	0.482	0.331	0.458
5	toluene	1	5	0.0112	0.0010	0.536	0.387	0.620
6	ethanol	0.12	0	0.0045	0.0002	0.095	0.070	0.300
7	ethanol	0.12	0.6	0.0101	0.0016	0.205	0.126	0.572
8	ethanol	0.24	0.6	0.0126	0.0017	0.230	0.177	0.655

[a]DABCO used as ATA; CD injection technique; T_r:20 °C; p_{H2}: 50 bar; catalyst: 0.125g, 3,4-hexanedione: batch4, active carbon treated, in all experiments after 240 min the yield of hexanediols is less than 5 %.

The data in Table 2 show that with CD concentration around of 10^{-4} M, the addition of ATA resulted in slight increase in the reaction rate, but a decrease in the ee_{240} values was observed, with ee_{max} unchanged. By adding ATA, at [CD] = 10^{-5} M, beside increased activity shown by k_1 and k_2 about 25 % increase of ee value was observed. Different from the observation in the hydrogenation of Etpy, this ATA effect was observed in ethanol solvent. This may be explained as that, in case of enantioselective hydrogenation of diones, beside the shift in the cinchonidine monomer-dimer equilibrium other processes have to be taken into account to explain the ATA effect. The importance of other factors probably arises due to the fact that the strength of adsorption of diketones to the platinum surface is much stronger than that of the α-ketoesters [3]. We suppose that (1) the coadsorption of ATA diminishes the hydrogenation of the aromatic ring of

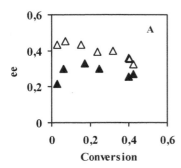 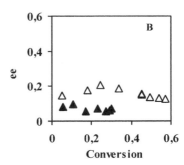

Figure 2 Conversion-enantioselectivity dependencies in the hydrogenation of 3,4–hexanedione; **A** - reaction in toluene, **B** - reaction in ethanol.
[CD] = 1.2×10^{-5} M, [substrate]$_0$ = 1 M, T_r = 20 °C, p_{H2} = 50 bar, catalyst = 0.125g, purification: using activated carbon; 7- in the absence of quinuclidine, Δ - in the presence of 6×10^{-5} M quinuclidine.

CD, which has a great importance at low CD concentration [9] and (2) the coadsoption of ATA decreases the poisoning effects of oligomeric by-products. We think that the latter plays a more significant role than the former.

Summary

In the asymmetric hydrogenation of BD and HD strong poisoning effect was observed. An accurate purification of the substrates is needed to get adequate kinetic data. Added achiral tertiary amines increase both the reaction rate and the enantioselectivity. Suggested explanation for this observation is that, beside the shift in the cinchonidine monomer-dimer equilibrium, the coadsorption of ATA diminished the hydrogenation of the aromatic ring of CD and decreased the poisoning effect of oligomeric by-products.

Acknowledgements

The authors thank to Engelhard Corp. providing the catalyst (E4759) as a gift. We also thank to Dr. Ágnes Gömöry for the MS measurements.

References

1. Y. Orito, S. Imai, and S. Niva, *J. Chem. Soc. Jpn.*, **8**, 1118 (1979).
2. H. U. Blaser, H. P. Jalett, D. M. Monti, J. F. Reber, J. T. and Wehrli, *Stud. Surf. Sci. Catal.*, **42**, 153 (1988).
3. J. A. Slipszenko, S. P. Griffiths, P. Johnston, K. E. Simons, W. A. H. Vermeer, and P. B. Wells, *J. Catal.*, **179**, 267 (1998).
4. M. Studer, V. Okafor, V., H. U. and Blaser, *Chem. Commun.*,1053 (1998).
5. X. Zuo, H. Liu, J. and Tian, *J. Mol. Catal. A: Chem.*, **157**, 217 (2000).
6. J. L. Margitfalvi, E. Tálas, and M. Hegedüs, *Chem. Commun.*, 645 (1999).
7. J. L. Margitfalvi, E. Tálas, E. Tfirst, C. V. Kumar, A. and Gergely, *Appl. Catal. A: General*, **191**, 177 (2000).
8. J. L. Margitfalvi, E. Tálas, and Á. Gömöry, *to be published.*
9. W.-R. Huck, T. Bürgi, T. Mallat, A. and Baiker, *J. Catal.*, **216**, 276 (2003).

61. Enantioselective Hydrogenation of Diketones on Pt Supported Catalysts

N. Marín-Astorga, G. Pecchi, and P. Reyes
Departamento de Fisicoquímica, Facultad de Ciencias, Universidad de Concepción, Casilla 160-C, Concepción, Chile; nmarin@udec.cl

J. L. Margitfalvi, E. Tálas, and O. Egyed
Chemical Research Center, Hungarian Academy of Sciences, 1025 Budapest, Pusztaszeri út 59-67 Hungary; joemarg@chemres.hu

Abstract

Enantioselective hydrogenation of 2,3-butanedione and 3,4-hexanedione has been studied over different type of supported Pt catalysts (Pt/Al_2O_3, Pt/SiO_2, Pt/MCM-41) in the presence of cinchonidine (CD). Kinetic results confirmed that (i) 2,3-butanedione is more reactive than 3,4-hexanedione over all catalysts studied and (ii) both substrates have a strong poisoning effect. The kinetic results confirmed also that CD concentration close to 10^{-3} M is necessary to achieve both high reaction rate and enantioselectivity in the range of 55-65 %. NMR results confirmed that substrate-modifier interaction takes also place in the liquid phase.

Introduction

The asymmetric hydrogenation (AH) of prochiral compounds, in particular diketones (see Figure 1) and α-ketoesters, is of significant importance since they are convenient building blocks in the asymmetric synthesis of biologically active compounds. [1-4]. The most frequently studied reaction of this class is the hydrogenation of α-ketoesters to the corresponding α-hydroxy esters [1]. The best catalyst system of this reaction is Pt/Al_2O_3 modified with cinchonidine (see Figure 2). High performances have been reported for Pt-cinchona catalyst systems in which Pt has been supported on silica, alumina, and carbon [5] or deposited in zeolites [6] and in mesopores of MCM-41 [7].

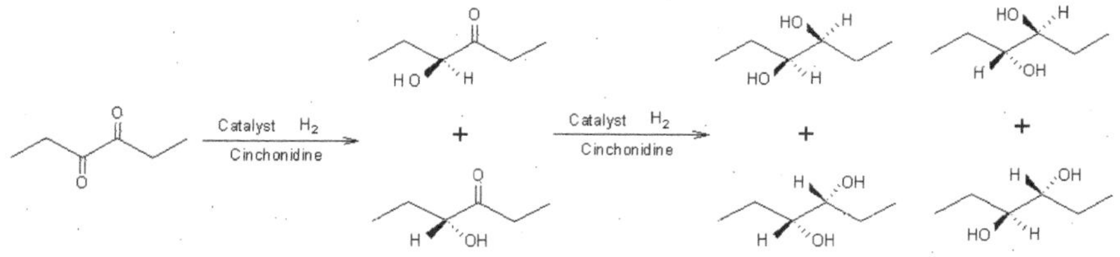

Figure 1 Reaction scheme for hydrogenation of 3,4-hexanedione.

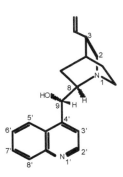

Figure 2 Structure of cinchonidine.

There are two views on the origin of enantiodifferentiation (ED) using Pt-cinchona catalyst system. In the classical approach it has been proposed that the ED takes place on the metal crystallite of sufficient size required for the adsorption of the chiral modifier, the reactant and hydrogen [8]. Contrary to that the "shielding effect model" suggest the formation of "substrate-modifier complex" in the liquid phase and its hydrogenation over Pt sites [9].

The aim of this study is to get further information with respect to the character of interactions involved in ED using α,β–diketones as substrates.

Experimental Section

Catalyst preparation and characterisation. Three different supports were used: 1% on MCM-41 and SiO_2, 5% on Al_2O_3 (Engelhard 4759). The synthesis of MCM-41 was performed using a standard procedure [10-11]. The catalysts were prepared by impregnation using a slight excess of solution required to fill the pore volume of the supports with a solution of $Pt(acac)_2$ in toluene, in the required amount to get 1wt.% of Pt. All catalysts were reduced *in situ* in hydrogen at 573 K for 1 h prior to their characterisation or catalytic test. Specific surface area and porosity were measured with an automatic Micromeritics system (Model ASAP 2010). The metallic dispersion was determined by H_2 chemisorption at 343 K in the same equipment. TGA of the supports were carried out in a Mettler Toledo TGA/SDTA851e apparatus. XRD patterns were obtained on a Rigaku diffractometer using a Ni filter and Cu $K\alpha_1$ radiation. TEM micrographs were obtained in a Jeol Model JEM-1200 EXII System.

Kinetic investigations. Enantioselective hydrogenation of 2,3-butanedione and 3,4-hexanedione has been investigated using cinchonidine (CD) as chiral modifier. To introduce chiral modifier injection and premixing techniques were used [12]. The reaction temperature and the hydrogen pressure was 23 °C and 50 bar in all experiments. The ee_{max} means the highest enantiomeric excess value measured in a given reaction. The $conv_{240}$ and ee_{240} values were measured after 4 h of reaction. The kinetic curves of the enantioselective hydrogenation had the characteristic two step behaviour described earlier (4,13). First order rate constants measured in the first hour of reaction and between 1 and 4 hours (k_1

and k_2, respectively) were used to compare the activity of catalysts under various experimental condition. All reagents used were purchased from Fluka.

NMR measurements. All spectra were measured at 30 °C The spectra were taken using a Varian Unity Inova 400 spectrometer. Chemical shifts (δ) are reported relative to internal TMS. 512 transients were accumulated to achieve the appropriate signal/noise ratio.

Results and Discussion

Characteristic features of catalysts used are given in Table 1. As emerges from chemisorption data all catalysts within the experimental error have similar mean particle size. This fact has also been proved by TEM for Pt/SiO$_2$ and Pt/MCM-41 catalysts. The channel size of MCM-41 was 2.4 nm.

Table 1 Characterisation data obtained for Pt-supported catalysts.

Catalysts	S_{BET} (m^2 g^{-1})	H/Pt
Pt/SiO$_2$	85	0.22
Pt/MCM-41	984	0.21
Pt/Al$_2$O$_3$	200	0.22

Results of kinetic measurements over Pt/Al$_2$O$_3$ catalyst under different reaction conditions are summarised in Table 2. A remarkable enhancement in the initial reaction rate was observed upon increasing the concentration of CD, while the increases of the concentration of the substrate resulted in a substantial drop in the reaction rate. The letter finding has been interpreted as a strong substrate poisoning. This interpretation has been further supported by the finding that $k_1 \gg k_2$ in all experiments. It is also known that the rate of hydrogenation of diketones is lower than that of the α-ketoesters [14]. This fact can be attributed to the higher ability of diketones to form oligomers and polymers, which are considered as strong catalyst poisons. Kinetic results confirmed that 2,3-butanedione is more reactive than 3,4-hexanedione over all catalysts studied. In case of 2,3-butanedione measurable amount of dihydroxy products also appeared, the kinetic behaviour was similar as described earlier [15]. No significant difference was found in toluene or dichloromethane solvents. The results confirmed also that CD concentration close to 10^{-3} M is necessary to achieve both high reaction rate and ee in the range of 0.5-0.6.

Hydrogenation of 3,4-hexanedione was used to compare the behaviour of different supported platinum catalysts. The highest rate has been obtained over Pt/MCM-41 catalyst. It was the only catalysts, where the rate constant k_2 exceeded k_1, i.e., there was no catalyst deactivation during the catalytic run. It is likely that the low acidity of MCM-41 can be responsible for the suppression of undesired side reactions. However, the e.e. values on this catalyst were only 25 %, while the highest e.e. values were obtained on Pt/Al_2O_3 catalyst.

Table 2 Kinetic data of hydrogenation of diketones over alumina supported Pt catalyst.

No	Substrate	$[S]_0$ M	$[CD]$ 10^{-3} M	Solvent	k_1	k_2	ee_{max}	ee_{240}	$conv_{240}$
1	3,4-HD	1	no	toluene	0.0015	0.0003	no	no	0.183
2	3,4-HD	1	0.05	toluene	0.0086	0.0005	0.449	0.430	0.383
3	3,4-HD	1	0.50	toluene	0.0049	0.0012	0.607	0.571	0.480
4	3,4-HD	1	1	toluene	0.0065	0.0014	0.613	0.569	0.433
5	3,4-HD	0.5	1	toluene	0.0151	0.0042	0.545	0.536	0.823
6	3,4-HD	0.1	1	toluene	0.0218	0.0044	0.552	0.552	0.845
7	3,4-HD	1	1	CH_2Cl_2	0.0077	0.0014	0.538	0.538	0.400
8^a	2,3-BD	1	1	toluene	0.0162	0.0058	0.510	0.510	0.659
9^b	2,3-BD	1	1	CH_2Cl_2	0.0198	0.0018	0.522	0.497	0.803

S: substrate; CD: cinchonidine; 3,4-HD: 3,4-hexanedione; 2,3-BD: 2,3-butane-dione;CD injection technique, T_r:20 °C; p_{H2}: 50 bar.

Table 3 Kinetic data of enantioselective hydrogenation of 3,4-hexanedione over different types of supported Pt catalysts.

No	$[CD]$ 10^{-3} M	S	Catalyst	k_1	k_2	ee_{max}	ee_{240}	$conv_{240}$
1	1	3,4-HD	Pt/Al_2O_3	0.0094	0.0009	0.643	0.562	0.551
2	1	3,4-HD	Pt/SiO_2	0.0058	0.0019	0.348	0.320	0.521
3	1	3,4-HD	Pt/MCM-41	0.0056	0.0128	0.255	0.247	0.932
4	no	3,4-HD	Pt/MCM-41					

CD: cinchonidine; S: substrate; 3,4-HD: 3,4-hexanedione; premixing technique; T_r:20 °C; p_{H2}: 50 bar; amount of catalysts used was calculated to have equal amount of surface platinum (Pt_s) in each reaction.

Table 4 Effect of diketones on H9 and aromatic protons of cinchonidine(CD) in CD_2Cl_2.

Substrate	Protons							
	H2' d	H5' d,d	H8' t,d	H6' t,d	H7' t,d	H3' d	H9	
no	8.86	8.15	8.09	7.69	7.54	7.54	5.59	d
Methyl pyruvate	8.89	8.09	8.07	7.66	7.51	7.76	6.40	s
2,3-Butanedione	8.84	8.11	8.07	7.69	7.57	7.65	6.02	d
2,3-Pentanedione	8.82	8.12	8.10	7.71	7.59	7.76	6.41	s
3,4-Hexanedione	8.88	8.11	8.09	7.70	7.58	7.75	6.42	s
1-Ph-1,2-PD*	8.92	8.15	8.12	n.m.	n.m.	7.76	6.49	s
Ethyl acetate	8.84	8.16	8.09	7.68	7.54	7.53	5.53	d
Acetophenone	8.88	8.11	n.m.	n.m.	n.m.	n.m.	5.91	d
Acetone	8.84	8.16	8.07	7.68	7.54	7.53	5.54	d

[CD]: 1×10^{-3} M; [substrate]: 1 M; room temperature, spectra measured after 30 min, d-doublet, s-singlet; *1-Phenyl-1,2-propanedione.

Results of liquid phase NMR measurements (Table 4) show that only the real substrates influences the proton shift of H3' and H9 protons of CD. These data confirmed the liquid phase interaction between the diketones and the chiral modifier. No effect of dummy substrates (ethyl acetate, acetone, etc.) was observed. No direct connection was found between kinetic data (reaction rate and optical yield) and NMR proton shift. Liquid phase NMR measurements confirmed the interaction of both 2,3-butandione and 3,4-hexanedione with the alkaloid used.

Acknowledgements

The authors thank Millennium Scientific Nucleus ICM P99-92, CONICYT (FONDECYT Grant 1030670). Mr. Marin-Astorga thanks CONICYT for a graduate fellowship. Thanks to Engelhard Corp. providing the catalyst (E4759) as a gift. Thanks to J. L. G. Fierro (Instituto de Catálisis y Petroleoquímica, CSIC, Madrid, Spain) for the characterization of MCM-41.

References

1. Y. Orito, S. Imai and S. Nina, *J. Chem. Soc. Jpn.,* **8**, 1118 (1979).
2. S. Bailey and F. King, in Fine Chemicals through Heterogeneous Catalysis, R. A. Sheldon and H. Van Bekkum, Eds., WILEY-VCH, Federal Republic of Germany.
3. Y. Izumi, in Advances in Catalysis, D. D. Eley, H. Pines, and P. B. Weisz, Eds., Academic Press, San Diego.
4. H. U. Blaser, *Tetrahedron: Asymm., 2*, 843 (1991).
5. G. Webb and P. B. Wells, *Catal. Today,* **12**, 319 (1992).
6. K. E. Simon, G. Wang, T. Heinz, T. Giger, T. Mallat, A. Pfaltz and A. Baiker, *Tetrahedron: Asymm.,* **6**, 505 (1995).
7. T. Hall, J. E. Halder, G. J. Hutchings, R. L. Jenkins, P. Johnston, P. McMorn, P. B. Wells and R. P. K. Wells, *Top. Catal.,* **11**, 351 (2000).
8. K. E. Simons, P. A. Meheux, S. P. Griffiths, I. M. Sutherland, P. Johnston, P. B. Wells, A. F. Carley, M. K. Rajumon, M. W. Roberts and A. Ibbotson, *Rec. Trav. Chim. Pays-Bas*, **113**, 465 (1994).
9. J. L. Margitfalvi, M. Hegedűs, and E. Tfirts, *Stud. Surf. Sci. Catal.*, **101**, 241 (1996).
10. J. Beck, J. Vartuli, W. Roth, M. Leonowicz, C. Kresge, K. Schmitt, C. Chu, D. Olson, E. Sheppard, S. McCullen, J. Higgins and J. Schlenker, *J. Am. Chem. Soc.*, **114**, 10834 (1992).
11. C. T Kresge, M. E. Leonowicz, W. J. Roth, J. C. Vartuli, and J. S. Beck, Nature, **359**, 710 (1992).

12. J. L. Margitfalvi, E. Tálas, E. Tfirst, C. V. Kumar, A. and Gergely, *Appl. Catal. A: General*, **191**, 177 (2000).
13. N. Marin-Astorga, G. Pecchi, J. L. G. Fierro and P. Reyes, *Catal. Lett.*, **91**, 115 (2003).
14. J. A. Slipszenko, S. P. Griffiths, P. Johnston, K. E. Simons, W. A. H. Vermeer and P. B. Wells, *J. Catal.*, **179**, 267 (1998).
15. M. Studer, V. Okafor and H.U. Blaser, *Chem. Commun.*, 1053 (1998).

62. Enantioselective Hydrogenation of α,β-Unsaturated Acids or Esters

S. Franceschini and A. Vaccari*

Dipartimento di Chimica Industriale e dei Materiali,
Alma Mater Studiorum—University of Bologna, INSTM-UdR di Bologna,
Viale del Risorgimento 4, 40136 Bologna, Italy

Abstract

The enantioselective hydrogenation of α,β–unsaturated acids or esters, using 5wt% Pt/Al_2O_3 or Pd/Al_2O_3 commercial catalysts doped with cinchonidine (CD), was deeply investigated to evidence the specific activity of Pd or Pt and the role of the reaction parameters and solvent polarity. Finally, the steric and electronic effects of different substituent groups were also studied.

Introduction

The enantioselective hydrogenation of α,β–unsaturated acids (or their esters) and α-ketoesters, mainly pyruvates, (Figure 1) is a subject of high industrial relevance in the pharmaceutical and agrochemical areas, considering the very different activity of pure enantiomers (1,2). However, the former reaction has been up to today less investigated, evidencing a lower enantioselectivity (maximum ee 38% in comparison to 90% for the ethyl pyruvate) (3,4).

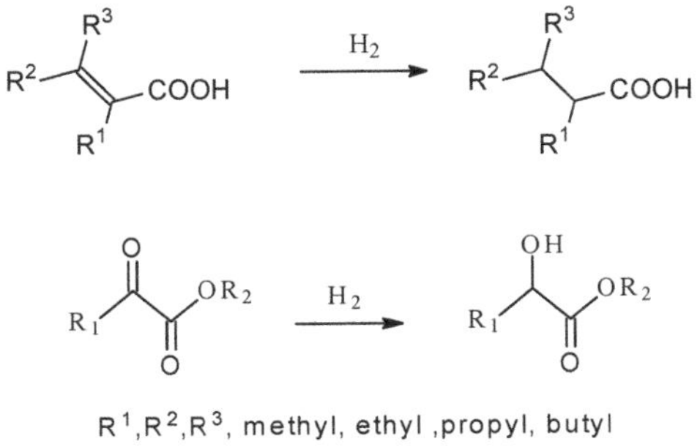

Figure 1 Hydrogenation of α,β-unsaturated acids and α-ketoesters.

Chiral heterogeneous catalysts, although have lower enantioselectivity and stability than homogeneous catalysts, are often preferable because of their handling and separation properties (5). Aim of this work was to shed light on the enantioselective hydrogenation of α,β-unsaturated acids or their esters using commercial noble-metal based heterogeneous catalysts in order to evidence the specific features of this reaction and the role of the different factors involved (nature of the metal, catalyst pre-activation, nature of the solvent, etc.). Finally, the effect of electron-donating or electron-withdrawing substituents in the enantioselective hydrogenation of α,β-unsaturated acids was also studied.

Experimental Section

Commercial 5wt% Pt or Pd supported on γ-Al_2O_3 catalysts were supplied by Engelhard and used as such or after reduction in H_2 flow at 673K for 3h (4). Cinchonidine (CD) was used as chiral modifier and added to the catalyst under an oxidizing atmosphere (6). The catalytic tests were carried out for 2 h, operating in liquid phase at 4.0MPa of H_2 and 298K and using a fully stirred Parr autoclave (54µmol of Pd or Pt, 25mmol of organic substrate, 170µmol of CD, 75ml of solvent). Two probe molecules, representative of each class of compounds, were used: trans-2-methyl-2-butenoic acid (tiglic acid) or its methyl ester and ethyl pyruvate. The products were analyzed using a Perkin Elmer Autosystem XL gas chromatograph, equipped with FID and a wide bore PS086 column (25m x 0.25mm, film width 0.25µm).

Results and Discussion

The heterogeneous enantioselective hydrogenation of α-ketoester or α,β–unsaturated substrates (acids or esters) may be catalysed using both Pt and Pd-based catalysts. However, feeding ethyl pyruvate only the CD-doped Pt/γ-Al_2O_3 catalysts gave rise to complete conversion with enantiomeric excesses (ee) up to 74%. The catalytic performances of the Pt-based catalysts were affected by the metal dispersion and, mainly, pre-treatment in H_2 flow (4). In fact, both conversion and ee increased with the metal dispersion, but only catalysts with Pt crystallites $\geq$ 3-4 nm gave rise to high optic yields. Lower crystallites opposed to the adsorption of CD, probably due to the hydrogenation of its quinoline ring (7). Even more relevant is the effect of the pre-treatment in H_2 flow: in absence of this step, using the same Pt/γ-Al_2O_3 catalyst the conversion decreased form 100% to 70% and the ee from 74% to 30%. On the contrary, CD-doped Pd/γ-Al_2O_3 catalysts showed very low activity and enantioselectivity values (for example, 10% conversion and 17% ee).

On the contrary, all the commercial CD-doped Pt/γ-Al_2O_3 catalysts investigated showed, in the same reaction conditions used for ethyl pyruvate hydrogenation, very poor performances feeding the tiglic acid. The conversion values were lower than 60% and ee values negligible (< 1%). These data were not improved by treating previously the catalysts in H_2 flow. With tiglic acid, the best performances were obtained with CD-doped Pd/γ-Al_2O_3 catalysts, reaching the complete conversion and an ee of 45%, in agreement with the best literature data (3). However, also for these catalysts the pre-treatment in H_2 flow did not improve the activity and/or enantioselectivity. Thus, a well-defined correlation exists between the active metal (Pt or Pd) and the organic substrate fed. However, feeding the corresponding ester (methyl tiglate) also the best CD-doped Pd/γ-Al_2O_3 catalyst gave rise to complete conversion, but with an ee of 2%. The ee was low probably since methyl tiglate does not form a dimer unlike the corresponding unsaturated acid. In tiglic acid the carboxylic groups can form dimer via H bonding, favouring the interaction between the substrate and CD and, consequently, the enantioselectivity of the reaction.

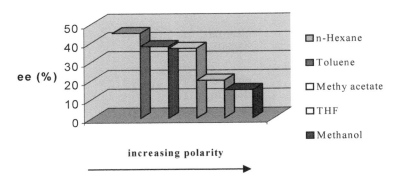

Figure 2 Enantiomeric excess (%) as a function of the solvent polarity in the hydrogenation of tiglic acid using a CD-doped Pd/γ-Al_2O_3 catalyst.

Another important factor in the hydrogenation of α,β-unsaturated acids was the solvent: by increasing the polarity of the solvent used (n-hexane $<$ toluene $<$ methyl acetate $<$ THF $<$ methanol) a progressive decrease in ee was observed. This was probably due to the fact that H_2 and the α,β-unsaturated acids were more soluble in the apolar solvents and also that CD was present in the open conformation, the enantioselective conformation (8). In fact, the highest conversion and ee were obtained with n-hexane (Figure 2).

Finally, the role of different substituents near the double bond in the hydrogenation of α,β-unsaturated acids using a CD-doped Pd/γ-Al_2O_3 was investigated; comparing the catalytic data with those collected feeding tiglic acid (Table 1). The presence of methyl or ethyl groups in position 2 or 3 strongly affected the catalytic performances, with an effect due to both the electron-donating and the steric hindrance, this latter hampering the right interaction between substrate, metal active sites and chiral modifier, worsening the ee. On the other hand, the presence of an electron-attracting

Table 1 Catalytic data for tiglic acid and other differently substituted α,β-unsaturated acids.

Substrate	Substituent properties	Conversion (%)	Yield (%)	ee (%)
(COOH, H_3C, CH_3)	--	99	99	45
(COOH, F_3C, CH_3)	electron-withdrawing	97	97	26
(F_3C, COOH, H_3C)	electron-withdrawing	100	99	< 1
(H_3C, COOH, H_3C)	electron-donating	26	21	--
(COOH, H_5C_2)	electron-donating	99	94	--
(COOH, H_5C_2, CH_3)	electron-donating	48	48	39

group (for example, -CF$_3$) favoured the conversion, whereas worsened the ee significantly. In fact, although the presence of electron-withdrawing groups weakened the double bond attracting the π–electrons, their steric hindrance again hampered the stereospecific interaction between catalyst-substrate-CD.

Conclusions

The enantioselective hydrogenation of α,β-unsaturated acids or ester is more complex than the analogous hydrogenation of ethyl pyruvate, giving rise to significantly lower ee values. Pd is better catalyst than Pt for this reaction, although with α,β-unsaturated esters negligible ee values were obtained. By decreasing the polarity of the solvent, a progressive increase in ee was detected, due to the higher solubility of H$_2$ and α,β-unsaturated acids. The substituents near the double bond affect the catalytic performances, with an effect due to both electron-donating and steric hindrance.

Acknowledgements

Thanks are due to American Chemical Society-Petroleum Research Foundation (ACS-PRF) for the financial support and Engelhard for providing the commercial catalysts.

References

1. S. C. Stinson, *Chem. Eng. News,* **28,** 46 (1992).
2. H. U. Blaser, *Catal. Today*, **60,** 161 (2000).
3. K. Borszeky, T. Bürgi, Z. Zhaohui, T. Mallat, and A. Baiker, *J. Catal.*, **187**, 160 (1999).
4. M. Bartok, G. Szöllösi, K. Balazsik, and T. Bartok, *J. Mol. Catal. A,* **177**, 299 (2002).
5. H. U. Blaser, *Tetrahedron Asymm.,* **2**, 843 (1991).
6. S. P. Griffiths, P. Johnston, and P. B. Wells, *Appl.Catal. A*, **191**,193 (2000).
7. P. B. Wells and A. Wilkinson, *Top. Catal.*, **5**, 39 (1998).
8. E. Toukoniitty, T. Salmi, and Yu. Murzin, *J. Mol. Catal. A*, **192**, 135 (2003).

63.

The Effects of Pt/Al₂O₃ Structure and Stability on the *O*-Methyl-10,11-dihydrocinchonidine Modified Enantioselective Hydrogenation of Activated Ketones

Daniel J. Ostgard[a], Rolf Hartung[a], Jürgen G. E. Krauter[a], Steffen Seebald[a], Pavel Kukula[b], Ulrike Nettekoven[b], Martin Studer[b], and Hans-Ulrich Blaser[b]

[a]*Degussa AG, Exclusive Synthesis & Catalysts, Rodenbacher Chaussee 4, D-63457 Hanau (Wolfgang), Germany*
[b]*Solvias AG, P.O. Box, CH-4002 Basel, Switzerland*

dan.ostgard@degussa.com

Abstract

Commercially available catalysts were characterized and tested for the enantioselective hydrogenation of activated ketones. It was found that the catalyst with a lower reducible residue level (cat*A*S*ium® F214; 5%Pt) and a narrower Pt crystal size distribution gave the best performance. Reductive pretreatment (RPT) of Pt/Al₂O₃ catalysts in hydrogen at 400°C improved their activity per surface Pt and their enantiomeric excesses (ee) at the expense of Pt dispersion (%D). Although the improvement in ee and activity over compensate for the slight loss in Pt dispersion from the RPT, cat*A*S*ium® F214 already performs excellently without a RPT. The RPT was most needed for the catalyst with the highest reducible residue level (E 4759 5%Pt) even though it also suffers the highest loss of Pt surface via sintering during RPT.

Introduction

Since the initial work of Orito et al. (1) a considerable amount of work has been performed to improve our understanding of the enantioselective hydrogenation of activated ketones over cinchona-modified Pt/Al₂O₃ (2, 3). Moderate to low dispersed Pt on alumina catalysts have been described as the catalysts of choice and pre-reducing them in hydrogen at 300-400°C typically improves their performance (3, 4). Recent studies have questioned the need for moderate to low dispersed Pt, since colloidal catalysts with Pt crystal sizes of <2 nm have also been found to be effective (3). A key role is ascribed to the effects of the catalyst support structure and the presence of reducible residues on the catalytic surface. Support structures that avoid mass transfer limitations and the removal of reducible residues obviously improve the catalyst performance. This work shows that creating a catalyst on an open porous support without a large concentration of reducible residues on the Pt surface not only leads to enhanced activity and ee, but also reduces the need for the pretreatment step. One factor not mentioned in the literature is the effect of Pt crystal size uniformity on catalyst performance. Since higher modifier-to-surface Pt atomic ratios shift the equilibrium from the desired face-adsorbed to the less effective edge-adsorbed cinchonidine species (5), a narrower Pt crystal size distribution can improve the control of modifier adsorption over the complete range of Pt crystals leading to higher reaction rates and ees.

Results and Discussion

Figure 1 clearly shows that the cat*A*Sium® F214 consists of uniform, finely divided Pt crystals supported on a needle-like alumina with an open pore structure allowing better flow of materials to and from the active sites. The E 4759 support is spherical and its Pt particles are lower dispersed (Figure 2) with a broader Pt crystal size distribution (Figure 3). The CO chemisorption data of Table 1 confirm the lower %D_{CO} of E 4759 and the excellent agreement between the %D_{TEM} obtained from the transmission electron microscope (TEM) and %D_{CO} verifies the accuracy of these measurements. During RPT the Pt dispersion of both catalysts drops due to sintering but the amount of sintering is roughly 5 times greater for E 4759. Crotonic acid (CA) hydrogenation (Table 2, for structures of modifier and test substrates see Figure 4) shows that removing reducible residues increases its turnover frequencies (TOF here defined as reactions per sec per surface Pt), while the activity per weight remains the same for cat*A*Sium® F214 and is roughly 24% lower for E 4759 due to their respective Pt sintering levels. The *O*-methyl-10,11-dihydrocinchonidine (MeOHCd) assisted enantioselective hydrogenation of activated ketones is more complex than CA reduction because the more effective face-adsorbed MeOHCd requires a larger ensemble of reduced residue-free Pt atoms than the less effective edge-adsorbed one that tends to form at higher MeOHCd-to-surface Pt ratios. Hence, the amount of surface reducible residues and the resulting Pt dispersion are critical parameters for this enantioselective hydrogenation. As seen in the hydrogenation of ethyl pyruvate (EtPy, Table 3), pyruvaldehyde dimethyl acetal (PADA, Table 4) and ethyl-2-oxo-4-phenylbutanoate (EOPB, Table 5), the RPT enhances the TOF of these reductions more than in the case of CA. Although the activity gain from RPT is higher for E 4759 than cat*A*Sium® F214, E 4759 also suffers more Pt sintering (Table 6), lower activities per weight and lower TOF in comparison to cat*A*Sium® F214. This suggests that the level of reducible residues and their possible retention after RPT are higher for E 4759. It is conceivable that higher reducible residue levels also favor Pt sintering. Only in the case of PADA is the TOF slightly higher for the E 4759 after RPT, but this is

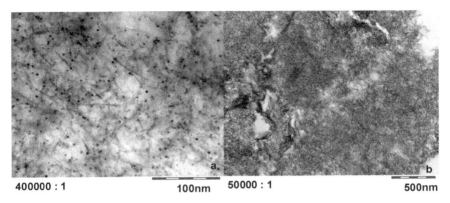

Figure 1 High (a) and low (b) magnification TEM images of cat*AS*ium® F214.

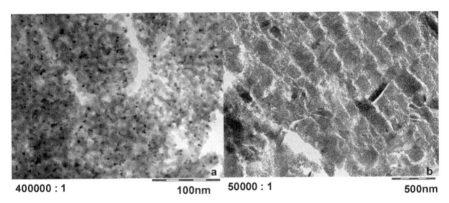

Figure 2 High (a) and low (b) magnification TEM images of E 4759.

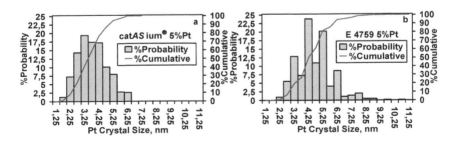

Figure 3 Pt crystal size distributions of cat*AS*ium® F214 (a) and E 4759 (b).

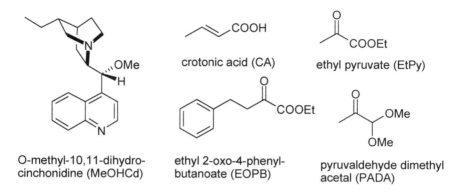

O-methyl-10,11-dihydro-
cinchonidine (MeOHCd)

ethyl 2-oxo-4-phenyl-
butanoate (EOPB)

pyruvaldehyde dimethyl
acetal (PADA)

Figure 4 Structures and abbreviations of modifier and test substrates.

Table 1 Characterization of the Pt/Al₂O₃ catalysts.

Catalyst	RPT[a]	CO Chemisorption Data		TEM Data	
		%D_{CO}[b]	Calc. PS_{SA}, nm[c]	%D_{TEM}[d]	PS_{SA}, nm[e]
cat*A*Sium® F214	No	26.3	4.54	27.8	4.3
cat*A*Sium® F214	Yes	24.4	4.89	---	---
E 4759	No	20.4	5.85	22.5	5.3
E 4759	Yes	12.9	9.25	---	---

[a] Reductive pretreatment.
[b] Pt %D from CO chemisorption data.
[c] Surface area average Pt particle size calculated from CO chemisorption data.
[d] Pt %D calculated from transmission electronmicroscopy (TEM) data.
[e] Surface area average Pt particle size calculated from TEM data.

Table 2 Crotonic acid (CA) hydrogenation data (atmospheric pressure).

Catalyst	RPT	Initial Reaction Rate ml H₂ / min · g_{cat}	TOF[a] sec⁻¹Pt$_{Surf.}$⁻¹
cat*A*Sium® F214	No	192	1.94
cat*A*Sium® F214	Yes	198	2.16
E 4759	No	105	1.37
E 4759	Yes	80	1.65

[a] Turnover frequency = number of reactions per sec per surface Pt atom.

Table 3 Ethyl Pyruvate (EtPy) hydrogenation data (60 bar).

Catalyst	RPT	Initial Reaction Rate ml H_2 / min · g_{cat}	TOF $sec^{-1}Pt_{Surf.}^{-1}$	Conv.[a] %	ee %
cat*A*Sium® F214	No	223	55.1	95	94
cat*A*Sium® F214	Yes	297	79.2	98	95
E 4759	No	76	24.2	44	90
E 4759	Yes	128	64.5	73	95

[a] % conversion.

Table 4 Pyruvaldehyde dimethyl acetal (PADA) hydrogenation data (60 bar).

Catalyst	RPT	Initial Reaction Rate ml H_2 / min · g_{cat}	TOF $sec^{-1}Pt_{Surf.}^{-1}$	Conv. %	ee %
cat*A*Sium® F214	No	52.3	12.9	80	95
cat*A*Sium® F214	Yes	67.4	18.0	89	96
E 4759	No	22.1	7.0	37	91
E 4759	Yes	38.7	19.5	57	96

Table 5 Ethyl-2-oxo-4-phenylbutanoate (EOPB) hydrogenation data (60 bar).

Catalyst	RPT	Initial Reaction Rate ml H_2 / min · g_{cat}	TOF $sec^{-1}Pt_{Surf.}^{-1}$	Conv. %	ee %
cat*A*Sium® F214	No	50.9	12.6	60	86
cat*A*Sium® F214	Yes	72.3	19.3	80	90
E 4759	No	21.6	6.9	41	79
E 4759	Yes	31.6	15.9	57	90

Table 6 Effect of the reductive pretreatment (RPT).

Catalyst	CO Chemisorption Data		% increase in Reaction TOF			
	%drop in %D	%incr. PS_{SA}[a]	EtPy	PADA	EOPB	CA
cat*A*Sium® F214	7.2	7.8	43.5	38.9	53.1	11.1
E 4759	36.8	58.1	166	177	131	20.5

[a] % increase of the surface area average Pt particle size calculated from CO chemisorption measurements.

balanced out by a far lower activity per weight and lower conversion in comparison to the cat*A*Sium® F214.

It is not always possible to calculate TOF for reactions with surface blocking modifiers. However, MeOHCd is reversibly adsorbed on the surface (4) and it also increases the TOF due to the favorable orientation of the substrate to the catalyst. Thus, the use of TOFs is more helpful than initial activities for understanding the optimal modifier conditions while taking into consideration the Pt crystal size. Improved Pt crystal size uniformity enhances the enantioselectivity of cat*A*Sium® F214 due to less variation in the modifier-to-surface Pt ratios over the complete catalyst in comparison to the E 4759 with its ~ 35% broader Pt crystal size range. This and the lower level of reducible residues make RPT practically unnecessary for cat*A*Sium® F214 for the enantioselective reduction of EtPy. The dimethyl acetal moiety of PADA does not activate the keto group as much as the ester functionality of EtPy and in this case, the RPT of cat*A*Sium® F214 improved its conversion to some extent. A similar effect could be seen for EOPB hydrogenation, where the presence of a CH₂Ph group might inhibit the optimal substrate orientation. Consequently, the RPT of cat*A*Sium® F214 increased the conversion of EOPB from 60 to 80%.

Conclusions

The RPT of Pt/Al₂O₃ catalysts improves the hydrogenation TOF and ee while sintering the Pt crystallites. The degree of TOF and ee enhancement induced by RPT increases with higher levels of reducible residues due to the poor performance of the initial catalyst. Pt sintering and potential residue retention also increase if higher reducible residue contents are present. Such catalysts tend to perform worse, even after RPT, than those with lower amounts of reducible residues. A cleaner catalyst will have a higher concentration of large residue-free Pt ensembles necessary for the formation of the more effective face-adsorbed MeOHCd. The ratio of the face- to the undesired edge-adsorbed MeOHCd is also dependent on the overall MeOHCd-to-surface Pt ratio and this is easier to control when the distribution of Pt crystal sizes is narrower. The cat*A*Sium® F214 meets the above mentioned requirements and has been found to exhibit very high enantioselectivity for the hydrogenation of activated ketones.

Experimental Section

The catalysts were obtained from their respective commercial sources and for some of these tests they were subjected to RPT by heating them to 400°C for 2 hours under a flow of hydrogen. The enantioselective ketone hydrogenations were carried out in a 50 ml stainless steel autoclave stirred with a magnetic stirring bar with 10 to 30 mg of catalyst, 10 to 30 mg of MeOHCd, 5 ml substrate and 20 ml AcOH at 60 bar and 25°C for 30 minutes. The crotonic acid hydrogenations were carried out in an ethanolic solution at atmospheric pressure and room temperature stirred at 2000 rpm with a hollow shaft bubbling stirrer.

CO chemisorption was measured via the pulse method and the TEM samples were prepared by embedding the catalyst in an epoxide followed by microtomically cutting it into thin slices. The diameters (dia) of at least 600 Pt crystals were measured at the magnifications ranging from 400,000 to 600,000 X and the surface area average particle size (PS_{SA}) was calculated from the following equation, $PS_{SA} = \Sigma n_i \cdot dia_i^3 / \Sigma n_i \cdot dia_i^2$, where n is the number of Pt crystals at that particular diameter. The calculation of %D from TEM data or PS_{SA} from CO chemisorption information was carried out with the standard formula, %D = $600/[PS_{SA} \cdot$ Pt density (surface area/g Pt)], where the Pt density is 2.14×10^{-20} g/nm^3 and the surface area/g Pt is 2.35×10^{20} nm^2/g.

References

1. Y. Orito, S. Imai, S. Niwa, G.-H. Nguyen, *J. Synth. Org. Chem. Jpn.*, **37**, 173 (1979).
2. A. Baiker, *J. Mol. Catal. A: Chemical*, **163**, 205 (2000).
3. M. Studer, H. U. Blaser and C. Exner, *Adv. Synth. Catal.*, **345**, 45 (2003).
4. H.U. Blaser, H.P. Jalett, M. Müller, and M. Studer, *Catal. Today,* **37**, 441 (1997).
5. J. Kubota and F. Zaera, *J. Am. Chem. Soc.*, **498**, 212 (2001).

Author Index

Keyword Index